UNITEXT

La Matematica per il 3+2

Volume 179

Editor-in-Chief

Alfio Quarteroni, Politecnico di Milano, Milan, Italy
  École Polytechnique Fédérale de Lausanne (EPFL), Lausanne, Switzerland

Series Editors

Luigi Ambrosio, Scuola Normale Superiore, Pisa, Italy

Paolo Biscari, Politecnico di Milano, Milan, Italy

Ciro Ciliberto, Università di Roma "Tor Vergata", Rome, Italy

Camillo De Lellis, Institute for Advanced Study, Princeton, USA

Lorenzo Rosasco, DIBRIS, Università degli Studi di Genova, Genova, Italy
  Center for Brains Mind and Machines, Massachusetts Institute of Technology,
  Cambridge, Massachusetts, US
  Istituto Italiano di Tecnologia, Genova, Italy

The **UNITEXT - La Matematica per il 3+2** series is designed for undergraduate and graduate academic courses, and also includes books addressed to PhD students in mathematics, presented at a sufficiently general and advanced level so that the student or scholar interested in a more specific theme would get the necessary background to explore it.

Originally released in Italian, the series now publishes textbooks in English addressed to students in mathematics worldwide.

Some of the most successful books in the series have evolved through several editions, adapting to the evolution of teaching curricula.

Submissions must include at least 3 sample chapters, a table of contents, and a preface outlining the aims and scope of the book, how the book fits in with the current literature, and which courses the book is suitable for.

For any further information, please contact the Editor at Springer: francesca.bonadei@springer.com.

THE SERIES IS INDEXED IN SCOPUS.

Carlo Presilla

Complex Functions
of a Variable

An Introduction with Solved Exercises

Carlo Presilla (iD)
Dipartimento di Matematica
Università degli Studi di Roma
"La Sapienza"
Roma, Roma, Italy

ISSN 2038-5714   ISSN 2532-3318   (electronic)
UNITEXT
ISSN 2038-5722   ISSN 2038-5757   (electronic)
La Matematica per il 3+2
ISBN 978-3-032-12493-7   ISBN 978-3-032-12494-4   (eBook)
https://doi.org/10.1007/978-3-032-12494-4

This Springer imprint is published by the registered company Springer Nature Switzerland AG
The registered company address is: Gewerbestrasse 11, 6330 Cham, Switzerland

If disposing of this product, please recycle the paper.

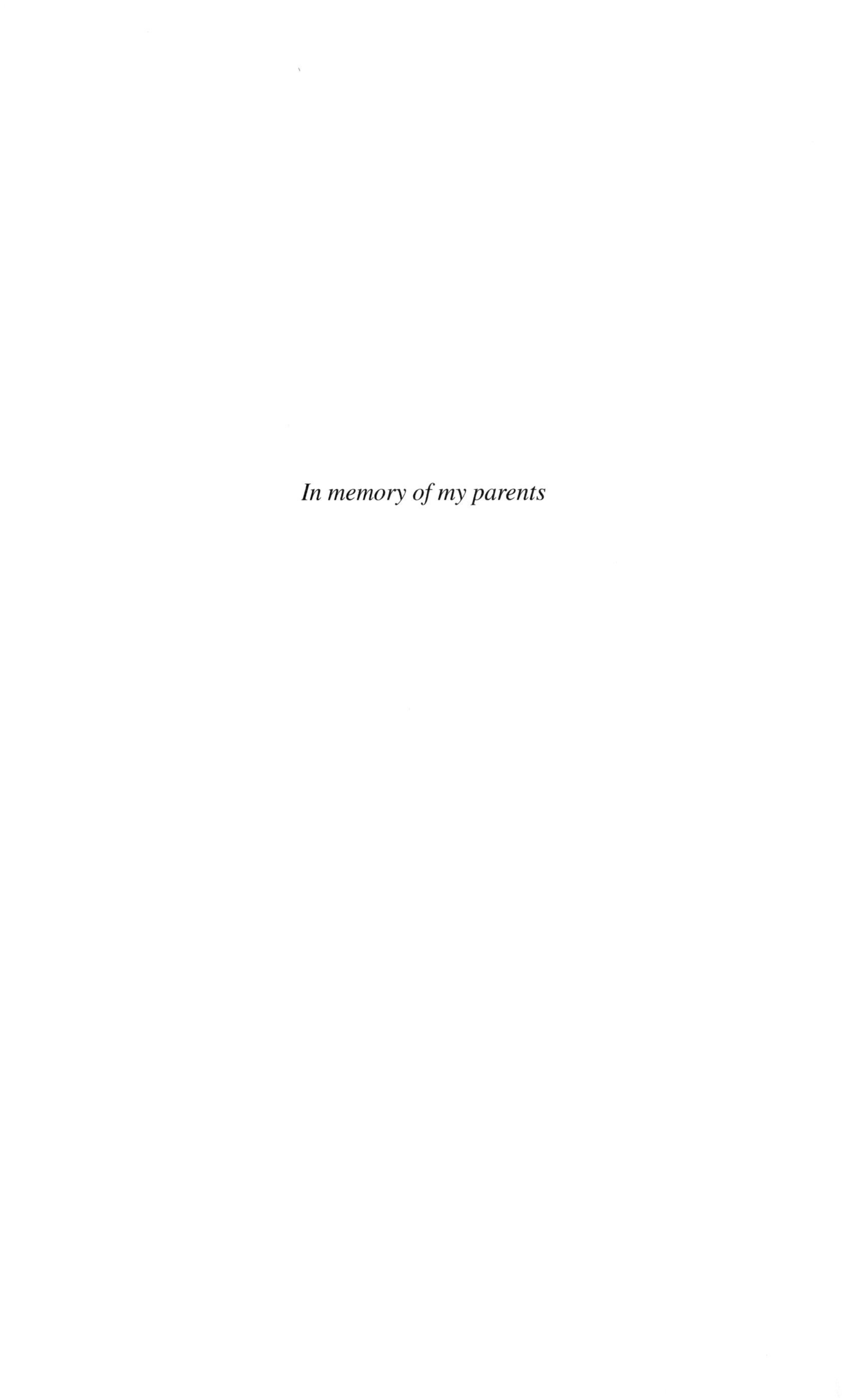

In memory of my parents

Preface

This book is intended as a first course in the theory of functions of a single complex variable for students who have already acquired the fundamental notions of real mathematical analysis. It is the expanded English version of an older text, published by Springer in Italian, based on the lectures I gave for several years at the University of Rome "La Sapienza" and at the Scuola Superiore di Catania.

The claim of mathematical rigor is combined with the idea that students should develop, at the same time, know-how skills. In fact, half of the book is devoted to the detailed solution of about 300 exercises illustrating, also with figures, the theory.

It's a pleasure to thank Prof. Filippo Cesi for countless discussions and suggestions. I also thank Prof. Alberto Perelli for his critical observations on the second edition of the Italian book, especially regarding the proof of the Cauchy-Goursat Theorem. Finally, I acknowledge the invaluable help of various students in highlighting misprints and errors.

Rome, Italy   Carlo Presilla
September 2025

Declarations

Competing Interests The author has no competing interests to declare that are relevant to the content of this manuscript.

Contents

Chapter 1
Complex Numbers

Abstract Axiomatic foundations: definition of complex numbers, their equality, sum and product. The field $\mathbb{C}$ of complex numbers. The imaginary unit. Geometric representation of complex numbers. Moduli and conjugates. Triangle inequality. Polar form of complex numbers: cis function or symbolic exponential, de Moivre's formula. Roots of complex numbers. Regions in the complex plane. The extended complex plane. Point at infinity.

1.1 Axiomatic Foundations

Definition 1.1 (Complex Numbers) We call $\mathbb{C}$, complex numbers, the set of all ordered pairs (x, y) with $x, y \in \mathbb{R}$.

For each complex number $z \in \mathbb{C}$ with $z = (x, y)$, the real number x is called the real part of z and is denoted $x = \operatorname{Re} z$ while the real number y is called the imaginary part of z and is denoted $y = \operatorname{Im} z$. Evidently, we have $\mathbb{R} \subset \mathbb{C}$.

Definition 1.2 (Equality, Sum and Product of Complex Numbers) Let z_1 and z_2 be two arbitrary complex numbers with $z_1 = (x_1, y_1)$ and $z_2 = (x_2, y_2)$, we define

(a) $z_1 = z_2$ if and only if $x_1 = x_2$ and $y_1 = y_2$;
(b) $z_1 + z_2 = (x_1 + x_2, y_1 + y_2)$;
(c) $z_1 z_2 = (x_1 x_2 - y_1 y_2, x_1 y_2 + y_1 x_2)$.

For $y_1 = 0$ and $y_2 = 0$, the previous definition reduces to that of equality, sum, and product of real numbers.

Theorem 1.3 *The set of complex numbers $\mathbb{C}$ with the Definition 1.2 forms a field, that is, the following properties hold:*

(a) $\forall z_1, z_2 \in \mathbb{C},\ z_1 + z_2 \in \mathbb{C}$ *and* $z_1 z_2 \in \mathbb{C}$ *(closure)*;
(b) $\forall z_1, z_2 \in \mathbb{C},\ z_1 + z_2 = z_2 + z_1$ *and* $z_1 z_2 = z_2 z_1$ *(commutative)*;
(c) $\forall z_1, z_2, z_3 \in \mathbb{C},\ (z_1 + z_2) + z_3 = z_1 + (z_2 + z_3)$ *and* $(z_1 z_2) z_3 = z_1 (z_2 z_3)$ *(associative)*;

© The Author(s), under exclusive license to Springer Nature Switzerland AG 2026
C. Presilla, *Complex Functions of a Variable*, La Matematica per il 3+2 179,
https://doi.org/10.1007/978-3-032-12494-4_1

(d) $\forall z_1, z_2, z_3 \in \mathbb{C}$, $z_1(z_2 + z_3) = z_1 z_2 + z_1 z_3$ *(distributive)*;

(e) $\exists! 0 \in \mathbb{C}$ *such that* $z + 0 = 0 + z = z\ \forall z \in \mathbb{C}$ *and* $\exists! 1 \in \mathbb{C}$ *such that* $z1 = 1z = z\ \forall z \in \mathbb{C}$ *(existence and uniqueness of identity). It turns out* $0 = (0,0)$ *and* $1 = (1,0)$;

(f) $\forall z \in \mathbb{C}$, $\exists! - z \in \mathbb{C}$ *such that* $-z + z = z + (-z) = 0$ *and, if* $z \neq 0$, $\exists! z^{-1} \in \mathbb{C}$ *such that* $z^{-1}z = zz^{-1} = 1$ *(existence and uniqueness of inverse). Let* $z = (x, y)$, *then* $-z = (-x, -y)$ *and* $z^{-1} = (x/(x^2 + y^2), -y/(x^2 + y^2))$.

Proof Elementary, using the Definition 1.2 and the fact that $\mathbb{R}$ is a field. $\square$

The existence and uniqueness of the inverse with respect to the sum and the product allow us to define in the usual way the difference and the ratio of two complex numbers z and w

$$z - w = z + (-w), \tag{1.1}$$

$$z/w = zw^{-1}, \quad \text{if } w \neq 0. \tag{1.2}$$

Definition 1.4 (Integer Powers) Let z be an arbitrary complex number and n be a positive integer, we define the n-th power of z as

$$z^n = zz \ldots z \ (n \text{ times}). \tag{1.3}$$

If $z \neq 0$, we then set $z^{-n} = 1/z^n$ and $z^0 = 1$.

1.2 The Imaginary Unit

The complex number $i = (0, 1)$ is called the imaginary unit. It is the solution, in the complex field, of the equation $x^2 + 1 = 0$, which does not admit a solution in the real field. By Definition 1.2 we have $i^2 = (0, 1)(0, 1) = (-1, 0)$. The integer powers of i have the following recurrence property:

$$i = (0, 1), \qquad i^2 = -1, \qquad i^3 = -i, \qquad i^4 = 1,$$

$$i^5 = i, \qquad\qquad i^6 = -1, \qquad i^7 = -i, \qquad i^8 = 1,$$

$$\vdots$$

By the Definition 1.2, every complex number $z = (x, y)$ can be written in the form $z = x + iy$, meaning with this that

$$(x, y) = (x, 0) + (0, 1)(y, 0).$$

The sum and product of two complex numbers are easily computed in terms of the imaginary unit by applying the usual rules of algebra

$$z_1 + z_2 = (x_1 + iy_1) + (x_2 + iy_2) = (x_1 + x_2) + i(y_1 + y_2),$$
$$z_1 z_2 = (x_1 + iy_1)(x_2 + iy_2) = (x_1 x_2 - y_1 y_2) + i(x_1 y_2 + y_1 x_2).$$

1.3 Modules and Conjugates

Definition 1.5 (Module) Each $z \in \mathbb{C}$ with $z = (x, y)$ and $x, y \in \mathbb{R}$ is associated with a non-negative real number, called the modulus of z, defined

$$|z| = \sqrt{x^2 + y^2}. \tag{1.4}$$

Note that $|z| = 0$ if and only if $z = 0$.

Definition 1.6 (Conjugate) Each $z \in \mathbb{C}$ with $z = (x, y)$ and $x, y \in \mathbb{R}$ is associated with a complex number, called the conjugate of z, defined

$$\bar{z} = (x, -y) = x - iy. \tag{1.5}$$

Note that $z\bar{z} = |z|^2$. Therefore, the modulus of a complex number is easily calculated by the formula

$$|z| = \sqrt{z\bar{z}}. \tag{1.6}$$

If $z \neq 0$ we also have

$$z^{-1} = \frac{\bar{z}}{|z|^2}. \tag{1.7}$$

The following properties immediately follow:

$$\operatorname{Re} z = \frac{1}{2}(z + \bar{z}), \qquad \operatorname{Im} z = \frac{1}{2i}(z - \bar{z}), \tag{1.8}$$

$$\overline{z \pm w} = \bar{z} \pm \bar{w}, \qquad \overline{zw} = \bar{z}\,\bar{w}, \qquad \overline{z/w} = \bar{z}/\bar{w}, \quad (w \neq 0), \tag{1.9}$$

$$|zw| = |z|\,|w|, \qquad |z/w| = |z|\,/\,|w|, \quad (w \neq 0), \tag{1.10}$$

$$\bar{\bar{z}} = z, \qquad |\bar{z}| = |z|. \tag{1.11}$$

1.4 Triangle Inequality

By definition we have $|z|^2 = (\operatorname{Re} z)^2 + (\operatorname{Im} z)^2$ and therefore $|\operatorname{Re} z| \le |z|$ as well as $|\operatorname{Im} z| \le |z|$ for every $z \in \mathbb{C}$. The following important inequality is then obtained:

Theorem 1.7 (Triangle Inequality) *For any two complex numbers z and w, we have*

$$|z + w| \le |z| + |w|. \tag{1.12}$$

Proof For the properties of modules and conjugates, we have

$$\begin{aligned}
|z + w|^2 &= (z + w)\overline{(z + w)} \\
&= (z + w)(\overline{z} + \overline{w}) \\
&= z\overline{z} + z\overline{w} + w\overline{z} + w\overline{w} \\
&= |z|^2 + 2\operatorname{Re}(z\overline{w}) + |w|^2 \\
&\le |z|^2 + 2|z\overline{w}| + |w|^2 \\
&= |z|^2 + 2|z|\,|w| + |w|^2 \\
&= (|z| + |w|)^2.
\end{aligned}$$

Taking the square root of both sides we get the result. $\square$

The equality sign in (1.12) holds only when $\operatorname{Re}(z\overline{w}) = |z\overline{w}|$, that is, when $\operatorname{Im}(z\overline{w}) = 0$ and, excluding the trivial cases in which one or both of the numbers z and w are 0, $\operatorname{Re}(z\overline{w}) > 0$. Given $z = x + iy$ and $w = u + iv$, these two conditions are equivalent to $xv = yu$ and $xu + yv > 0$.

From the triangle inequality it follows, in turn, that

$$|z + w| \ge \big||z| - |w|\big|. \tag{1.13}$$

In fact, assuming $|z| \ge |w|$ we have

$$|z| = |(z + w) + (-w)| \le |z + w| + |-w| = |z + w| + |w|,$$

or $|z + w| \ge |z| - |w| = \big||z| - |w|\big|$. The same result is reached by exchanging z with w in the case $|w| \ge |z|$. In conclusion, the modulus of the sum or difference of two arbitrary complex numbers z and w is bounded by

$$\big||z| - |w|\big| \le |z \pm w| \le |z| + |w|. \tag{1.14}$$

1.5 Geometric Representation of Complex Numbers

In the complex plane, also known as the Argand plane, each complex number $z = (x, y)$ can be associated with a vector that joins the origin $(0, 0)$ to the point (x, y). The vectorial interpretation of the operations of sum and difference of complex numbers as well as that of modulus and conjugate are immediate. The triangle inequality is nothing but the well-known property of triangles in which the length of one side is less than or equal to the sum of the lengths of the other two.

For a geometric interpretation of the multiplication and ratio of complex numbers see [4, 9]. These operations are more easily expressed in the so-called polar form of complex numbers.

1.6 Polar Form of Complex Numbers

In the complex plane each number $z = (x, y) \neq (0, 0)$ can also be represented by assigning the polar coordinates (r, θ), defined by the relations $x = r \cos \theta$, $y = r \sin \theta$. Evidently we have

$$r = |z| = \sqrt{x^2 + y^2}, \tag{1.15}$$

while θ is the angle between the positive real semi-axis and the segment that goes from 0 to z. The angle θ is called the argument of z and is denoted $\theta = \arg z$. The argument of a complex number is defined up to multiples of 2π. The value of the angle θ included in $(-\pi, \pi]$ is called the principal argument of z and is indicated with $\operatorname{Arg} z$

$$-\pi < \operatorname{Arg} z \leq \pi, \qquad \arg z = \operatorname{Arg} z + 2\pi n, \quad n \in \mathbb{Z}. \tag{1.16}$$

We now express the principal argument of z in terms of x and y by inverting the relation $\tan(\operatorname{Arg} z) = y/x$. Observing that the function arctan is conventionally a bijective transformation of the interval $(-\infty, +\infty)$ into the interval $(-\pi/2, \pi/2)$, we have

$$\operatorname{Arg} z = \begin{cases} \arctan(y/x) - \pi & x < 0, \ y < 0 \\ \arctan(y/x) & x \geq 0, \ (x, y) \neq (0, 0) \\ \arctan(y/x) + \pi & x < 0, \ y \geq 0 \end{cases} . \tag{1.17}$$

According to this formula we have $\operatorname{Arg} z = \operatorname{sgn}(y)\pi/2$ for $x = 0$ while $\operatorname{Arg} z$ is not defined for $z = 0$. The plot of Eq. (1.17) is shown in Fig. 1.1.

It is convenient to introduce the function $\operatorname{cis}(\theta) = \cos \theta + i \sin \theta$. For reasons that will be obvious later, this function is also indicated with the symbol $\operatorname{cis}(\theta) = e^{i\theta}$.

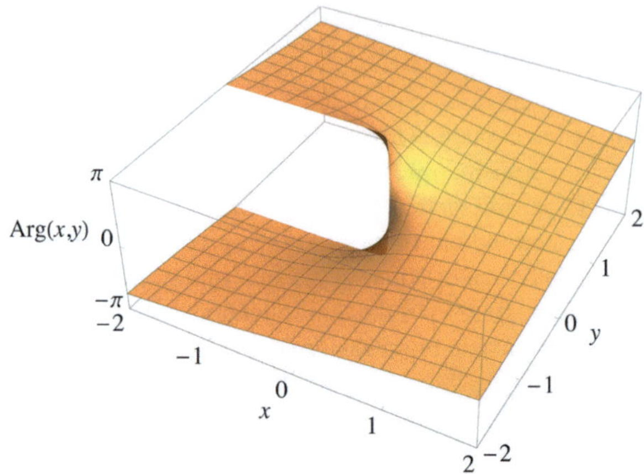

Fig. 1.1 Graph of Arg(z) with $z = (x, y)$. Note the discontinuity along the negative real semi-axis. The function is not defined at the origin

Given two complex numbers $z_1 = r_1 \operatorname{cis}(\theta_1)$ and $z_2 = r_2 \operatorname{cis}(\theta_2)$ their product is

$$
\begin{aligned}
z_1 z_2 &= r_1 r_2 [(\cos\theta_1 \cos\theta_2 - \sin\theta_1 \sin\theta_2) + i(\sin\theta_1 \cos\theta_2 - \cos\theta_1 \sin\theta_2)] \\
&= r_1 r_2 [\cos(\theta_1 + \theta_2) + i \sin(\theta_1 + \theta_2)] \\
&= r[\cos(\theta) + i \sin(\theta)].
\end{aligned}
$$

So in polar form the product $z = z_1 z_2$ is given by the simple rule $r = r_1 r_2$ and $\theta = \theta_1 + \theta_2$, with $z = r \operatorname{cis}(\theta)$. The rule extends to the product of an arbitrary number of factors

$$
|z_1 z_2 \ldots z_n| = |z_1| \, |z_2| \ldots |z_n|,
$$

$$
\arg(z_1 z_2 \ldots z_n) = \arg z_1 + \arg z_2 + \cdots + \arg z_n.
$$

Similarly, for the ratio of two complex numbers z_1 and $z_2 \neq 0$ we get

$$
|z_1/z_2| = |z_1| \, / \, |z_2|,
$$

$$
\arg(z_1/z_2) = \arg z_1 - \arg z_2.
$$

Note that in general $\operatorname{Arg}(z_1 z_2) \neq \operatorname{Arg} z_1 + \operatorname{Arg} z_2$ as well as $\operatorname{Arg}(z_1/z_2) \neq \operatorname{Arg} z_1 - \operatorname{Arg} z_2$. Consider, for example, $z_1 = -1$ and $z_2 = i$ for which we have $\operatorname{Arg} z_1 = \pi$, $\operatorname{Arg} z_2 = \pi/2$ and therefore $\operatorname{Arg} z_1 + \operatorname{Arg} z_2 = 3\pi/2$ while $\operatorname{Arg}(z_1 z_2) = \operatorname{Arg}(-i) = -\pi/2$.

The n-th power of $z = r(\cos\theta + i \sin\theta)$ is given by

$$
z^n = r^n [\cos(n\theta) + i \sin(n\theta)], \qquad n \in \mathbb{Z}. \tag{1.18}
$$

For $r = 1$ we obtain the de Moivre's formula

$$(\cos\theta + i\sin\theta)^n = \cos(n\theta) + i\sin(n\theta). \tag{1.19}$$

1.7 Roots of Complex Numbers

Given the complex number $z \neq 0$, we call n-th root of z, with $n \geq 2$ integer, every complex number w such $w^n = z$. Setting $w = re^{i\theta}$, by (1.18) we have $r^n = |z|$ and $n\theta = \operatorname{Arg} z + 2\pi k$ and, therefore, $r = \sqrt[n]{|z|}$ and $\theta = (\operatorname{Arg} z + 2\pi k)/n$ with $k \in \mathbb{Z}$. Of the infinite solutions thus found, only n are distinct, for example those obtained for $k = 0, 1, \ldots, n - 1$,

$$w_k = \sqrt[n]{|z|}\,e^{i(\operatorname{Arg} z + 2\pi k)/n}, \qquad k = 0, 1, \ldots, n - 1. \tag{1.20}$$

Note in fact that for every $k = 0, 1, \ldots, n - 1$, we have $w_k = w_{k+pn}$ with $p = \pm 1, \pm 2, \ldots$. In conclusion, every complex number $z \neq 0$ admits n distinct n-th roots given by the formula (1.20).

Example 1.8 The n-th roots of unity are the complex numbers

$$c_k = \operatorname{cis}(2\pi k/n), \qquad k = 0, 1, \ldots, n - 1,$$

lying on the circle of radius 1 centered at the origin equally spaced by angles $2\pi/n$ starting from the positive real semi-axis. If w is any n-th root of the number $z \neq 0$, then the n distinct n-th roots of z are $wc_0 = w, wc_1, wc_2, \ldots, wc_{n-1}$.

Example 1.9 The second-degree equation $az^2 + bz + c = 0$, with $a, b, c \in \mathbb{C}$, admits the solution

$$z = \frac{-b + \sqrt{b^2 - 4ac}}{2a}, \tag{1.21}$$

as it is immediate to verify by substitution

$$a\frac{(-b + \sqrt{b^2 - 4ac})^2}{4a^2} + b\frac{-b + \sqrt{b^2 - 4ac}}{2a} + c$$

$$= \frac{2b^2 - 4ac - 2b\sqrt{b^2 - 4ac}}{4a} + \frac{-b^2 + b\sqrt{b^2 - 4ac}}{2a} + c$$

$$= 0.$$

Let $b^2 - 4ac = re^{i\theta}$, (1.21) corresponds to the two solutions, which are distinct if $r \neq 0$,

$$z_1 = \frac{-b + \sqrt{r}e^{i\theta/2}}{2a}, \qquad z_2 = \frac{-b - \sqrt{r}e^{i\theta/2}}{2a}. \tag{1.22}$$

By the Fundamental Theorem of Algebra 7.40 these are the only two solutions of the equation considered.

1.8   Regions in the Complex Plane

Let λ be the straight line segment in $\mathbb{C}$ joining the points z_1 and z_2. From geometry we have that λ is the set of points of $\mathbb{C}$ parametrically defined by

$$\lambda(t) = z_1 + (z_2 - z_1)t, \qquad 0 \leq t \leq 1. \tag{1.23}$$

Therefore, the set of points

$$\mu(t) = z_1 + (z_2 - z_1)t, \qquad -\infty < t < \infty \tag{1.24}$$

represents the line μ that passes through the points z_1 and z_2. Obviously there are infinite other parametrizations that represent the same line. A non-parametric equation of the line passing through z_1 and z_2 is obtained by observing that for $z_1 \neq z_2$ we have $t = (\mu(t) - z_1)/(z_2 - z_1)$ and since $t \in \mathbb{R}$ it must be $\mathrm{Im}((\mu(t) - z_1)/(z_2 - z_1)) = 0$. Therefore, the non-parametric representation of the line for z_1 and z_2 is

$$\{\mu\} = \{z \in \mathbb{C} : \mathrm{Im}\,\frac{z - z_1}{z_2 - z_1} = 0\}. \tag{1.25}$$

It is possible to show that the sets

$$\{z \in \mathbb{C} : \mathrm{Im}\,\frac{z - z_1}{z_2 - z_1} < 0\}, \qquad \{z \in \mathbb{C} : \mathrm{Im}\,\frac{z - z_1}{z_2 - z_1} > 0\} \tag{1.26}$$

represent respectively the half-planes to the right and left of the oriented line μ when this is traveled from z_1 towards z_2.

Let γ be the boundary of a disk in $\mathbb{C}$ of radius R centered at the point z_0, namely, the circle of radius R centered at z_0. From geometry we have that γ is the set of points of $\mathbb{C}$ defined by

$$\{\gamma\} = \{z \in \mathbb{C} : |z - z_0| = R\}. \tag{1.27}$$

A possible parametric representation of this boundary is $\gamma(\theta) = z_0 + R\,\mathrm{cis}(\theta)$, $0 \leq \theta \leq 2\pi$. The sets

$$\{z \in \mathbb{C} : |z - z_0| < R\}, \qquad \{z \in \mathbb{C} : |z - z_0| > R\} \tag{1.28}$$

represent respectively the portion of $\mathbb{C}$ internal and external to the circle with center z_0 and radius R.

1.9 The Extended Complex Plane

It is convenient to define an extended complex plane $\mathbb{C}_\infty = \mathbb{C} \cup \{\infty\}$ that includes, in addition to $\mathbb{C}$, the point at infinity indicated with the symbol ∞. This point at infinity can be visualized by means of the stereographic projection of $\mathbb{C}$ on the Riemann sphere (Fig. 1.2). Consider in $\mathbb{R}^3$ the sphere with center $O = (0, 0, 0)$ and radius 1 and make $\mathbb{C}$ coincide with the plane $x_3 = 0$. The ray coming out from the point $N = (0, 0, 1)$ and passing through the generic point $P_z = (x, y, 0)$ intercepts on the sphere a point Q called the stereographic projection of the complex number $z = (x, y)$. The point N is the stereographic projection of the point at infinity.

It is easy to find the coordinates of the point Q given z. In $\mathbb{R}^3$ the half-line coming out of N and passing through P_z has parametric equation $\mu(t) = N + (P_z - N)t$, with $t \geq 0$, that is, it is the subset of $\mathbb{R}^3$

$$\{(xt, yt, 1 - t) \in \mathbb{R}^3 : t \geq 0\}. \tag{1.29}$$

The coordinates of the point Q correspond to the value of t such that

$$1 = (xt)^2 + (yt)^2 + (1 - t)^2. \tag{1.30}$$

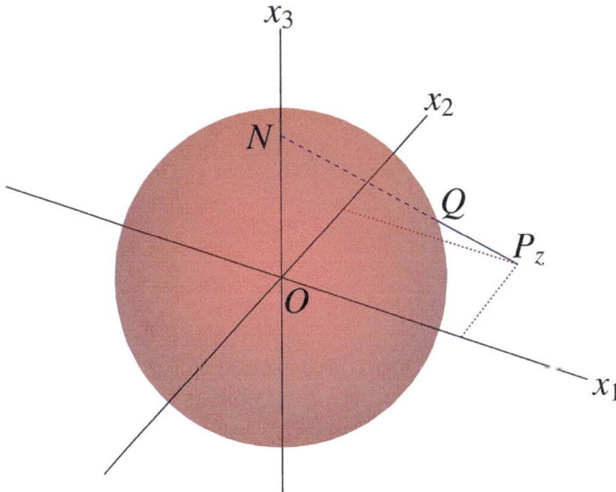

Fig. 1.2 Stereographic projection of the complex plane on the Riemann sphere. To each complex number $z = (x, y)$ of $\mathbb{C}$ (point P_z of the plane $x_3 = 0$) corresponds on the sphere the point $Q = (2x/(1 + r^2), 2y/(1 + r^2), (r^2 - 1)/(1 + r^2))$, where $r^2 = x^2 + y^2$

We obtain $t = 2/(1 + |z|^2)$ and therefore, according to (1.29), the coordinates of Q are

$$x_1 = \frac{2\operatorname{Re} z}{1 + |z|^2}, \qquad x_2 = \frac{2\operatorname{Im} z}{1 + |z|^2}, \qquad x_3 = \frac{|z|^2 - 1}{1 + |z|^2}. \tag{1.31}$$

Conversely, given the coordinates (x_1, x_2, x_3) of Q, the value of z is obtained by placing $1 - t = x_3$ in (1.29). From the same equation we have $x = x_1/t$ and $y = x_2/t$, that is,

$$z = \frac{x_1 + ix_2}{1 - x_3}. \tag{1.32}$$

Exercises

1.1 Calculate the following quantities:

(a) $\operatorname{Im}\left(i^{37}\right)$, (b) $\operatorname{Re}\left(\dfrac{2}{1 - 3i}\right)$, (c) $\operatorname{Im}\left(\left(1 + i\sqrt{3}\right)^3 e^{i\frac{\pi}{4}}\right)$.

1.2 Calculate the following quantities:

(a) $\dfrac{i^{101} - 3}{1 - 4i}$, (b) $\operatorname{Im}\dfrac{3i}{2 + 2i}$, (c) $\operatorname{Im}\left(e^{i\pi/4}\operatorname{Re}\left(e^{i\pi/4}\right)\right)$.

1.3 Calculate the following quantities:

(a) $\operatorname{Arg}(\operatorname{cis}(17\pi/5))$, (b) $\displaystyle\sum_{k=1}^{19}(1 + i)^k$, (c) $\left|\dfrac{ie^{-5i}(1 + i)^3}{1 + 3i}\right|$.

1.4 Let $z \in \mathbb{C}$ be such that $\operatorname{Re}(z^n) \geq 0$ for every positive integer n. Show that z is a non-negative real number.

1.5 Show that every complex number z of unit modulus, except $z = -1$, can be written in the form

$$z = \frac{1 + it}{1 - it}, \qquad t \in \mathbb{R}, \qquad |z| = 1, \ z \neq -1.$$

1.6 Show that for every complex number $z \neq 1$ and for every integer n we have

$$1 + z + z^2 + \cdots + z^n = \frac{1 - z^{n+1}}{1 - z}.$$

1.7 Let n and k be two positive integers and $z \in \mathbb{C}$ such that $z^n = 1$. Compute for any value of k the sum

$$S_k = \sum_{j=0}^{n-1} z^{jk}.$$

1.8 Assuming that $|z| < 1$, find an upper bound of $|\mathrm{Im}(1 + 2\bar{z} + z^2)|$.

1.9 Determine $a \in \mathbb{R}^+$ such that we have

$$\left| \mathrm{Re}\left(i + 2\bar{z} + z^2 \right) \right| < 3, \qquad |z| < a.$$

1.10 Prove that for every integer n the following inequality holds

$$|z_1 + z_2 + \cdots + z_n| \leq |z_1| + |z_2| + \cdots + |z_n|.$$

1.11 Prove that $\forall z \in \mathbb{C}$ we have

$$|\mathrm{Re}\, z| + |\mathrm{Im}\, z| \leq \sqrt{2}\, |z|.$$

1.12 Find all distinct values of the following roots and graph them:

$$\text{(a)} \ \sqrt[3]{-8}, \qquad \text{(b)} \ \sqrt[2]{1 - i}.$$

1.13 Find all distinct values of the following roots and graph them:

$$\text{(a)} \ \sqrt[2]{i^3}, \qquad \text{(b)} \ \sqrt[3]{8 + i}.$$

1.14 Find all the solutions of the following equations:

$$\text{(a)} \ z^4 + z^2 + 1 = 0, \qquad \text{(b)} \ z^3 - i = 0.$$

1.15 Find all the solutions of the equation

$$z^5 - (1 + i)z = 0.$$

1.16 Find all the solutions of the following equations:

$$\text{(a)} \ 3z + i\bar{z} = 4, \qquad \text{(b)} \ \bar{z}^4 = 1 + i.$$

1.17 Find all the solutions of the equation

$$z^4 - \frac{1 + i}{z} = 0.$$

1.18 Determine all the solutions of the following equations:

$$\text{(a) } z^2 + iz + 2 = 0, \qquad \text{(b) } \bar{z}^4 + \bar{z}^2 + 1 = 0.$$

1.19 Let $a, b, c \in \mathbb{C}$. Establish under which condition the equation

$$az + b\bar{z} + c = 0$$

has one and only one solution and determine that solution.

1.20 Calculate the value of $\cos(3\arccos(1/7))$.

1.21 Calculate the value of $\cos(4\arctan(3))$.

1.22 Calculate the value of $\sin(5\arccos(1/\sqrt{17}))$.

1.23 Calculate the sum

$$S = \cos\frac{\pi}{11} + \cos\frac{3\pi}{11} + \cos\frac{5\pi}{11} + \cos\frac{7\pi}{11} + \cos\frac{9\pi}{11}.$$

1.24 Calculate the sum

$$S = \sin\frac{\pi}{9} + \sin\frac{3\pi}{9} + \sin\frac{5\pi}{9} + \sin\frac{7\pi}{9}.$$

1.25 Let $\sum_{k=0}^{\infty} a_k z^k = f(z)$ for $|z| < R$. Prove that $\forall n \in \mathbb{N}$ and $\forall j \in \{0, 1, 2, \ldots, n-1\}$ we have

$$\frac{1}{n}\sum_{m=0}^{n-1} w^{-jm} f(w^m z) = \sum_{k=0}^{\infty} a_{j+nk} z^{j+nk},$$

where $w = e^{i2\pi/n}$ and $|z| < R$. Hint: Explicitly compute the left-hand side.

1.26 Describe in words and draw the set of points $z \in \mathbb{C}$ that satisfy the following conditions:

$$\text{(a) } 7 \geq |z - i + 3| > 5, \qquad \text{(b) } |z - i| = |z + 1|.$$

1.27 Determine and draw the locus of points $z \in \mathbb{C}$ such that

$$\left|\frac{z - 3}{z + 3}\right| < 2.$$

1.28 Determine and graph the locus of points $z \in \mathbb{C}$ such that

$$0 \leq \text{Arg}\left((z - 1)^2\right) \leq \pi/2.$$

Chapter 2
Metric Spaces

Abstract Metric spaces, distance. Diameter of a metric space. Bounded, unbounded metric spaces. Open, closed balls. Open, closed sets. Union and intersection of open or closed sets. Definition of interior, closure and boundary of a set and their properties. Metric spaces and connected sets. Connected sets in $\mathbb{R}$. Polygonal. Open connected sets in $\mathbb{C}$. Convergent sequences, limit points. The closure of a set coincides with its limit points. Dense sets. Cauchy sequences. Convergent sequences are Cauchy sequences. A Cauchy sequence that admits a convergent subsequence is convergent. Metric spaces and complete sets. Completeness of $\mathbb{C}$ (assuming $\mathbb{R}$ complete). A subset of a complete metric space is complete if and only if it is closed. Metric spaces and (sequentially) compact sets. A compact metric space is complete. Totally bounded metric spaces. A totally bounded metric space is bounded. A metric space is compact if and only if it is complete and totally bounded. A subset of $\mathbb{R}^n$ is compact if and only if it is closed and bounded.

2.1 Distance and Metric Spaces

Definition 2.1 (Metric Space) A metric space is the pair (S, d), where S is a set and $d : S \times S \mapsto \mathbb{R}$ a function, called distance, that for arbitrary $x, y, z \in S$ satisfies the following properties:

(a) $d(x, y) \geq 0$;
(b) $d(x, y) = 0$ if and only if $x = y$;
(c) $d(x, y) = d(y, x)$;
(d) $d(x, y) \leq d(x, z) + d(z, y)$

Definition 2.2 (Diameter) Let (S, d) be a metric space. The diameter of S is the quantity

$$\operatorname{diam} S = \sup_{x, y \in S} d(x, y). \tag{2.1}$$

© The Author(s), under exclusive license to Springer Nature Switzerland AG 2026

C. Presilla, *Complex Functions of a Variable*, La Matematica per il 3+2 179,

https://doi.org/10.1007/978-3-032-12494-4_2

A metric space is said to be bounded if diam $S \in \mathbb{R}$, unbounded if diam $S = \infty$.

The diameter of a space depends on the distance, in particular a space can be bounded with a distance and unbounded with another.

Example 2.3 (Complex Metric Space) Unless otherwise specified, from now on we will mean by complex metric space the pair $(\mathbb{C}, d)$ where $\mathbb{C}$ is the set of complex numbers and d the Euclidean distance

$$d(z, z') = |z - z'| = \sqrt{(\operatorname{Re} z - \operatorname{Re} z')^2 + (\operatorname{Im} z - \operatorname{Im} z')^2}, \qquad z, z' \in \mathbb{C}.$$

With this distance we have diam $\mathbb{C} = \infty$. It is possible to equip $\mathbb{C}$ with other distances. For example, consider

$$d(z, z') = \sqrt{(x_1 - x_1')^2 + (x_2 - x_2')^2 + (x_3 - x_3')^2} \tag{2.2}$$

equal to the Euclidean distance in $\mathbb{R}^3$ between the points $Q = (x_1, x_2, x_3)$ and $Q' = (x_1', x_2', x_3')$, stereographic projections of z and z' onto the Riemann sphere. Using the fact that Q and Q' lie on a sphere of unit radius centered at the origin, we have $d(z, z')^2 = 2 - 2(x_1 x_1' + x_2 x_2' + x_3 x_3')$ and so by (1.31)

$$d(z, z') = \frac{2|z - z'|}{\sqrt{(1 + |z|^2)(1 + |z'|^2)}}, \qquad z, z' \in \mathbb{C}. \tag{2.3}$$

With this distance we have diam $\mathbb{C} = 2$. The function d just discussed can be used to define the extended complex metric space $(\mathbb{C}_\infty, d)$. Similarly to what we saw for two points $z, z' \in \mathbb{C}$, the distance between $z \in \mathbb{C}$ and ∞ is

$$d(z, \infty) = \frac{2}{\sqrt{1 + |z|^2}}, \qquad z \in \mathbb{C}. \tag{2.4}$$

Definition 2.4 (Open Ball, Closed Ball) Let x be a point of the metric space (S, d) and $r > 0$. We call open ball and closed ball with center x and radius r, respectively, the sets

$$B(x, r) = \{y \in S : d(x, y) < r\}, \tag{2.5}$$

$$\overline{B}(x, r) = \{y \in S : d(x, y) \leq r\}. \tag{2.6}$$

2.2 Open and Closed Sets

Definition 2.5 (Open Set) Let (S, d) be a metric space. The set $G \subset S$ is said to be open in S if $\forall x \in G \, \exists \varepsilon > 0$ such that $B(x, \varepsilon) \subset G$.

Theorem 2.6 *An open ball in the metric space (S, d) is an open set in S.*

Proof Let $B(x, r)$ with $x \in S$ and $r > 0$ be an open ball in (S, d). Let y be a generic point of $B(x, r)$. By definition, $d(y, x) < r$. Let us show that it is possible to construct an open ball $B(y, \varepsilon) \subset B(x, r)$. Let $\varepsilon = r - d(y, x) > 0$, $\forall z \in B(y, \varepsilon)$ we have $d(z, y) < \varepsilon$ from which it follows, by the triangle property, $d(z, x) \leq d(z, y) + d(y, x) < \varepsilon + (r - \varepsilon) = r$, that is, $z \in B(x, r)$. From the arbitrariness of z we conclude that $B(y, \varepsilon) \subset B(x, r)$. $\qquad\square$

Theorem 2.7 *Let (S, d) be a metric space. Then*

(a) *the sets S and $\emptyset$ are open in S;*
(b) *if $(G_1, G_2, \ldots, G_n)$ is a finite collection of open sets in S, then $\bigcap_{k=1}^{n} G_k$ is open in S;*
(c) *if $(G_\alpha)_{\alpha \in I}$ is an arbitrary collection of open sets in S, then $\bigcup_{\alpha \in I} G_\alpha$ is open in S.*

Proof

(a) Let x be an arbitrary point of S. By definition of open ball $B(x, \varepsilon) \subset S$ with arbitrary ε. On the other hand $\emptyset$ has no points, so for each of its points it is possible to find an open ball, which is empty, contained in $\emptyset$.
(b) Let x be an arbitrary point of $\bigcap_{k=1}^{n} G_k$. It follows that $x \in G_k, k = 1, 2, \ldots, n$, and, since G_k is open, $\exists \varepsilon_k > 0$ such that $B(x, \varepsilon_k) \subset G_k$. Setting $\varepsilon = \min(\varepsilon_1, \varepsilon_2, \ldots, \varepsilon_n)$ and noticing that $\varepsilon > 0$, we have $B(x, \varepsilon) \subset B(x, \varepsilon_k) \subset G_k$ for every $k = 1, 2, \ldots, n$. Therefore $B(x, \varepsilon) \subset \bigcap_{k=1}^{n} G_k$.
(c) Let x be an arbitrary point of $\bigcup_{\alpha \in I} G_\alpha$. It follows that $\exists \alpha_0 \in I$ such that $x \in G_{\alpha_0}$. Since G_{α_0} is open $\exists \varepsilon > 0$ such that $B(x, \varepsilon) \subset G_{\alpha_0}$. Therefore $B(x, \varepsilon) \subset \bigcup_{\alpha \in I} G_\alpha$.

$\qquad\square$

Definition 2.8 (Closed Set) Let (S, d) be a metric space. The set $F \subset S$ is said to be closed in S if its complement $F^c = S \setminus F$ is open in S.

Theorem 2.9 *A closed ball in the metric space (S, d) is a closed set in S.*

Proof Let $\overline{B}(x, r)$ with $x \in S$ and $r > 0$ be a closed ball in (S, d). Let y be a generic point of $\overline{B}(x, r)^c$. By definition we have $d(y, x) > r$. Let us show that it is possible to construct an open ball $B(y, \varepsilon) \subset \overline{B}(x, r)^c$. Set $\varepsilon = d(y, x) - r > 0$, $\forall z \in B(y, \varepsilon)$ we have $d(z, y) < \varepsilon$ from which it follows, by the triangle property, $r + \varepsilon = d(y, x) \leq d(y, z) + d(z, x) < \varepsilon + d(z, x)$. This implies $d(z, x) > r$ and therefore $z \in \overline{B}(x, r)^c$. From the arbitrariness of z we conclude that $B(y, \varepsilon) \subset \overline{B}(x, r)^c$. $\qquad\square$

Theorem 2.10 *Let (S, d) be a metric space. Then*

(a) *the sets S and $\emptyset$ are closed in S;*
(b) *if $(F_1, F_2, \ldots, F_n)$ is a finite collection of closed sets in S, then $\bigcup_{k=1}^{n} F_k$ is closed in S;*
(c) *if $(F_\alpha)_{\alpha \in I}$ is an arbitrary collection of closed sets in S, then $\bigcap_{\alpha \in I} F_\alpha$ is closed in S.*

Proof It follows immediately from Theorem 2.7 using the definition of closed set and the de Morgan laws $\left(\bigcup F \right)^c = \bigcap F^c$ and $\left(\bigcap F \right)^c = \bigcup F^c$. $\qquad\square$

Definition 2.11 (Interior, Closure, Boundary) Let (S, d) be a metric space and A a subset of S. Let the interior, closure and boundary of A be the sets respectively indicated by A°, $\overline{A}$ and ∂A and defined as

$$A^\circ = \bigcup_{\text{open } G \subset A} G, \tag{2.7}$$

$$\overline{A} = \bigcap_{\text{closed } F \supset A} F, \tag{2.8}$$

$$\partial A = \overline{A} \setminus A^\circ = \overline{A} \cap (A^\circ)^c. \tag{2.9}$$

Since the union of an arbitrary collection of open sets is an open set and the intersection of an arbitrary collection of closed sets is a closed set, the interior of a subset A of S is an open set in S and its closure is a closed set in S. Furthermore, we always have $A^\circ \subset A \subset \overline{A}$. In particular, one can have $A^\circ = \emptyset$ and $\overline{A} = S$ as in the case $S = \mathbb{R}$ and $A = \mathbb{Q}$.

Theorem 2.12 *Let (S, d) be a metric space, $A \subset S$ and $B \subset S$. Then*

(a) *A is open in S if and only if $A = A^\circ$;*
(b) *A is closed in S if and only if $A = \overline{A}$;*
(c) *$A^\circ = (\overline{A^c})^c$;*
(d) *$\overline{A} = ((A^c)^\circ)^c$;*
(e) *$\partial A = \overline{A} \cap \overline{A^c}$;*
(f) *$(A \cap B)^\circ = A^\circ \cap B^\circ$;*
(g) *$\overline{A \cup B} = \overline{A} \cup \overline{B}$.*

Proof

(a) Let A be open in S. By definition we have $A^\circ \subset A$. On the other hand, A, being contained in itself, is one of the open sets whose union defines A° and therefore $A \subset A^\circ$. We conclude that $A = A^\circ$. Conversely, suppose that $A = A^\circ$. Since A° is open in S, A is also open in S.
(b) Let A be closed in S. By definition we have $A \subset \overline{A}$. On the other hand, A, containing itself, is one of the closed sets whose intersection defines $\overline{A}$ and therefore $\overline{A} \subset A$. We conclude that $A = \overline{A}$. Conversely, suppose that $A = \overline{A}$. Since $\overline{A}$ is closed in S, A is also closed in S.

(c) Using the definitions of interior and closure and de Morgan's laws, we have

$$
A^\circ = \bigcup_{\text{open } G \subset A} G = \left(\bigcap_{\text{open } G \subset A} G^c \right)^c = \left(\bigcap_{\text{closed } G^c \supset A^c} G^c \right)^c
$$

$$
= \left(\bigcap_{\text{closed } F \supset A^c} F \right)^c = (\overline{A^c})^c,
$$

where we exploited the fact that the correspondence between the open sets G contained in A and the closed sets G^c containing A^c is biunivocal.

(d) Reason in a similar way to the previous point. Alternatively, take the complement of the previous relation with the substitution $A \to A^c$.

(e) Follows from the definition of frontier and from point (c).

(f) Let us prove separately the two inclusions $(A \cap B)^\circ \subset A^\circ \cap B^\circ$ and $A^\circ \cap B^\circ \subset (A \cap B)^\circ$. Since $A \cap B \subset A$ it follows $(A \cap B)^\circ \subset A^\circ$. Similarly $(A \cap B)^\circ \subset B^\circ$ and therefore $(A \cap B)^\circ \subset A^\circ \cap B^\circ$. On the other hand, $A^\circ \cap B^\circ$ is an open set with the property $A^\circ \cap B^\circ \subset A \cap B$. Since $(A \cap B)^\circ$ represents the union of all open sets contained in $A \cap B$, it must be $A^\circ \cap B^\circ \subset (A \cap B)^\circ$.

(g) Reason in a similar way to the previous point. Alternatively, using the identity $\overline{A} = ((A^c)^\circ)^c$, the de Morgan's laws and the previous result, we have

$$
\overline{A \cup B} = (((A \cup B)^c)^\circ)^c = ((A^c \cap B^c)^\circ)^c = ((A^c)^\circ \cap (B^c)^\circ)^c
$$

$$
= ((A^c)^\circ)^c \cup ((B^c)^\circ)^c = \overline{A} \cup \overline{B}.
$$

$\square$

Theorem 2.13 *Let (S, d) be a metric space, $A \subset S$ and $x \subset S$. Then*

(a) $x \in A^\circ$ *if and only if* $\exists \varepsilon > 0$ *such that* $B(x, \varepsilon) \subset A$;
(b) $x \in \overline{A}$ *if and only if* $\forall \varepsilon > 0$ $B(x, \varepsilon) \cap A \neq \emptyset$;
(c) $x \in \partial A$ *if and only if* $\forall \varepsilon > 0$ $B(x, \varepsilon) \cap A \neq \emptyset$ *and* $B(x, \varepsilon) \cap A^c \neq \emptyset$.

Proof

(a) Let $x \in A^\circ$. Since A° is open in S then $\exists \varepsilon > 0$ such that $B(x, \varepsilon) \subset A^\circ \subset A$. Conversely, suppose that $\exists \varepsilon > 0$ such that $B(x, \varepsilon) \subset A$. Since the ball $B(x, \varepsilon)$ is an open set, from the definition of A° we have $B(x, \varepsilon) \subset A^\circ$ and therefore $x \in A^\circ$.

(b) Since $\overline{A} = ((A^c)^\circ)^c$, $x \in \overline{A}$ if and only if $x \notin (A^c)^\circ$. For the previous point, that is true if and only if $\forall \varepsilon > 0$ $B(x, \varepsilon) \not\subset A^c$, that is, if and only if $\forall \varepsilon > 0$ $B(x, \varepsilon) \cap A \neq \emptyset$.

(c) By definition $x \in \partial A = \overline{A} \cap (A^\circ)^c$ if and only if $x \in \overline{A}$ and $x \notin A^\circ$. By the two previous points this is true if and only if $\forall \varepsilon > 0$ both properties $B(x, \varepsilon) \cap A \neq \emptyset$ and $B(x, \varepsilon) \cap A^c \neq \emptyset$ hold.

$\square$

A subset $A \subset S$ in the metric space (S, d) is not necessarily open or closed. It may be that A is neither open nor closed, or both open and closed. Furthermore, the closure of an open ball does not necessarily coincide with the corresponding closed ball.

Example 2.14 Let (S, d) be the metric space in which $S = S_1 \cup S_2$ with

$$S_1 = \{z \in \mathbb{C} : d(z, 0) \leq 1\}, \qquad S_2 = \{z \in \mathbb{C} : d(z, 3) < 1\},$$

and d is the usual Euclidean distance $d(z, w) = |z - w|$. In (S, d) the open ball $B(3, 2)$ and the closed ball $\overline{B}(3, 2)$ correspond to the sets

$$B(3, 2) = \{z \in S : d(z, 3) < 2\} = S_2,$$

$$\overline{B}(3, 2) = \{z \in S : d(z, 3) \leq 2\} = S_2 \cup \{1\}.$$

Evidently $B(3, 2)$ is an open set in S. On the other hand $B(3, 2)$ is also a closed set in S since its complement $B(3, 2)^c = S_1$ is an open set in S. In fact, $\forall z \in S_1$ we have, for example, $B(z, 1/2) \subset S_1$. So $B(3, 2)$ is a proper subset of S (different from S and $\emptyset$) that is simultaneously open and closed in S. It is also an example of an open ball whose closure, coinciding with $B(3, 2)$, is different from the closed ball $\overline{B}(3, 2) = B(3, 2) \cup \{1\}$.

2.3 Connected Metric Spaces

Definition 2.15 (Connected Metric Space, Connected Subset) A metric space (S, d) is said to be connected if the only subsets of S that are simultaneously open and closed are $\emptyset$ and S. If $A \subset S$, then A is a connected subset of S if the metric space (A, d) is connected.

According to the previous definition, the metric space S is not connected if in S there exist two disjoint open sets A and B, both nonempty, such that $A \cup B = S$. In this case, in fact, $A = S \setminus B$ is also closed. The metric space of Example 2.14 is not connected.

Theorem 2.16 *A set $S \subset \mathbb{R}$ is connected if and only if S is an interval.*

Proof Let us limit ourselves to showing that a connected set $S \subset \mathbb{R}$ is necessarily an interval (for the proof of the inverse implication see [5]). So let $S \subset \mathbb{R}$ be connected. If, by contradiction, S were not an interval, then $\exists a, b \in S$ and $\exists t \notin S$ such that $a < t < b$. Let $A = (-\infty, t) \cap S$ and $B = (t, \infty) \cap S$. The two sets A and B are open in S, moreover $A \neq \emptyset$ since $a \in A$, and $B \neq \emptyset$ since $b \in B$. It is evident that $A \cap B = \emptyset$ and, since $t \notin S$, $A \cup B = S$. We conclude with the absurdity that S is not connected. Therefore S must be an interval. $\qquad\square$

Definition 2.17 (Polygonal) If $z, w \in \mathbb{C}$ let us indicate the straight line segment from z to w as

$$[z, w] = \{z + t(w - z),\ 0 \le t \le 1\}.$$

We call polygonal from z to w a set $P = \cup_{k=1}^{n-1}[z_k, z_{k+1}]$, where $z_k \in \mathbb{C}$, $k = 1, \ldots, n \in \mathbb{N}$, with $z_1 = z$ and $z_n = w$.

Theorem 2.18 *An open set $G \subset \mathbb{C}$ is connected if and only if $\forall z, w \in G$ there exists a polygonal from z to w that lies entirely in G.*

Proof Suppose that $\forall z, w \in G$ there exists a polygonal from z to w which lies entirely in G. If, by contradiction, G were not connected then we would have $G = Z \cup W$ with Z and W open and closed, non-empty and disjoint. Consider two points $z \in Z$ and $w \in W$ and let P be a polygonal from z to w such that $P \subset G$. For simplicity we assume that $P = [z, w]$; if P were composed of several intervals we would reason in a similar way considering that segment $[z_k, z_k+1]$ whose endpoints are such that $z_k \in Z$ and $z_{k+1} \in W$. Define $S = \{s \in [0, 1] : z+(w-z)s \in Z\}$ and $T = \{t \in [0, 1] : z+(w-z)t \in W\}$, it follows that $S \cap T = \emptyset$ and $S \cup T = [0, 1]$. Furthermore, S and T are open in $[0, 1]$ and not empty since $0 \in S$ and $1 \in T$. We have thus arrived at a contradiction with the fact that the interval $[0, 1]$ is connected. We must therefore admit that G is connected.

Conversely, suppose that G is open and connected. Let z be a generic point of G and define

$$W = \{w \in G : \text{ there exists a polygonal } P \subset G \text{ from } z \text{ to } w\}.$$

It is enough to show that W is simultaneously open and closed in G. In fact, if this is true, since $W \ne \emptyset$, indeed $z \in W$, and G is connected, it must be $W = G$.

To show that W is open in G, consider a point $w \in W$ and let P be a polygonal from z to w with $P \subset G$. Since G is open $\exists \varepsilon > 0$ such that $B(w, \varepsilon) \subset G$. For every $s \in B(w, \varepsilon)$ we have $[w, s] \subset B(w, \varepsilon) \subset G$. So the polygonal $P \cup [w, s]$ that connects z to s is still contained in G. We conclude that $B(w, \varepsilon) \subset W$ and therefore W is open in G.

To show that W is closed in G we prove that $W^c = G \setminus W$ is open in G. Note that $\forall w \in W^c$, since $w \in G$ and G is open, there exists $\varepsilon > 0$ such that $B(w, \varepsilon) \subset G$. Necessarily $B(w, \varepsilon) \cap W = \emptyset$, otherwise it would be possible to construct a polygonal from z to w contained in G. This means that $B(w, \varepsilon) \subset W^c$, that is W^c is open in G. $\qquad\square$

Note that the previous theorem is valid for open sets $G \subset \mathbb{C}$ and that, more precisely, this property is exploited only in the second part of the proof. Consider the case of a non-open G consisting of two disjoint disks connected by a non-rectilinear arc: according to Definition 2.15 G is a connected set but not all pairs of points of G can be connected by a polygonal entirely lying in G.

2.4 Convergence of Sequences

Definition 2.19 (Convergent Sequence) Let $(x_n)_{n=1}^{\infty}$, or, as it will be often more briefly indicated, (x_n), be a sequence in the metric space (S, d). We say that such a sequence converges to x, and we write

$$\lim_{n \to \infty} x_n = x \qquad \text{or} \qquad x_n \xrightarrow{n \to \infty} x, \tag{2.10}$$

if $\forall \varepsilon > 0 \exists N(\varepsilon)$ integer such that $d(x, x_n) < \varepsilon \forall n \geq N$. In other words, we have $\lim_{n \to \infty} x_n = x$ if and only if $\lim_{n \to \infty} d(x, x_n) = 0$.

Theorem 2.20 *Let (S, d) be a metric space. A set $F \subset S$ is closed in S if and only if for every sequence (x_n) in F convergent to x we have $x \in F$.*

Proof Let F be closed in (S, d). If (x_n) is a generic sequence in F convergent to x, then by definition $\forall \varepsilon > 0 \; \exists N(\varepsilon)$ integer such that $d(x, x_n) < \varepsilon \; \forall n \geq N$. Therefore, $\forall \varepsilon > 0 \; B(x, \varepsilon) \cap F \neq \emptyset$. By Theorem 2.13 we conclude that $x \in \overline{F}$ and therefore, since $F = \overline{F}$, we have that $x \in F$. Conversely, suppose that for every sequence (x_n) in F convergent to x we have $x \in F$. If, by contradiction, F were not closed, there would exist $y \in \overline{F}$ such that $y \notin F$. Since $y \in \overline{F}$, $\forall \varepsilon > 0 \; B(y, \varepsilon) \cap F \neq \emptyset$. Then, setting $\varepsilon = 1/n$, we have that $\exists y_n \in F$ such that $d(y_n, y) < 1/n$. We have thus constructed a sequence (y_n) in F that converges to $y \notin F$. This contradicts the hypothesis and therefore F must be closed. $\qquad \square$

Definition 2.21 (Limit Point) Let A be a subset of the metric space (S, d). A point $x \in S$ is called a limit point (or point of closure or adherent point) of A if there exists a sequence (x_n) of elements $x_n \in A$ that converges to x.

Let $L(A)$ denote the set of all limit points of A. Obviously $A \subset L(A)$.

Theorem 2.22 *Let A be a subset of the metric space (S, d). Then $\overline{A} = L(A)$.*

Proof We show that $\overline{A} \subset L(A)$ and then that $L(A) \subset \overline{A}$. Let $x \in \overline{A}$. Then $\forall \varepsilon > 0$ $B(x, \varepsilon) \cap A \neq \emptyset$. Choosing $\varepsilon = 1/n$ we conclude that $\exists x_n \in A$ such that $d(x_n, x) < 1/n$, that is, the sequence (x_n) of points of A converges to x. Therefore $x \in L(A)$. Conversely, let $x \in L(A)$. Then there exists a sequence (x_n) of points of A that converges to x. Therefore $\forall \varepsilon > 0 \exists N(\varepsilon)$ integer such that $d(x, x_n) < \varepsilon \; \forall n \geq N$. Since every $x_n \in A$ we conclude that $\forall \varepsilon > 0 \; B(x, \varepsilon) \cap A \neq \emptyset$, that is, $x \in \overline{A}$. $\quad \square$

Definition 2.23 (Dense Set) Let (S, d) be a metric space and A and B two subsets of S such that $A \subset B$. The set A is said to be dense in B if $B \subset \overline{A}$. In other words, A is dense in B if $\forall x \in B$ and $\forall \varepsilon > 0 \; \exists y \in A$ such that $d(x, y) < \varepsilon$.

Definition 2.24 (Cauchy Sequence) Let (S, d) be a metric space and (x_n) be a sequence of points $x_n \in S$. The sequence is called a Cauchy sequence if $\forall \varepsilon > 0 \exists N(\varepsilon)$ integer such that $d(x_n, x_m) < \varepsilon \forall n, m \geq N$.

Theorem 2.25 *Every convergent sequence is a Cauchy sequence.*

Proof Let (x_n) be a sequence in the metric space (S, d) convergent to $x \in S$. This means that $\forall \varepsilon > 0 \exists N(\varepsilon/2)$ integer such that $d(x, x_n) < \varepsilon/2 \forall n \geq N$. From the triangle inequality satisfied by the distance d, it follows that $d(x_n, x_m) \leq d(x_n, x) + d(x, x_m) < \varepsilon/2 + \varepsilon/2 = \varepsilon \ \forall n, m \geq N$, that is, (x_n) is a Cauchy sequence. $\square$

A Cauchy sequence is not necessarily convergent.

Example 2.26 Consider the sequence (x_n) defined by

$$x_{n+1} = \frac{1}{2}\left(x_n + \frac{s}{x_n}\right)$$

with $s > 0$ and $x_0 \neq 0$ arbitrary. This sequence, thought of as a sequence in $\mathbb{R}$, converges to $\sqrt{s}$ and constitutes the so-called Babylonian method for estimating the square root. Now consider the metric space $(\mathbb{Q}, d)$ with $d(x, y) = |x - y|$. For $x_0 \in \mathbb{Q}$ and $s \in \mathbb{Q}$, it is clear that (x_n) is a Cauchy sequence in the metric space $(\mathbb{Q}, d)$. If $\sqrt{s} \notin \mathbb{Q}$, consider for example $s = 2$, the sequence is not convergent in $\mathbb{Q}$.

Theorem 2.27 *If (x_n) is a Cauchy sequence in the metric space (S, d) which admits a subsequence (x_{n_k}) convergent in S, then (x_n) is convergent in S.*

Proof Since the sequence (x_n) is Cauchy $\forall \varepsilon > 0 \ \exists N'(\varepsilon/2)$ such that $d(x_n, x_m) < \varepsilon/2 \ \forall n, m \geq N'$. Since the subsequence (x_{n_k}) is convergent in S, say to $x \in S$, $\forall \varepsilon > 0 \ \exists N''(\varepsilon/2)$ such that $d(x_{n_k}, x) < \varepsilon/2 \ \forall n_k \geq N''$. Put $N = \max(N', N'')$ and choose $n_k \geq N$, from the triangle property of the distance it follows that $\forall n \geq N$ $d(x_n, x) \leq d(x_n, x_{n_k}) + d(x_{n_k}, x) \leq \varepsilon/2 + \varepsilon/2 = \varepsilon$, that is $x_n \xrightarrow{n \to \infty} x$. $\square$

2.5 Complete Metric Spaces

Definition 2.28 (Complete Metric Space, Complete Subset) A metric space (S, d) is said to be complete if all Cauchy sequences in S converge in S. If $A \subset S$, then A is a complete subset of S if the metric space (A, d) is complete.

One can show that $(\mathbb{R}, d)$ with d Euclidean distance is complete. On the other hand, $(\mathbb{Q}, d)$ is not complete. For the field of complex numbers we have:

Theorem 2.29 *$(\mathbb{C}, d)$ with d Euclidean distance is a complete metric space.*

Proof Assume that we have already proved that $(\mathbb{R}, d)$ with d Euclidean distance is complete. Let (z_n) with $z_n = x_n + iy_n$ be a generic Cauchy sequence in $\mathbb{C}$. It follows that (x_n) and (y_n) are Cauchy sequences in $\mathbb{R}$, in fact, it is enough to observe that $|x_n - x_m| \leq |z_n - z_m|$ and $|y_n - y_m| \leq |z_n - z_m|$ whatever the indices n and m are. Since $\mathbb{R}$ is complete, $x_n \xrightarrow{n \to \infty} x \in \mathbb{R}$ and $y_n \xrightarrow{n \to \infty} y \in \mathbb{R}$. It follows that $z_n \xrightarrow{n \to \infty} x + iy \in \mathbb{C}$, namely, $\mathbb{C}$ is complete. $\square$

From the completeness of a metric space follows an important criterion to establish whether a subset of it is complete or not.

Theorem 2.30 *Let (S, d) be a complete metric space and $F \subset S$. The metric space (F, d) is complete if and only if F is closed in S.*

Proof Suppose that (F, d) is complete and let us show that F is closed in S. It is sufficient to prove that for every sequence (x_n) of points $x_n \in F$ convergent to x we have $x \in F$. This is exactly what happens because (x_n), being a convergent sequence, is also a Cauchy sequence and (F, d) is complete. Conversely, suppose that F is closed in S and let us show that (F, d) is complete. Consider a generic Cauchy sequence (x_n) in F. Obviously this is also a Cauchy sequence in S which is complete. Therefore, $\exists x \in S$ such that $x_n \xrightarrow{n \to \infty} x$. Since F is closed in S, it must be $x \in F$, so (x_n) converges in F and (F, d) is complete. $\qquad\square$

The closure property of F in S is obviously a necessary condition for (F, d) to be complete. In general it is not a sufficient condition if the metric space (S, d) is not complete. For example, consider the metric space $(\mathbb{Q}, d)$ with d Euclidean distance and take $F = [0, 1] \cap \mathbb{Q}$. It is immediately shown that F is closed in $\mathbb{Q}$ but (F, d) is not complete.

2.6 Compact Metric Spaces

Definition 2.31 ((Sequentially) Compact Metric Space, (Sequentially) Compact Subset) A metric space (S, d) is said to be sequentially compact if every sequence in S has a convergent subsequence in S. If $A \subset S$, then A is a sequentially compact subset of S if the metric space (A, d) is sequentially compact.

It can be shown that every sequentially compact metric space is also compact (every covering has a finite subcover) and vice versa. Therefore, from now on we will not distinguish between compactness and sequential compactness.

Corollary 2.32 *Every (sequentially) compact metric space is complete.*

Proof It follows immediately from Theorem 2.27 and Definition 2.31. $\qquad\square$

Definition 2.33 (Totally Bounded Metric Space, Totally Bounded Subset) A metric space (S, d) is said to be totally bounded if $\forall \varepsilon > 0$ there exists a finite set of points of S, say $\{x_1, x_2, \ldots, x_n\}$, such that $S = \bigcup_{i=1}^{n} B(x_i, \varepsilon)$. If $A \subset S$, then A is a totally bounded subset of S if the metric space (A, d) is totally bounded.

Theorem 2.34 *A totally bounded metric space (S, d) is bounded.*

Proof According to Definition 2.33 $\forall \varepsilon > 0$ there exists a finite set of points of S, say $\{x_1, x_2, \ldots, x_n\}$, such that $S = \bigcup_{i=1}^{n} B(x_i, \varepsilon)$. Then $\forall x, y \in S$ it is possible to find two of these points, say x_j and x_k, such that $x \in B(x_j, \varepsilon)$ and $y \in B(x_k, \varepsilon)$. It follows that

$$d(x, y) \leq d(x, x_j) + d(x_j, x_k) + d(x_k, y) \leq \varepsilon + d(x_j, x_k) + \varepsilon$$

and therefore

$$\operatorname{diam} S = \sup_{x, y \in S} d(x, y) \leq 2\varepsilon + \max_{j,k=1,n} d(x_j, x_k) < \infty,$$

from which we conclude that S is bounded. $\qquad\square$

Theorem 2.35 *A metric space (S, d) is compact if and only if it is complete and totally bounded.*

Proof Let us start by assuming that (S, d) is compact and show that it is complete and totally bounded. The completeness of (S, d) follows from Corollary 2.32. If by contradiction (S, d) is not totally bounded then there exists $\varepsilon_0 > 0$ such that for no finite set of points of S $\{x_1, x_2, \ldots, x_n\}$ we have $S = \cup_{i=1}^{n} B(x_i, \varepsilon_0)$. Choosing an arbitrary point $x_1 \in S$ we then have $B(x_1, \varepsilon_0) \neq S$. Therefore $\exists x_2 \in S$ such that $d(x_1, x_2) \geq \varepsilon_0$. On the other hand, we have $B(x_1, \varepsilon_0) \cup B(x_2, \varepsilon_0) \neq S$ and therefore $\exists x_3 \in S$ such that $d(x_1, x_3) \geq \varepsilon_0$ and $d(x_2, x_3) \geq \varepsilon_0$. Continuing in this way we construct a sequence (x_n) of points of S with the property that $d(x_i, x_j) \geq \varepsilon_0$ if $i \neq j$. Such a sequence does not admit any convergent subsequence and this contradicts the hypothesis that (S, d) is sequentially compact.

Conversely, suppose that (S, d) is complete and totally bounded and let us show that it is sequentially compact. Let (x_n) be a generic sequence of points $x_n \in S$ and set $\varepsilon_k = 2^{-k}$. Since (S, d) is totally bounded, we can write S as the union of a finite number of balls of radius ε_1 centered on as many points of S. At least one of these balls, say the one centered on y_1, must contain infinite elements of the sequence (x_n). In other words, (x_n) admits a subsequence (x_{n_1}) whose elements are all contained in $B(y_1, \varepsilon_1)$. Observing that $(B(y_1, \varepsilon_1), d)$ is also totally bounded, we can write it as the union of a finite number of balls of radius ε_2 centered in as many points of $B(y_1, \varepsilon_1)$. Therefore $\exists y_2 \in B(y_1, \varepsilon_1)$ and a subsequence (x_{n_2}) of the sequence (x_{n_1}) whose elements are all contained in $B(y_2, \varepsilon_2)$. Continuing in this way we construct the sequence (z_k) where z_k is the first element of the subsequence (x_{n_k}) contained in the ball $B(y_k, \varepsilon_k)$. Evidently $d(z_n, z_m) < 2^{-\min(n,m)}$, so the sequence (z_n) is a Cauchy sequence in S. By the completeness of (S, d) this sequence is convergent in S. In conclusion, from the arbitrary sequence (x_n) of S we have extracted the subsequence (z_n) convergent in S, which means that (S, d) is sequentially compact. $\qquad\square$

Theorem 2.36 (Heine-Borel) *A subset $K \subset \mathbb{R}^n$ is compact if and only if K is closed and bounded.*

Proof Suppose, to begin with, that $K \subset \mathbb{R}^n$ is compact and let us show that it is closed and bounded. By Theorem 2.35 K is complete and totally bounded. Therefore, by Theorem 2.30 K is closed in $\mathbb{R}^n$ and by Theorem 2.34 bounded.

Conversely, suppose that K is closed and bounded and let us show that it is compact. Since K is bounded $\exists a_1, \ldots, a_n \in \mathbb{R}$ and $\exists b_1, \ldots, b_n \in \mathbb{R}$ such that $K \subset F = [a_1, b_1] \times \cdots \times [a_n, b_n]$. Since F is closed in $\mathbb{R}^n$ and $\mathbb{R}^n$ is complete, it follows that F is complete. On the other hand, F is also totally bounded[1] and therefore by Theorem 2.35 it is compact. In conclusion, K is a closed subset of the compact F and according to Definition 2.31 it is itself compact. $\qquad\square$

The compact subsets of $\mathbb{C}$ are all and only those that are closed and bounded. While in $\mathbb{R}^n$ boundedness and total boundedness coincide, in general total boundedness is a stronger property than boundedness. For this reason the Heine-Borel theorem does not apply to a generic metric space in which it is possible to find closed and bounded subsets that are not compact.

Exercises

2.1 Let S be an arbitrary nonempty set and $d : S \times S \mapsto \mathbb{R}$ the function

$$d(x, y) = \begin{cases} 0 & x = y \\ 1 & x \neq y \end{cases}.$$

Show that (S, d) is a metric space (called a space of isolated points).

2.2 Let $S = \mathbb{R}$ and consider the function $d : S \times S \mapsto \mathbb{R}$

$$d(x, y) = \frac{|x - y|}{1 + |x - y|}.$$

Show that (S, d) is a metric space and evaluate its diameter. Hint: note that $g(t) = t/(1 + t)$ is monotone increasing for $t \in [0, \infty)$.

2.3 Let $d : \mathbb{C} \times \mathbb{C} \mapsto \mathbb{R}$ be the function defined by

$$d(z_1, z_2) = |\mathrm{Re}(z_1 - z_2)| + |\mathrm{Im}(z_1 - z_2)|, \qquad z_1, z_2 \in \mathbb{C},$$

prove that $(\mathbb{C}, d)$ is a metric space and draw the open ball $B(1 + i3, 1)$.

2.4 Let $d : \mathbb{C} \times \mathbb{C} \mapsto \mathbb{R}$ be the function defined by

$$d(z_1, z_2) = \max\left(|\mathrm{Re}(z_1 - z_2)|, |\mathrm{Im}(z_1 - z_2)|\right), \qquad z_1, z_2 \in \mathbb{C},$$

prove that $(\mathbb{C}, d)$ is a metric space and draw the closed ball $\overline{B}(3 + i, 2)$.

---

[1] $\forall \varepsilon > 0$ consider a grid of F in n-dimensional cubes of side $\leq \varepsilon$: the corresponding finite number of vertices can be used as the finite set of points $\{x_1, x_2, \ldots, x_k\}$, such that $S = \cup_{i=1}^{k} B(x_i, \varepsilon)$

2.5 Prove the following statements, if true, or provide a counterexample, if false:

(a) the union of a countable infinity of open sets is an open set;
(b) the intersection of a countable infinity of open sets is an open set;
(c) the union of a countable infinity of closed sets is a closed set;
(d) the intersection of a countable infinity of closed sets is a closed set.

2.6 Prove the following statements if true, or provide a counterexample if false:

(a) $(A \cup B)^\circ = A^\circ \cup B^\circ$;
(b) $(A \cap B)^\circ = A^\circ \cap B^\circ$;
(c) $\overline{A \cup B} = \overline{A} \cup \overline{B}$;
(d) $\overline{A \cap B} = \overline{A} \cap \overline{B}$.

2.7 Prove the following statement, if true, or provide a counterexample, if false. Let A be a subset of the metric space (S, d), then diam $A = $ diam $\overline{A}$.

2.8 Let A and B be two subsets of a metric space (S, d). Prove that if $A \subset B$, then $\overline{A} \subset \overline{B}$ and $A^\circ \subset B^\circ$.

2.9 Let A be a subset of a metric space (S, d). Prove that, in general, $\overline{A^\circ} \neq \overline{A}$ and $(\overline{A})^\circ \neq A^\circ$.

2.10 Consider the set $A = \{x \in \mathbb{Q} : \sqrt{2} < x < \sqrt{3}\}$. Determine whether, in the metric space $(\mathbb{Q}, d)$ with $d(x, y) = |x - y|$, the set A is: (a) open; (b) closed; (c) totally bounded; (d) compact.

Chapter 3
Limits and Continuity

Abstract Definition of convergence for functions between metric spaces. Uniqueness of the limit and characterization in terms of convergent sequences. Limits of composite functions. The case of complex functions: relationship with the limits of real part and imaginary part functions, limit of the sum, product and ratio of two functions. Limits involving the point at infinity. Functions continuous at a point, continuous functions. A function is continuous if and only if the inverse function transforms open sets into open sets or closed sets into closed sets. The composition of continuous functions is continuous. The case of complex functions: relationship with the continuity of real part and imaginary part functions, continuity of the sum, product and ratio of two continuous functions. Uniformly continuous and Lipschitz continuous functions: mutual implications. A continuous function transforms compact sets into compact sets and connected sets into connected sets. A continuous function on a compact set with values in $\mathbb{R}$ has absolute maximum and minimum. The modulus of a continuous function on a compact set with values in $\mathbb{C}$ has absolute maximum and minimum. A function with values in $\mathbb{C}$ which is continuous and non-zero at a point is non-zero in a neighborhood of that point. A continuous function on a compact set is uniformly continuous.

3.1 Limits

Definition 3.1 (Limit) Let (S, d) and (Ω, ρ) be two metric spaces and consider a function $f : S \mapsto \Omega$. If $a \in S$ and $\omega \in \Omega$, then $\lim_{x \to a} f(x) = \omega$ is equivalent to saying that $\forall \varepsilon > 0 \exists \delta(\varepsilon) > 0$ such that $f(x) \in B(\omega, \varepsilon) \ \forall x \in B(a, \delta) \setminus \{a\}$. In other words, $\forall \varepsilon > 0 \exists \delta(\varepsilon) > 0$ such that $\rho(f(x), \omega) < \varepsilon$ whenever $0 < d(x, a) < \delta$.

Note that if $f : D \mapsto \Upsilon$ with $D \subset S$ and $\Upsilon \subset \Omega$, it could be $a \notin D$ and $\omega \notin \Omega$, however it must be $a \in L(D)$ and $\omega \in L(\Upsilon)$. Even if it were $a \in D$ and $\omega \in \Upsilon$ it does not necessarily have to be $f(a) = \omega$.

Theorem 3.2 (Uniqueness of the Limit) *Let* $f : (S, d) \mapsto (\Omega, \rho)$, $a \in S$ *and* $\omega \in \Omega$. *If* $\lim_{x \to a} f(x) = \omega$ *then* ω *is unique.*

© The Author(s), under exclusive license to Springer Nature Switzerland AG 2026
C. Presilla, *Complex Functions of a Variable*, La Matematica per il 3+2 179,
https://doi.org/10.1007/978-3-032-12494-4_3

Proof Suppose that there exists $\omega' \in \Omega$ such that $\lim_{x \to a} f(x) = \omega'$. Then $\forall \varepsilon > 0$ $\exists \delta'(\varepsilon/2) > 0$ such that $\rho(f(x), \omega') < \varepsilon/2$ whenever $0 < d(x, a) < \delta'$. Since by hypothesis we also have $\lim_{x \to a} f(x) = \omega$, correspondingly to the same ε $\exists \delta(\varepsilon/2) > 0$ such that $\rho(f(x), \omega) < \varepsilon/2$ whenever $0 < d(x, a) < \delta$. Therefore, if $x \neq a$ is a point of S such that $d(x, a) < \min(\delta, \delta')$, by the triangle property of the distance ρ we have $\rho(\omega, \omega') \leq \rho(\omega, f(x)) + \rho(f(x), \omega') < \varepsilon/2 + \varepsilon/2 = \varepsilon$. From the arbitrariness of ε it follows $\rho(\omega, \omega') = 0$ and hence $\omega' = \omega$. $\square$

Theorem 3.3 *Let* $f : (S, d) \mapsto (\Omega, \rho)$, $a \in S$ *and* $\omega \in \Omega$. *Then* $\lim_{x \to a} f(x) = \omega$ *if and only if for every sequence* (x_n) *of points* $x_n \in S \setminus \{a\}$ *such that* $\lim_{n \to \infty} x_n = a$ *we have* $\lim_{n \to \infty} f(x_n) = \omega$.

Proof Suppose that $\lim_{x \to a} f(x) = \omega$. Then $\forall \varepsilon > 0$ $\exists \delta(\varepsilon) > 0$ such that $\rho(f(x), \omega) < \varepsilon$ whenever $0 < d(x, a) < \delta$. On the other hand, if (x_n) is a generic sequence of points $x_n \in S \setminus \{a\}$ such that $\lim_{n \to \infty} x_n = a$, then $\exists N(\delta)$ integer such that $0 < d(x_n, a) < \delta$ whenever $n \geq N$. Combining the two properties we find that $\rho(f(x_n), \omega) < \varepsilon$ whenever $n \geq N$, that is, $\lim_{n \to \infty} f(x_n) = \omega$.

Conversely, suppose that for every sequence (x_n) of points $x_n \in S \setminus \{a\}$ such that $\lim_{n \to \infty} x_n = a$ we have $\lim_{n \to \infty} f(x_n) = \omega$. Suppose, for contradiction, that it is false that $\lim_{x \to a} f(x) = \omega$. In that case there would exist $\varepsilon > 0$ such that $\forall \delta > 0 \exists x \in B(a, \delta) \setminus \{a\}$ with the property $\rho(f(x), \omega) \geq \varepsilon$. Choose $\delta = 1/n$, and call x_n the point (one of the points) in $B(a, 1/n) \setminus \{a\}$ for which $\rho(f(x_n), \omega) \geq \varepsilon$. We have thus constructed a sequence (x_n) in $S \setminus \{a\}$ such that $\lim_{n \to \infty} x_n = a$ but for which $f(x_n)$ does not converge to ω. We must therefore admit that $\lim_{x \to a} f(x) = \omega$. $\square$

Theorem 3.4 *Let* $f : (S_1, d_1) \mapsto (S_2, d_2)$ *and* $g : (S_2, d_2) \mapsto (S_3, d_3)$ *be two functions such that* $\lim_{x \to a} f(x) = b$ *and* $\lim_{y \to b} g(y) = c$, *with* $a \in S_1$, $b \in S_2$ *and* $c \in S_3$. *Then* $\lim_{x \to a}(g \circ f)(x) = c$ *if and only if at least one of the following two properties is satisfied*

(a) $\exists r > 0$ *such that* $f(x) \neq b$ $\forall x \in B(a, r) \setminus \{a\}$;
(b) $g(b) = c$.

Proof First, we show that only one of the above two properties is sufficient to have $\lim_{x \to a}(g \circ f)(x) = \lim_{x \to a} g(f(x)) = c$.

Suppose that property (a) is satisfied. Since $\lim_{y \to b} g(y) = c$, $\forall \varepsilon > 0 \exists \delta_g(\varepsilon) > 0$ such that $g(B(b, \delta_g) \setminus \{b\}) \subset B(c, \varepsilon)$, and since $\lim_{x \to a} f(x) = b$, $\forall \eta > 0 \exists \delta_f(\eta) > 0$ such that $f(B(a, \delta_f) \setminus \{a\}) \subset B(b, \eta)$. Choose $\eta = \delta_g$ and set $\delta = \min\{\delta_f, r\}$, we have $g(f(B(a, \delta) \setminus \{a\})) \subset g(B(b, \delta_g) \setminus \{b\}) \subset B(c, \varepsilon)$ with $\varepsilon > 0$ arbitrary and $\delta(\varepsilon) > 0$. We conclude that $\lim_{x \to a}(g \circ f)(x) = c$.

Suppose that property (b) is satisfied. Since $\lim_{y \to b} g(y) = c$ and $g(b) = c$, $\forall \varepsilon > 0 \exists \delta_g(\varepsilon) > 0$ such that $g(B(b, \delta_g)) \subset B(c, \varepsilon)$, and since $\lim_{x \to a} f(x) = b$, $\forall \eta > 0 \exists \delta_f(\eta) > 0$ such that $f(B(a, \delta_f) \setminus \{a\}) \subset B(b, \eta)$. Choose $\eta = \delta_g$ and set $\delta = \delta_f$, we have $g(f(B(a, \delta) \setminus \{a\})) \subset g(B(b, \delta_g)) \subset B(c, \varepsilon)$ with $\varepsilon > 0$ arbitrary and $\delta(\varepsilon) > 0$. We conclude that $\lim_{x \to a}(g \circ f)(x) = c$.

Conversely, let $\lim_{x \to a}(g \circ f)(x) = c$ and suppose, for the sake of argument, that both properties (a) and (b) are false. If property (a) is false then $\forall r > 0 \; \exists x \in B(a, r) \setminus \{a\}$ such that $f(x) = b$. Let $r = 1/n$, $n \in \mathbb{N}$, and call this point (one of these points) x_n, we conclude that there exists a sequence (x_n) of points $x_n \in S_1 \setminus \{a\}$ that converges to a and such that $f(x_n) = b$ for every n. Now consider the sequence $(g(f(x_n)))$ in S_3. For every n we have $g(f(x_n)) = g(b)$ with $g(b) \neq c$ since property (b) is supposed to be false. Therefore, the sequence $(g(f(x_n)))$ does not converge to c and by Theorem 3.3 this is in contradiction with the hypothesis $\lim_{x \to a} g(f(x)) = c$. $\qquad\square$

Note that if neither of the two properties 1 and 2 is satisfied, two cases may arise: the limit for $x \to a$ of $g \circ f$ exists but is different from c, or such a limit does not exist.

Example 3.5 Consider the functions $f : (\mathbb{R}, d) \mapsto (\mathbb{R}, d)$ and $g : (\mathbb{R}, d) \mapsto (\mathbb{R}, d)$, with d Euclidean distance, defined by

$$f(x) = 1,$$

$$g(y) = \begin{cases} 2 & y \neq 1 \\ 0 & y = 1 \end{cases}.$$

We have $\lim_{x \to 0} f(x) = 1$ and $\lim_{y \to 1} g(y) = 2$ with f not satisfying property 1 and g not satisfying property 2. The limit for $x \to 0$ of $g \circ f$ exists but is equal to $\lim_{x \to 0} g(f(x)) = 0$.

Example 3.6 Consider the functions $f : (\mathbb{R}, d) \mapsto (\mathbb{R}, d)$ and $g : (\mathbb{R}, d) \mapsto (\mathbb{R}, d)$, with d Euclidean distance, defined by

$$f(x) = \begin{cases} x \sin(\pi/x) & x \neq 0 \\ 0 & x = 0 \end{cases},$$

$$g(y) = \begin{cases} y & y \neq 0 \\ 1 & y = 0 \end{cases}.$$

We have $\lim_{x \to 0} f(x) = 0$ and $\lim_{y \to 0} g(y) = 0$ with f not satisfying property 1 and g not satisfying property 2. The limit for $x \to 0$ of $g \circ f$ does not exist since

$$g(f(x)) = \begin{cases} x \sin(\pi/x) & x \in \mathbb{R} \setminus Z \\ 1 & x \in Z \end{cases},$$

where $Z = \{0, \pm 1, \pm \frac{1}{2}, \pm \frac{1}{3}, \dots\}$ is the set of the solutions of $f(x) = 0$.

In the case of complex functions, to the properties of the limit above derived from Definition 3.1, we can add the next one.

Theorem 3.7 *Let $f : \mathbb{C} \mapsto \mathbb{C}$ and $a, w \in \mathbb{C}$. Then*

$$\lim_{z \to a} f(z) = w \quad \textit{if and only if} \quad \begin{cases} \lim_{z \to a} \mathrm{Re}\, f(z) = \mathrm{Re}\, w \\ \lim_{z \to a} \mathrm{Im}\, f(z) = \mathrm{Im}\, w \end{cases}.$$

Proof Elementary, taking advantage of the definition of limit and observing that $\forall z \in G$ the following inequalities hold

$$|\mathrm{Re}\, f(z) - \mathrm{Re}\, w| = |\mathrm{Re}(f(z) - w)| \le |f(z) - w|,$$

$$|\mathrm{Im}\, f(z) - \mathrm{Im}\, w| = |\mathrm{Im}(f(z) - w)| \le |f(z) - w|,$$

and the triangle inequality

$$|f(z) - w| = |\mathrm{Re}(f(z) - w) + i\,\mathrm{Im}(f(z) - w)|$$

$$\le |\mathrm{Re}\, f(z) - \mathrm{Re}\, w| + |\mathrm{Im}\, f(z) - \mathrm{Im}\, w|.$$

$\square$

Theorem 3.8 *Let $f : (S, d) \mapsto \mathbb{C}$, $g : (S, d) \mapsto \mathbb{C}$, $a \in S$ and $w_f, w_g \in \mathbb{C}$. If*

$$\lim_{x \to a} f(x) = w_f \quad e \quad \lim_{x \to a} g(x) = w_g,$$

then

$$\lim_{x \to a} (f \pm g)(x) = w_f \pm w_g,$$

$$\lim_{x \to a} (fg)(x) = w_f w_g,$$

$$\lim_{x \to a} (f/g)(x) = w_f / w_g, \qquad \textit{if } w_g \ne 0.$$

Proof Elementary, using the definition of limit. $\square$

In the extended complex plane the open ball with center ∞ and radius r is defined as the set $B(\infty, r) = \{z \in \mathbb{C} : 1/|z| < r\}$. As a result, the limits involving the point at infinity can be rewritten in terms of suitable ordinary limits.

Theorem 3.9 *Let $f : \mathbb{C} \mapsto \mathbb{C}$ and $a, w \in \mathbb{C}$. Then*

$$\lim_{z \to a} f(z) = \infty \quad \textit{if and only if} \quad \lim_{z \to a} \frac{1}{f(z)} = 0, \tag{3.1}$$

$$\lim_{z \to \infty} f(z) = w \quad \textit{if and only if} \quad \lim_{z \to 0} f\left(\frac{1}{z}\right) = w, \tag{3.2}$$

$$\lim_{z \to \infty} f(z) = \infty \quad \textit{if and only if} \quad \lim_{z \to 0} \frac{1}{f\left(\frac{1}{z}\right)} = 0. \tag{3.3}$$

Proof Elementary, using the definition of limit and that of open ball of center ∞.
□

Example 3.10 Consider the following three limits:

$$\lim_{z \to 1} \frac{1}{(z-1)^3} = \infty \quad \text{since} \quad \lim_{z \to 1} \left(\frac{1}{(z-1)^3} \right)^{-1} = \lim_{z \to 1} (z-1)^3 = 0,$$

$$\lim_{z \to \infty} \frac{4z^2}{(z-1)^2} = 4 \quad \text{since} \quad \lim_{z \to 0} \frac{4z^{-2}}{(z^{-1}-1)^2} = \lim_{z \to 0} \frac{4}{(1-z)^2} = 4,$$

$$\lim_{z \to \infty} \frac{z^2+1}{z-1} = \infty \quad \text{since} \quad \lim_{z \to 0} \left(\frac{z^{-2}+1}{z^{-1}-1} \right)^{-1} = \lim_{z \to 0} \frac{z-z^2}{1+z^2} = 0.$$

3.2 Continuity

Definition 3.11 (Continuous Function) Let $f : (S, d) \mapsto (\Omega, \rho)$. The function f is said to be continuous at the point $a \in S$ if $\lim_{x \to a} f(x) = f(a)$, that is, if $\forall \varepsilon > 0 \exists \delta(\varepsilon) > 0$ such that $f(x) \in B(f(a), \varepsilon) \forall x \in B(a, \delta)$. The function f is said to be continuous if it is continuous at every point of S, that is, if $\forall a \in S$ and $\forall \varepsilon > 0$ $\exists \delta(\varepsilon, a) > 0$ such that $\rho(f(x), f(a)) < \varepsilon \ \forall x \in S$ such that $d(x, a) < \delta$.

The continuity of a function at a point or in the whole domain is characterized by the following properties.

Theorem 3.12 *Let $f : (S, d) \mapsto (\Omega, \rho)$ and $a \in S$. The following statements are equivalent:*

(a) *f is continuous at a;*
(b) *$\forall \varepsilon > 0$ $f^{-1}(B(f(a), \varepsilon))$ contains a ball with center a;*
(c) *if $\lim_{n \to \infty} x_n = a$, then $\lim_{n \to \infty} f(x_n) = f(a)$.*

Proof Elementary.
□

Example 3.13 The distance function $d : S \times S \mapsto \mathbb{R}$ is continuous. We assume that $S \times S$ is equipped with a distance $D : (S \times S) \times (S \times S) \to \mathbb{R}$ such that the property $\lim_{(x', y') \to (x, y)} D((x', y'), (x, y)) = 0$ is equivalent to saying that $\lim_{x' \to x} d(x', x) = 0$ and $\lim_{y' \to x} d(y', y) = 0$.

By the previous theorem, it is enough to show that, if $((x_n, y_n))$ is any sequence convergent in $S \times S$ to (x, y), then $\lim_{n \to \infty} d(x_n, y_n) = d(x, y)$. We have

$$|d(x, y) - d(x_n, y_n)| = |d(x, y) - d(x_n, y) + d(x_n, y) - d(x_n, y_n)|$$

$$\leq |d(x, y) - d(x_n, y)| + |d(x_n, y) - d(x_n, y_n)|.$$

For the triangle property of the distance d we can write

$$d(x, y) \le d(x, x_n) + d(x_n, y), \qquad d(x_n, y) \le d(x_n, x) + d(x, y),$$

which implies $|d(x, y) - d(x_n, y)| \le d(x, x_n)$. Similarly,

$$d(x_n, y) \le d(x_n, y_n) + d(y_n, y), \qquad d(x_n, y_n) \le d(x_n, y) + d(y, y_n),$$

which implies $|d(x_n, y) - d(x_n, y_n)| \le d(y, y_n)$. We thus obtain

$$|d(x, y) - d(x_n, y_n)| \le d(x, x_n) + d(y, y_n).$$

On the other hand, by the properties of D $\lim_{n \to \infty} d(x, x_n) = \lim_{n \to \infty} d(y, y_n) = 0$ and therefore the statement is proved.

Theorem 3.14 *Let* $f : (S, d) \mapsto (\Omega, \rho)$. *The following statements are equivalent:*

(a) *f is continuous;*
(b) *if Γ is open in Ω then $f^{-1}(\Gamma)$ is open in S;*
(c) *if Γ is closed in Ω then $f^{-1}(\Gamma)$ is closed in S.*

Proof We prove in order the implications (a) $\Rightarrow$ (b), (b) $\Rightarrow$ (c) and finally (c) $\Rightarrow$ (a).

(a) $\Rightarrow$ (b) Suppose that f is continuous. Let $\Gamma \subset \Omega$ be open and consider a point $x \in f^{-1}(\Gamma) \subset S$. Since $f(x) \in \Gamma$ and Γ is open, $\exists \varepsilon > 0$ such that $B(f(x), \varepsilon) \subset \Gamma$. By point 2 of Theorem 3.12 $\exists \delta > 0$ such that $B(x, \delta) \subset f^{-1}(\Gamma)$. From the arbitrariness of x it follows that $f^{-1}(\Gamma)$ is open.

(b) $\Rightarrow$ (c) Suppose that every open set in Ω is transformed by f^{-1} into an open set in S. If $\Gamma \subset \Omega$ is closed, then $\Omega \setminus \Gamma$ is open and hence also $f^{-1}(\Omega \setminus \Gamma) = S \setminus f^{-1}(\Gamma)$ is open. Therefore, $f^{-1}(\Gamma)$ is closed.

(c) $\Rightarrow$ (a) Suppose that every closed set in Ω is transformed by f^{-1} into a closed set in S. If, by contradiction, f is not continuous at some point $x \in S$, then there exists $\varepsilon > 0$ and a sequence (x_n) such that for every n we have $\rho(f(x_n), f(x)) \ge \varepsilon$ while $\lim_{n \to \infty} x_n = x$. Let $\Gamma = \Omega \setminus B(f(x), \varepsilon)$. By construction Γ is closed and every $x_n \in f^{-1}(\Gamma)$. Since $f^{-1}(\Gamma)$ is closed, we must have $x \in f^{-1}(\Gamma)$. This implies $\rho(f(x), f(x)) \ge \varepsilon$, which is absurd, so f must be continuous. $\square$

Theorem 3.15 *Let* $f : (S_1, d_1) \mapsto (S_2, d_2)$ *and* $g : (S_2, d_2) \mapsto (S_3, d_3)$ *continue. Then* $g \circ f : (S_1, d_1) \mapsto (S_3, d_3)$ *is continuous.*

Proof If A is open in (S_3, d_3), then $g^{-1}(A)$ is open in (S_2, d_2) and $f^{-1}(g^{-1}(A))$ is open in (S_1, d_1). Observing that $(g \circ f)(x) = g(f(x))$ and therefore $f^{-1}(g^{-1}(x)) = (g \circ f)^{-1}(x)$, by point 2 of Theorem 3.14 we conclude that $g \circ f$ is continuous. $\square$

In the case of complex functions, the following properties also hold.

Theorem 3.16 *Let $f : \mathbb{C} \mapsto \mathbb{C}$ and $a \in \mathbb{C}$. The function f is continuous at a if and only if the component functions $\operatorname{Re} f, \operatorname{Im} f : \mathbb{R}^2 \mapsto \mathbb{R}$ are continuous at $(\operatorname{Re} a, \operatorname{Im} a)$.*

Proof Elementary, using the definition of continuity and Theorem 3.7. □

Theorem 3.17 *Let $f : (S, d) \mapsto \mathbb{C}$ and $g : (S, d) \mapsto \mathbb{C}$ be continuous. Then the functions $f \pm g$ and fg are continuous. The function f/g is continuous at every point $x \in S$ such that $g(x) \neq 0$.*

Proof Elementary, using the definition of continuity. □

Recall that a function $f : (S, d) \mapsto (\Omega, \rho)$ is said to be continuous in S if $\forall x \in S$ and $\forall \varepsilon > 0$ there exists $\delta(\varepsilon, x) > 0$ such that $\rho(f(x), f(y)) < \varepsilon \; \forall y \in B(x, \delta)$. We give the following further definitions.

Definition 3.18 (Uniformly Continuous Function) Let $f : (S, d) \mapsto (\Omega, \rho)$. The function f is said to be uniformly continuous in S if $\forall \varepsilon > 0 \; \exists \delta(\varepsilon) > 0$ such that $\rho(f(x), f(y)) < \varepsilon \; \forall x, y \in S$ with $d(x, y) < \delta$.

Definition 3.19 (Continuous Lipschitz Function) Let $f : (S, d) \mapsto (\Omega, \rho)$. The function f is said to be Lipschitz continuous in S if $\exists M > 0$ such that $\forall x, y \in S$ $\rho(f(x), f(y)) < M d(x, y)$.

It is evident that a uniformly continuous function is continuous and a Lipschitz continuous function is uniformly continuous. In both cases, the vice versa is generally false.

Example 3.20 The function $f(x) = x^2$ with $x \in [0, \infty)$ is continuous but not uniformly continuous. In fact, given a real number $a > 0$ and choosing $\varepsilon > 0$ such that $f(a) - \varepsilon > 0$, let $\delta(\varepsilon, a)$ be the radius of the ball with center a such that $f(B(a, \delta)) \subset B(f(a), \varepsilon)$. We have

$$\delta(\varepsilon, a) = \min \left(f^{-1}(f(a) + \varepsilon) - a, a - f^{-1}(f(a) - \varepsilon) \right)$$

$$= \min \left(\sqrt{a^2 + \varepsilon} - a, a - \sqrt{a^2 - \varepsilon} \right)$$

$$= \sqrt{a^2 + \varepsilon} - a.$$

Since $\lim_{a \to \infty} \sqrt{a^2 + \varepsilon} - a = 0$, it turns out

$$\inf_{a \in [0, \infty)} \delta(\varepsilon, a) = 0,$$

so the function f is not uniformly continuous in $[0, \infty)$ (Fig. 3.1).

Example 3.21 The function $f(x) = \sqrt{x}$ with $x \in [0, \infty)$ is uniformly continuous but not Lipschitz continuous. To demonstrate that f is uniformly continuous, reason as in the previous example. Given a real number $a > 0$ and choosing $\varepsilon > 0$ such

Fig. 3.1 Graph of $f(x) = x^2$. Constant increments of the function $f(a) \pm \varepsilon$ correspond to increments of the variable $a - \delta_-$ and $a + \delta_+$ that differ depending on the value of a. Setting $\delta = \min(\delta_-, \delta_+)$ we have $f(B(a, \delta)) \subset B(f(a), \varepsilon)$. Note that $\delta(\varepsilon, a)$ decreases as a increases

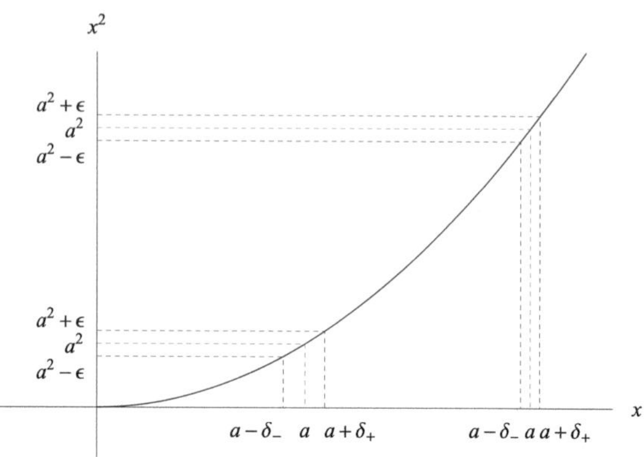

that $f(a) - \varepsilon > 0$, let $\delta(\varepsilon, a)$ be the radius of the ball with center a such that $f(B(a, \delta)) \subset B(f(a), \varepsilon)$. We have

$$\delta(\varepsilon, a) = \min\left(f^{-1}(f(a) + \varepsilon) - a, a - f^{-1}(f(a) - \varepsilon) \right)$$

$$= \min\left((\sqrt{a} + \varepsilon)^2 - a, a - (\sqrt{a} - \varepsilon)^2 \right)$$

$$= 2\varepsilon\sqrt{a} - \varepsilon^2$$

$$= \varepsilon(\sqrt{a} + (\sqrt{a} - \varepsilon)) > \varepsilon\sqrt{a} > \varepsilon^2,$$

having exploited the fact that $f(a) - \varepsilon > 0$, that is $\sqrt{a} > \varepsilon$. Since

$$\inf_{a \in [0, \infty)} \delta(\varepsilon, a) \geq \varepsilon^2 > 0,$$

the choice $\delta(\varepsilon) = \varepsilon^2$ ensures the uniform continuity of f in $[0, \infty)$. On the other hand, f is not Lipschitz continuous because, considering the pair of points x and 0, the inequality $\left| \sqrt{x} - \sqrt{0} \right| \leq M |x - 0|$ implies $M \geq 1/\sqrt{x}$ which cannot be satisfied for any value of the real constant M when $x \in (0, \infty)$ (Fig. 3.2).

If f and g are two uniformly (Lipschitz) continuous functions from (S, d) to $\mathbb{C}$, their sum $f + g$ is still a uniformly (Lipschitz) continuous function while the product fg in general is not.

Example 3.22 The function $f(x) = x^{2/3}$ with $x \in [0, \infty)$ is uniformly continuous while $(ff)(x) = x^{4/3}$ is only continuous. In fact, generalizing the two previous examples, we have that a function of the type $f(x) = x^\alpha$ is uniformly continuous in $[0, \infty)$ if $0 < \alpha < 1$, while for $\alpha > 1$ in the same domain it is only continuous.

Example 3.23 The function $f(x) = (1 + x^2)^{1/2}$ with $x \in \mathbb{R}$ is Lipschitz continuous with constant $M = 1$ while $(ff)(x) = 1 + x^2$ is only continuous.

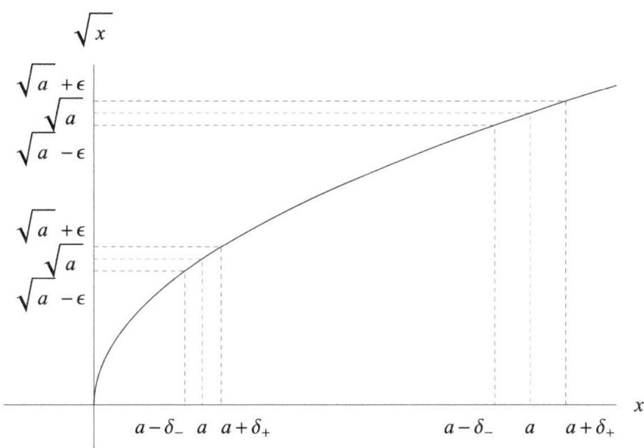

Fig. 3.2 As in the previous figure in the case of $f(x) = \sqrt{x}$. Note that $\delta(\varepsilon, a)$ decreases as a decreases

Theorem 3.24 *Let $f : (S, d) \mapsto (\Omega, \rho)$ be continuous. Then*

(a) *if S is compact, $f(S)$ is a compact subset of Ω;*
(b) *if S is connected, $f(S)$ is a connected subset of Ω.*

Proof

(a) Let (ω_n) be a generic sequence of points $\omega_n \in f(S) \subset \Omega$. For every ω_n $\exists x_n \in S$ such that $\omega_n = f(x_n)$. Consider the sequence (x_n) in S. Since S is compact, (x_n) admits a subsequence (x_{n_k}) convergent to $x \in S$. Set $f(x) = \omega \in f(S)$, since f is continuous $\lim_{n_k \to \infty} \omega_{n_k} = \omega$. We conclude that $f(S)$ is sequentially compact.
(b) Let $\Gamma \subset f(S)$ be a subset of $f(S) \subset \Omega$ that is nonempty and at the same time open and closed in Ω. Since $\Gamma \neq \emptyset$, it must also be $f^{-1}(\Gamma) \neq \emptyset$. Furthermore, since Γ is both open and closed in Ω and f is continuous, $f^{-1}(\Gamma)$ is both open and closed in S. Since S is connected, it must be $f^{-1}(\Gamma) = S$ and hence $\Gamma = f(S)$. We conclude that $f(S)$ is connected.

$\square$

Corollary 3.25 *Let $f : (S, d) \mapsto (\Omega, \rho)$ be continuous and $K \subset S$. Then*

(a) *if K is compact, $f(K)$ is compact in Ω;*
(b) *if K is connected, $f(K)$ is connected in Ω.*

Theorem 3.26 *Let $f : (S, d) \mapsto \mathbb{R}$ be continuous and $K \subset S$ compact. Then f assumes absolute maximum and minimum in K, that is $\exists x_a, x_b \in K$ such that $\forall x \in K$ we have $f(x_a) \leq f(x) \leq f(x_b)$.*

Proof Set $\alpha = \inf_{x \in K} f(x)$ and $\beta = \sup_{x \in K} f(x)$. Since f is continuous and K is compact, the subset $f(K) \subset \mathbb{R}$ is compact and hence, by the Heine-Borel theorem, closed and bounded. Therefore, $\alpha, \beta \in f(K)$, that is, $\exists x_a, x_b \in K$ such that $f(x_a) = \alpha$ and $f(x_b) = \beta$. $\square$

Corollary 3.27 *Let $f : (S, d) \mapsto \mathbb{C}$ be continuous and $K \subset S$ be compact. Then $|f|$ assumes in K an absolute maximum and minimum, that is, $\exists x_a, x_b \in K$ such that $\forall x \in K$ we have $|f(x_a)| \leq |f(x)| \leq |f(x_b)|$.*

Proof The function $|f(x)| : (S, d) \mapsto \mathbb{R}$ is continuous.   □

Theorem 3.28 *Let $f : (S, d) \mapsto \mathbb{C}$. If f is continuous at $a \in S$ and $f(a) \neq 0$, then $\exists \delta > 0$ such that $f(x) \neq 0 \; \forall x \in B(a, \delta)$.*

Proof Since f is continuous at a, $\forall \varepsilon > 0$ there exists $\delta(\varepsilon) > 0$ such that $|f(x) - f(a)| < \varepsilon \; \forall x \in B(a, \delta)$. Choose $\varepsilon = |f(a)|/2$. Then $|f(x) - f(a)| < |f(a)|/2 \forall x \in B(a, \delta)$. If $f(x) = 0$ at some $x \in B(a, \delta)$ it would be $|f(a)| < |f(a)|/2$. Therefore, we must have $f(x) \neq 0 \; \forall x \in B(a, \delta)$.   □

Theorem 3.29 *Let $f : (S, d) \mapsto (\Omega, \rho)$ be continuous and S compact, then f is uniformly continuous.*

Proof Suppose by contradiction that f is not uniformly continuous. Then $\exists \varepsilon > 0$ such that $\forall \delta > 0$ for at least one pair of points $x, y \in S$ with distance $d(x, y) < \delta$ we have $\rho(f(x), f(y)) \geq \varepsilon$. Choosing $\delta = 1/n$, this means that $\forall n \geq 1, \exists x_n, y_n \in S$ such that $d(x_n, y_n) < 1/n$ and $\rho(f(x_n), f(y_n)) \geq \varepsilon$. Since S is compact, it is possible to extract from (x_n) a subsequence (x_{n_k}) convergent to some point $x \in S$. If $\lim_{n_k \to \infty} x_{n_k} = x$, then also $\lim_{n_k \to \infty} y_{n_k} = x$. Indeed,

$$d(x, y_{n_k}) \leq d(x, x_{n_k}) + d(x_{n_k}, y_{n_k}) < d(x, x_{n_k}) + 1/n_k \xrightarrow{n_k \to \infty} 0.$$

Since f is continuous, $\lim_{n_k \to \infty} f(x_{n_k}) = \lim_{n_k \to \infty} f(y_{n_k}) = f(x)$ and therefore

$$\rho(f(x_{n_k}), f(y_{n_k})) \leq \rho(f(x_{n_k}), f(x)) + \rho(f(x), f(y_{n_k})) \xrightarrow{n_k \to \infty} 0$$

in contradiction with the property $\rho(f(x_{n_k}), f(y_{n_k})) \geq \varepsilon \forall n_k \geq 1$.   □

Exercises

3.1 Determine, if they exist, the following limits justifying the answer:

$$\text{(a)} \quad \lim_{z \to \infty} \frac{6(z^3 - 2)^2}{(z - 1)^4}, \qquad \text{(b)} \quad \lim_{z \to 1} \frac{z^{101} - 1}{z - 1}, \qquad \text{(c)} \quad \lim_{z \to i} \frac{(z^2 + 1)^2}{(z - i)^2}.$$

3.2 Determine, if they exist, the following limits justifying the answer:

$$\text{(a)} \quad \lim_{z \to i} \frac{z}{z^2 + 1}, \qquad \text{(b)} \quad \lim_{z \to i} \frac{\sqrt{z}}{z}, \qquad \text{(c)} \quad \lim_{z \to i} \frac{z^4 - z^3 - z^2 + z}{z + 1}.$$

Assume $\sqrt{w} = \sqrt{|w|} e^{\frac{i}{2} \operatorname{Arg} w}$.

3.3 Determine, if they exist, the following limits justifying the answer:

$$\text{(a)} \quad \lim_{z \to 1} \operatorname{Re} \frac{1+z}{z-3i}, \qquad \text{(b)} \quad \lim_{z \to -3} |z|^{\frac{1}{2}} \, e^{\frac{i}{2} \operatorname{Arg} z}, \qquad \text{(c)} \quad \lim_{z \to 0} \left(\frac{z}{\bar{z}} \right)^3 .$$

3.4 Determine, if they exist, the following limits justifying the answer:

$$\text{(a)} \quad \lim_{z \to 0} \frac{z^2}{|z|^2}, \qquad \text{(b)} \quad \lim_{z \to \infty} \left(\sqrt{z+i} - \sqrt{z-i} \right) \sqrt{z}, \qquad \text{(c)} \quad \lim_{z \to \infty} \frac{1 - |z|^2}{z^2 + i}.$$

Assume $\sqrt{w} = \sqrt{|w|} e^{\frac{i}{2} \operatorname{Arg} w}$.

3.5 Determine, if they exist, the following limits justifying the answer:

$$\text{(a)} \quad \lim_{z \to 1+i} \frac{z^2 - 2i}{z^2 - 2z + 2}, \qquad \text{(b)} \quad \lim_{z \to -3} \frac{\sqrt{z}}{z}, \qquad \text{(c)} \quad \lim_{z \to \infty} (\sqrt{z+2i} - \sqrt{z+i}).$$

Assume $\sqrt{w} = \sqrt{|w|} e^{\frac{i}{2} \operatorname{Arg} w}$.

3.6 Prove that $\lim_{z \to z_0} f(z) = w_0$ implies $\lim_{z \to z_0} |f(z)| = |w_0|$. Show with an example that the vice versa is false if $w_0 \neq 0$. What happens if $w_0 = 0$?

3.7 Study the convergence of the following sequences $(z_n)_{n=1}^{\infty}$:

$$\text{(a)} \quad z_n = \frac{i^n}{n}, \qquad \text{(b)} \quad z_n = \frac{(1+i)^n}{n}, \qquad \text{(c)} \quad z_n = \sqrt{n} \left(\sqrt{n+2i} - \sqrt{n+i} \right).$$

Assume $\sqrt{w} = \sqrt{|w|} e^{\frac{i}{2} \operatorname{Arg} w}$.

3.8 Find the points of discontinuity of the following functions:

$$1. \quad f(z) = \operatorname{Arg} z, \qquad 2. \quad f(z) = \frac{z}{z^3 + 27}, \qquad 3. \quad f(z) = \sqrt{|z|} e^{\frac{i}{2} \operatorname{Arg} z}.$$

3.9 Show that the function $f(z) = 1/z$ is not uniformly continuous in $A = \{z \in \mathbb{C} : 0 < |z| \leq 2\}$.

3.10 Show that the function $f(z) = z^{-2}$ is not uniformly continuous in the domain $D = \{z \in \mathbb{C} : 0 < |z| \leq 3\}$. Propose a modification of D such that in the modified domain f is uniformly continuous.

3.11 Let $f : (S, d) \mapsto \mathbb{C}$ and $g : (S, d) \mapsto \mathbb{C}$ be two functions that are uniformly continuous in S. Show that $f + g : (S, d) \mapsto \mathbb{C}$ is uniformly continuous in S.

3.12 Let $f : (S, d) \mapsto \mathbb{C}$ and $g : (S, d) \mapsto \mathbb{C}$ be Lipschitz continuous functions in S. Prove that $f + g : (S, d) \mapsto \mathbb{C}$ is Lipschitz continuous in S.

Chapter 4
Sequences and Series of Functions

Abstract Convergent and uniformly convergent sequences of functions. A uniformly convergent sequence of continuous functions converges to a continuous function. Partial sums of a sequence of functions with values in $\mathbb{C}$, series of functions. Convergence, uniform convergence and absolute convergence of a series of functions. A series of functions absolutely convergent is convergent. Weierstrass criterion for the uniform convergence of a series of functions. Upper and lower limits of a real numerical sequence. Power series. The geometric series. Radius of convergence, Abel's theorem. Ratio criterion.

4.1 Sequences and Series of Functions

Let S be a set and (Ω, ρ) be a metric space. Consider the sequence of functions $(f_n)_{n=1}^{\infty}$ with $f_n : S \mapsto (\Omega, \rho)$.

Definition 4.1 (Convergent Sequence of Functions) The sequence of functions (f_n) converges to $f : S \mapsto (\Omega, \rho)$, and we write

$$\lim_{n \to \infty} f_n(x) = f(x) \qquad \text{or} \qquad f_n(x) \xrightarrow{n \to \infty} f(x),$$

if $\forall x \in S$ and $\forall \varepsilon > 0 \; \exists N(\varepsilon, x)$ integer such that $\rho(f_n(x), f(x)) < \varepsilon \; \forall n \geq N$.

Definition 4.2 (Uniformly Convergent Sequence of Functions) The sequence of functions (f_n) converges uniformly to $f : S \mapsto (\Omega, \rho)$ if $\forall \varepsilon > 0 \; \exists N(\varepsilon)$ integer such that $\rho(f_n(x), f(x)) < \varepsilon \; \forall n \geq N$ and $\forall x \in S$.

If S is equipped with a distance d the following theorem holds.

Theorem 4.3 *Let (f_n) be a sequence of continuous functions $f_n : (S, d) \mapsto (\Omega, \rho)$ uniformly convergent to $f : (S, d) \mapsto (\Omega, \rho)$, then f is continuous.*

Proof Take arbitrary $x \in S$ and $\varepsilon > 0$. Since the sequence (f_n) converges uniformly to f, $\exists n$ such that $\rho(f(y), f_n(y)) < \varepsilon/3 \; \forall y \in S$. On the other hand, since f_n is

© The Author(s), under exclusive license to Springer Nature Switzerland AG 2026

C. Presilla, *Complex Functions of a Variable*, La Matematica per il 3+2 179,

https://doi.org/10.1007/978-3-032-12494-4_4

continuous, $\exists \delta(\varepsilon) > 0$ such that $\rho(f_n(y), f_n(x)) < \varepsilon/3 \; \forall x, y \in S$ with $d(x, y) < \delta$. By the triangle property of the distance, we then have

$$\rho(f(y), f(x)) \le \rho(f(y), f_n(y)) + \rho(f_n(y), f_n(x)) + \rho(f_n(x), f(x))$$

$$< \varepsilon/3 + \varepsilon/3 + \varepsilon/3 = \varepsilon,$$

$\forall y \in S$ with $d(y, x) < \delta$. From the arbitrariness of ε follows the continuity of f at x and from the arbitrariness of x the continuity of f in S. $\qquad \square$

One may suspect that, if the functions $f_n : (S, d) \mapsto (\Omega, \rho)$ converging to $f : (S, d) \mapsto (\Omega, \rho)$ are continuous and S is compact, then the convergence is always uniform. In general, this is not true. However, if the convergence is not uniform, the limit function may be continuous.

Example 4.4 Consider the sequence of functions $f_n(x) : [0, 1] \mapsto [0, 1]$ with

$$f_n(x) = \begin{cases} 1 - n \, |x - 1/n| & 0 \le x \le 2/n \\ 0 & 2/n < x \le 1 \end{cases}.$$

It is immediate to show that this sequence converges, pointwise but not uniformly, to the continuous function $f(x) = 0$.

Consider now the special case in which $\Omega = \mathbb{C}$. If (f_n) is a sequence of functions $f_n : (S, d) \mapsto \mathbb{C}$ (we omit to indicate the distance of $\mathbb{C}$, which is chosen as the usual one if not differently specified) it is possible to construct the sequence (s_n) of the n-th partial sum functions

$$s_n(x) = f_1(x) + f_2(x) + \cdots + f_n(x) : (S, d) \mapsto \mathbb{C}.$$

The sequence of functions (s_n) is called the series of functions associated with the sequence of functions (f_n).

Definition 4.5 (Convergent Series of Functions) The series of functions associated with the sequence of functions (f_n) with $f_n(x) : (S, d) \mapsto \mathbb{C}$ is said to converge to the function $s(x) : S \mapsto \mathbb{C}$, and we write

$$s(x) = \sum_{k=1}^{\infty} f_k(x),$$

if $\forall x \in S$

$$s(x) = \lim_{n \to \infty} s_n(x) = \lim_{n \to \infty} \sum_{k=1}^{n} f_k(x).$$

A necessary condition for a series of functions to converge is that $\forall x \in S$ we have $\lim_{n\to\infty} f_n(x) = 0$. In fact, if the series of functions converges, $\forall x \in S$ we have

$$\lim_{n\to\infty} f_n(x) = \lim_{n\to\infty} (s_n(x) - s_{n-1}(x)) = \lim_{n\to\infty} s_n(x) - \lim_{n\to\infty} s_{n-1}(x) = 0.$$

A series of functions that does not converge is said to be divergent. In other words, a series of functions is divergent if for some $x \in S$ we have $s(x) = \infty$ or $\nexists \lim_{n\to\infty} s_n(x)$.

Definition 4.6 (Uniformly Convergent Series of Functions) The series of functions $\sum_{k=1}^{\infty} f_k(x)$ is said to converge uniformly to $s(x)$ if the sequence of n-th partial sums (s_n) converges uniformly to s.

Definition 4.7 (Absolutely Convergent Series of Functions) The series of functions $\sum_{k=1}^{\infty} f_k(x)$ is said to converge absolutely if the series of functions $\sum_{k=1}^{\infty} |f_k(x)|$ is convergent.

Theorem 4.8 *Let (f_n) be a sequence of functions $f_n : (S, d) \mapsto \mathbb{C}$. If the series of functions $\sum_{k=1}^{\infty} f_k(x)$ is absolutely convergent, then this series is convergent.*

Proof Let $s_n(x) = f_1(x) + f_2(x) + \cdots + f_n(x)$. By hypothesis $\exists g : (S, d) \mapsto \mathbb{R}$ such that $\sum_{k=1}^{\infty} |f_k(x)| = g(x)$. In other words $\forall x \in S$ and $\forall \varepsilon > 0\ \exists N(\varepsilon, x)$ integer such that $\left| g(x) - \sum_{k=1}^{n} |f_k(x)| \right| = \sum_{k=n+1}^{\infty} |f_k(x)| < \varepsilon\ \forall n \geq N$. Then $\forall n, m \geq N$, assuming $m > n$, we have

$$|s_m(x) - s_n(x)| = \left| \sum_{k=n+1}^{m} f_k(x) \right| \leq \sum_{k=n+1}^{m} |f_k(x)| \leq \sum_{k=n+1}^{\infty} |f_k(x)| < \varepsilon.$$

For any fixed $x \in S$ we thus find that $(s_n(x))$ is a Cauchy sequence in $\mathbb{C}$. Since $\mathbb{C}$ is complete, $\exists s(x) \in \mathbb{C}$ such that $\lim_{n\to\infty} s_n(x) = s(x)$. $\square$

Theorem 4.9 (Weierstrass Criterion) *Let (f_n) be a sequence of functions $f_n : (S, d) \mapsto \mathbb{C}$. If for every integer $n\ \exists M_n > 0$ with the properties $|f_n(x)| \leq M_n$ $\forall x \in S$ and $\sum_{k=1}^{\infty} M_k < \infty$, then the series of functions $\sum_{k=1}^{\infty} f_k(x)$ is uniformly convergent.*

Proof Let $s_n(x) = f_1(x) + f_2(x) + \cdots + f_n(x)$. For each pair of integers m, n, suppose $m > n$, and $\forall x \in S$ we have

$$|s_m(x) - s_n(x)| = \left| \sum_{k=n+1}^{m} f_k(x) \right| \leq \sum_{k=n+1}^{n} |f_k(x)| \leq \sum_{k=n+1}^{n} M_k \leq \sum_{k=n+1}^{\infty} M_k.$$

Since $\sum_{k=1}^{\infty} M_k = M \in \mathbb{R}$, we have

$$\lim_{n \to \infty} \sum_{k=n+1}^{\infty} M_k = M - \lim_{n \to \infty} \sum_{k=1}^{n} M_k = 0.$$

This implies that $\forall \varepsilon > 0 \; \exists N(\varepsilon)$ integer such that $\sum_{k=n+1}^{\infty} M_k < \varepsilon \; \forall n \geq N$. Therefore $\forall \varepsilon > 0 \; \exists N(\varepsilon)$ integer such that

$$|s_m(x) - s_n(x)| < \varepsilon, \tag{4.1}$$

$\forall m, n \geq N$ and $\forall x \in S$. In other words, for every fixed $x \in S$ the sequence $(s_n(x))$ is a Cauchy sequence in $\mathbb{C}$. For the completeness of $\mathbb{C}$ this sequence is convergent. Let $s(x) = \lim_{n \to \infty} s_n(x)$. We have thus defined the limit function $s : (S, d) \mapsto \mathbb{C}$. Taking the limit $m \to \infty$ in (4.1), we obtain finally that $\forall \varepsilon > 0 \; \exists N(\varepsilon)$ integer such that $|s(x) - s_n(x)| \leq \varepsilon \; \forall n \geq N$ and $\forall x \in S$, that is, the sequence of functions $(s_n(x))$ converges to $s(x)$ uniformly in S. $\square$

4.2 Upper and Lower Bounds of a Real Sequence

Definition 4.10 (Upper Bound, Lower Bound) Given a sequence (a_n), we define the upper and lower bounds

$$\limsup_{n \to \infty} a_n = \lim_{n \to \infty} \sup_{k \geq n} a_k, \tag{4.2}$$

$$\liminf_{n \to \infty} a_n = \lim_{n \to \infty} \inf_{k \geq n} a_k. \tag{4.3}$$

Obviously this definition makes sense only for sequences in ordered sets where the concepts of sup and inf are defined. In the following we will limit ourselves to considering the case in which $a_n \in \mathbb{R}$.

The limits defined in (4.2) and (4.3) always exist. In the case of $\limsup$ set $a_n^{(s)} = \sup_{k \geq n} a_k$, it turns out that $(a_n^{(s)})$ is a non-increasing monotone sequence of real numbers, $a_{n+1}^{(s)} \leq a_n^{(s)}$, or $a_n^{(s)} = +\infty \forall n$. It follows that $\lim_{n \to \infty} a_n^{(s)} = a^{(s)}$ with $a^{(s)} \in \mathbb{R}$ or $a^{(s)} = \pm\infty$. In the case of $\liminf$ place $a_n^{(i)} = \inf_{k \geq n} a_k$, it turns out that $(a_n^{(i)})$ is a non-decreasing monotone sequence of real numbers, $a_{n+1}^{(i)} \geq a_n^{(i)}$, or $a_n^{(i)} = -\infty \; \forall n$. It follows that $\lim_{n \to \infty} a_n^{(i)} = a^{(i)}$ with $a^{(i)} \in \mathbb{R}$ or $a^{(i)} = \pm\infty$.

The properties of $\limsup$ and $\liminf$ and the relationship with the ordinary $\lim$ are established by the following theorems.

Theorem 4.11 *Let (a_n) be a sequence of real numbers. Then*

(a) $\limsup_{n\to\infty} a_n = a^{(s)} \in \mathbb{R}$ *if and only if* $\forall \varepsilon > 0 \; \exists N(\varepsilon)$ *integer such that* $a_n < a^{(s)} + \varepsilon \; \forall n \geq N$ *and furthermore* $a_n > a^{(s)} - \varepsilon$ *for infinite values of n;*

(b) $\limsup_{n\to\infty} a_n = +\infty$ *if and only if* a_n *assumes arbitrarily large values for infinite values of n;*

(c) $\limsup_{n\to\infty} a_n = -\infty$ *if and only if* $\lim_{n\to\infty} a_n = -\infty$;

(d) $\liminf_{n\to\infty} a_n = a^{(i)} \in \mathbb{R}$ *if and only if* $\forall \varepsilon > 0 \; \exists N(\varepsilon)$ *integer such that* $a_n > a^{(i)} - \varepsilon \; \forall n \geq N$ *and furthermore* $a_n < a^{(i)} + \varepsilon$ *for infinite values of n;*

(e) $\liminf_{n\to\infty} a_n = +\infty$ *if and only if* $\lim_{n\to\infty} a_n = +\infty$;

(f) $\liminf_{n\to\infty} a_n = -\infty$ *if and only if* a_n *assumes arbitrarily small values for infinite values of n.*

Proof Let's see in detail the three cases relating to $\limsup$:

(a) $\lim_{n\to\infty} \sup_{k\geq n} a_k = a^{(s)} \in \mathbb{R}$ is equivalent to saying that $\forall \varepsilon > 0 \; \exists N(\varepsilon)$ integer such that $|a^{(s)} - \sup_{k\geq n} a_k| < \varepsilon$, that is $a^{(s)} - \varepsilon < \sup_{k\geq n} a_k < a^{(s)} + \varepsilon$, $\forall n \geq N$. The above two inequalities in turn are equivalent to the two conditions $a_n < a^{(s)} + \varepsilon \; \forall n \geq N$ and $a^{(s)} - \varepsilon < a_n$ for infinite values of n;

(b) $\lim_{n\to\infty} \sup_{k\geq n} a_k = +\infty$ is equivalent to saying that $\forall \varepsilon > 0 \; \exists N(\varepsilon)$ integer such that $\sup_{k\geq n} a_k > 1/\varepsilon \; \forall n \geq N$. In turn this is equivalent to saying that a_n takes arbitrarily large values for infinite values of n;

(c) $\lim_{n\to\infty} \sup_{k\geq n} a_k = -\infty$ is equivalent to saying that $\forall \varepsilon > 0 \; \exists N(\varepsilon)$ integer such that $\sup_{k\geq n} a_k < -1/\varepsilon$, or $a_n < -1/\varepsilon$, $\forall n \geq N$. In turn this is equivalent to saying that $\lim_{n\to\infty} a_n = -\infty$.

A similar reasoning applies to $\liminf$. $\square$

Example 4.12 Consider the following sequences (a_n) of real numbers:

(a) $a_n - n$

$a_n^{(s)} = \sup_{k\geq n} a_k = +\infty$ therefore $\lim_{n\to\infty} a_n^{(s)} = +\infty$,

$a_n^{(i)} = \inf_{k\geq n} a_k = n$ therefore $\lim_{n\to\infty} a_n^{(i)} = +\infty$;

(b) $a_n = 1 - 1/n$

$a_n^{(s)} = \sup_{k\geq n} a_k = 1$ therefore $\lim_{n\to\infty} a_n^{(s)} = 1$,

$a_n^{(i)} = \inf_{k\geq n} a_k = 1 - 1/n$ therefore $\lim_{n\to\infty} a_n^{(i)} = 1$;

(c) $a_n = 1 + 1/n$

$a_n^{(s)} = \sup_{k\geq n} a_k = 1 + 1/n$ therefore $\lim_{n\to\infty} a_n^{(s)} = 1$,

$a_n^{(i)} - \inf_{k\geq n} a_k - 1$ therefore $\lim_{n\to\infty} a_n^{(i)} = 1$;

(d) $a_n = -n$

$a_n^{(s)} = \sup_{k\geq n} a_k = -n$ therefore $\lim_{n\to\infty} a_n^{(s)} = -\infty$,

$a_n^{(i)} = \inf_{k\geq n} a_k = -\infty$ therefore $\lim_{n\to\infty} a_n^{(i)} = -\infty$;

(e) $a_n = (-1)^n$

$a_n^{(s)} = \sup_{k\geq n} a_k = 1$ so $\lim_{n\to\infty} a_n^{(s)} = 1$,

$a_n^{(i)} = \inf_{k>n} a_k = -1$ so $\lim_{n\to\infty} a_n^{(i)} = -1$.

The previous example suggests that when the ordinary limit of the sequence (a_n) exists then the values of $\lim$, $\limsup$ and $\liminf$ coincide. This is better specified by the following theorems.

Theorem 4.13 *Let (a_n) be a sequence of real numbers. Then*

$$\liminf_{n\to\infty} a_n \leq \limsup_{n\to\infty} a_n. \tag{4.4}$$

Proof The obvious inequality $\inf_{k\geq n} a_k \leq \sup_{k\geq n} a_k$ is valid for every value of n and thus remains valid taking the limit $n \to \infty$. $\qquad\square$

Theorem 4.14 *Let (a_n) be a sequence of real numbers. Then $\lim_{n\to\infty} a_n = a$ if and only if $\liminf_{n\to\infty} a_n = \limsup_{n\to\infty} a_n = a$.*

Proof If, by hypothesis, $\lim_{n\to\infty} a_n = a \in \mathbb{R}$, then $\forall \varepsilon > 0 \; \exists N(\varepsilon)$ integer such that $|a_n - a| < \varepsilon/2$, that is, $a - \varepsilon/2 < a_n < a + \varepsilon/2$, $\forall n \geq N$. As a consequence $a - \varepsilon/2 < \sup_{k\geq n} a_k \leq a + \varepsilon/2$, that is $|\sup_{k\geq n} a_k - a| \leq \varepsilon/2 < \varepsilon$, $\forall n \geq N$. So $\limsup_{n\to\infty} a_n = a$. Similarly, $a - \varepsilon/2 \leq \inf_{k\geq n} a_k < a + \varepsilon/2$, that is $|\inf_{k\geq n} a_k - a| \leq \varepsilon/2 < \varepsilon$, $\forall n \geq N$. So $\liminf_{n\to\infty} a_n = a$.

Conversely, suppose that $\limsup_{n\to\infty} a_n = \liminf_{n\to\infty} a_n = a \in \mathbb{R}$. This means that $\forall \varepsilon > 0 \; \exists N_s(\varepsilon)$ and $\exists N_i(\varepsilon)$ integers such that $|\sup_{k\geq n} a_k - a| < \varepsilon$ $\forall n \geq N_s$ and $|\inf_{k\geq n} a_k - a| < \varepsilon$ $\forall n \geq N_i$. Put $N = \max(N_s, N_i)$, from the above relations it follows that $a - \varepsilon < a_n < a + \varepsilon$ $\forall n \geq N$, that is $\lim_{n\to\infty} a_n = a$.

The cases $a = \pm\infty$ follow immediately from the points (c) and (e) of Theorem 4.11 and from Theorem 4.13. $\qquad\square$

Theorem 4.15 *Let (a_n) and (b_n) be two bounded sequences of real numbers. Then*

$$\liminf_{n\to\infty} a_n + \liminf_{n\to\infty} b_n \leq \liminf_{n\to\infty}(a_n + b_n)$$

$$\leq \liminf_{n\to\infty} a_n + \limsup_{n\to\infty} b_n$$

$$\leq \limsup_{n\to\infty}(a_n + b_n)$$

$$\leq \limsup_{n\to\infty} a_n + \limsup_{n\to\infty} b_n. \tag{4.5}$$

Proof Note that $\forall k \geq n$ we have $a_k + \inf_{k'\geq n} b_{k'} \leq a_k + b_k \leq a_k + \sup_{k'\geq n} b_{k'}$. Taking the lower bound of this expression for $k \geq n$ and observing that $\inf_{k\geq n}(a_k + c) = \inf_{k\geq n} a_k + c$, we have

$$\inf_{k\geq n}\left(a_k + \inf_{k'\geq n} b_{k'} \right) = \inf_{k\geq n} a_k + \inf_{k'\geq n} b_{k'}$$

$$\leq \inf_{k\geq n}(a_k + b_k)$$

$$\leq \inf_{k \geq n} \left(a_k + \sup_{k' \geq n} b_{k'} \right)$$

$$= \inf_{k \geq n} a_k + \sup_{k' \geq n} b_{k'}.$$

We conclude taking the limit $n \to \infty$

$$\liminf_{n \to \infty} a_n + \liminf_{n \to \infty} b_n \leq \liminf_{n \to \infty}(a_n + b_n) \leq \liminf_{n \to \infty} a_n + \limsup_{n \to \infty} b_n.$$

To demonstrate the remaining inequalities, note that $\forall k \geq n$ we have $\inf_{k' \geq n} a_{k'} + b_k \leq a_k + b_k \leq \sup_{k' \geq n} a_{k'} + b_k$. Taking the upper bound of this expression for $k \geq n$ and observing that $\sup_{k \geq n}(c + b_k) = c + \sup_{k \geq n} b_k$, we have

$$\sup_{k \geq n} \left(\inf_{k' \geq n} a_{k'} + b_k \right) = \inf_{k' \geq n} a_{k'} + \sup_{k \geq n} b_k$$

$$\leq \sup_{k \geq n} (a_k + b_k)$$

$$\leq \sup_{k \geq n} \left(\sup_{k' \geq n} a_{k'} + b_k \right)$$

$$= \sup_{k' \geq n} a_{k'} + \sup_{k \geq n} b_k.$$

Taking again the limit $n \to \infty$

$$\liminf_{n \to \infty} a_n + \limsup_{n \to \infty} b_n \leq \limsup_{n \to \infty}(a_n + b_n) \leq \limsup_{n \to \infty} a_n + \limsup_{n \to \infty} b_n.$$

$\square$

Corollary 4.16 *Let (a_n) and (b_n) be two sequences of real numbers and let $\lim_{n \to \infty} a_n = a \in \mathbb{R}$. Then*

$$\limsup_{n \to \infty}(a_n + b_n) = a + \limsup_{n \to \infty} b_n, \tag{4.6}$$

$$\liminf_{n \to \infty}(a_n + b_n) = a + \liminf_{n \to \infty} b_n. \tag{4.7}$$

Proof By Theorem 4.15 we have

$$\liminf_{n \to \infty} a_n + \limsup_{n \to \infty} b_n \leq \limsup_{n \to \infty}(a_n + b_n) \leq \limsup_{n \to \infty} a_n + \limsup_{n \to \infty} b_n,$$

$$\liminf_{n \to \infty} a_n + \liminf_{n \to \infty} b_n \leq \liminf_{n \to \infty}(a_n + b_n) \leq \limsup_{n \to \infty} a_n + \liminf_{n \to \infty} b_n.$$

Relations (4.6) and (4.7) follow immediately from these observing that $\liminf_{n\to\infty} a_n = \limsup_{n\to\infty} a_n = \lim_{n\to\infty} a_n = a$. $\qquad\square$

4.3   Power Series

Definition 4.17 (Power Series) Let $z, z_0, a_n \in \mathbb{C}$ with $n \geq 0$ integer. We call power series around z_0 the series of functions

$$\sum_{n=0}^{\infty} a_n (z - z_0)^n. \tag{4.8}$$

Example 4.18 A notable power series is the geometric series of ratio z

$$\sum_{n=0}^{\infty} z^n. \tag{4.9}$$

Using the identity $(1 + z + z^2 + \cdots + z^n)(1 - z) = 1 - z^{n+1}$, for $z \neq 1$ we have

$$\sum_{k=0}^{n} z^k = \frac{1 - z^{n+1}}{1 - z}.$$

If $|z| < 1$, then $\lim_{n\to\infty} z^n = 0$ and the geometric series converges to $1/(1 - z)$. This convergence is also absolute as can be seen immediately by changing $z \to |z|$ in the previous reasoning. If instead $|z| > 1$, we have $\lim_{n\to\infty} |z|^n = \infty$ and therefore the series diverges.

The geometric series (4.9) converges to $1/(1 - z)$ uniformly for $z \in \overline{B}(0, r)$ with arbitrary $0 < r < 1$. In fact, $\forall z \in \overline{B}(0, r)$ we have

$$\left| \sum_{k=0}^{n} z^k - \frac{1}{1 - z} \right| = \frac{|z|^{n+1}}{|1 - z|} \leq \frac{r^{n+1}}{1 - r}.$$

Since $0 < r < 1$, $\forall \varepsilon > 0$ it is possible to find an integer $N(\varepsilon)$ large enough, for example

$$N(\varepsilon) = \left\lceil \frac{\ln(\varepsilon(1 - r))}{\ln r} \right\rceil - 1,$$

such that $\forall n > N(\varepsilon)$ we have $r^{n+1}/(1 - r) < \varepsilon$ and therefore

$$\left| \sum_{k=0}^{n} z^k - \frac{1}{1 - z} \right| < \varepsilon, \qquad \forall z \in \overline{B}(0, r).$$

Note that the convergence of the geometric series to $1/(1-z)$ is not uniform in $B(0,1)$. In fact, choosing $0 < \varepsilon < 1$, for every integer n it is possible to find $z \in B(0,1)$ such that $\left| \sum_{k=0}^{n} z^k - 1/(1-z) \right| = |z|^{n+1}/|1-z| > \varepsilon$.

The result of the previous example is an archetype of the convergence scheme of a generic power series and is used in the proof of the following theorem.

Theorem 4.19 (Abel) *Given the power series $\sum_{n=0}^{\infty} a_n (z - z_0)^n$ with $a_n, z, z_0 \in \mathbb{C}$, define the number $R \in [0, \infty]$ by Hadamard's formula*

$$\frac{1}{R} = \limsup_{n \to \infty} |a_n|^{1/n} . \tag{4.10}$$

Then

(a) *if $|z - z_0| < R$ the series converges absolutely;*
(b) *if $|z - z_0| > R$ the series diverges (the terms of the series become unbounded);*
(c) *if $0 < r < R$ the series converges uniformly in $\overline{B}(z_0, r)$.*

Proof For simplicity of notation, set $w = z - z_0$.

(a) We exclude the trivial case $R = 0$ and suppose $0 \leq 1/R < \infty$. If $|w| < R$, $\exists r$ such that $|w| < r < R$. Since $1/r > 1/R = \limsup |a_n|^{1/n}$, from definition of $\limsup$ we obtain that $\exists N$ such that $|a_n|^{1/n} < 1/r \ \forall n \geq N$. The last inequality can be rewritten as $|a_n| < 1/r^n$ and so $|a_n w^n| < (|w|/r)^n \ \forall n \geq N$. Since $|w|/r < 1$, the real series $\sum_{n=0}^{\infty} (|w|/r)^n$ converges and therefore the series $\sum_{n=0}^{\infty} a_n w^n$ converges absolutely;
(b) We exclude the trivial case $R = \infty$ and suppose $0 < 1/R \leq \infty$. If $|w| > R$, $\exists r$ such that $|w| > r > R$. Since $1/r < 1/R = \limsup |a_n|^{1/n}$, from the definition of $\limsup$ we obtain that $|a_n|^{1/n} > 1/r$ for infinite values of n. The last inequality can be rewritten as $|a_n| > 1/r^n$ and therefore $|a_n w^n| > (|w|/r)^n$ for infinite values of n. Since $|w|/r > 1$, the terms of the series $\sum_{n=0}^{\infty} a_n w^n$ become unbounded for $n \to \infty$;
(c) As in point (a), we exclude the trivial case $R = 0$ and suppose $0 \leq 1/R < \infty$. If $0 < r < R$, $\exists \rho$ such that $r < \rho < R$. Since $1/\rho > 1/R = \limsup |a_n|^{1/n}$, from the definition of $\limsup$ we obtain that $\exists N$ such that $|a_n|^{1/n} < 1/\rho \ \forall n \geq N$. The last inequality can be rewritten as $|a_n| < 1/\rho^n$ and so $\forall w \in \overline{B}(0, r)$ we get $|a_n w^n| < (|w|/\rho)^n \leq (r/\rho)^n \ \forall n \geq N$. Since $r/\rho < 1$, the real series $\sum_{n=0}^{\infty} (r/\rho)^n$ converges and therefore by the Weierstrass criterion the series $\sum_{n=0}^{\infty} a_n w^n$ converges uniformly in $\overline{B}(0, r)$.

$\square$

Theorem 4.20 (Ratio Criterion) *Let $\sum_{n=0}^{\infty} a_n (z - z_0)^n$ be a power series with radius of convergence R. Then*

$$R = \lim_{n \to \infty} \left| \frac{a_n}{a_{n+1}} \right| \tag{4.11}$$

if this limit exists.

Proof Suppose that the limit of $|a_n/a_{n+1}|$ for $n \to \infty$ exists and let α denote its value. Define $w = z - z_0$ and consider $|w| < r < \alpha$. By hypothesis $\exists N$ integer such that $|a_n/a_{n+1}| > r \ \forall n \geq N$. Setting $A = |a_N| r^N$ and observing that

$$|a_{N+k}| r^{N+k} = |a_{N+k-1}| r^{N+k-1} \frac{|a_{N+k}|}{|a_{N+k-1}|} r$$

$$< |a_{N+k-1}| r^{N+k-1}, \qquad k = 1, 2, \ldots,$$

we have $|a_n| r^n \leq A \ \forall n \geq N$. Then $|a_n w^n| = |a_n| r^n (|w|/r)^n \leq A(|w|/r)^n$ $\forall n \geq N$. Since $|w| < r$ the geometric series $\sum_{n=0}^{\infty} (|w|/r)^n$ converges and therefore the series $\sum_{n=0}^{\infty} |a_n w^n|$ converges too. Since $r < \alpha$ is arbitrary and the series $\sum_{n=0}^{\infty} a_n w^n$ converges absolutely for $|w| < r$, the convergence radius of such series must be $R \geq \alpha$.

Now consider $|w| > r > \alpha$. As $\lim_{n \to \infty} |a_n/a_{n+1}| = \alpha$, $\exists N$ integer such that $|a_n/a_{n+1}| < r \ \forall n \geq N$. Setting $A = |a_N| r^N$ and observing that

$$|a_{N+k}| r^{N+k} = |a_{N+k-1}| r^{N+k-1} \frac{|a_{N+k}|}{|a_{N+k-1}|} r$$

$$> |a_{N+k-1}| r^{N+k-1}, \qquad k = 1, 2, \ldots,$$

we have $|a_n| r^n \geq A \ \forall n \geq N$. Then $|a_n w^n| = |a_n| r^n (|w|/r)^n \geq A(|w|/r)^n$ $\forall n \geq N$. Since $|w| > r$, we have $\lim_{n \to \infty} |a_n w^n| = \infty$. Since $r > \alpha$ is arbitrary and the series $\sum_{n=0}^{\infty} a_n w^n$ diverges for $|w| > r$, the convergence radius of this series must be $R \leq \alpha$.

We conclude that $R = \alpha$. $\qquad \square$

Example 4.21 Consider the power series

$$\sum_{n=0}^{\infty} \frac{z^n}{n!}. \tag{4.12}$$

The n-th coefficient of the series is $a_n = 1/n!$ and its radius of convergence is easily calculated with the ratio criterion

$$R = \lim_{n \to \infty} \left| \frac{a_n}{a_{n+1}} \right| = \lim_{n \to \infty} \frac{(n+1)!}{n!} = \lim_{n \to \infty} (n+1) = \infty.$$

The series converges on the whole complex plane. In the particular case $z = i\theta$, $\theta \in \mathbb{R}$, we have

$$\sum_{n=0}^{\infty} \frac{(i\theta)^n}{n!} = \sum_{k=0}^{\infty} \frac{(-1)^k \theta^{2k}}{(2k)!} + i \sum_{k=0}^{\infty} \frac{(-1)^k \theta^{2k+1}}{(2k+1)!} = \cos\theta + i\sin\theta.$$

Thus we obtain Euler's formula $e^{i\theta} = \cos\theta + i\sin\theta$ provided that we define the sum of the series (4.12) as the complex exponential $\exp z$.

Exercises

4.1 Consider the sequence of functions $f_n(z) : \mathbb{C} \mapsto \mathbb{C}$ with $f_n(z) = z^n$ and $n \in \mathbb{N}$. Show that for $z \in \overline{B}(0, r)$ with $0 < r < 1$ this sequence converges uniformly to $f(z) = 0$. Show also that the convergence is not uniform in $B(0, 1)$.

4.2 Prove that if $\sum_{n=0}^{\infty} a_n = A$ and $\sum_{n=0}^{\infty} b_n = B$ with $a_n, b_n, A, B \in \mathbb{C}$, then $\sum_{n=0}^{\infty}(a_n + b_n) = A + B$.

4.3 Find the radius of convergence of the following power series:

$$\text{(a) } \sum_{n=0}^{\infty} \cosh(n)z^n, \qquad \text{(b) } \sum_{n=0}^{\infty}(n + c^n)z^n, \qquad \text{(c) } \sum_{n=1}^{\infty} 3z^{\lfloor n \ln \ln n \rfloor},$$

where $c \in \mathbb{C}$ and $\lfloor x \rfloor$ with $x \in \mathbb{R}$ denoting the integer part of x.

4.4 Find the radius of convergence of the following power series:

$$\text{(a) } \sum_{n=0}^{\infty} \frac{2^n}{n^3} z^n, \qquad \text{(b) } \sum_{n=0}^{\infty} i \sinh(n)z^n, \qquad \text{(c) } \sum_{n=0}^{\infty} n^3 4^n z^n.$$

4.5 Find the radius of convergence of the following power series:

$$\text{(a) } \sum_{n=0}^{\infty} n e^n z^n, \qquad \text{(b) } \sum_{n=0}^{\infty} z^{n!+n}, \qquad \text{(c) } \sum_{n=0}^{\infty} n^n \left(\frac{1+i}{2}\right)^{n^2} z^n.$$

4.6 Find the radius of convergence of the following power series:

$$\text{(a) } \sum_{n=0}^{\infty} \cosh(n^2)z^n, \qquad \text{(b) } \sum_{n=0}^{\infty} \frac{n^{3n}}{(3n)!} z^n.$$

4.7 Find the radius of convergence of the following power series:

$$\text{(a) } \sum_{n=0}^{\infty} \frac{3^n}{(n+1)^3} z^n, \qquad \text{(b) } \sum_{n=1}^{\infty} \left(1 + n^6\right) z^{n^2}.$$

4.8 Find the radius of convergence of the following power series:

$$\text{(a) } \sum_{n=0}^{\infty} n^3 e^{3n} z^n, \qquad \text{(b) } \sum_{n=1}^{\infty} \frac{(\ln n)^n}{(1+i)^{n^2}} z^n, \qquad \text{(c) } \sum_{n=0}^{\infty} (-1)^n z^{n^3}.$$

4.9 Find the radius of convergence of the following series of powers:

$$\text{(a) } \sum_{n=0}^{\infty} \frac{(3n)! + 4^n}{(3n+1)!} z^n, \qquad \text{(b) } \sum_{n=1}^{\infty} i^{-in} z^n, \qquad \text{(c) } \sum_{n=1}^{\infty} \log(in) z^{2n}.$$

Set $\log(in) = \ln(n) + i\pi/2$ and observe that $i^{-in} = e^{n\pi/2}$.

4.10 Find the radius of convergence of the following power series:

$$\text{(a) } \sum_{n=0}^{\infty} (-1)^n z^{1+n+n^2+n^3}, \qquad \text{(b) } \sum_{n=0}^{\infty} n^2 3^n z^n.$$

4.11 Find the radius of convergence of the following power series:

$$\text{(a) } \sum_{n=0}^{\infty} n^3 2^n z^n, \qquad \text{(b) } \sum_{n=0}^{\infty} (-5)^{n^3} z^{1+n^3}.$$

4.12 Show that the power series

$$\text{(a) } \sum_{n=0}^{\infty} a_n z^n, \qquad \text{(b) } \sum_{n=0}^{\infty} a_n n^c z^n, \quad c \in \mathbb{C},$$

have the same radius of convergence and determine this radius in the case $a_n = e^{-n}$.

4.13 Find the radius of convergence of the following power series:

$$\text{(a) } \sum_{n=1}^{\infty} e^{n!} z^{n!}, \qquad \text{(b) } \sum_{n=1}^{\infty} e^n z^{n!}, \qquad \text{(c) } \sum_{n=1}^{\infty} e^{n!} z^n.$$

4.14 Find the radius of convergence of the following power series:

$$\text{(a) } \sum_{n=0}^{\infty} n^2 \pi^n [(1 - (-1)^n)/2] z^n, \qquad \text{(b) } \sum_{n=0}^{\infty} n^{2+i} \pi^n z^n.$$

Let $n^{2+i} = n^2 e^{i \ln n}$.

4.15 Find the radius of convergence of the following power series:

$$\text{(a) } \sum_{n=0}^{\infty} \frac{n^3}{e^{3n}} z^n, \qquad \text{(b) } \sum_{n=1}^{\infty} \frac{(-1)^n}{n} z^{n(n+1)}.$$

4.16 Knowing that the power series $\sum_{n=0}^{\infty} c_n z^n$ has radius of convergence R_0, determine the radius of convergence of the following series:

$$\text{(a)} \ \sum_{n=0}^{\infty} c_n^3 z^n, \qquad \text{(b)} \ \sum_{n=0}^{\infty} c_n z^{3n}.$$

4.17 Find the radius of convergence of the following power series:

$$\text{(a)} \ \sum_{n=0}^{\infty} (n + c^n) z^n, \quad c \in \mathbb{C}, \qquad \text{(b)} \ \sum_{n=1}^{\infty} 5^n z^{n!}.$$

4.18 Find the radius of convergence of the following power series:

$$\text{(a)} \ \sum_{n=0}^{\infty} n e^{2n - \sqrt{n}} z^n, \qquad \text{(b)} \ \sum_{n=0}^{\infty} (\ln(1 + n))^4 z^n, \qquad \text{(c)} \ \sum_{n=0}^{\infty} \frac{n^2 (2n)!}{(n!)^2} z^n.$$

4.19 Determine for which values of $z \in \mathbb{C}$ the series

$$\sum_{n=0}^{\infty} e^{nz}$$

is convergent. Put $e^z = e^{\operatorname{Re} z} e^{i \operatorname{Im} z}$.

4.20 Find the radius of convergence of the following power series:

$$\text{(a)} \ \sum_{n=1}^{\infty} c^{n^2} z^n, \quad c \in \mathbb{C}, \qquad \text{(b)} \ \sum_{n=1}^{\infty} c^n z^{n^2}, \quad c \in \mathbb{C}.$$

4.21 Determine for which values of $z \in \mathbb{C}$ the series

$$\sum_{n=0}^{\infty} \left(\frac{z^n}{n!} + \frac{n^2}{z^n} \right)$$

is convergent.

4.22 Let $(a_n)_{n=0}^{\infty}$ be a sequence of complex numbers $a_n \neq 0$ such that

$$\sum_{k=0}^{\infty} a_k \ \text{diverge}, \qquad \sum_{k=1}^{\infty} |a_k - a_{k-1}| < \infty.$$

Show that the power series $\sum_{k=0}^{\infty} a_k z^k$ has radius of convergence $R = 1$. Does the result change if $a_n \neq 0$ not always but for infinite values of the index n?

Chapter 5
Derivatives and Analytic Functions

Abstract Derivatives of complex functions. A differentiable function is continuous. Derivative formulas for the sum, product, ratio, composition of differentiable functions. Derivative of the inverse function. Cauchy-Riemann equations. Sufficient conditions for the existence of the derivative. Analytic functions. Singular points. If f is analytic and $f' = 0$ in D open and connected then f is constant in D. If f and $\overline{f}$ are analytic in D open and connected then f is constant in D. If f is analytic in D open and connected and $|f|$ is constant in D then f is constant in D. Derivatives of complex functions of real variables. Conformal transformations. An analytic function f is conformal at all points where $f' \neq 0$.

5.1 Derivatives of Complex Functions

Definition 5.1 (Differentiable Function) Let $f : G \mapsto \mathbb{C}$ with $G \subset \mathbb{C}$ open. The function f is differentiable at a point $a \in G$ if there exists in $\mathbb{C}$ the limit

$$\lim_{z \to a} \frac{f(z) - f(a)}{z - a}. \tag{5.1}$$

The value of this limit is indicated by $f'(a)$ or by $\mathrm{d}f(z)/\mathrm{d}z|_{z=a}$ and is called the derivative of f at a. If f is differentiable at every point of G then f is said to be differentiable in G.

Note that the definition of differentiability at a point assumes that the function is defined in a whole neighborhood of that point. Defined $\Delta z = z - a$, one can also write

$$f'(a) = \lim_{\Delta z \to 0} \frac{f(a + \Delta z) - f(a)}{\Delta z}. \tag{5.2}$$

If f is differentiable everywhere in G, the derivative of f defines a new function $f' : G \mapsto \mathbb{C}$. If f' is differentiable in G we say that f is twice differentiable in G, et cœtera.

C. Presilla, *Complex Functions of a Variable*, La Matematica per il 3+2 179,
https://doi.org/10.1007/978-3-032-12494-4_5

Example 5.2 Consider $f(z) = z^2$. This function is differentiable everywhere in $\mathbb{C}$, in fact $\forall z \in \mathbb{C}$ we have

$$\lim_{\Delta z \to 0} \frac{(z + \Delta z)^2 - z^2}{\Delta z} = \lim_{\Delta z \to 0} (2z + \Delta z) = 2z$$

and therefore $f'(z) = 2z$.

Example 5.3 Consider $f(z) = |z|^2$. Since

$$\frac{|z + \Delta z|^2 - |z|^2}{\Delta z} = \frac{(z + \Delta z)(\bar{z} + \overline{\Delta z}) - z\bar{z}}{\Delta z} = \bar{z} + \overline{\Delta z} + z\frac{\overline{\Delta z}}{\Delta z},$$

for $z \neq 0$ the limit $\Delta z \to 0$ of this incremental ratio does not exist. In fact, with the choice $\Delta z = (\Delta x, 0)$ for $\Delta x \to 0$ we obtain the value $\bar{z} + z$ while with the choice $\Delta z = (0, \Delta y)$ for $\Delta y \to 0$ we have $\bar{z} - z$ and for $z \neq 0$ these two values do not coincide. The function f is instead differentiable at $z = 0$ and $f'(0) = 0$.

The function just considered $f(z) = |z|^2$ is an example of a continuous but not differentiable function. The converse is always true.

Theorem 5.4 *Let $f : G \mapsto \mathbb{C}$ with $G \subset \mathbb{C}$ open. If f is differentiable at $a \in G$, then f is continuous at a.*

Proof Using the properties of the limit and the existence of $f'(a)$, we have

$$\lim_{z \to a} (f(z) - f(a)) = \lim_{z \to a} (z - a) \frac{f(z) - f(a)}{z - a}$$

$$= \lim_{z \to a} (z - a) \lim_{z \to a} \frac{f(z) - f(a)}{z - a}$$

$$= 0.$$

$\square$

Theorem 5.5 (Derivative of Sum, Product, Ratio) *Let $f : G \mapsto \mathbb{C}$ and $g : G \mapsto \mathbb{C}$, with $G \subset \mathbb{C}$ open, be two differentiable functions in G, then the functions $f \pm g$, fg and f/g are also differentiable in G, in the last case excluding the points $z \in G$ where $g(z) = 0$. Furthermore, the usual derivative rules hold:*

(a) $(f \pm g)'(z) = f'(z) \pm g'(z)$;
(b) $(fg)'(z) = f'(z)g(z) + f(z)g'(z)$;
(c) $(f/g)'(z) = (f'(z)g(z) - f(z)g'(z))/g(z)^2$.

Proof Elementary, follows from the definition of derivative and the properties of the limit. $\square$

A constant function $f(z) = c$ with $c \in \mathbb{C}$ has zero derivative $f'(z) = 0$ throughout $\mathbb{C}$. For the function $f(z) = z$ we have $f'(z) = 1 \forall z \in \mathbb{C}$. For integer powers $f(z) = z^n$ with $n \in \mathbb{Z}$, it follows $f'(z) = nz^{n-1}$ assuming $z \neq 0$ in the case in which $n \leq 0$.

Theorem 5.6 (Derivative of Composite Function) *Let $f : G \mapsto \mathbb{C}$ with $G \subset \mathbb{C}$ open and let $g : F \mapsto \mathbb{C}$ with $F \subset \mathbb{C}$ open such that $F \supset f(G)$. If f is differentiable at $a \in G$ and g is differentiable at $f(a) \in F$, then $g \circ f$ is differentiable at a and its derivative is*

$$(g \circ f)'(a) = g'(f(a))f'(a). \tag{5.3}$$

Proof Since G is open, $\exists r > 0$ such that $B(a, r) \subset G$. To prove the statement, it is enough to prove that, given any sequence (z_n) with $z_n \in B(a, r)$ and $\lim_{n \to \infty} z_n = a$, the following limit exists and has the specified value

$$\lim_{n \to \infty} \frac{(g \circ f)(z_n) - (g \circ f)(a)}{z_n - a} = g'(f(a))f'(a). \tag{5.4}$$

Let's start by assuming that for each n we have $f(z_n) \neq f(a)$. In this case we can write

$$\frac{(g \circ f)(z_n) - (g \circ f)(a)}{z_n - a} = \frac{g(f(z_n)) - g(f(a))}{f(z_n) - f(a)} \frac{f(z_n) - f(a)}{z_n - a}. \tag{5.5}$$

Since f is continuous at a we have $\lim_{n \to \infty} f(z_n) = f(a)$ and therefore, being f and g differentiable at a and $f(a)$ respectively, both fractions in the right-hand side of (5.5) have limit for $n \to \infty$ and the product of their limits is $g'(fa))f'(a)$.

Now suppose that for infinitely many values of the index n we have $f(z_n) = f(a)$ and for still infinitely many values of n we have $f(z_n) \neq f(a)$. The sequence (z_n) is then the union of two sequences (s_n) and (w_n) such that $f(s_n) = f(a)$ and $f(w_n) \neq f(a)$ for each value of the respective indices n. To demonstrate that (5.4) is valid, it is sufficient to show that it is true for both the sequences (s_n) and (w_n). First of all, observe that, since f is differentiable at a, it follows that $f'(a) = \lim_{n \to \infty}(f(s_n) - f(a))/(s_n - a) = 0$. Since $g'(f(a)) \in \mathbb{C}$, the right-hand side of (5.4) is 0. In the case of the sequence (s_n), the left-hand side of (5.4) is $\lim_{n \to \infty}(g(f(s_n)) - g(f(a)))/(s_n - a) = 0$. In the case of the sequence (w_n), by what has already been demonstrated we have $\lim_{n \to \infty}(g(f(w_n)) - g(f(a)))/(w_n - a) = g'(f(a))f'(a) = 0$. In conclusion, (5.4) is still valid with both sides equal to 0.

Finally, for sequences (z_n) in which $f(z_n) = f(a)$ or $f(z_n) \neq f(a)$ only for a finite number of values of n, the proof of (5.4) follows immediately from the cases already considered. $\qquad\qquad\square$

Theorem 5.7 (Derivative of Inverse Function) *Let $f : G \mapsto \mathbb{C}$ with $G \subset \mathbb{C}$ open. If $f(z) = w$ is bijective and its inverse function $f^{-1}(w) = z$ is continuous* *in the open set $f(G)$, then, if f is differentiable at $a \in G$ and $f'(a) \neq 0$, f^{-1} is differentiable at $w = f(a)$ and its derivative is*

$$\left. \frac{\mathrm{d} f^{-1}(w)}{\mathrm{d} w} \right|_{w=f(a)} = \frac{1}{f'(a)}. \tag{5.6}$$

Proof Since f is bijective, if $w = f(z)$ is different from $f(a)$ then $z \neq a$. Therefore, the incremental ratio of f^{-1} at point $f(a)$ can be written as

$$\frac{f^{-1}(w) - f^{-1}(f(a))}{w - f(a)} = \frac{z - a}{f(z) - f(a)} = \frac{1}{(f(z) - f(a))/(z - a)}.$$

Since f^{-1} is continuous in $f(G)$, we have $\lim_{w \to f(a)} f^{-1}(w) = f^{-1}(f(a))$, which means that $z \to a$ when $w \to f(a)$. In conclusion

$$\lim_{w \to f(a)} \frac{f^{-1}(w) - f^{-1}(f(a))}{w - f(a)} = \lim_{z \to a} \frac{1}{(f(z) - f(a))/(z - a)}$$

$$= \frac{1}{f'(a)}$$

exists and formula (5.6) is established. $\qquad\qquad\qquad\qquad\qquad\qquad\qquad\qquad \square$

5.2 Cauchy-Riemann Equations

Theorem 5.8 *Let $f : G \mapsto \mathbb{C}$ with $G \subset \mathbb{C}$ open and set $f(z) = u(x, y) + iv(x, y)$ with $z = x + iy$ and $u, v : \mathbb{R}^2 \mapsto \mathbb{R}$. If f is differentiable at $z_0 = x_0 + iy_0 \in G$, then the functions u and v admit partial first derivatives at (x_0, y_0) and these satisfy the Cauchy-Riemann equations*

$$u_x(x_0, y_0) = v_y(x_0, y_0), \tag{5.7}$$

$$u_y(x_0, y_0) = -v_x(x_0, y_0). \tag{5.8}$$

Furthermore $f'(z_0) = u_x(x_0, y_0) + iv_x(x_0, y_0)$.

Proof Define $\Delta z = \Delta x + i \Delta y$ and choose real increments Δx and Δy small enough so that $B(z_0, |\Delta z|) \subset G$. The corresponding increment of f at z_0 is

$$\Delta f = f(z_0 + \Delta z) - f(z_0)$$

$$= [u(x_0 + \Delta x, y_0 + \Delta y) - u(x_0, y_0)]$$

$$+ i[v(x_0 + \Delta x, y_0 + \Delta y) - v(x_0, y_0)]$$

and the derivative of f at z_0 can be written as

$$f'(z_0) = \lim_{\Delta z \to 0} \frac{\Delta f}{\Delta z}$$

$$= \lim_{\Delta z \to 0} \text{Re}\, \frac{\Delta f}{\Delta z} + i \lim_{\Delta z \to 0} \text{Im}\, \frac{\Delta f}{\Delta z}. \tag{5.9}$$

Since both the limits of the real and imaginary parts of $\Delta f/\Delta z$ exist, their value is independent of the way $\Delta z \to 0$. If $\Delta z = (\Delta x, 0)$ we have

$$\lim_{\Delta z \to 0} \text{Re}\, \frac{\Delta f}{\Delta z} = \lim_{\Delta x \to 0} \frac{u(x_0 + \Delta x, y_0) - u(x_0, y_0)}{\Delta x} = u_x(x_0, y_0),$$

$$\lim_{\Delta z \to 0} \text{Im}\, \frac{\Delta f}{\Delta z} = \lim_{\Delta x \to 0} \frac{v(x_0 + \Delta x, y_0) - v(x_0, y_0)}{\Delta x} = v_x(x_0, y_0),$$

while for $\Delta z = (0, \Delta y)$ we obtain

$$\lim_{\Delta z \to 0} \text{Re}\, \frac{\Delta f}{\Delta z} = \lim_{\Delta y \to 0} \frac{v(x_0, y_0 + \Delta y) - v(x_0, y_0)}{\Delta y} = v_y(x_0, y_0),$$

$$\lim_{\Delta z \to 0} \text{Im}\, \frac{\Delta f}{\Delta z} = \lim_{\Delta y \to 0} \frac{-u(x_0, y_0 + \Delta y) + u(x_0, y_0)}{\Delta y} = -u_y(x_0, y_0).$$

We conclude that the partial derivatives with respect to x and y of u and v exist at (x_0, y_0) and are related by (5.7) and (5.8). Finally by (5.9) we have $f'(z_0) = u_x(x_0, y_0) + iv_x(x_0, y_0)$. $\qquad\square$

Example 5.9 Consider the function $f(z) = z^2$. Setting $z = x + iy$ and $f = u + iv$, we have $u(x, y) = x^2 - y^2$ and $v(x, y) = 2xy$. We know that f is differentiable throughout $\mathbb{C}$, therefore $\forall (x, y) \in \mathbb{R}^2$ we have $u_x(x, y) = v_y(x, y) = 2x$ and $u_y(x, y) = -v_x(x, y) = -2y$. Moreover, $u_x(x, y) + iv_x(x, y) = 2x + i2y = 2z = f'(z)$.

Example 5.10 Consider the function $f(z) = |z|^2$. We have $u(x, y) = x^2 + y^2$ and $v(x, y) = 0$. Therefore, $u_x(x, y) = 2x$, $u_y(x, y) = 2y$, $v_x(x, y) = v_y(x, y) = 0$. The Cauchy-Riemann equations are not satisfied for $(x, y) \neq (0, 0)$ and therefore $f(z)$ is not differentiable at $z \neq 0$. At $z = 0$, as seen previously, f is differentiable. It follows that the Cauchy-Riemann equations for u and v are satisfied at $(0, 0)$.

The Cauchy-Riemann equations (5.7) and (5.8) represent a necessary condition, in general not sufficient, for the function $f(z)$ to be differentiable at $z_0 = x_0 + iy_0$.

Example 5.11 Consider the function $f(z) = \sqrt{|\text{Re}\, z\, \text{Im}\, z|}$ intending the root to be the positive branch. The component functions $u(x, y) = \sqrt{|xy|}$ and $v(x, y) = 0$ have at $(0, 0)$ partial first derivatives $u_x(0, 0) = u_y(0, 0) = v_x(0, 0) = v_y(0, 0) = 0$. The Cauchy-Riemann equations are therefore satisfied at $(0, 0)$ but f is not differentiable at $z = 0$. In fact, the incremental ratio $(f(0 + \Delta z) - f(0))/\Delta z$ is $1/(1 \pm i)$ depending on whether $\Delta z = t \pm it$ with $t > 0$.

Theorem 5.12 *Let $f : G \mapsto \mathbb{C}$ with $G \subset \mathbb{C}$ open and set $f(z) = u(x, y) + iv(x, y)$ with $z = x + iy$ and $u, v : \mathbb{R}^2 \mapsto \mathbb{R}$. If the functions u and v admit partial first derivatives in G and these are continuous at $(x_0, y_0) \in G$ and satisfy the Cauchy-Riemann equations $u_x(x_0, y_0) = v_y(x_0, y_0)$ and $u_y(x_0, y_0) = -v_x(x_0, y_0)$, then f is differentiable at $z_0 = x_0 + iy_0$ and we have $f'(z_0) = u_x(x_0, y_0) + iv_x(x_0, y_0)$.*

Proof Define $\Delta z = \Delta x + i \Delta y$ and choose real increments Δx and Δy small enough so that $B(z_0, |\Delta z|) \subset G$. We write the corresponding increment of f at $z_0 = x_0 + iy_0$ as $\Delta f = \Delta u + i \Delta v$ with $\Delta u = u(x_0 + \Delta x, y_0 + \Delta y) - u(x_0, y_0)$ and $\Delta v = v(x_0 + \Delta x, y_0 + \Delta y) - v(x_0, y_0)$. Since u_x, u_y, v_x, v_y exist and are continuous at (x_0, y_0), the functions u and v are differentiable at (x_0, y_0) and the following formulas hold [12]

$$\Delta u = u_x(x_0, y_0)\Delta x + u_y(x_0, y_0)\Delta y + \varepsilon_u(\Delta x, \Delta y)\sqrt{\Delta x^2 + \Delta y^2},$$

$$\Delta v = v_x(x_0, y_0)\Delta x + v_y(x_0, y_0)\Delta y + \varepsilon_v(\Delta x, \Delta y)\sqrt{\Delta x^2 + \Delta y^2},$$

where $\varepsilon_u(\Delta x, \Delta y) \xrightarrow{\Delta x, \Delta y \to 0} 0$ and $\varepsilon_v(\Delta x, \Delta y) \xrightarrow{\Delta x, \Delta y \to 0} 0$. From these formulas and using the Cauchy-Riemann equations, we obtain

$$\Delta f = [u_x(x_0, y_0) + iv_x(x_0, y_0)](\Delta x + i\Delta y)$$

$$+ [\varepsilon_u(\Delta x, \Delta y) + i\varepsilon_v(\Delta x, \Delta y)]\sqrt{\Delta x^2 + \Delta y^2}$$

and therefore

$$\frac{\Delta f}{\Delta z} = u_x(x_0, y_0) + iv_x(x_0, y_0) + [\varepsilon_u(\Delta x, \Delta y) + i\varepsilon_v(\Delta x, \Delta y)]\frac{|\Delta z|}{\Delta z}.$$

Since $||\Delta z|/\Delta z| = 1$ and $\varepsilon_u + i\varepsilon_v \xrightarrow{\Delta z \to 0} 0$, the limit of $\Delta f/\Delta z$ for $\Delta z \to 0$ exists and furthermore we have $f'(z_0) = u_x(x_0, y_0) + iv_x(x_0, y_0)$. $\square$

Sometimes it is convenient to express the function f in terms of the polar coordinates r, θ rather than the Cartesian coordinates x, y. Let $f(z) = u(r, \theta) + iv(r, \theta)$ with $z = r\,\mathrm{cis}(\theta)$, it is easy to show that the Cauchy-Riemann equations (5.7) and (5.8) at the point (x_0, y_0) are equivalent to

$$ru_r(r_0, \theta_0) = v_\theta(r_0, \theta_0), \tag{5.10}$$

$$u_\theta(r_0, \theta_0) = -rv_r(r_0, \theta_0), \tag{5.11}$$

where (r_0, θ_0) are the coordinates connected to (x_0, y_0) by the transformation $x = r\cos\theta$, $y = r\sin\theta$. Finally we get $f'(z_0) = (u_r(r_0, \theta_0) + iv_r(r_0, \theta_0))e^{-i\theta_0}$.

5.3 Analytic Functions

Definition 5.13 (Analytic Function) Let $f : G \mapsto \mathbb{C}$ with $G \subset \mathbb{C}$ open. The function f is said to be analytic in G if $\exists f'(z)\ \forall z \in G$. If G is closed, f analytic in G means that f is analytic in some open set containing G. In particular, f analytic at z_0 means that $\exists r > 0$ such that f is analytic in $B(z_0, r)$.

Definition 5.14 (Entire Function) A function is said to be entire if it is analytic throughout $\mathbb{C}$.

Example 5.15 The function $f(z) = |z|^2$ is not analytic at any point because it is differentiable only at $z = 0$.

Definition 5.16 (Singular Point) A point z_0 is called a singular point of f if f is not analytic at z_0 but $\forall \varepsilon > 0\ f$ is analytic at some point of $B(z_0, \varepsilon)$.

Example 5.17 The function $f(z) = 1/z$ is analytic in $\mathbb{C} \setminus \{0\}$ and has at $z = 0$ a singular (isolated) point. The function $f(z) = \operatorname{Arg} z - \mathrm{i} \ln |z|$ is analytic in $\mathbb{C} \setminus \{z = -t,\ t \in [0, \infty)\}$. Every point of the negative real semi-axis, including the origin, is a singular point of f.

It is evident that if f and g are two analytic functions in G, also the functions $f \pm g$, fg and f/g are analytic in G, with the exception, in the case of the ratio, of the points $z \in G$ where $g(z) = 0$.

Since the constant functions and the identity function are analytic in $\mathbb{C}$, it follows that polynomials are entire functions while rational functions are analytic in the complement of the set of zeros of the denominator.

Theorem 5.18 *Let $f : D \mapsto \mathbb{C}$ with $D \subset \mathbb{C}$ open and connected. If f is analytic in D and $f'(z) = 0\ \forall z \in D$, then f is constant in D.*

Proof Since $f(z) = u(x, y) + \mathrm{i}v(x, y)$, with $z = x + \mathrm{i}y$, is differentiable in D, in D there exist the first partial derivatives of the component functions u and v. Furthermore, since the Cauchy-Riemann conditions $u_x = v_y$ and $u_y = -v_x$ hold, and since $f' = u_x + \mathrm{i}v_x = 0$ by hypothesis, we have $u_x = u_y = v_x = v_y = 0$ throughout D. Let z and w be two any points of D. Since $D \subset \mathbb{C}$ is open and connected, such points can be joined by a polygonal of vertices $z_1, z_2, \ldots, z_n$ with $z_1 = z$ and $z_n = w$ contained in D. Consider, to begin, the segment $[z_1, z_2]$. Any point of this segment has real and imaginary parts that we parametrize as

$$x(t) = x_1 + t(x_2 - x_1), \quad y(t) = y_1 + t(y_2 - y_1), \qquad 0 \le t \le 1,$$

where $x_1 + \mathrm{i}y_1 = z_1$ and $x_2 + \mathrm{i}y_2 = z_2$. The total derivative of u with respect to the parameter t is

$$\frac{\mathrm{d}u}{\mathrm{d}t} = u_x(x(t), y(t))\frac{\mathrm{d}x(t)}{\mathrm{d}t} + u_y(x(t), y(t))\frac{\mathrm{d}y(t)}{\mathrm{d}t} = 0,$$

therefore

$$u(x_2, y_2) - u(x_1, y_1) = \int_0^1 \frac{du}{dt} dt = 0.$$

Moving on to consider the subsequent segments $[z_{k-1}, z_k]$, $k = 3, \ldots, n$, we find $u(x_k, y_k) = u(x_{k-1}, y_{k-1})$ and in conclusion $u(x_1, y_1) = u(x_n, y_n)$. Similarly $v(x_1, y_1) = v(x_n, y_n)$ and therefore $f(z) = f(w)$. From the arbitrariness of z and w follows the statement. $\qquad\square$

Theorem 5.19 *Let $f : D \mapsto \mathbb{C}$ with $D \subset \mathbb{C}$ open and connected. If f and $\overline{f}$ are analytic in D, then f and $\overline{f}$ are constant in D.*

Proof Let $f(z) = u(x, y) + iv(x, y)$ and $\overline{f}(z) = U(x, y) + iV(x, y)$ with $z = x + iy$. Since f and $\overline{f}$ are analytic in D, the functions u and v and the functions U and V satisfy the Cauchy-Riemann conditions at every point $(x, y) \in D$

$$u_x = v_y \qquad u_y = -v_x,$$
$$U_x = V_y \qquad U_y = -V_x.$$

The conditions for U and V are equivalent to

$$u_x = -v_y \qquad u_y = v_x$$

since, being $\overline{f}(z) = \overline{f(z)}$, we have $U(x, y) = u(x, y)$ and $V(x, y) = -v(x, y)$. Necessarily then it is $u_x = u_y = v_x = v_y = 0 \ \forall (x, y) \in D$ and therefore $f'(z) = 0$ $\forall z \in D$. Since D is open and connected, by Theorem 5.18 it follows that f is constant in D. $\qquad\square$

Theorem 5.20 *Let $f : D \mapsto \mathbb{C}$ with $D \subset \mathbb{C}$ open and connected. If f is analytic in D and $|f|$ is constant in D, then f is constant in D.*

Proof Let $|f(z)| = c \ \forall z \in D$. If $c = 0$, then $f(z) = 0 \ \forall z \in D$. If $c > 0$, then $f(z) \neq 0 \ \forall z \in D$ and we have $\overline{f}(z) = c^2/f(z)$. It follows that f and $\overline{f}$ are both analytic in D and by Theorem 5.19 they are constants in D. $\qquad\square$

5.4 Derivatives of Complex Functions of Real Variable

It is very useful to consider complex functions $f(t)$ of real variable t. The differentiability properties of such functions are strictly related to those of the component functions $\operatorname{Re} f(t)$ and $\operatorname{Im} f(t)$.

Definition 5.21 A function $f : [a, b] \mapsto \mathbb{C}$, with $a, b \in \mathbb{R}$ and $a < b$, is differentiable at the point $t \in (a, b)$ if the limit

$$\lim_{h \to 0} \frac{f(t + h) - f(t)}{h}$$

exists and is finite. The value of this limit is denoted by $f'(t)$. If $t = a$ or $t = b$, the definition is modified by taking the right or left limit respectively. If f is differentiable at every point of $[a, b]$ then f is said to be differentiable.

Theorem 5.22 *A function* $f : [a, b] \mapsto \mathbb{C}$ *is differentiable if and only if the component functions* $\mathrm{Re}\, f, \mathrm{Im}\, f : [a, b] \mapsto \mathbb{R}$ *are differentiable. Furthermore* $\forall t \in [a, b]$ *we have* $f'(t) = (\mathrm{Re}\, f)'(t) + \mathrm{i}(\mathrm{Im}\, f)'(t)$.

Proof Elementary.   □

Theorem 5.23 *If* $f : [a, b] \mapsto \mathbb{C}$ *is differentiable and* $f'(t) = 0 \; \forall t \in [a, b]$, *then* f *is constant.*

Proof Since $f'(t) = 0 \; \forall t \in [a, b]$ then $(\mathrm{Re}\, f)'(t) = (\mathrm{Im}\, f)'(t) = 0 \; \forall t \in [a, b]$. It follows that $\mathrm{Re}\, f$ and $\mathrm{Im}\, f$ are constant and therefore f is constant.   □

Theorem 5.24 *Let* $f : [a, b] \mapsto \mathbb{C}$ *and* $g : [a, b] \mapsto \mathbb{C}$ *be two functions differentiable in* $[a, b]$, *then the functions* $f \pm g$, fg *and* f/g *are differentiable in* $[a, b]$, *in the latter case* $\forall t \in [a, b]$ *such that* $g(t) \neq 0$. *Furthermore, the usual derivative rules hold*

(a) $(f \pm g)'(t) = f'(t) \pm g'(t)$;
(b) $(fg)'(t) = f'(t)g(t) + f(t)g'(t)$;
(c) $(f/g)'(t) = (f'(t)g(t) - f(t)g'(t))/g(t)^2$.

Theorem 5.25 *Let* $f : [a, b] \mapsto [c, d]$ *and* $g : [c, d] \mapsto \mathbb{C}$ *be differentiable, then* $g \circ f : [a, b] \mapsto \mathbb{C}$ *is differentiable and the derivative is*

$$(g \circ f)'(t) = g'(f(t))f'(t).$$

Proof Since $(g \circ f)(t) = g(f(t)) = \mathrm{Re}\, g(f(t)) + \mathrm{i}\, \mathrm{Im}\, g(f(t))$, the proof follows from the analogous theorem for real functions.   □

Not all properties of real functions extend to complex-valued functions of one real variable. For example, the mean value theorem does not hold in general. In fact, according to this theorem if $f : [a, b] \mapsto \mathbb{R}$ is differentiable, then $\exists c \in (a, b)$ such that

$$f(b) - f(a) = f'(c)(b - a).$$

However, consider the complex-valued function $f(t) = \cos(t) + \mathrm{i}\sin(t)$ with $t \in [0, 2\pi]$. We have $f(2\pi) - f(0) = 0$ while $f'(t) = -\sin(t) + \mathrm{i}\cos(t) \neq 0 \; \forall t \in (0, 2\pi)$ since in this entire interval we have $\left| f'(t) \right| = 1$.

5.5 Conformal Transformations

A path in the complex plane is a continuous function $\gamma : [a, b] \mapsto \mathbb{C}$, where $a, b \in \mathbb{R}$ with $a < b$. If $\forall t \in [a, b]$ there exists

$$\gamma'(t) = \lim_{h \to 0} \frac{\gamma(t + h) - \gamma(t)}{h},$$

meaning that $h \to 0^+$ if $t = a$ and $h \to 0^-$ if $t = b$, and furthermore $\gamma' : [a, b] \mapsto \mathbb{C}$ is continuous, then γ is called a regular path. It is evident that, if γ is a regular path and for $t = t_0$ we have $\gamma'(t_0) \neq 0$, then γ has a tangent line at the point $z_0 = \gamma(t_0)$ whose angular coefficient is $\operatorname{Im}\gamma'(t_0)/\operatorname{Re}\gamma'(t_0)$. In other words, $\arg(\gamma'(t_0))$ is, modulo 2π, the angle formed with the real axis by the tangent line to γ at $z_0 = \gamma(t_0)$.

Example 5.26 Consider the semicircle $\gamma(t) = e^{it}$, $0 \leq t \leq \pi$. This is a regular path with $\gamma'(t) = ie^{it} \neq 0$ $\forall t \in [0, \pi]$ that admits a tangent at each of its points. For $0 \leq t \leq \pi$, the tangent line to γ forms with the real axis an angle given by $\arg(\gamma'(t)) = \arg(e^{i(t+\pi/2)}) = t + \pi/2 + 2\pi k$, $k \in \mathbb{Z}$.

Definition 5.27 (Conformal Transformation) Let $f : G \to \mathbb{C}$ with $G \subset \mathbb{C}$ open. The function f is said to be a conformal transformation at $z_0 \in G$ if

(a) the angle formed by two regular paths that intersect at z_0 is equal to the angle formed by their images through f at the intersection point $f(z_0)$;
(b) there exists, finite and non-zero, the limit

$$\lim_{z \to z_0} \frac{|f(z) - f(z_0)|}{|z - z_0|}. \tag{5.12}$$

The function f is a conformal transformation in G if it is a conformal transformation at every point of G (Fig. 5.1).

Note that a conformal transformation preserves both the magnitude and the direction of the rotation angle of a path. Moreover, the existence of the limit (5.12) ensures the existence, for z close to z_0, of a scale factor $|f(z) - f(z_0)| / |z - z_0|$. Even though the rotation angle and the scale factor change, in general, as the point z_0 varies, for reasons of continuity the image through f of a small region around z_0 will have, approximately, the same shape as the starting region. A transformation that preserves the magnitude of the angle between two regular paths but not the direction of that angle is called isogonal or conformal transformation of the second kind.

Example 5.28 Consider the transformation $f(z) = \bar{z}$. The two paths $\gamma_1(t) = t$ and $\gamma_2(t) = te^{i\pi/4}$, $0 \leq t \leq 1$, form an oriented angle, from path 1 to path 2, equal to $\pi/4$. The transformed paths, $\sigma_1(t) = f(\gamma_1(t)) = t$ and $\sigma_2(t) = f(\gamma_2(t)) = te^{-i\pi/4}$, instead form an angle equal to $-\pi/4$. The transformation is isogonal non-conformal.

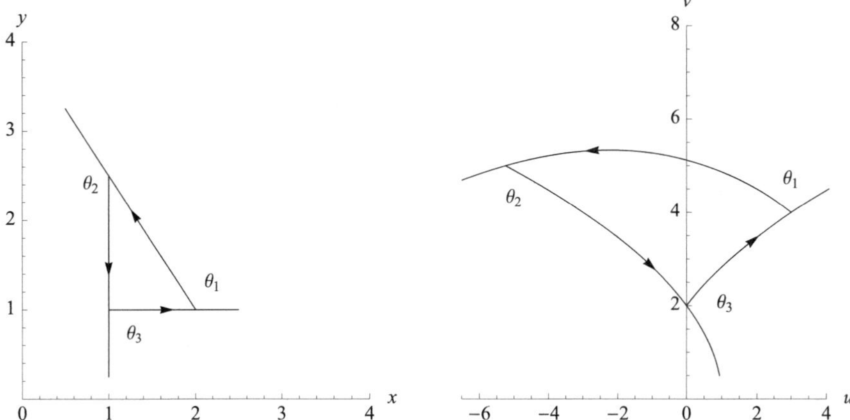

Fig. 5.1 Conformal transformation of three rectilinear paths that form a right-angled triangle (left) using the function $f(x + iy) = (x + iy)^2 = u(x, y) + iv(x, y)$ (right)

Theorem 5.29 *Let $f : G \to \mathbb{C}$ be analytic in $G \subset \mathbb{C}$ open. Then f is a conformal transformation at every point $z_0 \in G$ where $f'(z_0) \neq 0$.*

Proof Let z_0 be a point of G such that $f'(z_0) \neq 0$. Note that f is differentiable in a neighborhood of z_0. Let $\gamma_1(t)$, $a_1 \leq t \leq b_1$, and $\gamma_2(t)$, $a_2 \leq t \leq b_2$, be two arbitrary regular paths such that, for two suitable values t_1 and t_2 of the parameters, $\gamma_1(t_1) = \gamma_2(t_2) = z_0$ with $\gamma_1'(t_1) \neq 0$ and $\gamma_2'(t_2) \neq 0$. Let us indicate with $\sigma_1(t) = (f \circ \gamma_1)(t)$, $a_1 \leq t \leq b_1$, and $\sigma_2(t) = (f \circ \gamma_2)(t)$, $a_2 \leq t \leq b_2$, the paths transformed by f. It turns out $\sigma_1(t_1) = \sigma_2(t_2) = f(z_0)$, that is, the transformed paths intersect at the point $w_0 = f(z_0)$. At this point σ_1 and σ_2 are differentiable with non-zero derivative. In fact, by the derivative rule of composite functions we have $\sigma_1'(t_1) = f'(z_0)\gamma_1'(t_1) \neq 0$ and $\sigma_2'(t_2) = f'(z_0)\gamma_2'(t_2) \neq 0$. Therefore, there exist tangent lines to σ_1 and σ_2 at w_0 and such lines form with the real axis the angles $\arg \sigma_1'(t_1) = \arg f'(z_0) + \arg \gamma_1'(t_1)$ and $\arg \sigma_2'(t_2) = \arg f'(z_0) + \arg \gamma_2'(t_2)$. It is concluded that the oriented angle between the paths σ_1 and σ_2 that intersect at w_0 coincides with that between the paths γ_1 and γ_2 that intersect at z_0,

$$\arg \sigma_2'(t_2) - \arg \sigma_1'(t_1) = \arg \gamma_2'(t_2) - \arg \gamma_1'(t_1).$$

Finally, since there exists $f'(z_0) \neq 0$ then there also exists, finite and nonzero,

$$\lim_{z \to z_0} \frac{|f(z) - f(z_0)|}{|z - z_0|} = |f'(z_0)|.$$

We conclude that f is a conformal transformation at z_0. $\square$

Corollary 5.30 *Let $f : G \to \mathbb{C}$ with $G \subset \mathbb{C}$ open. Then f is analytic in G and $f'(z) \neq 0 \; \forall z \in G$ if and only if f is conformal in G.*

Proof The necessary condition follows from the previous theorem. For the sufficient condition note that, if f is a conformal transformation in G, by Definition 5.27 at every point $z \in G$ there must exist $\arg f'(z)$ and $|f'(z)|$ with, necessarily, $|f'(z)| \neq 0$. Therefore, $\forall z \in G$ there exists $f'(z) \neq 0$. $\square$

Example 5.31 Consider in $\mathbb{R}^2$ the regular contour lines $u(x, y) = c_1$ and $v(x, y) = c_2$, with $c_1, c_2 \in \mathbb{R}$, and suppose that the two curves intersect at the point (x_0, y_0). If the functions u and v constitute the real and imaginary parts of a complex function

$$f(z) = u(x, y) + iv(x, y), \qquad z = x + iy,$$

analytic at $z_0 = x_0 + iy_0$ with $f'(z_0) \neq 0$, then f conformally transforms the two contour lines into the straight lines $u = c_1$ and $v = c_2$. These straight lines intersect at the point $f(z_0)$ forming a right angle. It follows that the contour lines $u(x, y) = c_1$ and $v(x, y) = c_2$ intersect at (x_0, y_0) orthogonally.

Conformal transformations play a crucial role in the geometric approach to the theory of analytic functions. For a detailed discussion see [1, 11, 13].

Exercises

5.1 Prove that if the limit

$$\lim_{\Delta z \to 0} \operatorname{Re}\left(\frac{f(z + \Delta z) - f(z)}{\Delta z} \right),$$

with $f(z) = u(x, y) + iv(x, y)$ and $z = x + iy$, exists then the partial derivatives $u_x(x, y)$ and $v_y(x, y)$ exist too and we have $u_x(x, y) = v_y(x, y)$.

5.2 Prove that if the limit

$$\lim_{\Delta z \to 0} \operatorname{Im}\left(\frac{f(z + \Delta z) - f(z)}{\Delta z} \right),$$

with $f(z) = u(x, y) + iv(x, y)$ and $z = x + iy$, exists, then the partial derivatives $u_y(x, y)$ and $v_x(x, y)$ exist too and we have $u_y(x, y) = -v_x(x, y)$.

5.3 Let f and g be two functions such that $f(z_0) = g(z_0) = 0$, both differentiable at z_0 with $g'(z_0) \neq 0$. Prove that

$$\lim_{z \to z_0} \frac{f(z)}{g(z)} = \frac{f'(z_0)}{g'(z_0)}.$$

5.4 Show that the function

$$f(z) = \begin{cases} z^5 / |z|^4 & z \neq 0 \\ 0 & z = 0 \end{cases}$$

satisfies the Cauchy-Riemann conditions at $z = 0$ but at the same point it is not differentiable.

5.5 Let $f : G \mapsto \mathbb{C}$ be analytic in $G \subset \mathbb{C}$ open. Show that the function $f^*(z) = \overline{f(\bar{z})}$ is analytic in the open set $G^* = \{z \in \mathbb{C} : \bar{z} \in G\}$.

5.6 Determine, justifying your answer, the domains of continuity, differentiability and analyticity of the following functions:

$$\text{(a) } f(z) = x, \qquad \text{(b) } f(z) = e^x e^{iy}, \qquad \text{(c) } f(z) = x^2 - y^2 + i3xy,$$

where $z = x + iy$ with $x, y \in \mathbb{R}$.

5.7 Find the singular points of the following functions:

$$\text{(a) } f(z) = \frac{2z+1}{z(z^2+1)}, \qquad \text{(b) } f(z) = \frac{z^2+1}{(z-2)(z^2+2z+2)}.$$

5.8 Determine, justifying your answer, the domain of analyticity of the following functions and calculate their first derivative:

$$\text{(a) } f(z) = z^{-4}, \qquad \text{(b) } f(z) = (\cos(\ln r) + i \sin(\ln r)) \, e^{-\theta},$$

where $z = re^{i\theta}$ with $\theta \in (-\pi, \pi]$ and $r \geq 0$.

5.9 Determine, justifying your answer, the domain of analyticity of the following functions and calculate the first derivative:

$$\text{(a) } f(z) = \sqrt{r} e^{i\theta/2}, \qquad \text{(b) } f(z) = (\ln r)^2 - \theta^2 + i2\theta \ln r,$$

where $z = re^{i\theta}$ with $\theta \in (-\pi, \pi]$ and $r \geq 0$.

5.10 Consider the function $f(z) = (3z^2 - \bar{z}^2)\bar{z}/2$. Determine, justifying your answer, the domains of continuity, differentiability and analyticity of f.

5.11 A vector field $F : \mathbb{R}^2 \mapsto \mathbb{C}$ with $F(x, y) = U(x, y) + iV(x, y)$ is said to be irrotational if $U_y = V_x$ and solenoidal if $U_x = -V_y$. Show that if $f : D \mapsto \mathbb{C}$ is analytic in D then $\bar{f}$ is irrotational and solenoidal in D.

5.12 Determine, giving reasons for your answer, the domains of continuity, differentiability and analyticity of the following functions:

$$\text{(a) } f(z) = y, \qquad \text{(b) } f(z) = e^x e^{-iy}, \qquad \text{(c) } f(z) = x^2 - y^2 + i2xy,$$

where $z = x + iy$ with $x, y \in \mathbb{R}$.

5.13 Consider the function $f : D \mapsto \mathbb{C}$, with $D = \mathbb{C} \setminus \{0\}$, defined by $f(z) = \text{Arg}(z^2) - i2a \ln |z|$ with $a \in \mathbb{R}$. Determine, justifying your answer, the domains of continuity, differentiability and analyticity of f as the parameter a varies.

Chapter 6
Elementary Functions

Abstract Exponential. Logarithm: branches of the logarithm. Powers with complex exponent. Exponentials with complex base. Trigonometric functions. Hyperbolic functions. Inverse trigonometric and hyperbolic functions.

6.1 Exponential

Definition 6.1 (Exponential) Given an arbitrary $z \in \mathbb{C}$ and setting $\operatorname{Re} z = x$ and $\operatorname{Im} z = y$, the exponential of z is defined as the complex number

$$\exp z = e^x \left(\cos y + i \sin y \right), \tag{6.1}$$

where e^x is the usual real exponential of the real number x (Fig. 6.1).

With deliberate abuse of notation we will also indicate $\exp z$ as e^z. For $y = 0$ Definition 6.1 is reduced to that of the real exponential e^x. For $x = 0$ the symbolic notation $e^{iy} = \operatorname{cis}(y)$ is justified. For $z = 0$ we have $e^0 = 1$.

Theorem 6.2 *Let* $z_1, z_2 \in \mathbb{C}$, *then*

$$\exp(z_1) \exp(z_2) = \exp(z_1 + z_2), \tag{6.2}$$

$$\frac{\exp(z_1)}{\exp(z_2)} = \exp(z_1 - z_2). \tag{6.3}$$

Proof Set $z_1 = x_1 + iy_1$ and $z_2 = x_2 + iy_2$ with $x_1, y_1, x_2, y_2 \in \mathbb{R}$. According to Definition 6.1 and observing that $\operatorname{Re}(z_1 + z_2) = x_1 + x_2$ and $\operatorname{Im}(z_1 + z_2) = y_1 + y_2$, it turns out

$$\exp(z_1) \exp(z_2) = e^{x_1} (\cos y_1 + i \sin y_1) e^{x_2} (\cos y_2 + i \sin y_2)$$

$$= e^{x_1} e^{x_2} ((\cos y_1 \cos y_2 - \sin y_1 \sin y_2)$$

$$+ i(\cos y_1 \sin y_2 + \sin y_1 \cos y_2))$$

C. Presilla, *Complex Functions of a Variable*, La Matematica per il 3+2 179,
https://doi.org/10.1007/978-3-032-12494-4_6

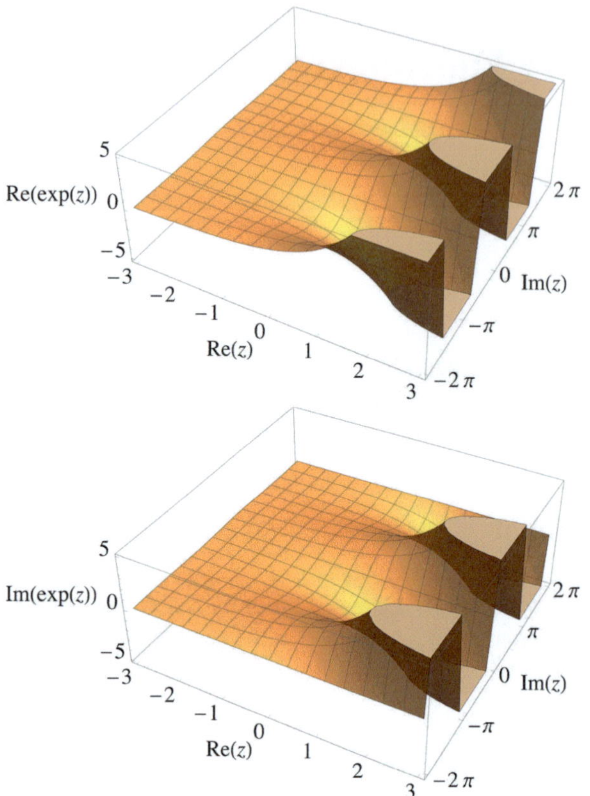

Fig. 6.1 Real and imaginary parts of $\exp(z)$

$$= e^{x_1+x_2}(\cos(y_1 + y_2) + i \sin(y_1 + y_2))$$

$$= \exp(z_1 + z_2),$$

having used the corresponding property, $e^{x_1}e^{x_2} = e^{x_1+x_2}$, valid for the real exponential. A similar reasoning applies to $\exp(z_1)/\exp(z_2)$.

By what has just been demonstrated, we have $1/\exp(z) = \exp(-z)$.

Theorem 6.3 *The function* $\exp z$ *is entire and* $\forall z \in \mathbb{C}$ *its derivative is*

$$\frac{\mathrm{d}}{\mathrm{d}z} \exp z = \exp z. \tag{6.4}$$

Proof Set $\exp z = u(x, y) + iv(x, y)$ where $z = x + iy$ with $x, y \in \mathbb{R}$. It turns out $u(x, y) = e^x \cos y$ and $v(x, y) = e^x \sin y$. By Theorem 5.12, whose hypotheses are satisfied $\forall (x, y) \in \mathbb{R}^2$, the function $\exp z$ is differentiable throughout $\mathbb{C}$ and its derivative is

$$\frac{d}{dz} \exp z = u_x(x, y) + iv_x(x, y) = e^x \cos y + ie^x \sin y = \exp z .$$

The polar form of $\exp z$ is immediately obtained from the definition $e^z = e^x e^{iy}$,

$$|\exp z| = e^{\operatorname{Re} z}, \tag{6.5}$$

$$\arg(\exp z) = \operatorname{Im} z + 2\pi n, \qquad n \in \mathbb{Z}. \tag{6.6}$$

From (6.5) it follows that $\exp z \neq 0 \ \forall z \in \mathbb{C}$. It is easy to verify that $\exp z$ is a periodic function of period $2\pi i$, which means that $\forall z \in \mathbb{C}$

$$\exp(z + 2\pi i) = \exp z. \tag{6.7}$$

Example 6.4 Determine all the solutions of the equation $\exp z = -1$. Given $z = x + iy$ with $x, y \in \mathbb{R}$ and noting that $-1 = 1e^{i\pi}$, by (6.5) and (6.6) we have

$$\begin{cases} e^x = 1, \\ y + 2\pi n = \pi, \qquad n \in \mathbb{Z}. \end{cases}$$

Therefore, we have infinite solutions $z = i\pi(2n + 1)$ with $n \in \mathbb{Z}$.

6.2 Logarithm

Definition 6.5 (Logarithm) Let $z \in \mathbb{C}$ with $z \neq 0$, we define the logarithm of z as any complex number

$$\log z = \ln |z| + i \arg z = \ln |z| + i(\operatorname{Arg} z + 2\pi n), \qquad n \in \mathbb{Z}, \tag{6.8}$$

where $\ln(\cdot)$ denotes the usual real natural logarithm.

The complex logarithm is a multi-valued function, or multi-function for brevity: for each value of z, $\log z$ assumes infinite values whose imaginary parts differ by an integer of 2π. Definition 6.5 is equivalent to requiring that $\log z = w$ be a number $w \in \mathbb{C}$ such that $\exp w = z$, in other words $\log$ is the inverse function of $\exp$. The exclusion of the point $z = 0$ corresponds to the fact that the equation $\exp w = 0$ does not admit solutions. Consistently, $\ln |z|$ is well defined in the region $\mathbb{C} \setminus \{0\}$. Note that the composition

$$\exp(\log(z)) = \exp(\ln |z| + i \arg z) = e^{\ln |z|} e^{i \arg z} = |z| e^{i \arg z} = z \tag{6.9}$$

is a single-valued function, while the inverse composition is multi-valued

$$\log(\exp(z)) = \log(|\exp z| \, e^{i \arg(\exp z)})$$

$$= \ln(e^{\operatorname{Re} z}) + i(\operatorname{Im} z + 2\pi n)$$

$$= z + i2\pi n, \qquad n \in \mathbb{Z}. \tag{6.10}$$

Definition 6.6 (Principal Value of the Logarithm) Let $z \in \mathbb{C}$ with $z \neq 0$, we define principal value of $\log z$, hereafter indicated as $\operatorname{Log} z$, the complex number

$$\operatorname{Log} z = \ln |z| + i \operatorname{Arg} z, \qquad -\pi < \operatorname{Arg} z \leq \pi. \tag{6.11}$$

Example 6.7 Determine $\log(-1)$. Observing that $-1 = 1e^{i\pi}$, from Definition 6.5 we have the infinite values

$$\log(-1) = \ln 1 + i(\pi + 2\pi n) = i\pi(2n + 1), \qquad n \in \mathbb{Z},$$

while $\operatorname{Log}(-1) = i\pi$.

Example 6.8 Determine all the solutions of the equation $\log z = a + ib$, where $a, b \in \mathbb{R}$. From Definition 6.5 we have

$$\ln |z| + i \arg z = a + ib,$$

it follows that $|z| = e^a$ and $\arg z = b$. We therefore have only one solution $z = e^a e^{ib} = \exp(a + ib)$. Alternatively, by (6.9) we immediately have

$$z = \exp(\log z) = \exp(a + ib).$$

Note that the equality $\log(z_1 z_2) = \log z_1 + \log z_2$ is in general false if the three logarithms in this expression are assumed to have the same value, for example the principal one. Equality instead holds in the sense that for every value of $\log(z_1 z_2)$ there exist a value of $\log z_1$ and a value of $\log z_2$ such that their sum is equal to the chosen value of $\log(z_1 z_2)$.

Example 6.9 Find two complex numbers z_1 and z_2 such that $\operatorname{Log}(z_1 z_2) \neq \operatorname{Log}(z_1) \operatorname{Log}(z_2)$. Take $z_1 = z_2 = -1$. We have $\operatorname{Log}(z_1 z_2) = \operatorname{Log}(1) = 0$ while $\operatorname{Log}(z_1) = \operatorname{Log}(z_2) = \operatorname{Log}(-1) = i\pi$.

Definition 6.10 (Branch of the Logarithm) Let $D \subset \mathbb{C}$ be open and connected and let $f : D \mapsto \mathbb{C}$ be a continuous function such that $\forall z \in D$ we have $\exp(f(z)) = z$. The function f is called a branch of the logarithm on D.

Note that D cannot contain the point $z = 0$ because the function $\exp$ is never zero (Figs. 6.2 and 6.3).

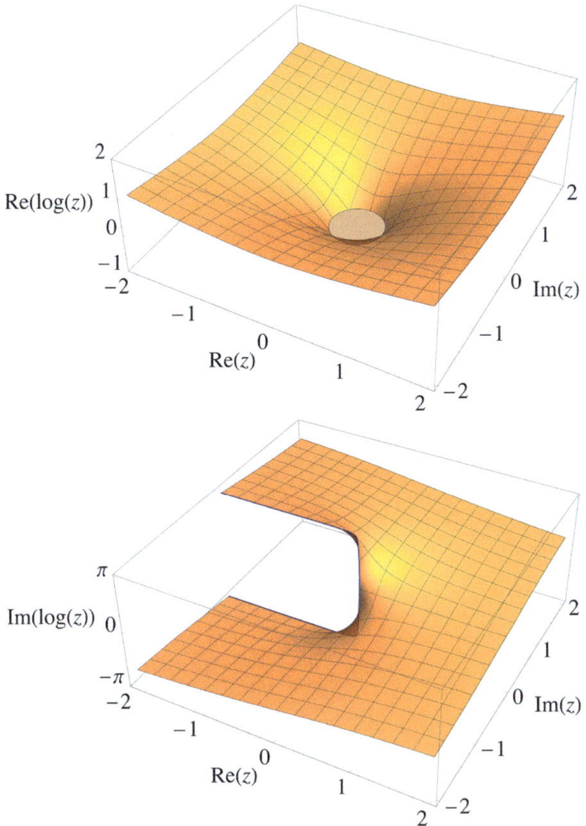

Fig. 6.2 Real and imaginary parts of the principal branch of $\log(z)$

Theorem 6.11 *Let $D \subset \mathbb{C}$ be open and connected and let $f : D \mapsto \mathbb{C}$ be a branch of $\log z$ on D. The totality of branches of $\log z$ on D are the functions $f(z) + \mathrm{i}2\pi n$, $n \in \mathbb{Z}$.*

Proof Let f be a branch of $\log z$ on D and let $g : D \mapsto \mathbb{C}$ be the continuous function defined as $g(z) = f(z) + \mathrm{i}2\pi n$ with $n \in \mathbb{Z}$. Since $\exp(g(z)) = \exp(f(z)) = z$, g is also a branch of $\log z$ on D. Conversely, if f and g are two branches of $\log z$ on D, then $\forall z \in D \ \exp(g(z) - f(z)) = \exp(g(z))/\exp(f(z)) = 1$. It follows that $g(z) - f(z) = \mathrm{i}2\pi n$, where n is a relative integer dependent, in principle, on z. We can show that in fact n is the same for any $z \in D$. Consider the function $h(z) = (g(z) - f(z))/(2\pi \mathrm{i})$ continuous in D. Since D is connected, $h(D) \subset \mathbb{Z}$ must also be connected. We conclude that there exists a unique integer $n \in \mathbb{Z}$ such that $\forall z \in D \ g(z) = f(z) + \mathrm{i}2\pi n$.

Theorem 6.12 *Let $D \subset \mathbb{C}$ be open and connected and let $f : D \mapsto \mathbb{C}$ be a branch of $\log z$ on D. The function f is analytic in D and its derivative is $f'(z) = 1/z$.*

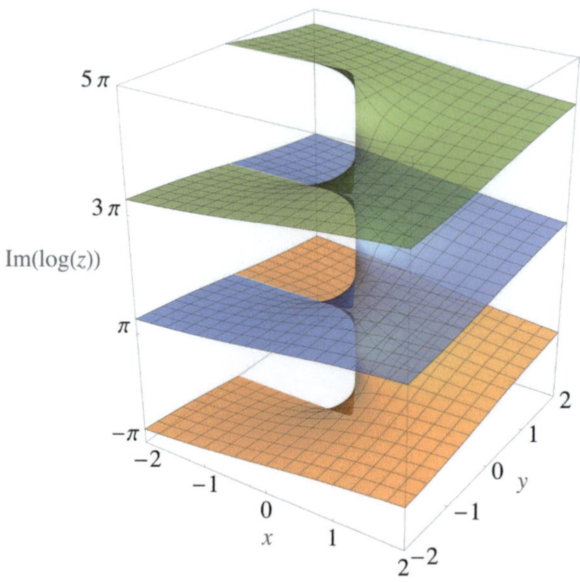

Fig. 6.3 Imaginary parts of three consecutive branches of $\log(z)$, the lowest one being the principal branch

Proof Observe that $\forall w \in f(D)\ \exists z \in D$ such that $f(z) = w$, therefore $f(\exp(w)) = f(\exp(f(z))) = f(z) = w$, that is, f is a continuous inverse function of $\exp$ in the open set $f(D)$. Since $\exp$ is differentiable in $f(D)$ and its derivative is never zero, by Theorem 5.7 also f is differentiable throughout D and its derivative is

$$\frac{\mathrm{d}}{\mathrm{d}z} f(z)\Big|_{z=\exp(w)} = \frac{1}{\exp'(w)} = \frac{1}{\exp(w)} = \frac{1}{z}. \tag{6.12}$$

Definition 6.13 (Principal Branch of the Logarithm) Let D be the open and connected subset of $\mathbb{C}$ defined by $D = \{z \in \mathbb{C} : z \neq -t,\ t \in [0, \infty)\}$. Each point $z \in D$ can be uniquely represented in the polar form $z = re^{i\theta}$ with $r > 0$ and $-\pi < \theta < \pi$. The function $f(z) = \ln r + i\theta$, continuous in D and such that $\exp(f(z)) = z$, is a branch of the logarithm on D. It is called the principal branch of the logarithm. The point $z = 0$ is called the branch point while the negative real semi-axis, including the origin, is called the branch line.

Note that the domain D of the principal branch corresponds to a maximal choice, in fact it is not possible to expand D by adding a single point of the branch line without losing the continuity of the function f defined by $\exp(f(z)) = z$. Furthermore, the domain D, or the branch line that defines it, is a natural choice that corresponds to identifying, for each $z \in D$, the value assumed by the principal branch of $\log z$ with the principal value $\mathrm{Log}\, z$. In the complex plane $w = \log z$, the image of the domain D is the strip $\log(D) = \{w \in \mathbb{C} : \ -\pi < \mathrm{Im}\, w < \pi\}$. The infinite other branches of the logarithm associated to D correspond to the transformation of D into the strips $-\pi + 2\pi k < \mathrm{Im}\, w < \pi + 2\pi k$, with $k = \pm 1, \pm 2, \ldots$. The boundaries of these strips are the paths $\Gamma_k(t) = \ln|\gamma(t)| + i(\mathrm{Arg}(\gamma(t)) + 2\pi k)$, where $\gamma(t) = -t$, $t \in [0, \infty)$, is the path that represents the branch line. The strip corresponding to the principal branch has as boundaries Γ_0 and Γ_{-1}.

Example 6.14 For some purposes it is convenient to define branches of the logarithm on more general maximal domains. Connect the point $z = 0$ to the point $z = \infty$ by means of a completely arbitrary line as long as it is continuous and never intersects and assume this as the branch line. Consider, for example, the path shown in the upper panel of Fig. 6.4 $\gamma(t) = (e^{-(1+t^2)/t} + t^2/1000)e^{i\alpha(t)}$ with $\alpha(t) = \pi \cos(t/5)$, $t \in [0, \infty)$. Note that $|\gamma(t)|$ is non-monotone. In fact, it is immediately clear that there exist two numbers $0 < r_{\min} < r_{\max}$ such that $|\gamma(t)| = r$ has only one solution for $r < r_{\min}$ or $r > r_{\max}$ and three solutions for $r_{\min} < r < r_{\max}$. Analogously to what we saw in the case $\gamma(t) = -t$, the domain $D = \{z \in \mathbb{C} : z \neq \gamma(t),\ t \in [0, \infty)\}$ is transformed into strips of the complex plane

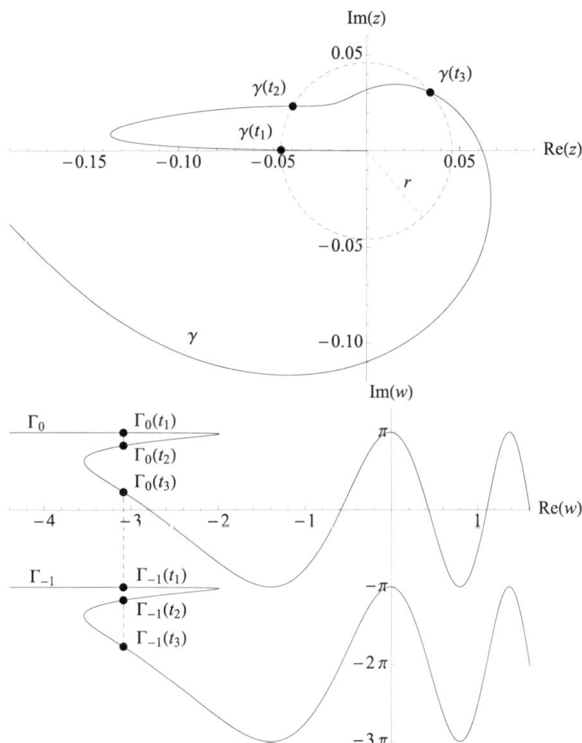

Fig. 6.4 Construction of the branch of $\log(z)$ associated with the branch line $\gamma(t) = (e^{-(1+t^2)/t} + t^2/1000)e^{i\alpha(t)}$ with $\alpha(t) = \pi \cos(t/5)$ and having as boundaries $\Gamma_0(t) = \ln|\gamma(t)| + i\,\mathrm{Arg}(\gamma(t))$ and $\Gamma_{-1}(t) = \ln|\gamma(t)| + i(\mathrm{Arg}(\gamma(t)) - 2\pi)$

$w = \log z$ whose boundaries are $\Gamma_k(t) = \ln|\gamma(t)| + i(\mathrm{Arg}(\gamma(t)) + 2\pi k)$, $k \in \mathbb{Z}$. The strip with boundaries Γ_0 and Γ_{-1} is shown in the lower panel of Fig. 6.4. The shape of this strip differs significantly from that of the principal branch of the logarithm and becomes equal to it only for $t \to 0$ when $\gamma(t) \simeq -t$.

Coming back to the z plane, we can explicitly define the branch of the logarithm corresponding to the w-plane strip with boundaries Γ_0 and Γ_{-1}. Let $z = re^{i\theta}$. For $r < r_{\min}$ or $r > r_{\max}$, we have $\log z = \ln r + i\theta$ with $\alpha(t_0) - 2\pi < \theta < \alpha(t_0)$ where t_0 is the unique solution of the equation $|\gamma(t_0)| = r$. For $r_{\min} < r < r_{\max}$, let $\gamma(t_1)$, $\gamma(t_2)$ and $\gamma(t_3)$, with $t_1 < t_2 < t_3$, be the three intersections of the circle $|z| = r$ with the branch line γ. We can still write $\log z = \ln r + i\theta$ but now θ ranges in three different intervals when we move counterclockwise along the three arcs $\gamma(t_1)\gamma(t_3)$, $\gamma(t_3)\gamma(t_2)$ and $\gamma(t_2)\gamma(t_1)$ of the circle $|\gamma| = r$.

$\gamma(t_1)\gamma(t_3)$ The points of this arc correspond to the segment from $\Gamma_{-1}(t_1)$ to $\Gamma_0(t_3)$ in the w plane, which implies $\alpha(t_1) - 2\pi < \theta < \alpha(t_3)$;

$\gamma(t_3)\gamma(t_2)$ The points of this arc correspond to the segment from $\Gamma_{-1}(t_3)$ to $\Gamma_{-1}(t_2)$ in the w plane, which implies $\alpha(t_3) - 2\pi < \theta < \alpha(t_2) - 2\pi$;

$\gamma(t_2)\gamma(t_1)$ The points of this arc correspond to the segment from $\Gamma_0(t_2)$ to $\Gamma_0(t_1)$ in the w plane, which implies $\alpha(t_2) < \theta < \alpha(t_1)$.

Note that we have a discontinuity in the value of θ whenever the branch line intersects the circle $|z| = r$. The discontinuity amounts to -2π when the branch line goes from inside to outside the circle and $+2\pi$ in the other direction.

Example 6.15 Determine the domain of analyticity of the principal branch of $\log(z^2)$. Let $w = z^2$, according to Definition 6.13 the principal branch of $\log(w)$ is an analytic function at all points $w \neq -t$ with $t \in [0, \infty)$. Therefore, the principal branch of $\log(z^2)$ is an analytic function everywhere in $\mathbb{C}$ except the points $z(t)^2 = -t$, that is, $z(t) = \pm it$, with $t \geq 0$. In conclusion, the domain of analyticity of $\log(z^2)$ is $\mathbb{C}$ except the imaginary axis. Note that if one had, erroneously, written $\log(z^2) = 2\log z$, one would have deduced an analytic domain equal to $\mathbb{C}$ minus the negative real semi-axis.

6.3 Powers with Complex Exponent

Definition 6.16 Let $c \in \mathbb{C}$, a power with complex exponent c is defined as the function

$$z^c = \exp(c \log z), \qquad z \neq 0. \tag{6.13}$$

This function, except for the case of real and integer c, is multi-valued and all the considerations made in the case of the logarithm apply to it. In particular, the principal branch of z^c is defined in terms of the principal branch of $\log z$ and is an analytic function in the domain $D = \{z \in \mathbb{C} : z \neq -t, \ t \in [0, \infty)\}$. For each $z \in D$ we also have

$$\frac{\mathrm{d}}{\mathrm{d}z} z^c = \frac{\mathrm{d}}{\mathrm{d}z} \exp(c \log z) = \frac{c}{z} \exp(c \log z) = c \frac{\exp(c \log z)}{\exp(\log z)}$$

$$= c \exp((c-1) \log z) = c z^{c-1}. \tag{6.14}$$

Example 6.17 Calculate the principal value of $(-\mathrm{i})^{\mathrm{i}}$. Based on Definitions 6.16 and 6.6, we have

$$(-\mathrm{i})^{\mathrm{i}} = \exp(\mathrm{i} \operatorname{Log}(-\mathrm{i})) = \exp(\mathrm{i}(\ln 1 - \mathrm{i}\pi/2)) = \mathrm{e}^{\pi/2}.$$

Definition 6.16 also applies in the case where c is an integer n or a fractional number $1/n$. Let's see how the results already known for integer or fractional powers are re-obtained in these cases. If $c = n \in \mathbb{N}$, setting $z = r\,\mathrm{e}^{\mathrm{i}\theta}$, we have

$$z^n = \exp(n \log z)$$

$$= \exp(n(\ln r + \mathrm{i}(\theta + 2\pi k)))$$

$$= r^n \mathrm{e}^{\mathrm{i}n\theta}.$$

The multiplicity of values of the logarithm corresponding to the different choices of $k \in \mathbb{Z}$ is completely removed by the subsequent application of the exponential function: the product $nk \in \mathbb{Z}$ and hence $\mathrm{e}^{\mathrm{i}2\pi nk} = 1$. We thus obtain the univocal result of Definition 1.4. Furthermore, since for integer n we have $\mathrm{e}^{\mathrm{i}n\pi} = \mathrm{e}^{-\mathrm{i}n\pi}$, the domain of analyticity extends to the entire $\mathbb{C}$ in accordance with the previously established result that the powers z^n are entire functions. For $c = 1/n$, we have instead

$$z^{1/n} = \exp((1/n) \log z)$$

$$= \exp((1/n)(\ln r + \mathrm{i}(\theta + 2\pi k)))$$

$$= r^{1/n} \mathrm{e}^{\mathrm{i}(\theta + 2\pi k)/n}.$$

The infinite multiplicity with respect to $k \in \mathbb{Z}$ is now only partially removed by the exponential function. In accordance with (1.20), we obtain n distinct values for $z^{1/n}$, for example those relating to $k = 0, 1, \ldots, n-1$. Starting from the principal branch of the logarithm, we can therefore identify n distinct functions, which we will call the branches of $z^{1/n}$, each of which has as its domain of analyticity $D = \{z \in \mathbb{C} : z \neq -t, \ t \in [0, \infty)\}$. The function corresponding to $k = 0$ is the principal branch of $z^{1/n}$ (Fig. 6.5).

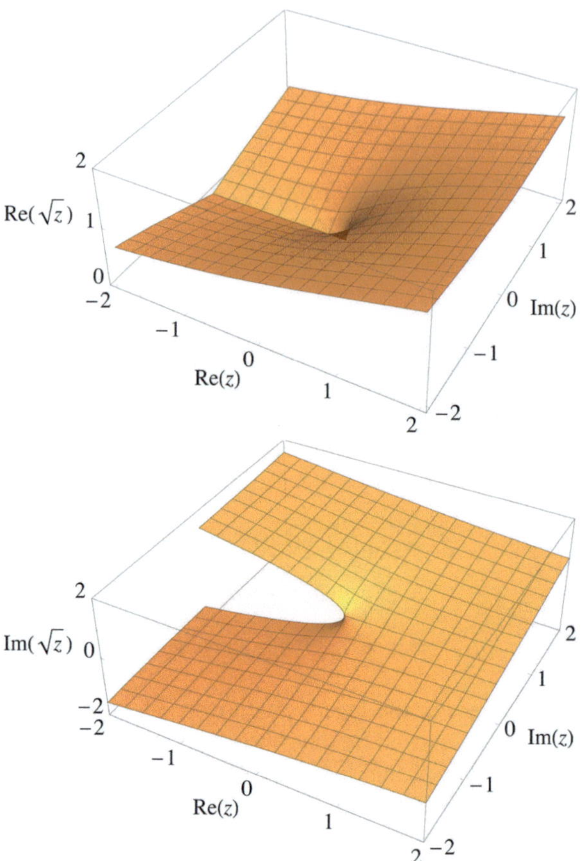

Fig. 6.5 Real and imaginary parts of the principal branch of $\sqrt{z}$. The corresponding secondary branch is $-\sqrt{z}$

6.4   Exponentials with Complex Base

Definition 6.18 Let $c \in \mathbb{C}$, we define exponential with complex base c any function

$$c^z = \exp(z \log c) \tag{6.15}$$

corresponding to the possible values of $\log c$.

For a fixed value of $\log c$, the function $\exp(z \log c)$ is entire and its derivative is

$$\frac{\mathrm{d}}{\mathrm{d}z} c^z = \frac{\mathrm{d}}{\mathrm{d}z} \exp(z \log c) = \exp(z \log c) \log c = c^z \log c. \tag{6.16}$$

Example 6.19 Determine the domain of analyticity of the principal branch of z^{z^2} and the value of its derivative. Based on the previous definitions we must set

$$z^{z^2} = \exp(z^2 \log z)$$

whose principal branch corresponds to taking the principal branch of $\log z$. The domain of analyticity of z^{z^2} is therefore all $\mathbb{C}$ except for the negative real semi-axis. In this domain the derivative is

$$\frac{d}{dz} z^{z^2} = \frac{d}{dz} \exp(z^2 \log z) = \exp(z^2 \log z)(2z \log z + z^2/z)$$

$$= z^{z^2}(2z \log z + z).$$

6.5 Trigonometric Functions

Definition 6.20 (Trigonometric Functions) Let $z \in \mathbb{C}$, the complex functions sine and cosine are defined by the following generalization of Euler's formulas:

$$\sin z = \frac{e^{iz} - e^{-iz}}{2i}, \tag{6.17}$$

$$\cos z = \frac{e^{iz} + e^{-iz}}{2}. \tag{6.18}$$

As linear combinations of entire functions, $\sin z$ and $\cos z$ are entire functions and their derivatives are given by

$$\frac{d}{dz} \sin z = \frac{d}{dz} \frac{e^{iz} - e^{-iz}}{2i} = \frac{1}{2i}\left(ie^{iz} + ie^{-iz}\right) = \cos z, \tag{6.19}$$

$$\frac{d}{dz} \cos z = \frac{d}{dz} \frac{e^{iz} + e^{-iz}}{2} = \frac{1}{2}\left(ie^{iz} - ie^{-iz}\right) = -\sin z. \tag{6.20}$$

For $z = x$ with $x \in \mathbb{R}$, (6.17) and (6.18) reduce to the corresponding real trigonometric functions. For $z = iy$ with $y \in \mathbb{R}$, the following connection with real hyperbolic functions holds:

$$\sin(iy) = i \sinh y, \qquad \cos(iy) = \cosh y. \tag{6.21}$$

Like the corresponding real functions, (6.17) and (6.18) are respectively odd and even functions

$$\sin(-z) = -\sin z, \qquad \cos(-z) = \cos z, \tag{6.22}$$

periodic of period 2π

$$\sin(z + 2\pi) = \sin z, \qquad \cos(z + 2\pi) = \cos z, \tag{6.23}$$

anti-periodic of period π

$$\sin(z + \pi) = -\sin z, \qquad \cos(z + \pi) = -\cos z. \tag{6.24}$$

Theorem 6.21 *Let* $z, z_1, z_2 \in \mathbb{C}$ *with* $z = x + iy$ *and* $x, y \in \mathbb{R}$. *The following identities hold:*

$$\sin^2 z + \cos^2 z = 1, \tag{6.25}$$

$$\sin(z_1 \pm z_2) = \sin z_1 \cos z_2 \pm \cos z_1 \sin z_2, \tag{6.26}$$

$$\cos(z_1 \pm z_2) = \cos z_1 \cos z_2 \mp \sin z_1 \sin z_2, \tag{6.27}$$

$$\sin z = \sin x \cosh y + i \cos x \sinh y, \tag{6.28}$$

$$\cos z = \cos x \cosh y - i \sin x \sinh y, \tag{6.29}$$

$$|\sin z|^2 = \sin^2 x + \sinh^2 y, \tag{6.30}$$

$$|\cos z|^2 = \cos^2 x + \sinh^2 y. \tag{6.31}$$

Proof The identities (6.25), (6.26), and (6.27) are verified in an elementary way starting from Definition 6.20. Identities (6.28) and (6.29) follow from (6.26) and (6.27) by setting $z_1 = x$ and $z_2 = iy$ and using the relations (6.21). Equations (6.30) and (6.31) follow from (6.28) and (6.29) using the fundamental trigonometric identity $\sin^2 x + \cos^2 x = 1$ and the corresponding hyperbolic identity $\cosh^2 y - \sinh^2 y = 1$.

From the value of the moduli (6.30) and (6.31) it is evident that $\sin z$ and $\cos z$ are not bounded functions in the complex field (Fig. 6.6).

Example 6.22 Determine all the solutions of the equation $\cos z = 0$. By (6.29) the given equation is equivalent to the system

$$\begin{cases} \cos x \cosh y = 0 \\ \sin x \sinh y = 0 \end{cases}.$$

Since $\cosh y \neq 0 \ \forall y \in \mathbb{R}$, the first equation of the system reduces to $\cos x = 0$ which gives $x = \pi/2 + \pi k$ with $k \in \mathbb{Z}$. For these values of x we have $\sin x \neq 0$, so the second equation reduces to $\sinh y = 0$ whose only solution is $y = 0$. In conclusion, $\cos z = 0$ for $z = \pi/2 + \pi k$ with $k \in \mathbb{Z}$.

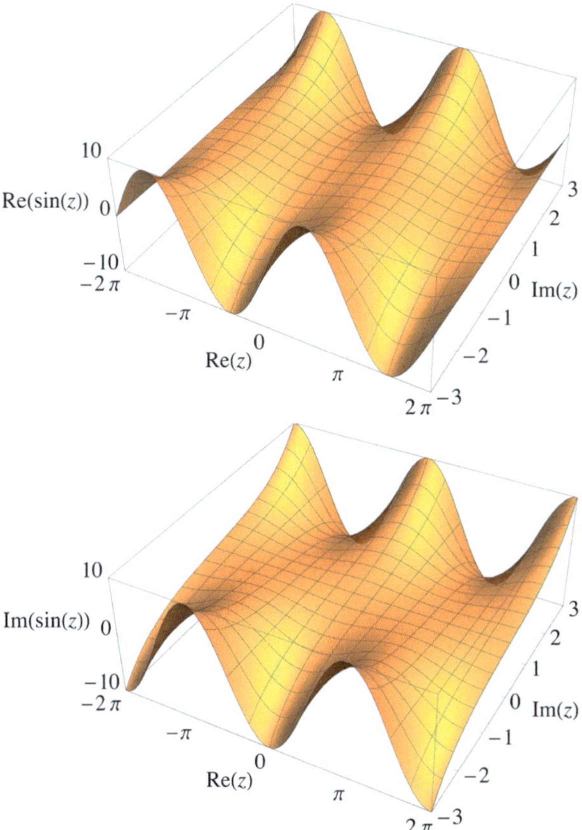

Fig. 6.6 Real and imaginary parts of $\sin(z)$

Example 6.23 Find the locus of points z of the complex plane such that $\cos z = a$, with $a \in \mathbb{R}$, $|a| > 1$. Using the definition $\cos z = (e^{iz} + e^{-iz})/2$, the equation can be rewritten as

$$e^{2iz} - 2ae^{iz} + 1 = 0,$$

whose solutions are

$$e^{iz} = a \pm \sqrt{a^2 - 1}.$$

Note that for $|a| > 1$ the radicand of this expression is positive. If $a > 1$ the solutions are

$$iz = \log\left(a \pm \sqrt{a^2 - 1}\right)$$

$$= \ln\left(a \pm \sqrt{a^2 - 1}\right) + i2\pi k, \qquad k \in \mathbb{Z},$$

or

$$z = -i \ln \left(a \pm \sqrt{a^2 - 1} \right) + 2k\pi, \qquad k \in \mathbb{Z},$$

and represent, as a varies in $(1, \infty)$, the imaginary axes passing through $z = 2k\pi$ with the exception of the points $z = 2k\pi$. If $a < -1$ the solutions are instead

$$iz = \log \left[-\left(a \pm \sqrt{a^2 - 1} \right) e^{i\pi} \right]$$

$$= \ln \left(-a \mp \sqrt{a^2 - 1} \right) + i(\pi + 2\pi k), \qquad k \in \mathbb{Z},$$

that is to say

$$z = -i \ln \left(-a \mp \sqrt{a^2 - 1} \right) + (2k + 1)\pi, \qquad k \in \mathbb{Z},$$

and represent, as a varies in $(-\infty, -1)$, the imaginary axes passing through $z = (2k + 1)\pi$ with the exception of points $z = (2k + 1)\pi$.

It is useful to define other non-fundamental trigonometric functions, for example $\tan z = \sin z / \cos z$. As seen with regard to the zeroes of $\cos z$, the function $\tan z$ is analytic everywhere in the complex plane except at the points $z = \pi/2 + n\pi$ with $n \in \mathbb{Z}$. In this domain its derivative is $d \tan z / dz = 1 / \cos^2 z$.

6.6 Hyperbolic Functions

Definition 6.24 (Hyperbolic Functions) Let $z \in \mathbb{C}$, the complex hyperbolic functions are defined analogously to the real case (Fig. 6.7)

$$\sinh z = \frac{e^z - e^{-z}}{2}, \tag{6.32}$$

$$\cosh z = \frac{e^z + e^{-z}}{2}. \tag{6.33}$$

As linear combinations of entire functions, $\sinh z$ and $\cosh z$ are entire functions and their derivatives are

$$\frac{d}{dz} \sinh z = \frac{d}{dz} \frac{e^z - e^{-z}}{2} = \frac{1}{2} \left(e^z + e^{-z} \right) = \cosh z, \tag{6.34}$$

$$\frac{d}{dz} \cosh z = \frac{d}{dz} \frac{e^z + e^{-z}}{2} = \frac{1}{2} \left(e^z - e^{-z} \right) = \sinh z. \tag{6.35}$$

Fig. 6.7 Real and imaginary
parts of $\sinh(z)$

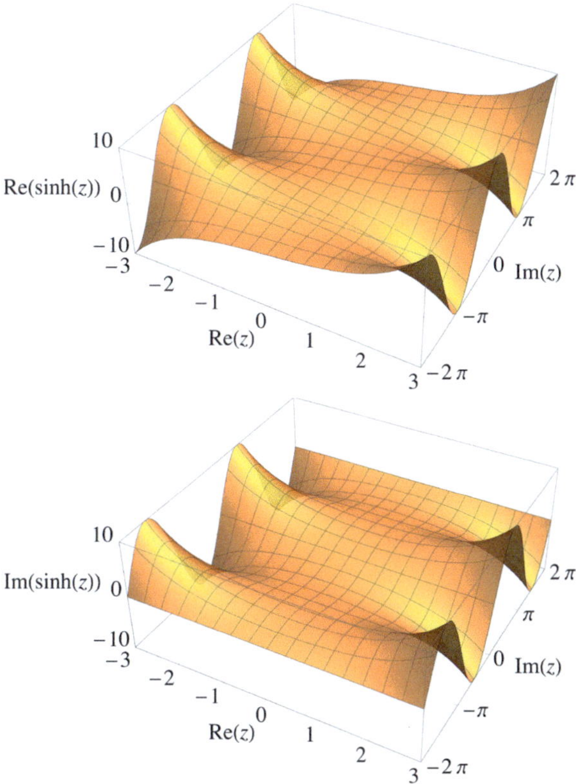

Like the corresponding real functions, (6.32) and (6.33) are respectively odd and
even functions

$$\sinh(-z) = -\sinh z, \qquad \cosh(-z) = \cosh z, \tag{6.36}$$

periodic of period $2\pi i$

$$\sinh(z + 2\pi i) = \sinh z, \qquad \cosh(z + 2\pi i) = \cosh z, \tag{6.37}$$

anti-periodic of period πi

$$\sinh(z + \pi i) = -\sinh z, \qquad \cosh(z + \pi i) = -\cosh z. \tag{6.38}$$

The following relationships with the trigonometric functions hold

$$\sinh z = -i\sin(iz), \qquad \sinh(iz) = i\sin z, \tag{6.39}$$

$$\cosh z = \cos(iz), \qquad \cosh(iz) = \cos z. \tag{6.40}$$

Theorem 6.25 *Let* $z, z_1, z_2 \in \mathbb{C}$ *with* $z = x + iy$ *and* $x, y \in \mathbb{R}$. *The following identities hold:*

$$\cosh^2 z - \sinh^2 z = 1, \tag{6.41}$$

$$\sinh(z_1 \pm z_2) = \sinh z_1 \cosh z_2 \pm \cosh z_1 \sinh z_2, \tag{6.42}$$

$$\cosh(z_1 \pm z_2) = \cosh z_1 \cosh z_2 \pm \sinh z_1 \sinh z_2, \tag{6.43}$$

$$\sinh z = \sinh x \cos y + i \cosh x \sin y, \tag{6.44}$$

$$\cosh z = \cosh x \cos y + i \sinh x \sin y, \tag{6.45}$$

$$|\sinh z|^2 = \sinh^2 x + \sin^2 y, \tag{6.46}$$

$$|\cosh z|^2 = \cosh^2 x + \sin^2 y. \tag{6.47}$$

Proof The identities (6.41), (6.42), and (6.43) are verified in an elementary way starting from Definition 6.24. Identities (6.44) and (6.45) follow from (6.42) and (6.43) by setting $z_1 = x$ and $z_2 = iy$ and using the relations (6.39) and (6.40). Equations (6.46) and (6.47) follow from (6.44) and (6.45) using the fundamental trigonometric identity $\sin^2 y + \cos^2 y = 1$ and the corresponding hyperbolic identity $\cosh^2 x - \sinh^2 x = 1$.

Example 6.26 Determine all solutions of the equation $\cosh z = 0$. By (6.45) the given equation is equivalent to the system

$$\begin{cases} \cosh x \cos y = 0 \\ \sinh x \sin y = 0 \end{cases}.$$

Since $\cosh x \neq 0 \ \forall x \in \mathbb{R}$, the first equation of the system reduces to $\cos y = 0$ which gives $y = \pi/2 + \pi k$ with $k \in \mathbb{Z}$. For these values of y we have $\sin y \neq 0$, so the second equation reduces to $\sinh x = 0$ whose only solution is $x = 0$. In conclusion, $\cosh z = 0$ for $z = i(\pi/2 + \pi k)$ with $k \in \mathbb{Z}$.

It is useful to define other non-fundamental hyperbolic functions, for example $\tanh z = \sinh z / \cosh z$. As seen about the zeros of $\cosh z$, the function $\tanh z$ is analytic everywhere in the complex plane except at the points $z = i(\pi/2 + n\pi)$ with $n \in \mathbb{Z}$. In this domain its derivative is $\mathrm{d}\tanh z / \mathrm{d}z = 1/\cosh^2 z$.

6.7 Inverse Trigonometric and Hyperbolic Functions

The definition of the inverse trigonometric and hyperbolic functions in the complex field follows from the corresponding trigonometric and hyperbolic functions introduced previously. For example, we set $\arcsin z = w$ with $\sin w = z$. Using (6.17) this last equation is equivalent to

$$(e^{iw})^2 - 2ize^{iw} - 1 = 0$$

whose solution is $e^{iw} = iz + \sqrt{(iz)^2 + 1}$, namely, $iw = \log(iz + \sqrt{1 - z^2})$. In the other cases we reason similarly and arrive at the following expressions.

Definition 6.27 (Inverse Trigonometric Functions) Let $z \in \mathbb{C}$, the inverse trigonometric functions in the complex field are defined by

$$\arcsin z = -i \log \left(iz + \sqrt{1 - z^2} \right), \tag{6.48}$$

$$\arccos z = -i \log \left(z + i\sqrt{1 - z^2} \right), \tag{6.49}$$

$$\arctan z = -\frac{i}{2} \log \left(\frac{i - z}{i + z} \right). \tag{6.50}$$

Definition 6.28 (Inverse Hyperbolic Functions) Let $z \in \mathbb{C}$, the inverse hyperbolic functions in the complex field are defined by

$$\operatorname{arcsinh} z = \log \left(z + \sqrt{z^2 + 1} \right), \tag{6.51}$$

$$\operatorname{arccosh} z = \log \left(z + \sqrt{z^2 - 1} \right), \tag{6.52}$$

$$\operatorname{arctanh} z = \frac{1}{2} \log \left(\frac{1 + z}{1 - z} \right). \tag{6.53}$$

All the previous functions are multi-valued and the considerations made in the case of the logarithm apply to them.

Example 6.29 Determine the domain of analyticity of the principal branch of the function $\arcsin z$ and the value of its derivative in this domain. The principal branch of $\arcsin z$ is defined by taking the principal branches of the root and the logarithm that appear in Definition (6.48). The principal branch of $\sqrt{1 - z^2} = \exp[\frac{1}{2} \log(1 - z^2)]$ is a function analytic everywhere in $\mathbb{C}$ except at the points z such that $1 - z^2 = -u$ with $u \in [0, \infty)$. Such points,

$$z(u) = \pm\sqrt{1 + u}, \qquad u \in [0, \infty),$$

represent the real half-lines $(-\infty, -1]$ and $[1, +\infty)$. The principal branch of $\log(iz + \sqrt{1 - z^2})$ turns out to be non-analytic at the points that satisfy $iz + \sqrt{1 - z^2} = -t$ with $t \in [0, \infty)$. The solution of this equation obtained by squaring the equivalent expression $\sqrt{1 - z^2} = -(t + iz)$ gives

$$z(t) = i\frac{t^2 - 1}{2t}, \qquad t \in [0, \infty),$$

which represents the entire imaginary axis. However, the points $z(t)$ thus found do not satisfy the starting equation in which the principal branch is considered for the

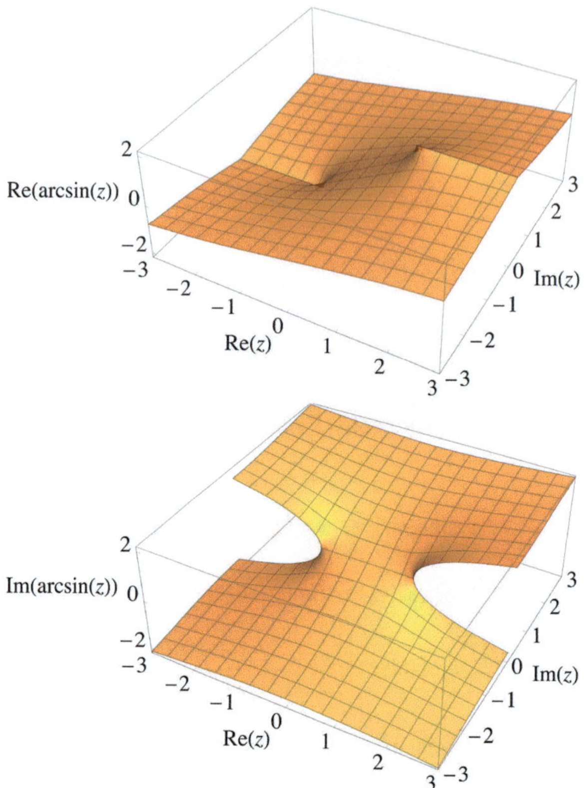

Fig. 6.8 Real and imaginary parts of the principal branch of $\arcsin(z)$

square root. In fact

$$\sqrt{1 - z(t)^2} = \sqrt{1 + (t^2 - 1)^2/(4t^2)} = \sqrt{(t^2 + 1)^2/(4t^2)} > 0,$$

while $-(t + iz(t)) = -(t^2 + 1)/2t < 0$. We have therefore found a spurious solution, to be discarded, introduced by the squaring operation. In conclusion, the domain of analyticity of $\arcsin z$ is the entire complex plane except for the real half-lines $(-\infty, -1]$ and $[1, +\infty)$ (Fig. 6.8). In the region of analyticity the derivative is

$$\frac{d}{dz}\arcsin z = -i\frac{1}{iz + \sqrt{1 - z^2}}\left(i + \frac{1}{2}\frac{-2z}{\sqrt{1 - z^2}}\right) = \frac{1}{\sqrt{1 - z^2}}. \tag{6.54}$$

Example 6.30 Determine the domain of analyticity of the principal branch of the function $\arctan z$ and the value of its derivative in that domain. Let us first see how the Definition (6.50) is obtained. Setting $\arctan z = w$, we have $\tan w = z$, that is

$$\frac{e^{iw} - e^{-iw}}{e^{iw} + e^{-iw}} = iz.$$

Multiplying by e^{iw} gives $e^{2iw} = (1+iz)/(1-iz)$ and from this, taking the logarithm, we arrive at

$$\arctan z = -\frac{i}{2} \log\left(\frac{i-z}{i+z}\right).$$

The principal branch of $\arctan z$ is defined by taking the principal branch of the logarithm appearing in this expression. This is analytic everywhere in $\mathbb{C}$ except at the points the points z such that $(i - z)/(i + z) = -t$ with $t \in [0, \infty)$,

$$z(t) = i\frac{1+t}{1-t}, \qquad t \in [0, \infty).$$

These points represent the entire imaginary axis except for the segment $(-i, i)$ (Fig. 6.9). Within the above domain the derivative is

$$\frac{d}{dz} \arctan z = -\frac{i}{2}\frac{1-iz}{1+iz}\frac{i(1-iz) - (1+iz)(-i)}{(1-iz)^2} = \frac{1}{1+z^2}. \tag{6.55}$$

Exercises

6.1 Determine the real and imaginary parts of the following expressions:

$$\text{(a)} \;\; \cosh(2+i), \qquad \text{(b)} \;\; \cotan\left(\frac{\pi}{4} - i\ln 2\right).$$

6.2 Determine the principal value of

$$\arctan(2+i).$$

6.3 Find the imaginary part of the principal value of

$$\arctan(1+2i).$$

6.4 Find the principal value of

$$(4 - 4i)^{1+i}.$$

6.5 Find the modulus of the principal value of

$$(1+i)^{2-i}.$$

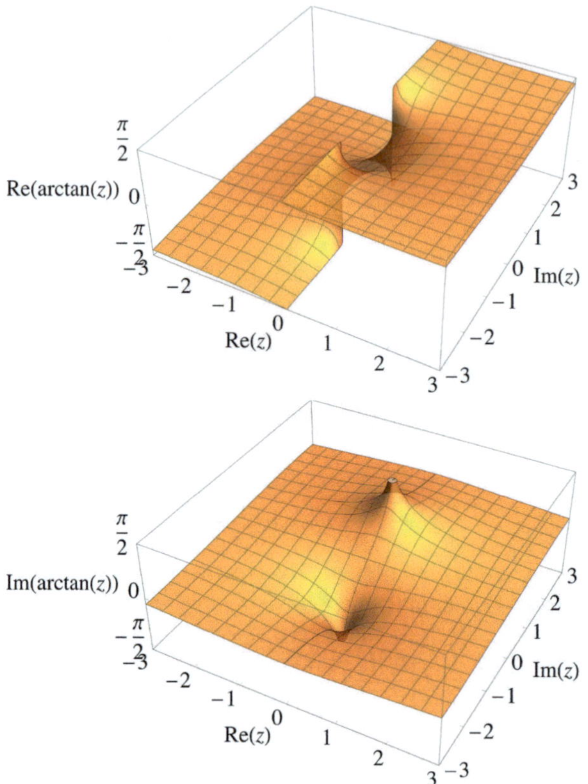

Fig. 6.9 Real and imaginary parts of the principal branch of $\arctan(z)$

6.6 Determine the principal argument of the principal value of

$$\left(\frac{e^{\pi}}{\sqrt{2}} + i\frac{e^{\pi}}{\sqrt{2}}\right)^{i+1}.$$

6.7 Determine the principal value of the following quantities:

$$\text{(a)}\quad \ln\left|(1-i)^{\sqrt{2}+i\pi}\right|, \qquad \text{(b)}\quad \operatorname{Im}\left(\arctan\left(\frac{i}{2}\right)\right).$$

6.8 Find all the solutions of the equation

$$\cos z = a, \qquad a > 1.$$

6.9 Find all the solutions of the equation

$$\cosh z = \frac{1}{2}.$$

6.10 Find all the solutions of the equation

$$\sin(z)\cos(z) = 2.$$

6.11 Find all the solutions of the equation

$$\cosh(z) + e^z = 2.$$

6.12 Find all the solutions of the equation

$$\cos(3z^2) = 0.$$

6.13 Find all the solutions of the equation

$$\sinh z = -i\pi.$$

6.14 Show that all the solutions of the following equation are real

$$\sin z = a, \qquad -1 \le a \le 1.$$

6.15 Find the locus of the points $z \in \mathbb{C}$ such that

$$\cosh z = a, \qquad 0 < a < 1.$$

6.16 Find all the solutions of the equation

$$\sinh z = i.$$

6.17 Find all the solutions of the equation

$$(\sin z)^2 = -4$$

and evaluate the limit points of these solutions in $\mathbb{C}_\infty$.

6.18 Find all the solutions of the equation

$$\exp\left(z^2\right) = -1.$$

6.19 Find all the solutions of the following equations:

$$\text{(a)} \ z^4 - i = 0, \qquad \text{(b)} \ z^i - 4 = 0.$$

6.20 Find all the solutions of the equation

$$\text{cotan}(\sqrt{z}) = 0.$$

6.21 Find all the solutions of the following equations:

$$\text{(a)} \ z^3 - 2i = 0, \qquad \text{(b)} \ z^{2i} - 3 = 0.$$

6.22 Determine the domain of analyticity of the following functions, considering the principal branch in the case of multi-valued functions:

$$\text{(a)} \ \log((2 - z)^2), \qquad \text{(b)} \ z^z.$$

6.23 Determine the domain of analyticity of the following functions, considering the principal branch in the case of multi-valued functions:

$$\text{(a)} \ z^{\sinh z}, \qquad \text{(b)} \ \log(\cos z).$$

6.24 Determine the domain of analyticity of the following functions, considering the principal branch in the case of multi-valued functions:

$$\text{(a)} \ \cosh(e^{\sinh z}), \qquad \text{(b)} \ \log(3 + z^2).$$

6.25 Determine the domain of analyticity of the following functions, considering the principal branch in the case of multi-valued functions:

$$\text{(a)} \ \exp\left(\frac{1}{1 + z^2}\right), \qquad \text{(b)} \ \log(z^3).$$

6.26 Determine the domain of analyticity of the following functions, considering the principal branch in the case of multi-valued functions:

$$\text{(a)} \ \left(1 + z^2\right)^z, \qquad \text{(b)} \ \frac{\sinh(\sin(z))}{z^2 + 9}.$$

6.27 Determine the domain of analyticity of the following functions, considering the principal branch in the case of multi-valued functions, and the value of the respective derivatives in this domain:

$$\text{(a)} \ \log(z^4), \qquad \text{(b)} \ \sin\sqrt{z}.$$

6.28 Determine the domain of analyticity of the following functions, considering the principal branch in the case of multi-valued functions, and the value of the respective derivatives in this domain. Specify how many distinct branches each function has.

$$\text{(a) } \sin(z^n), \qquad \text{(b) } \sin(z^{n/m}), \quad n, m \in \mathbb{N}, \qquad \text{(c) } \sin(z^c), \quad c \in \mathbb{R} \setminus \mathbb{Q}.$$

6.29 Determine the domain of analyticity of the principal branch of $\arccos z$ and the value of its derivative in this domain.

6.30 Determine the domain of analyticity of the following functions, considering the principal branch in the case of multi-valued functions:

$$\text{(a) } \sqrt{z^2 - 1}, \qquad \text{(b) } i\sqrt{1 - z^2}.$$

6.31 Establish how many distinct branches the following functions have and determine the domain of analyticity of the principal branch. In the particular case $n = 17, m = 3, c = 1/\pi$, calculate their principal value at the point $z = i$.

$$\text{(a) } z^n, \qquad \text{(b) } z^{n/m}, \quad n, m \in \mathbb{N}, \qquad \text{(c) } z^c, \quad c \in \mathbb{R} \setminus \mathbb{Q}.$$

6.32 Determine the domain of analyticity of the following functions, considering the principal branch in the case of multi-valued functions, and the value of the respective derivatives in this domain:

$$\text{(a) } \log((\log z)^2), \qquad \text{(b) } e^{z^2 + i}, \qquad \text{(c) } \frac{\sqrt{1 - z}}{z^2 + i}.$$

6.33 Determine the domain of analyticity of the principal branch of

$$\sqrt{4 + z^4}.$$

6.34 Determine the domain of analyticity of the principal branch of

$$f(z) = \frac{\sin z}{\sqrt{\cos z}}.$$

6.35 Determine the domain of analyticity of the function

$$f(z) = \frac{\sin z}{\sqrt{\cos z}}$$

assuming for $f(z)$ the branch corresponding to the branch of $\log z$ that has as its branch line the positive real semi-axis.

6.36 Consider the function $g(z) = \log((3z^2 - \bar{z}^2)\bar{z}/2)$, where $\log(\cdot)$ is the principal branch of the logarithm. Determine, justifying your answer, the domains of continuity, differentiability and analyticity of g.

6.37 Determine the domain of analyticity of the function

$$f(z) = \sqrt{\cos z}$$

assuming for $f(z)$ the branch corresponding to the principal branch of $\log z$ that has as its branch line the positive imaginary semi-axis.

6.38 Consider the power series

$$\sum_{n=0}^{\infty} a_n (z - z_0)^n$$

convergent to $f(z) = \tan(z) \log(z - \pi + \mathrm{i})$ for $|z - z_0| < R$ with $z_0 = 2 + \mathrm{i}$. Determine the radius of convergence R of this series. Assume the principal branch for multi-valued functions.

6.39 Provide an example of an analytic function that transforms biunivocally A in B, where

$$A = \{z \in \mathbb{C} : 1 \leq |z| \leq 2, \ 0 \leq \operatorname{Arg} z \leq \pi/2\},$$

$$B = \{z \in \mathbb{C} : 1 \leq \operatorname{Re} z \leq 2, \ 0 \leq \operatorname{Im} z \leq \pi/(2 \ln 2)\}.$$

6.40 Assuming the principal branch for multi-valued functions, determine for which values of $z \in \mathbb{C}$ the following series is convergent

$$\sum_{n=0}^{\infty} \mathrm{i}^{n z^2}.$$

Chapter 7
Integrals

Abstract Integrals of complex functions of a real variable. Paths, closed paths, simple closed paths. Piecewise regular paths. Trace of a path. Equivalent paths: curves. Length of a piecewise regular path and its invariance under reparametrization. Jordan curve theorem (not proved). Integrals of complex functions along piecewise regular paths. Invariance of the integral under reparametrization of the path. Upper bound of the modulus of an integral: Darboux inequality. Theorem on the existence of a primitive. Primitives of multi-valued functions and calculation of integrals on closed curves. Cauchy-Goursat theorem for simple closed curves. Homotopic curves. Closed curves homotopic to zero. Simply connected open sets. Extension of the Cauchy-Goursat theorem to closed curves in simply connected domains. An analytic function in a simply connected domain admits primitive. Principle of deformation of paths. Cauchy's integral formula for an analytic function. Existence and Cauchy's integral formula for the derivatives of any order of analytic functions. The derivative of any order of an analytic function is analytic. Morera's theorem. Cauchy's inequalities. Liouville's theorem. Fundamental theorem of algebra: decomposition of a polynomial.

7.1 Integrals of Complex Functions of a Real Variable

As an introduction to the integration of complex functions of a complex variable, in this section we extend the definition and properties of the integral of real functions to complex functions $f(t)$ of real variable t.

Definition 7.1 Let $f : [a, b] \mapsto \mathbb{C}$ be a continuous function, the integral of f between a and b is the complex number

$$\int_a^b f(t)\mathrm{d}t = \int_a^b \mathrm{Re}\, f(t)\mathrm{d}t + \mathrm{i} \int_a^b \mathrm{Im}\, f(t)\mathrm{d}t. \tag{7.1}$$

Definition 7.2 Let $f : [a, b] \mapsto \mathbb{C}$ and $F : [a, b] \mapsto \mathbb{C}$. If $F'(t) = f(t)\ \forall t \in [a, b]$, then F is a primitive of f in $[a, b]$.

C. Presilla, *Complex Functions of a Variable*, La Matematica per il 3+2 179,
https://doi.org/10.1007/978-3-032-12494-4_7

Theorem 7.3 (Fundamental Theorem of Calculus) *Every continuous function* $f : [a, b] \mapsto \mathbb{C}$ *has a primitive and any two primitives differ by a constant. If* F *is a primitive of* f *then*

$$\int_a^b f(t)\mathrm{d}t = F(b) - F(a). \tag{7.2}$$

Proof The real functions $\mathrm{Re}\,f$ and $\mathrm{Im}\,f$ are continuous and therefore admit primitives. If U and V are two such primitives with $U' = \mathrm{Re}\,f$ and $V' = \mathrm{Im}\,f$, then $F = U + iV$ is a primitive of f since $F' = f$. Equation (7.2) follows immediately.
$\square$

Theorem 7.4 *Let* $f : [a, b] \mapsto \mathbb{C}$ *be continuous. Then the following inequality holds*

$$\left| \int_a^b f(t)\mathrm{d}t \right| \le \int_a^b |f(t)|\,\mathrm{d}t. \tag{7.3}$$

Proof Setting $\int_a^b f(t)\mathrm{d}t = r\mathrm{e}^{i\theta}$ with r and θ real, we have

$$\left| \int_a^b f(t)\mathrm{d}t \right| = \mathrm{e}^{-i\theta} \int_a^b f(t)\mathrm{d}t$$

$$= \int_a^b \mathrm{e}^{-i\theta} f(t)\mathrm{d}t$$

$$= \int_a^b \mathrm{Re}\left(\mathrm{e}^{-i\theta} f(t) \right) \mathrm{d}t$$

$$\le \int_a^b \left| \mathrm{e}^{-i\theta} f(t) \right| \mathrm{d}t$$

$$= \int_a^b |f(t)|\,\mathrm{d}t,$$

having used $\mathrm{Re}\,z \le |z|$ valid $\forall z \in \mathbb{C}$.
$\square$

Definition (7.1), the fundamental theorem of calculus (7.2) and the inequality (7.3) extend immediately to piecewise continuous functions $f : [a, b] \mapsto \mathbb{C}$.

7.2 Paths, Path Traces, Curves

We now want to introduce the integral of a complex function $f(z)$ of a complex variable z along a curve in $\mathbb{C}$ having sufficient regularity. The meaning of this sentence is clarified by the following definitions partly anticipated when introducing the conformal transformations.

Definition 7.5 (Path) We call path in the complex plane any continuous function $\gamma : [a, b] \mapsto \mathbb{C}$, with $a, b \in \mathbb{R}$ and $a < b$. A path is said to be simple if γ is injective in $[a, b]$, that is, if $\gamma(t_1) \neq \gamma(t_2)$ for $t_1 \neq t_2$ with $t_1, t_2 \in [a, b]$. A path is said to be closed if $\gamma(a) = \gamma(b)$. A path is called simple closed if γ is injective in (a, b) and $\gamma(a) = \gamma(b)$.

If $\gamma : [a, b] \mapsto \mathbb{C}$ is a path, then the set $\{\gamma(t) : a \leq t \leq b\} \subset \mathbb{C}$ is called the trace of γ and denoted by the symbol $\{\gamma\}$. Note that the trace of a path is always a compact subset of $\mathbb{C}$ as it is the image of the compact $[a, b]$ through the continuous function γ. Different paths can have the same trace. Consider for example $\gamma_1(t) = e^{it}$ and $\gamma_2(t) = e^{2it}$ with $0 \leq t \leq 2\pi$. For both these paths the trace is the unit circle centered at the origin.

Definition 7.6 (Regular Path, Piecewise Regular Path) A regular path is a function $\gamma : [a, b] \mapsto \mathbb{C}$ that can be differentiated and has continuous first derivative. The path γ is said to be piecewise regular if there exists a partition of the interval $[a, b]$, $a = t_0 < t_1 < \cdots < t_{n-1} < t_n = b$, such that γ is regular in every subinterval $[t_{j-1}, t_j]$, $j = 1, \ldots, n$.

In the case of a regular (piecewise regular) path

$$\left|\gamma'(t)\right| = \sqrt{((\mathrm{Re}\gamma)'(t))^2 + ((\mathrm{Im}\gamma)'(t))^2}$$

is a continuous (piecewise continuous) function and therefore integrable between a and b. The integral

$$L_\gamma = \int_a^b \left|\gamma'(t)\right| \, dt \tag{7.4}$$

represents the length of the path γ.

Definition 7.7 (Equivalent Paths, Curve) Let $\gamma : [a, b] \mapsto \mathbb{C}$ and $\eta : [c, d] \mapsto \mathbb{C}$ be two paths. The path η is said to be equivalent to γ if there exists a continuous, strictly increasing function $\varphi : [c, d] \mapsto [a, b]$ with $\varphi(c) = a$ and $\varphi(d) = b$, such that $\eta = \gamma \circ \varphi$. The function φ is called reparametrization of the path. A curve is an equivalence class of paths. The trace of a curve is the trace of any one of the member paths of the class. A curve is regular (piecewise regular) if and only if at least one of the member paths of the class is regular (piecewise regular).

The trace of a curve should not be confused with the curve itself. The curve is a function (an equivalence class of functions), while the trace is its image. Different curves can have the same trace. For example, a circle traveled clockwise or counterclockwise corresponds to two different curves (there are no equivalent paths between the two cases) that have the circle as the same trace.

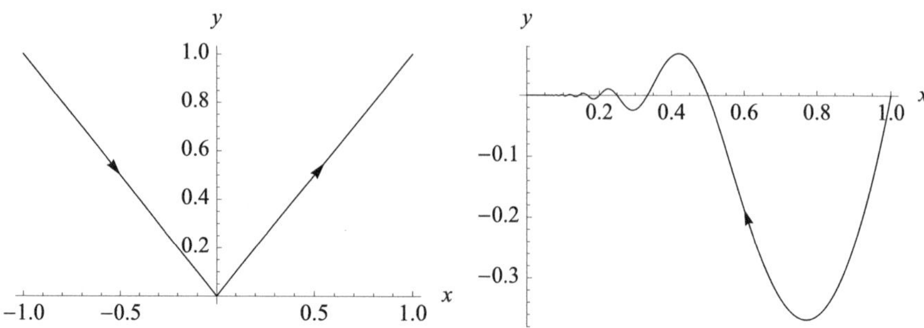

Fig. 7.1 Regular curves $y(x) = |x|$ with $-1 \le x \le 1$ (left) and $y(x) = x^3 \sin(\pi/x)$ with $0 \le x \le 1$ (right)

From now on we will not distinguish between curve and path belonging to the equivalence class that defines the curve since all the results we will obtain do not depend on the path chosen to represent the curve. For example, the length of a regular curve is invariant under reparametrization of the path. Let $\gamma : [a, b] \mapsto \mathbb{C}$ and $\eta : [c, d] \mapsto \mathbb{C}$ be two equivalent paths and $\varphi : [c, d] \mapsto [a, b]$ be the change of parameter. Observing that φ is differentiable almost everywhere[1] in $[c, d]$, almost everywhere in $[c, d]$ we have $\eta'(u) = \gamma'(\varphi(u))\varphi'(u)$. Let $t = \varphi(u)$, since $\varphi'(u) = |\varphi'(u)|$ at the points where the derivative of φ exists, we obtain

$$\int_c^d \left|\eta'(u)\right| du = \int_c^d \left|\gamma'(\varphi(u))\right| \varphi'(u)du = \int_a^b \left|\gamma'(t)\right| dt. \tag{7.5}$$

Example 7.8 The path $\gamma(t) = t^3 + \mathrm{i}\,|t|^3$ with $-1 \le t \le 1$ is regular. In fact, for $t < 0$ we have $\gamma'(t) = 3t^2 - \mathrm{i}3t^2$, while for $t > 0$ we have $\gamma'(t) = 3t^2 + \mathrm{i}3t^2$. It is easy to show using the definition of derivative that $\gamma'(0) = 0$. So γ' exists and is continuous in the whole interval $[-1, 1]$. The length of the associated curve is

$$L_\gamma = \int_{-1}^1 \sqrt{9t^4 + 9t^4}dt = 2 \int_0^1 3\sqrt{2}t^2 dt = 2\sqrt{2}.$$

Let $t^3 = x$, in the x, y plane the equation of the curve associated with the path considered is $y(x) = |x|$. Using the function $\varphi(u) = \mathrm{sgn}(u)\,|u|^{1/3}$ with $-1 \le u \le 1$ we can reparametrize the curve $y(x) = |x|$ in the path $\eta(u) = (\mathrm{sgn}(u)|u|^{1/3})^3 + \mathrm{i}|\mathrm{sgn}(u)|u|^{1/3}|^3 = u + \mathrm{i}\,|u|$. Note that η has no derivative at $u = 0$ while $\eta'(u)$ exists continuously for $u \in [-1, 0) \cup (0, 1]$, therefore η is a piecewise regular path. The curve is however said regular because a member of the equivalence class that defines it, the path γ just studied, is regular (Fig. 7.1).

---

[1] The set of points where the derivative does not exist has zero Lebesgue measure.

Example 7.9 The path $\gamma(t) = t + it^3 \sin(\pi/t)$ with $0 < t \le 1$ and $\gamma(0) = 0$ is regular. In fact for $t \in (0, 1]$ we immediately have $\gamma'(t) = 1 + i[3t^2 \sin(\pi/t) - \pi t \cos(\pi/t)]$, while applying the definition of derivative we find $\gamma'(0) = 1$. Therefore, γ' exists and is continuous in the whole interval $[0, 1]$. The path intersects the real axis infinitely many times at the points $x_k = k^{-1}, k = 1, 2, 3, \ldots$, and its length is

$$L_\gamma = \int_0^1 \sqrt{1 + [3t^2 \sin(\pi/t) - \pi t \cos(\pi/t)]^2}\, dt \simeq 1.50793.$$

In the x, y plane the associated curve has equation $y(x) = x^3 \sin(\pi/x)$ (Fig. 7.1).

Theorem 7.10 (Of the Jordan Curve) *Let γ be a simple closed curve (Jordan curve) in $\mathbb{R}^2$ or in $\mathbb{C}$. The complement of the trace of γ is the union of two distinct open and connected sets: a bounded one, the internal part $\mathrm{Int}(\gamma)$, and an unbounded one, the external part $\mathrm{Ext}(\gamma)$. Moreover, the trace of γ is the boundary of each of these two sets.*

$$\{\gamma\}^c = \mathrm{Int}(\gamma) \cup \mathrm{Ext}(\gamma), \qquad \mathrm{diamInt}(\gamma) < \infty, \qquad \mathrm{diamExt}(\gamma) = \infty,$$

$$\partial \mathrm{Int}(\gamma) = \partial \mathrm{Ext}(\gamma) = \{\gamma\}.$$

For a proof of the Jordan curve theorem see [14].

Definition 7.11 (Positively Oriented Simple Closed Curve) A simple closed curve γ in $\mathbb{R}^2$ or in $\mathbb{C}$ is said to be positively oriented if, walking it, the internal part of γ remains on the left.

7.3   Integrals of Complex Functions Along Piecewise Regular Curves

If $\gamma : [a, b] \mapsto \mathbb{C}$ is a piecewise regular path whose trace $\{\gamma\} \subset D \subset \mathbb{C}$ and $f : D \mapsto \mathbb{C}$ is a continuous function, then $(f \circ \gamma)\gamma' : [a, b] \mapsto \mathbb{C}$ is a piecewise continuous function. The following definition is then well-posed:

Definition 7.12 (Integral Along a Piecewise Regular Curve) Let $\gamma : [a, b] \mapsto \mathbb{C}$ be a piecewise regular path and $f : D \mapsto \mathbb{C}$ be a continuous function in $D \supset \{\gamma\}$, then the integral of f along γ is the complex number

$$\int_\gamma f(z)\,dz = \int_a^b f(\gamma(t))\gamma'(t)\,dt. \tag{7.6}$$

Actually, to grant the existence of the integral (7.6) it is sufficient that f is defined and continuous on the trace of γ up to a finite number of points. In fact, even in this case $f(\gamma(t))\gamma'(t)$ is piecewise continuous in $[a, b]$. It is immediate to verify that the result of the integral (7.6) is invariant under reparametrization of the path. Therefore, the index γ in the integral symbol stands for the equivalence class curve to which the specific path γ belongs. Finally, if the curve γ is split into two curves γ_1 and γ_2, we will symbolically indicate the split with $\gamma = \gamma_1 + \gamma_2$ and we will have

$$\int_{\gamma_1+\gamma_2} f(z)\mathrm{d}z = \int_{\gamma_1} f(z)\mathrm{d}z + \int_{\gamma_2} f(z)\mathrm{d}z. \tag{7.7}$$

Example 7.13 Compute the integral of the function $f(z) = \bar{z}$ along the path $\gamma(t) = 2e^{it}$ with $-\pi/2 \le t \le \pi/2$. By definition we have

$$\int_{\gamma} \bar{z}\mathrm{d}z = \int_{-\pi/2}^{\pi/2} \overline{\gamma(t)}\gamma'(t)\mathrm{d}t = \int_{-\pi/2}^{\pi/2} 2e^{-it}2e^{it}i\mathrm{d}t = 4\pi i.$$

Example 7.14 Consider the function $f(z) = |z|^2$ continuous in $\mathbb{C}$ and compute its integral along (a) the arc of the circle centered at the origin that goes from 1 to i, (b) along the segment $[1, i]$ and (c) along the polygonal $[1, 1+i, i]$.

(a) Let $\gamma_1(t) = e^{it}$ with $0 \le t \le \pi/2$, the integral of f along this arc of circle is

$$\int_{\gamma_1} |z|^2\,\mathrm{d}z = \int_0^{\pi/2} |\gamma_1(t)|^2 \gamma'(t)\mathrm{d}t$$

$$= \int_0^{\pi/2} \left|e^{it}\right|^2 ie^{it}\mathrm{d}t = e^{it}\Big|_0^{\pi/2}$$

$$= -1 + i.$$

(b) The segment $[1, i]$ is represented by the path $\gamma_2(t) = 1+(i-1)t$ with $0 \le t \le 1$, so

$$\int_{\gamma_2} |z|^2\,\mathrm{d}z = \int_0^1 |\gamma_2(t)|^2 \gamma_2'(t)\mathrm{d}t$$

$$= \int_0^1 |1 + (i-1)t|^2 (i-1)\mathrm{d}t$$

$$= (i-1) \int_0^1 ((1-t)^2 + t^2)\mathrm{d}t$$

$$= \frac{2}{3}(-1+i).$$

(c) Finally, we represent the polygonal $[1, 1 + i, i]$ by the two paths $\gamma_3(t) = 1 + it$ and $\gamma_4(t) = 1 + i - t$ with $0 \le t \le 1$. For the corresponding integrals we have

$$\int_{\gamma_3} |z|^2 \, dz = \int_0^1 |\gamma_3(t)|^2 \, \gamma_3'(t) dt$$

$$= \int_0^1 |1 + it|^2 \, i dt = i \int_0^1 (1 + t^2) dt = i \left(t + \frac{1}{3} t^3 \right) \Big|_0^1 = i \frac{4}{3},$$

$$\int_{\gamma_4} |z|^2 \, dz = \int_0^1 |\gamma_4(t)|^2 \, \gamma_4'(t) dt$$

$$= \int_0^1 |1 + i - t|^2 \, (-1) dt = - \int_0^1 ((1 - t)^2 + 1) dt = -\frac{4}{3},$$

and so

$$\int_{\gamma_3 + \gamma_4} |z|^2 \, dz = \int_{\gamma_3} |z|^2 \, dz + \int_{\gamma_4} |z|^2 \, dz$$

$$= \frac{4}{3} (-1 + i).$$

This example shows that the integral of a function f along different curves that connect the same initial point z_a to the same final point z_b takes, in general, different values. If it happens that the integral of f along curves from z_a to z_b is independent of the type of curve considered, then we will use the notation

$$\int_{z_a}^{z_b} f(z) dz$$

to indicate the integral along any of these curves from z_a to z_b.

If $\gamma : [a, b] \mapsto \mathbb{C}$ is a path, we indicate with $-\gamma$ the same path traveled in the opposite direction. Evidently a parametrization of $-\gamma$ is $\gamma(-t)$ with $-b \le t \le -a$. If $w \in \mathbb{C}$, the path $\gamma + w$ has trace obtained by translating the trace of γ by w. Next properties of the integral follow immediately from Definition 7.6 and the above observations.

Theorem 7.15 *Let $\gamma : [a, b] \mapsto \mathbb{C}$ be a piecewise regular path, f and g continuous functions on $\{\gamma\}$ and $w \in \mathbb{C}$ a constant, then*

$$\int_{\gamma} wf(z) dz = w \int_{\gamma} f(z) dz,$$

$$\int_{\gamma} (f(z) + g(z)) \, dz = \int_{\gamma} f(z) dz + \int_{\gamma} g(z) dz,$$

$$\int_{-\gamma} f(z)\mathrm{d}z = -\int_{\gamma} f(z)\mathrm{d}z,$$

$$\int_{\gamma+w} f(z)\mathrm{d}z = \int_{\gamma} f(z+w)\mathrm{d}z.$$

Theorem 7.16 (Darboux Inequality) *Let $\gamma : [a,b] \mapsto \mathbb{C}$ be a piecewise regular path of length L_γ and f a continuous function on $\{\gamma\}$, then*

$$\left| \int_{\gamma} f(z)\mathrm{d}z \right| \le L_\gamma \sup_{z\in\{\gamma\}} |f(z)| . \tag{7.8}$$

Proof Using the definition of integral and the inequality (7.3) we have

$$\left| \int_{\gamma} f(z)\mathrm{d}z \right| = \left| \int_a^b f(\gamma(t))\gamma'(t)\mathrm{d}t \right|$$

$$\le \int_a^b \left| f(\gamma(t))\gamma'(t) \right| \mathrm{d}t$$

$$\le \sup_{z\in\{\gamma\}} |f(z)| \int_a^b \left| \gamma'(t) \right| \mathrm{d}t$$

$$= L_\gamma \sup_{z\in\{\gamma\}} |f(z)| . \tag{7.9}$$

Note that since $f(\gamma(t))$ is continuous for $t \in [a,b]$, $|f(\gamma(t))|$ assumes an absolute maximum in $[a,b]$, that is, there exists $t_0 \in [a,b]$ such that $\sup_{z\in\{\gamma\}} |f(z)| = \sup_{t\in[a,b]} |f(\gamma(t))| = |f(\gamma(t_0))|$. $\square$

Example 7.17 Prove that, if $\gamma(\theta) = 2e^{i\theta}$ with $0 \le \theta \le \pi/2$, then

$$\left| \int_{\gamma} \frac{\mathrm{d}z}{z^2 - 1} \right| \le \frac{\pi}{3} .$$

The integration curve has length $L_\gamma = \pi$ and the modulus of the integrand function for $z \in \{\gamma\}$ can be upper bounded as follows

$$\left| \frac{1}{z^2 - 1} \right| = \frac{1}{|z^2 - 1|} \le \frac{1}{||z^2| - 1|} = \frac{1}{4 - 1} .$$

We conclude applying the Darboux's inequality.

Example 7.18 Prove that, if $\gamma_R(\theta) = Re^{i\theta}$ with $0 \le \theta \le \pi$, then

$$\lim_{R\to\infty} \int_{\gamma_R} \frac{\sqrt{z}}{z^2 + 1}\mathrm{d}z = 0,$$

where, setting $z = re^{i\theta}$, we assume

$$\sqrt{z} = \sqrt{r}e^{i\theta/2}, \qquad r \geq 0, \quad -\pi < \theta \leq \pi.$$

The length of the integration path γ_R is $L_{\gamma_R} = \pi R$, while the modulus of the integrand function for $z \in \{\gamma_R\}$ can be upper bounded as follows

$$\left| \frac{\sqrt{z}}{z^2 + 1} \right| = \frac{\sqrt{R}}{|z^2 + 1|} \leq \frac{\sqrt{R}}{||z|^2 - 1|} = \frac{\sqrt{R}}{R^2 - 1}.$$

By Darboux's inequality we have then

$$\left| \int_{\gamma_R} \frac{\sqrt{z}}{z^2 + 1} \mathrm{d}z \right| \leq \pi R \, \frac{\sqrt{R}}{R^2 - 1} \xrightarrow{R \to \infty} 0,$$

which implies $\lim_{R \to \infty} \int_{\gamma_R} \sqrt{z}/(z^2 + 1)\mathrm{d}z = 0$.

Definition 7.19 (Primitive) Let $f : D \mapsto \mathbb{C}$ and $F : D \mapsto \mathbb{C}$ with $D \subset \mathbb{C}$ open and connected. If $F'(z) = f(z) \, \forall z \in D$, then F is a primitive of f in D.

Theorem 7.20 (Existence of Primitive) *Let $f : D \mapsto \mathbb{C}$ be continuous in $D \subset \mathbb{C}$ open and connected. Then the following properties are equivalent:*

(a) *f has primitive in D;*
(b) *$\forall z_1, z_2 \in D$ and for every piecewise regular curve γ contained in D and joining z_1 to z_2 $\int_{\gamma} f(z)\mathrm{d}z = \int_{z_1}^{z_2} f(z)\mathrm{d}z$;*
(c) *for every closed piecewise regular curve γ contained in D $\int_{\gamma} f(z)\mathrm{d}z = 0$.*

Proof We will prove in order the implications (a) $\Rightarrow$ (b), (b) $\Rightarrow$ (a), (b) $\Rightarrow$ (c) and (c) $\Rightarrow$ (b).

(a) $\Rightarrow$ (b) By hypothesis $\exists F : D \mapsto \mathbb{C}$ such that $F'(z) = f(z) \, \forall z \in D$. Given two arbitrary points $z_1, z_2 \in D$, suppose to begin with that there exists a regular path $\gamma : [a, b] \mapsto \mathbb{C}$ such that $\gamma(a) = z_1$ and $\gamma(b) = z_2$. The integral of f along γ is

$$\int_{\gamma} f(z)\mathrm{d}z = \int_a^b f(\gamma(t))\gamma'(t)\mathrm{d}t = \int_a^b F'(\gamma(t))\mathrm{d}t = F(\gamma(b)) - F(\gamma(a))$$

$$= F(z_2) - F(z_1),$$

that is, it depends only on the endpoints z_1 and z_2. We reason in a similar way if γ is piecewise regular. Note that the path γ definitely exists because D is connected.

(b) $\Rightarrow$ (a) Given a point $z_0 \in D$, every point $z \in D$ can be connected to z_0 by a polygonal contained in D. It therefore makes sense to speak of an integral of f along a piecewise regular curve that goes from z_0 to z and is entirely contained in D. Since such integral is, by hypothesis, independent of the considered curve but depends only on the endpoints z_0 and z, we define

$$F(z) = \int_{z_0}^{z} f(w)\mathrm{d}w. \tag{7.10}$$

Since D is open $\exists \delta_1(z) > 0$ such that $B(z, \delta_1) \subset D$. Then, choosing Δz such that $|\Delta z| < \delta_1$, by the properties of the integral we have

$$F(z + \Delta z) - F(z) = \int_{z_0}^{z+\Delta z} f(w)\mathrm{d}w - \int_{z_0}^{z} f(w)\mathrm{d}w = \int_{z}^{z+\Delta z} f(w)\mathrm{d}w.$$

The curve that goes from z to $z + \Delta z$ is arbitrary. Choose for it the segment corresponding to the path $\gamma(t) = z + t\Delta z$ with $0 \leq t \leq 1$. Since then

$$\int_{z}^{z+\Delta z} \mathrm{d}w = \int_{0}^{1} \gamma'(t)\mathrm{d}t = \int_{0}^{1} \Delta z\, \mathrm{d}t = \Delta z,$$

we can write

$$\frac{F(z + \Delta z) - F(z)}{\Delta z} - f(z) = \frac{1}{\Delta z} \int_{z}^{z+\Delta z} (f(w) - f(z))\mathrm{d}w.$$

Since f is continuous at z, $\forall \varepsilon > 0\ \exists \delta_2(\varepsilon) > 0$ such that $|f(w) - f(z)| < \varepsilon$ whenever $|w - z| < \delta_2$. Therefore, if $|\Delta z| < \min(\delta_1, \delta_2)$, using Darboux's inequality we get

$$\left| \frac{F(z + \Delta z) - F(z)}{\Delta z} - f(z) \right| = \frac{1}{|\Delta z|} \left| \int_{z}^{z+\Delta z} (f(w) - f(z))\mathrm{d}w \right|$$

$$\leq \frac{1}{|\Delta z|} |\Delta z|\, \varepsilon = \varepsilon.$$

By the arbitrariness of ε we conclude that there exists the limit of $(F(z + \Delta z) - F(z))/\Delta z$ for $\Delta z \to 0$, namely, F is differentiable at z, and we have $F'(z) = f(z)$.

(b) $\Rightarrow$ (c) Let γ be an arbitrary piecewise regular closed curve with $\{\gamma\} \subset D$. Let z_1 and z_2 be two non-coincident points on the trace of γ. In this way we identify two curves, γ_1 and γ_2, which both join z_1 to z_2 and are such that $\gamma = \gamma_1 + (-\gamma_2)$. By hypothesis the integrals of f along γ_1 and γ_2 have the same value, therefore

$$\int_{\gamma} f(z)\mathrm{d}z = \int_{\gamma_1} f(z)\mathrm{d}z + \int_{-\gamma_2} f(z)\mathrm{d}z = \int_{\gamma_1} f(z)\mathrm{d}z - \int_{\gamma_2} f(z)\mathrm{d}z = 0.$$

(c) $\Rightarrow$ (b) Retrace the reasoning of the previous point backwards. $\quad\square$

Note the difference between the theorem just proved and the Fundamental Theorem of Calculus 7.3 for complex functions of a real variable. In fact, the continuity of $f : D \mapsto \mathbb{C}$ with $D \subset \mathbb{C}$ open and connected is not sufficient to conclude that there exists a primitive of f in D. In Example 7.14 we saw that for the function $f(z) = |z|^2$ continuous in $\mathbb{C}$, integrals along different curves joining the points 1 and i may have different values. By the previous theorem we conclude that $f(z) = |z|^2$ does not admit a primitive in $\mathbb{C}$.

Example 7.21 Compute integrals of the type

$$\int_\gamma f_n(z)\mathrm{d}z, \qquad f_n(z) = z^{-n}, \quad n \in \mathbb{N},$$

where γ is a piecewise regular simple closed path such that $0 \notin \{\gamma\}$.

For $n = 2, 3, \ldots$, the function f_n is continuous in $D = \mathbb{C} \setminus \{0\}$ open and connected and in D it admits the primitive $F_n(z) = -z^{-(n-1)}/(n-1)$. Therefore, the integral of f_n along any path γ such that $0 \notin \{\gamma\}$ is zero, both in the case in which $0 \in \mathrm{Int}(\gamma)$ and in the case in which $0 \in \mathrm{Ext}(\gamma)$.

For $n = 1$, the function $f_1(z) = 1/z$ is still continuous in $D = \mathbb{C} \setminus \{0\}$ but in D it does not admit a primitive. In fact, the primitive of f_1 can only be some branch of $F_1(z) = \log z$ and such a primitive exists only in the domain of analyticity $\tilde{D}$ of the logarithm. On the other hand, every maximal domain of analyticity of the logarithm is strictly contained in D. If $0 \in \mathrm{Ext}(\gamma)$, it is possible to choose $\tilde{D}$ in such a way that $\{\gamma\} \subset \tilde{D}$. In this case we obtain $\int_\gamma z^{-1}\mathrm{d}z = 0$. If instead $0 \in \mathrm{Int}(\gamma)$, the integral is not zero and can be computed in the following way. Suppose that γ is positively oriented. We choose a branch of the logarithm whose branch line σ has a unique point of intersection with γ. Let this point be $u_0 e^{\mathrm{i}\alpha}$, with $u_0, \theta \in \mathbb{R}$. We parametrize the path $\gamma : [a, b] \mapsto \mathbb{C}$ such that $\gamma(a) = \gamma(b) = u_0 e^{\mathrm{i}\alpha}$. We then define a new path $\gamma_\varepsilon : [a + \varepsilon, b - \varepsilon] \mapsto \mathbb{C}$ with $\varepsilon > 0$ such that $\gamma_\varepsilon(t) = \gamma(t) \; \forall t \in [a + \varepsilon, b - \varepsilon]$. We have

$$\int_\gamma \frac{1}{z}\mathrm{d}z = \lim_{\varepsilon \to 0} \int_{\gamma_\varepsilon} \frac{1}{z}\mathrm{d}z.$$

For each point $z \in \tilde{D} = \mathbb{C} \setminus \{\sigma\}$ the corresponding branch of the logarithm admits a derivative and $\mathrm{d}(\log z)/\mathrm{d}z = z^{-1}$. Therefore, in $\tilde{D}$ the function $\log z$ is a primitive of z^{-1} and since the trace of γ_ε is contained in $\tilde{D}$, we have

$$\int_{\gamma_\varepsilon} \frac{1}{z}\mathrm{d}z = \log(\gamma_\varepsilon(b - \varepsilon)) - \log(\gamma_\varepsilon(a + \varepsilon))$$

$$= \log(\gamma(b - \varepsilon)) - \log(\gamma(a + \varepsilon)).$$

For the present branch of $\log z$ we have

$$\lim_{\varepsilon \to 0} \log(\gamma(a + \varepsilon)) = \ln u_0 + \mathrm{i}\alpha,$$

$$\lim_{\varepsilon \to 0} \log(\gamma(b - \varepsilon)) = \ln u_0 + \mathrm{i}(\alpha + 2\pi),$$

and we conclude

$$\int_\gamma \frac{1}{z}\,dz = (\ln u_0 + i(\alpha + 2\pi)) - (\ln u_0 + i\alpha) = 2\pi i.$$

In the case of a closed but not simple path γ we reason in a similar way by considering γ as the sum of simple closed paths.

Example 7.22 Calculate the integral of the function of $z = re^{i\theta}$ defined by

$$\sqrt{z} = \sqrt{r}e^{i\theta/2}, \qquad r \geq 0, \quad \alpha < \theta \leq \alpha + 2\pi, \quad \alpha \in \mathbb{R},$$

along a piecewise regular, simple closed, positively oriented path $\gamma : [a, b] \mapsto \mathbb{C}$ which embodies the point $z = 0$ and intersects the ray $\sigma(u) = ue^{i\alpha}$, $u \in [0, \infty)$, at the point $u_0 e^{i\alpha}$. The integrand function is analytic in $D = \mathbb{C} \setminus \{\sigma\}$ and in D it admits as primitive the branch of $\frac{2}{3}z^{3/2}$ that has σ as branch line. Parametrized the path γ in such a way that $\gamma(a) = \gamma(b) = u_0 e^{i\alpha}$ and defined $\gamma_\varepsilon : [a + \varepsilon, b - \varepsilon] \mapsto \mathbb{C}$ with $\varepsilon > 0$ and $\gamma_\varepsilon(t) = \gamma(t) \; \forall t \in [a + \varepsilon, b - \varepsilon]$, we have

$$\int_\gamma \sqrt{z}\,dz = \lim_{\varepsilon \to 0} \int_{\gamma_\varepsilon} \sqrt{z}\,dz$$

$$= \lim_{\varepsilon \to 0} \frac{2}{3}z^{3/2} \Big|_{z=\gamma_\varepsilon(a+\varepsilon)}^{z=\gamma_\varepsilon(b-\varepsilon)}$$

$$= \frac{2}{3}u_0^{3/2}e^{i3(\alpha+2\pi)/2} - \frac{2}{3}u_0^{3/2}e^{i3\alpha/2}$$

$$= -\frac{4}{3}u_0^{3/2}e^{i3\alpha/2}.$$

7.4   Cauchy-Goursat Theorem

Definition 7.23 (Finite Cover of a Jordan Curve) Let γ be a simple closed curve (Jordan curve) in $\mathbb{C}$. Let us call finite cover of γ a set of closed squares Q_j, $j = 1, \ldots, N$, not necessarily equal to each other, with the properties

$$Q_j^\circ \cap Q_k^\circ = \emptyset, \qquad j, k = 1, \ldots, N, \quad j \neq k,$$

$$Q_j^\circ \cap \mathrm{Int}(\gamma) \neq \emptyset, \qquad j = 1, \ldots, N,$$

$$\mathrm{Int}(\gamma) \subset \bigcup_{j=1}^{N} Q_j^\circ.$$

Note that, since the squares Q_j are closed, we also have

$$\text{Int}(\gamma) \cup \{\gamma\} = \overline{\text{Int}(\gamma)} \subset \bigcup_{j=1}^{N} Q_j.$$

A finite covering of γ can be obtained, for example, in the following way. Consider the closed square whose internal part contains $\text{Int}(\gamma)$. Such a square exists because $\text{diamInt}(\gamma) < \infty$. Divide this square into four equal squares. One or more of the squares obtained is again divided into four equal squares. Iterate the procedure a finite number of times, taking care to discard those new squares whose internal part does not contain points of $\text{Int}(\gamma)$.

Lemma 7.24 *Let f be analytic on and within γ a piecewise regular simple closed curve. For every $\varepsilon > 0$ there exists a finite cover of γ such that for each of the squares Q_j, $j = 1, 2, \ldots, N$, that compose it, the following property holds*

$$\exists z_j \in Q_j \cap \overline{\text{Int}(\gamma)} :$$

$$\left| \frac{f(z) - f(z_j)}{z - z_j} - f'(z_j) \right| < \varepsilon, \qquad \forall z \in Q_j \cap \overline{\text{Int}(\gamma)} \setminus \{z_j\}. \tag{7.11}$$

Proof Suppose, by contradiction, that the statement of the Lemma is false. Then there exists $\varepsilon > 0$ such that every finite cover of γ contains at least one square in which (7.11) is false. Consider initially a covering with N_0 squares $Q_k^{(0)}$, $k = 1, 2, \ldots, N_0$, all equal and with side d. Divide each of the squares $Q_k^{(0)}$ in which (7.11) is false (there must be at least one) into four squares with side $d/2$. Taking care to discard among the new squares with side $d/2$ those whose internal part does not contain points of $\text{Int}(\gamma)$, we find a new covering of γ in squares $Q_k^{(1)}$, $k = 1, 2, \ldots, N_1$. Divide each of the squares $Q_k^{(1)}$ in which (7.11) is false (there must be at least one) into four squares with side $d/2^2$ and among these new squares discard those whose internal part does not contain points of $\text{Int}(\gamma)$. At the n-th iteration of this procedure we will have a covering of γ consisting of N_n squares $Q_k^{(n)}$, $k = 1, 2, \ldots, N_n$. The squares $Q_k^{(n)}$ are not all equal: the smallest have side $d/2^n$ and among these, and only among these, there is at least one in which the property (7.11) is false. Define

$$A_n = \bigcup_{k:\,(7.11)\text{ false}} Q_k^{(n)} \cap \overline{\text{Int}(\gamma)}, \qquad n = 0, 1, 2, \ldots$$

By construction it must be $A_n \supset A_{n+1}$ as well as $A_n \neq \emptyset$. Then choose a point $w_n \in A_n$ and consider the sequence (w_n). Since $w_n \in \overline{\text{Int}(\gamma)}$ and $\overline{\text{Int}(\gamma)}$ is a subset of $\mathbb{C}$ closed and bounded, hence compact, it is possible to extract from (w_n) a convergent subsequence (w_{n_j}). Let w be the corresponding limit point. We have $w \in \overline{\text{Int}(\gamma)}$ since $\overline{\text{Int}(\gamma)}$ is closed. Since $f(z)$ is analytic in $\overline{\text{Int}(\gamma)}$ and hence at w, it is possible to find a $\delta(\varepsilon) > 0$ such that

$$\left| \frac{f(z) - f(w)}{z - w} - f'(w) \right| < \varepsilon, \qquad \forall z \in B(w, \delta) \setminus \{w\}.$$

Every point of A_{n_j}, the set containing the n_j-th element of the subsequence convergent to w, is contained in a square of side $d/2^{n_j}$. Consider n_j so large that $\sqrt{2}d/2^{n_j} < \delta$. Since $A_{n_j} \supset A_{n_{j+1}}$, that is, contains the elements of a sequence convergent to w, and A_{n_j} is closed, then $w \in A_{n_j}$. It follows that there exists a covering square of order n_j, say $Q_{k_{n_j}}^{(n_j)} \subset A_{n_j}$, where (7.11) is false and which contains w. On the other hand, since $Q_{k_{n_j}}^{(n_j)} \subset B(w, \delta)$, we have

$$\exists w \in Q_{k_{n_j}}^{(n_j)} \cap \overline{\mathrm{Int}(\gamma)} :$$

$$\left| \frac{f(z) - f(w)}{z - w} - f'(w) \right| < \varepsilon, \qquad \forall z \in Q_{k_{n_j}}^{(n_j)} \cap \overline{\mathrm{Int}(\gamma)} \setminus \{w\},$$

in contradiction to what was supposed. $\qquad\qquad\qquad\qquad\qquad\qquad\qquad\square$

Theorem 7.25 (Cauchy-Goursat) *Let f be analytic on and within γ a piecewise regular simple closed curve. Then*

$$\int_\gamma f(z)\mathrm{d}z = 0. \tag{7.12}$$

Proof By Lemma 7.24, $\forall \varepsilon > 0$ it is possible to find a finite cover of $\mathrm{Int}(\gamma)$ in squares Q_j, $j = 1, 2, \ldots, N$, such that for each of these $\exists z_j \in Q_j \cap \overline{\mathrm{Int}(\gamma)}$ and an associated function $\delta_j(z)$ defined by

$$\delta_j(z) = \begin{cases} (f(z) - f(z_j))/(z - z_j) - f'(z_j) & z \neq z_j \\ 0 & z = z_j \end{cases},$$

which is continuous and enjoys the property $\left| \delta_j(z) \right| < \varepsilon \ \forall z \in Q_j \cap \overline{\mathrm{Int}(\gamma)}$. Let γ_j, $j = 1, 2, \ldots, N$, be the positively oriented path defined by the boundary of $Q_j \cap \overline{\mathrm{Int}(\gamma)}$. Using the identity $f(z) = f(z_j) + f'(z_j)(z - z_j) + \delta_j(z)(z - z_j)$, for each path γ_j we find

$$\int_{\gamma_j} f(z)\mathrm{d}z = (f(z_j) - z_j f'(z_j)) \int_{\gamma_j} \mathrm{d}z + f'(z_j) \int_{\gamma_j} z\mathrm{d}z$$

$$+ \int_{\gamma_j} \delta_j(z)(z - z_j)\mathrm{d}z.$$

We have $\int_{\gamma_j} \mathrm{d}z = 0$ and $\int_{\gamma_j} z\,\mathrm{d}z = 0$ since the integrand functions have primitives z and $z^2/2$ respectively and the path γ_j is closed. Therefore

$$\int_{\gamma_j} f(z)\mathrm{d}z = \int_{\gamma_j} \delta_j(z)(z - z_j)\mathrm{d}z. \tag{7.13}$$

Summing the integrals of f along all the paths γ_j, $j = 1, 2, \ldots, N$, we get

$$\sum_{j=1}^{N} \int_{\gamma_j} f(z)\mathrm{d}z = \int_{\gamma} f(z)\mathrm{d}z,$$

and therefore, by (7.13) and using the triangle and Darboux inequalities,

$$\left| \int_{\gamma} f(z)\mathrm{d}z \right| = \left| \sum_{j=1}^{N} \int_{\gamma_j} \delta_j(z)(z - z_j)\mathrm{d}z \right|$$

$$\leq \sum_{j=1}^{N} \left| \int_{\gamma_j} \delta_j(z)(z - z_j)\mathrm{d}z \right|$$

$$< \sum_{j=1}^{N} \varepsilon\sqrt{2}d_j(4d_j + L_j),$$

where d_j is the side of the square Q_j, and L_j is the length of $\{\gamma_j\} \cap \{\gamma\}$. Let $L = \sum_{j=1}^{N} L_j$ be the length of the path γ. Since γ is piecewise regular, L is finite. Furthermore, by Jordan's curve theorem, there must exist a square with finite side D that contains $\{\gamma\}$. It follows that $\sum_{j=1}^{N} d_j^2 < D^2$ and $\sum_{j=1}^{N} d_j L_j < D \sum_{j=1}^{N} L_j = DL$. In conclusion,

$$\left| \int_{\gamma} f(z)\mathrm{d}z \right| < \varepsilon\sqrt{2}D(4D + L)$$

and from the arbitrariness of ε follows the statement. $\square$

In the Cauchy-Goursat theorem just proved it is crucial that the closed curve is simple. Only in this case, in fact, it makes sense to speak of the internal part of the γ curve and therefore of the analyticity of f in that region. If $f : G \mapsto \mathbb{C}$ is an analytic function in $G \subset \mathbb{C}$, we can ask ourselves under what conditions for G the integral of f along any closed curve γ contained in G is zero. This property may be false for domains of the type $G = \mathbb{C} \setminus \{0\}$, as shown in Example 7.21 for the function $f(z) = 1/z$.

Definition 7.26 (Homotopic Curves) Let γ_0 and γ_1 be two curves in $G \subset \mathbb{C}$ represented by the paths $\gamma_0, \gamma_1 : [a, b] \mapsto G$. We say that γ_0 is homotopic to γ_1 in G, and we write $\gamma_0 \sim \gamma_1$ in G, if there exists a continuous function $\Gamma : [a, b] \times [0, 1] \mapsto G$ such that $\Gamma(s, 0) = \gamma_0(s)$ and $\Gamma(s, 1) = \gamma_1(s)$ $\forall s \in [a, b]$. If the two curves γ_0 and γ_1 have the same extreme points $z_a = \gamma_0(a) = \gamma_1(a)$ and $z_b = \gamma_0(b) = \gamma_1(b)$, we also impose that $\Gamma(a, t) = z_a$ and $\Gamma(b, t) = z_b$ $\forall t \in [0, 1]$. In particular, if γ_0 and γ_1 are closed curves, then $\Gamma(a, t) = \Gamma(b, t)$ $\forall t \in [0, 1]$.

Note that $\gamma_t : [a, b] \mapsto G$ defined by $\gamma_t(s) = \Gamma(s, t)$ represents a curve in G that, as t varies from 0 to 1, continuously deforms from γ_0 to γ_1. Homotopy (for brevity, if it does not cause confusion, we will omit specifying the domain G) is an equivalence relation. Every curve is homotopic to itself. If $\gamma_0 \sim \gamma_1$ and $\gamma_1 \sim \gamma_2$ then $\gamma_0 \sim \gamma_2$.

Definition 7.27 (Closed Curves Homotopic to Zero) Let γ be a closed curve in $G \subset \mathbb{C}$. We say that γ is homotopic to zero, and we write $\gamma \sim 0$, if γ is homotopic to a constant curve in G, that is, to a point of G.

A closed curve homotopic to zero in G is deformable with continuity at some point of G. For example, in $G = \mathbb{C} \setminus \{z_0\}$ every simple closed curve γ such that $z_0 \in \mathrm{Int}(\gamma)$ is not homotopic to zero since there is no point of G at which γ can deform continuously. If instead $z_0 \in \mathrm{Ext}(\gamma)$ then evidently $\gamma \sim 0$.

Definition 7.28 (Simply Connected Set) An open set $G \subset \mathbb{C}$ is said to be simply connected if G is connected and every closed curve contained in G is homotopic to zero. A connected set that is not simply connected is said multiply connected.

An open ball, all $\mathbb{C}$, the domain of analyticity of the principal branch of $\log z$ are simply connected sets, while the domain of analyticity of $1/z$ is multiply connected.

If G is simply connected, every closed curve γ contained in G can be deformed continuously into any point of G. In fact, let z be an arbitrary point of G, every closed curve γ contained in G is deformable with continuity into a point $w \in G$. Since G is open and connected, there exists a polygonal contained in G that connects the points z and w and such polygonal is deformable with continuity into its endpoint z.

Theorem 7.29 *Let $G \subset \mathbb{C}$ be open and simply connected. Then*

$$\int_\gamma f(z)\mathrm{d}z = 0 \tag{7.14}$$

for every function $f : G \mapsto \mathbb{C}$ analytic in G and for every piecewise regular closed curve γ contained in G.

Proof If γ is a simple closed curve, $\forall z \in \mathrm{Int}(\gamma)$ we have $z \in G$. In fact, if z does not belong to G, γ is not homotopic to zero, in contradiction with the hypothesis that G is simply connected. It follows that every function f analytic in G is analytic on and within γ and Theorem 7.25 applies. If γ is a closed curve that self-intersects $n - 1$ times, we write $\gamma = \sum_{k=1}^{n} \gamma_k$, where γ_k, $k = 1, \ldots, n$, are n simple closed curves. Since $\gamma \sim 0$, we also have $\gamma_k \sim 0$ for any k. It follows that every point within γ_k is contained in G and the Cauchy-Goursat theorem applies to every curve γ_k. We conclude

$$\int_\gamma f(z)\mathrm{d}z = \sum_{k=1}^{n} \int_{\gamma_k} f(z)\mathrm{d}z = \sum_{k=1}^{n} 0 = 0. \qquad (7.15)$$

The result still holds for $n \to \infty$ and in the case the curve self-intersects in a continuum of points. □

Corollary 7.30 *Let f be analytic in $G \subset \mathbb{C}$ open and simply connected, then f admits primitive in G.*

Since $\mathbb{C}$ is open and simply connected, the entire functions admit primitive in $\mathbb{C}$. The principal branch of $\log z$ admits primitive in its open and simply connected domain of analyticity $\{z \in \mathbb{C} : z \neq -t,\ t \in [0, \infty)\}$ and the primitive is $-z + z \log z$. Note that Corollary 7.30 provides only a sufficient condition for a function to admit primitive. The function $1/z^2$ analytic in $\mathbb{C} \setminus \{0\}$, open but not simply connected, nevertheless admits in this domain the primitive $-1/z$.

Using the definition of homotopy of paths, Theorem 7.29 can be stated in the following equivalent forms.

Theorem 7.31 *Let f be analytic in $G \subset \mathbb{C}$ open and let γ be a piecewise regular closed curve contained in G, homotopic to zero, then*

$$\int_\gamma f(z)\mathrm{d}z = 0. \qquad (7.16)$$

Theorem 7.32 *Let f be analytic in $G \subset \mathbb{C}$ open and let γ_1 and γ_2 be two piecewise regular open curves contained in G, homotopic with coinciding endpoints, then*

$$\int_{\gamma_1} f(z)\mathrm{d}z = \int_{\gamma_2} f(z)\mathrm{d}z. \qquad (7.17)$$

Finally, the following result is very useful in applications.

Theorem 7.33 (Path Deformation Principle) *Let γ and γ_1 be two piecewise regular simple closed curves positively oriented with $\{\gamma_1\} \subset \mathrm{Int}(\gamma)$. If f is analytic on γ, γ_1 and in the region between them, then*

$$\int_\gamma f(z)\mathrm{d}z = \int_{\gamma_1} f(z)\mathrm{d}z. \qquad (7.18)$$

Proof Join a point of γ to a point of γ_1 by a piecewise regular open curve λ contained in the region between γ and γ_1 and consider the closed curve $\eta = \gamma + \lambda + (-\gamma_1) + (-\lambda)$. The trace of this curve provides the boundary of a simply connected open set G in which f is analytic. By Theorem 7.29 the integral of f along any piecewise regular closed curve contained in G is zero. Let us consider a piecewise regular closed curve η_ε contained in G homotopic to η and such that $\lim_{\varepsilon \to 0} \eta_\varepsilon = \eta$. We have

$$\int_\eta f(z)\mathrm{d}z = \lim_{\varepsilon \to 0} \int_{\eta_\varepsilon} f(z)\mathrm{d}z = \lim_{\varepsilon \to 0} 0 = 0.$$

On the other hand

$$\int_\eta f(z)\mathrm{d}z = \int_\gamma f(z)\mathrm{d}z + \int_\lambda f(z)\mathrm{d}z - \int_{\gamma_1} f(z)\mathrm{d}z - \int_\lambda f(z)\mathrm{d}z$$

$$= \int_\gamma f(z)\mathrm{d}z - \int_{\gamma_1} f(z)\mathrm{d}z.$$

Combining the two results, the statement follows. $\square$

The path deformation principle is easily generalized to an arbitrary number of positively oriented simple closed curves, $\gamma_1, \gamma_2, \dots, \gamma_n$, all contained in γ and not intersecting each other. In this case

$$\int_\gamma f(z)\mathrm{d}z = \sum_{k=1}^{n} \int_{\gamma_k} f(z)\mathrm{d}z. \tag{7.19}$$

7.5   Cauchy Integral Formula

Theorem 7.34 (Cauchy Integral Formula) *Let f be analytic on and within γ a piecewise regular positively oriented simple closed curve. Then $\forall z \in \mathrm{Int}(\gamma)$ we have*

$$f(z) = \frac{1}{2\pi i} \int_\gamma \frac{f(w)}{w - z}\mathrm{d}w. \tag{7.20}$$

Proof Consider the path $\gamma_\rho(\theta) = z + \rho e^{i\theta}$ with $0 \le \theta \le 2\pi$ and ρ sufficiently small so that the circle γ_ρ is inside γ. Since $f(w)/(w - z)$ is a function of w that is analytic on γ, on γ_ρ and in the region between γ and γ_ρ, by the path deformation principle we have

$$\int_\gamma \frac{f(w)}{w - z}\mathrm{d}w = \int_{\gamma_\rho} \frac{f(w)}{w - z}\mathrm{d}w.$$

From this, and using $\int_{\gamma_\rho} (w - z)^{-1}\mathrm{d}w = 2\pi i$, follows

$$\int_\gamma \frac{f(w)}{w - z}\mathrm{d}w - 2\pi i f(z) = \int_{\gamma_\rho} \frac{f(w) - f(z)}{w - z}\mathrm{d}w.$$

Since f is analytic and hence continuous at z, $\forall \varepsilon > 0$ there exists $\delta(\varepsilon) > 0$ such that $|f(w) - f(z)| < \varepsilon$ whenever $|w - z| < \delta$. Then choosing $\rho < \delta$, by Darboux's inequality we have

$$\left| \int_{\gamma_\rho} \frac{f(w) - f(z)}{w - z} dw \right| < 2\pi\rho \, \frac{\varepsilon}{\rho} = 2\pi\varepsilon.$$

From the arbitrariness of ε it follows that $\int_{\gamma_\rho} (f(w) - f(z))/(w - z)dw = 0$ and hence the statement. $\square$

Theorem 7.35 (Derivatives of Analytic Functions) *Let $f : D \mapsto \mathbb{C}$ be analytic in $D \subset \mathbb{C}$ open, then f has derivatives of each order in D and $\forall z \in D$ the derivative of order n of f at z can be calculated as*

$$f^{(n)}(z) = \frac{n!}{2\pi i} \int_\gamma \frac{f(w)}{(w - z)^{n+1}} dw, \qquad n = 0, 1, 2, \ldots, \tag{7.21}$$

where γ is any piecewise regular positively oriented simple closed curve such that $z \in \mathrm{Int}(\gamma)$ and $\overline{\mathrm{Int}(\gamma)} \subset D$.

Proof Let us reason by induction. For $n = 0$ formula (7.21) is the Cauchy integral formula of Theorem 7.34. Let us then suppose that (7.21) is valid for a non-negative integer $n - 1$ and show that it is also valid for the next integer n. Given an arbitrary point $z \in D$ and an arbitrary piecewise regular positively oriented simple closed curve γ such that $z \in \mathrm{Int}(\gamma)$ and $\overline{\mathrm{Int}(\gamma)} \subset D$, we compute $f^{(n)}(z)$ as

$$f^{(n)}(z) = \lim_{\Delta z \to 0} \frac{f^{(n-1)}(z + \Delta z) - f^{(n-1)}(z)}{\Delta z}.$$

For the hypothesis made and using the path deformation principle, we have

$$\frac{f^{(n-1)}(z + \Delta z) - f^{(n-1)}(z)}{\Delta z}$$

$$= \frac{(n-1)!}{2\pi i \Delta z} \int_\gamma f(w) \left(\frac{1}{(w - z - \Delta z)^n} - \frac{1}{(w - z)^n} \right) dw$$

$$- \frac{(n-1)!}{2\pi i \Delta z} \int_{\gamma_\rho} f(w) \frac{(w - z)^n - (w - z - \Delta z)^n}{(w - z - \Delta z)^n (w - z)^n} dw$$

$$= \frac{(n-1)!}{2\pi i} \int_{\gamma_\rho} f(w) \frac{\sum_{k=0}^{n-1} (w - z)^{n-1-k}(w - z - \Delta z)^k}{(w - z - \Delta z)^n (w - z)^n} dw, \tag{7.22}$$

where γ_ρ is the circle $|w - z| = \rho$ oriented as γ and of radius ρ sufficiently small to have $\{\gamma_\rho\} \subset \mathrm{Int}(\gamma)$. Since we are interested in evaluating the limit $\Delta z \to 0$, without loss of generality we have assumed $z + \Delta z \in \mathrm{Int}(\gamma)$ and $|\Delta z| < \rho$. Finally, in the last line of (7.22) we have used the identity $a^n - b^n = (a - b) \sum_{k=0}^{n-1} a^{n-1-k} b^k$. Using again the path deformation principle, we can write

$$\frac{f^{(n-1)}(z + \Delta z) - f^{(n-1)}(z)}{\Delta z} - \frac{n!}{2\pi i} \int_\gamma \frac{f(w)}{(w - z)^{n+1}} dw$$

$$= \frac{(n - 1)!}{2\pi i} \int_{\gamma_\rho} f(w) h(w) dw,$$

where we have defined

$$h(w) = \frac{\sum_{k=0}^{n-1} \left[(w - z)^{n-k} (w - z - \Delta z)^k - (w - z - \Delta z)^n \right]}{(w - z - \Delta z)^n (w - z)^{n+1}}.$$

By Darboux's inequality, we have then

$$\left| \frac{f^{(n-1)}(z + \Delta z) - f^{(n-1)}(z)}{\Delta z} - \frac{n!}{2\pi i} \int_\gamma \frac{f(w)}{(w - z)^{n+1}} dw \right|$$

$$\leq \frac{(n - 1)!}{2\pi} 2\pi \rho \, M(\rho) \frac{\rho^{n-1} |\Delta z| \left[2^n (n + 1) - n^2 - n - 1 \right]}{(\rho - |\Delta z|)^n \rho^{n+1}}, \tag{7.23}$$

where $M(\rho) = \sup_{w \in \{\gamma_\rho\}} |f(w)|$ and we used the inequality

$$\left| \sum_{k=0}^{n-1} \left[(w - z)^{n-k} (w - z - \Delta z)^k - (w - z - \Delta z)^n \right] \right|$$

$$= \left| \sum_{k=0}^{n-1} \left[\sum_{j=0}^{k} \binom{k}{j} (w - z)^{n-j} \Delta z^j - \sum_{j=0}^{n} \binom{n}{j} (w - z)^{n-j} \Delta z^j \right] \right|$$

$$= \left| \sum_{k=0}^{n-1} \left[(n - k)(w - z)^{n-1} \Delta z \right. \right.$$

$$\left. + \sum_{j=2}^{k} \binom{k}{j} (w - z)^{n-j} \Delta z^j - \sum_{j=2}^{n} \binom{n}{j} (w - z)^{n-j} \Delta z^j \right]$$

$$\leq \sum_{k=0}^{n-1} \left[(n - k) \left| (w - z)^{n-1} \Delta z \right| \right.$$

$$\left. + \sum_{j=2}^{k} \binom{k}{j} \left| (w - z)^{n-j} \Delta z^j \right| + \sum_{j=2}^{n} \binom{n}{j} \left| (w - z)^{n-j} \Delta z^j \right| \right]$$

$$< \rho^{n-1} |\Delta z| \sum_{k=0}^{n-1} \left[(n-k) + (2^k - 1 - k) + (2^n - 1 - n) \right]$$

$$= \rho^{n-1} |\Delta z| \left[2^n (n+1) - n^2 - n - 1 \right],$$

which holds for $|\Delta z| < |w - z| = \rho$. The statement follows immediately by observing that the right-hand side of (7.23) vanishes for $\Delta z \to 0$.   □

Corollary 7.36 *Let f be analytic in $D \subset \mathbb{C}$ open, then its derivatives of every order $f^{(n)}$, $n = 0, 1, 2, \ldots$, are also analytic in D*

Theorem 7.37 (Morera) *Let $f : D \mapsto \mathbb{C}$ be continuous in $D \subset \mathbb{C}$ open and connected. If for every piecewise regular closed curve γ contained in D we have $\int_\gamma f(z)\mathrm{d}z = 0$, then f is analytic in D.*

Proof By Theorem 7.20 f admits in D a primitive F. Since $F'(z) = f(z) \ \forall z \in D$, the function F is analytic in D and by Theorem 7.36 also F', that is, f, is analytic in D.   □

Note that if f is analytic in D open and connected it is not in general true that $\int_\gamma f(z)\mathrm{d}z = 0$ for every closed curve γ contained in D, recall Example 7.21 with $f(z) = z^{-1}$ and $D = \mathbb{C} \setminus \{0\}$. Instead, according to Theorem 7.29, the integral is always zero if we add the hypothesis that D is simply connected.

The Cauchy integral formula is of central importance in the theory of analytic functions. Some first relevant applications are given below.

Theorem 7.38 (Cauchy Inequalities) *Let f be analytic in $B(z_0, R)$ and let $|f(z)| \leq M \ \forall z \in B(z_0, R)$. Then*

$$\left| f^{(n)}(z_0) \right| \leq \frac{n! \, M}{R^n}, \qquad n = 0, 1, 2, \ldots. \tag{7.24}$$

Proof Let γ_r be the circle of center z_0 and radius r. At any order n and for arbitrary $r < R$ Theorem 7.35 allows us to write

$$f^{(n)}(z_0) = \frac{n!}{2\pi \mathrm{i}} \int_{\gamma_r} \frac{f(w)}{(w - z_0)^{n+1}} \mathrm{d}w.$$

By Darboux's inequality

$$\left| f^{(n)}(z_0) \right| \leq 2\pi r \, \frac{n!}{2\pi} \frac{\sup_{w \in \{\gamma_r\}} |f(w)|}{r^{n+1}} \leq \frac{n! M}{r^n}.$$

Since $r < R$ is arbitrary, the statement follows by taking the limit $r \to R^-$.   □

Theorem 7.39 (Liouville) *Let f be entire and bounded in $\mathbb{C}$, then f is constant in $\mathbb{C}$.*

Proof Since, by hypothesis, the function f is analytic in the whole complex plane, $\forall z \in \mathbb{C}$ the Cauchy integral formula applies

$$f'(z) = \frac{1}{2\pi i} \int_{\gamma_R} \frac{f(w)}{(w-z)^2} \, dw,$$

where γ_R is a positively oriented circle, centered at z and of arbitrary radius R. On the other hand, since f is bounded in $\mathbb{C}$, $\exists M > 0$ such that $|f(z)| \leq M \ \forall z \in \mathbb{C}$. By Darboux's inequality we then obtain

$$|f'(z)| \leq 2\pi R \, \frac{1}{2\pi} \frac{M}{R^2} = \frac{M}{R}.$$

From the arbitrariness of R it follows that $f'(z) = 0 \ \forall z \in \mathbb{C}$. By Theorem 5.18 f is constant in $\mathbb{C}$. $\qquad\qquad\square$

Theorem 7.40 (Fundamental Theorem of Algebra) *Let* $P_n(z) = a_0 + a_1 z + a_2 z^2 + \cdots + a_n z^n$ *with* $z, a_0, a_1, a_2, \ldots, a_n \in \mathbb{C}$ *and* $a_n \neq 0$ *a polynomial of degree* $n \geq 1$. *Then there exists at least one point* $z_0 \in \mathbb{C}$ *such that* $P_n(z_0) = 0$.

Proof Suppose, by contradiction, that $P_n(z) \neq 0 \ \forall z \in \mathbb{C}$. It follows that the function $f(z) = 1/P_n(z)$ is entire. This function is also bounded. In fact

$$\lim_{z \to \infty} \frac{1}{P_n(z)} = \lim_{z \to 0} \frac{1}{P_n(1/z)} = \lim_{z \to 0} \frac{z^n}{a_0 z^n + a_1 z^{n-1} + \cdots + a_n} = 0,$$

therefore $\exists R > 0$ such that $|f(z)| < 1$ if $|z| > R$. On the other hand $|f(z)|$ is a continuous function and therefore bounded in the compact $|z| \leq R$. Since f is analytic and bounded in $\mathbb{C}$, by Liouville's theorem we arrive at the absurd that it is constant in $\mathbb{C}$. There must therefore exist at least one point $z_0 \in \mathbb{C}$ such that $P_n(z_0) = 0$. $\qquad\qquad\square$

A consequence of the fundamental theorem of algebra is that the polynomial P_n can be rewritten as

$$P_n(z) = a_n(z - z_1)(z - z_2) \ldots (z - z_n), \tag{7.25}$$

where $z_1, z_2, \ldots, z_n$ are n complex constants that are not necessarily distinct. In fact, by Theorem 7.40, $\exists z_1 \in \mathbb{C}$ such that $P_n(z_1) = 0$. Observing that for any integer k

$$z^k - z_1^k = (z - z_1)\left(z^{k-1} + z^{k-2}z_1 + z^{k-3}z_1^2 + \cdots + zz_1^{k-2} + z_1^{k-1}\right),$$

we have

$$P_n(z) = P_n(z) - P_n(z_1)$$

$$= \sum_{k=0}^{n} a_k z^k - \sum_{k=0}^{n} a_k z_1^k$$

$$= \sum_{k=1}^{n} a_k (z^k - z_1^k)$$

$$= (z - z_1) P_{n-1}(z),$$

where $P_{n-1}(z)$ is the polynomial of degree $n - 1$ defined by

$$P_{n-1}(z) = \sum_{k=1}^{n} a_k \sum_{j=0}^{k-1} z^{k-1-j} z_1^j = \sum_{m=0}^{n-1} a_{m+1} \sum_{j=0}^{m} z^{m-j} z_1^j.$$

Iterating the reasoning we arrive at writing $P_n(z) = (z - z_1)(z - z_2) \ldots (z - z_n) P_0$, where P_0 is a constant, necessarily equal to the coefficient a_n of z^n of the polynomial P_n. The complex numbers $z_1, z_2, \ldots, z_n$ are called the roots of the polynomial P_n. A polynomial of degree n can have at most n distinct roots.

Exercises

7.1 Calculate the integral

$$\int_\gamma \frac{z}{\bar{z}} dz,$$

where γ is the integration path, traveled counterclockwise, consisting of the semicircle $|z| = 2$ with $\mathrm{Im}\, z > 0$ and the segment of the real axis with endpoints $z = -2$ and $z = 2$.

7.2 Calculate the integral of the function $f(z) = (z - \bar{z})^2$ along the path γ represented by the boundary of the triangle with vertices $z_1 = (0, 0)$, $z_2 = (1, 0)$, $z_3 = (0, 1)$, traveled counterclockwise.

7.3 Calculate the integral

$$\int_\gamma \left(z^2 - \bar{z}^2 \right) dz,$$

where γ is the rectilinear path that goes from $z = 1$ to $z = i$.

7.4 Calculate the integral

$$\int_\gamma \frac{1}{\bar z}\,dz,$$

where γ is the closed path, traveled counterclockwise, formed by the segment γ_1 joining the points $\pm 1 - i$ and by the arc of the circle $\gamma_2(t) = \sqrt{2}e^{it}$, $-\pi/4 \le t \le 5\pi/4$.

7.5 Calculate the integral of the function $f(z) = \mathrm{Im}z - \mathrm{Re}z - i3(\mathrm{Re}z)^2$ along the curves γ_1, γ_2 and γ_3, respectively corresponding to the segments $[0, i]$, $[i, 1+i]$ and $[0, 1+i]$.

7.6 Let γ be the boundary of the square with vertices $1, i, -1, -i$, prove that

$$\left| \int_\gamma \frac{\sqrt{z}}{z^2}\,dz \right| \le 8\sqrt{2}.$$

7.7 Assuming the principal branch for multi-valued functions, prove that

$$\lim_{R\to\infty} \int_{|z|=R} z^{-3/2} \log z\,dz = 0.$$

7.8 Assuming $f(z)$ to be a continuous function for $\mathrm{Re}z \ge 0$ and $\lim_{z\to\infty} f(z) = 0$, prove that

$$\lim_{R\to\infty} \int_{\gamma_R} e^{az} f(z)\,dz = 0, \qquad a < 0,$$

where $\gamma_R(\theta) = Re^{i\theta}$ with $-\pi/2 \le \theta \le \pi/2$.

7.9 Let $a \ne 0$ be an arbitrary real. Calculate the integral of the principal branch of $f(z) = z^{a-1}$ along the positively oriented circle $|z| = R$, with arbitrary $R > 0$. Finally, determine the limit $a \to 0$ of the found result.

7.10 Let $f : D \mapsto \mathbb{C}$, with $D = \{z \in \mathbb{C} : 0 < |z| < \pi/2\}$, defined by

$$f(z) = \frac{\tan z}{z^2}.$$

Show that f has no primitive in D.

7.11 Calculate the integral

$$\int_\gamma z^{-1} \log z\,dz,$$

where $\gamma(\theta) = \exp(i\theta)$, with $0 \le \theta \le 2\pi$, and for the logarithm it is assumed the branch $\log z = \ln r + i\theta$, where $z = re^{i\theta}$ with $r > 0$ and $0 < \theta < 2\pi$.

7.12 Assuming the principal branch for multi-valued functions, calculate the integral

$$\int_{\gamma} z^{3/2}\mathrm{d}z,$$

where $\gamma(\theta) = Re^{i\theta}$, $-\pi \le \theta \le \pi$, $R > 0$.

7.13 Assuming the principal branch for multi-valued functions, calculate the integral

$$\int_{\gamma} z^{4/3}\mathrm{d}z,$$

where $\gamma(\theta) = Re^{i\theta}$, $-\pi \le \theta \le \pi$, $R > 0$.

7.14 Assuming the principal branch for multi-valued functions, calculate the integral

$$\int_{\gamma} z^z(1 + \log z)\mathrm{d}z,$$

where γ is the circle $|z| = R$ traveled counterclockwise.

7.15 Assuming the principal branch for multi-valued functions, calculate the integral

$$\int_{\gamma} \frac{\cos \sqrt{z}}{\sqrt{z}}\mathrm{d}z,$$

where γ is the circle $|z| = R$ traveled counterclockwise.

7.16 Assuming the principal branch for multi-valued functions, calculate the integral

$$\int_{\gamma} \frac{\log z}{z^2}\mathrm{d}z,$$

where γ is the circle $|z| = R$ traveled counterclockwise.

7.17 Assuming the principal branch for multi-valued functions, calculate the integral

$$\int_{\gamma} z \log z\,\mathrm{d}z,$$

where γ is the circle $|z| = R$ traveled counterclockwise.

7.18 Assuming the principal branch for multi-valued functions, calculate the integral

$$\int_{\gamma} \frac{1}{2\sqrt{z}\cos^2(\sqrt{z})}\,dz,$$

where γ is the circle $|z| = R$ traveled counterclockwise with $R \neq ((2n+1)\pi/2)^2$, $n = 0, 1, 2, \ldots$.

7.19 Assuming the principal branch for multi-valued functions, calculate the integral

$$\int_{\gamma} z^{1/3}\,dz,$$

where γ is the circle $|z| = R$ traveled counterclockwise.

7.20 Assuming the principal branch for multi-valued functions, calculate the integral

$$\int_{\gamma} z^{3/2}\,dz,$$

where γ is the boundary, positively oriented, of the square with vertices $L - iL$, $L + iL$, $-L + iL$, $-L - iL$, with $L > 0$.

7.21 Assuming the principal branch for multi-valued functions, calculate the integral

$$\int_{\gamma} (z - z_0)^{q-1}\,dz,$$

where γ is a piecewise regular positively oriented simple closed curve passing through the point $z_0 - R$, with $R > 0$ and $z_0 \in \mathrm{Int}(\gamma)$. Consider $q \neq 0$ arbitrary real.

7.22 Assuming the principal branch for multi-valued functions, calculate the integral

$$\int_{\gamma} \frac{\cos(\log z)}{z}\,dz,$$

where γ is the path, traveled counterclockwise, given by the boundary of the triangle with vertices b and $-a \pm ib$, with a and b real positive.

7.23 Assuming the principal branch for multi-valued functions, calculate the integral

$$\int_\gamma \frac{\sin(\log z)}{z}\,\mathrm{d}z,$$

where γ is the path, traveled counterclockwise, given by the boundary of the rhombus with vertices $\pm a$ and $\pm ib$, with a and b real positive.

7.24 Give an example of a function f that is analytic in $A = \{z \in \mathbb{C} : 1 < |z| < 2\}$, but not analytic in the closure of A, which does not admit a primitive in A.

7.25 Calculate the integral

$$\int_\gamma \frac{e^z}{z(1-z)^3}\,\mathrm{d}z,$$

in the following cases where γ is a positively oriented closed path of integration such that:

(a) contains the point $z = 0$ but does not contain the point $z = 1$;
(b) contains the point $z = 1$ but does not contain the point $z = 0$;
(c) contains both the points $z = 0$ and $z = 1$.

7.26 Calculate the integral

$$\int_\gamma \frac{z}{z^4 - 1}\,\mathrm{d}z,$$

where γ is the regular hexagon, inscribed in the circle of radius $a > 1$ and center $z = a$, with two sides parallel to the real axis and traveled counterclockwise.

7.27 Let $f(z)$ be analytic on and within the simple closed path γ and let z_0 be a point not belonging to $\{\gamma\}$. Prove that

$$\int_\gamma \frac{f(z)}{(z - z_0)^2}\,\mathrm{d}z = \int_\gamma \frac{f'(z)}{z - z_0}\,\mathrm{d}z.$$

7.28 Assuming the principal branch for multi-valued functions, calculate the integral

$$\int_\gamma \frac{\log(z + 2\pi)}{(z - i\pi)^3}\,\mathrm{d}z,$$

where γ is the circle $|z| = 4$ traveled counterclockwise.

7.29 Assuming the principal branch for multi-valued functions, calculate the integral

$$\int_\gamma \frac{e^z}{z^2\sqrt{3+z}}\,dz,$$

where γ is the path, traveled counterclockwise, given by the boundary of the square with vertices $1, i, -1, -i$.

7.30 Let f be a function analytic in $\mathbb{C}$ except for the isolated singularity, of unspecified nature, at $z = 2 + 2i$ and let γ be the positively oriented circle $|z - 3| = 2$. Calculate the following integrals, expressing the result, if necessary, in terms of the values assumed by f and its derivatives at suitable points. If the information is not sufficient to evaluate the integral, say that it is not computable.

$$\text{(a)} \int_\gamma \frac{f(z)\sqrt{z}}{(z+i)^2}\,dz, \qquad \text{(b)} \int_\gamma \frac{f(z)\sqrt{z}}{(z-1)^2}\,dz, \qquad \text{(c)} \int_\gamma \frac{f(z)\sqrt{z}}{(z-4)^2}\,dz.$$

7.31 Let $a, b \in \mathbb{C}$ and $p, q \in \mathbb{N}$. Establish a necessary and sufficient condition among the coefficients a, b, p, q such that the function

$$f(z) = a\frac{\sin z}{z^{p+1}} - b\frac{\cos z}{z^{q+1}}$$

has a primitive in $\mathbb{C}\backslash\{0\}$. Discuss explicitly the two cases $p = q = 0$ and $p = q = 1$.

7.32 Let $f(z)$ be analytic in $\mathbb{C}$ with $\left|\mathrm{Im}f'(z)\right|$ bounded in $\mathbb{C}$. Prove that $f(z) = a + bz$ with $a, b \in \mathbb{C}$.

7.33 Let f be an entire function with the property $\exists M > 0$ such that $\forall z \in \mathbb{C}$ we have $|f(z)| \le M + \sqrt{|z|}$. What can we state about f?

7.34 Let $f(z)$ be an entire function. Show that if $\exists a \in \mathbb{R}$ such that $\mathrm{Re}(f(z)) \le a$ $\forall z \in \mathbb{C}$, then $f(z)$ is constant in $\mathbb{C}$. Does this also hold if the imaginary part of f is bounded?

7.35 Let f be a function analytic in $\mathbb{C}$ except for a finite number of isolated singularities at points $z_1, z_2, \ldots, z_n$, all distinct from the origin. Let $R = \min_{k=1,\ldots,n} |z_k| > 0$. Estimate R knowing that $f^{(4)}(0) = 1 - i$ and $|f(z)| \le 5\,|z|^3$ $\forall z \in \mathbb{C}$.

Chapter 8
Taylor and Laurent Series

Abstract Taylor series expansion around the point z_0 of functions analytic in a ball centered at z_0. Integral along a curve of a power series multiplied by a continuous function. The sum of a power series is an analytic function within the circle of convergence. Derivative of a power series. Uniqueness of the Taylor series expansion. Notable examples of Taylor series expansions. Definition of annulus. Laurent series expansion around the point z_0 of functions analytic in an annulus centered at z_0. Integral along a curve of a generalized power series multiplied by a continuous function. The sum of a generalized power series is an analytic function within the convergence annulus. Derivative of a generalized power series. Uniqueness of the Laurent series expansion. Notable examples of Laurent series expansions. Multiplication and division of power series.

8.1 Taylor Series

Theorem 8.1 (Taylor Series Expansion) *Let f be analytic in $B(z_0, R)$, then $\forall z \in B(z_0, R)$ the following Taylor series expansion holds*

$$f(z) = \sum_{n=0}^{\infty} \frac{f^{(n)}(z_0)}{n!} (z - z_0)^n. \tag{8.1}$$

Proof Arbitrarily choose $z \in B(z_0, R)$ and set $\gamma(t) = z_0 + re^{it}$ with $0 \le t \le 2\pi$ and $|z - z_0| < r < R$, by the Cauchy integral formula we have

$$f(z) = \frac{1}{2\pi i} \int_\gamma \frac{f(w)}{w - z} dw = \frac{1}{2\pi i} \int_\gamma \frac{f(w)}{(w - z_0) - (z - z_0)} dw.$$

Using identity

$$\frac{1}{w - z} = \frac{1}{w} \frac{1}{1 - z/w}$$

© The Author(s), under exclusive license to Springer Nature Switzerland AG 2026
C. Presilla, *Complex Functions of a Variable*, La Matematica per il 3+2 179,
https://doi.org/10.1007/978-3-032-12494-4_8

$$= \frac{1}{w}\frac{1-(z/w)^n}{1-z/w} + \frac{1}{w}\frac{(z/w)^n}{1-z/w}$$

$$= \sum_{k=0}^{n-1} \frac{z^k}{w^{k+1}} + \frac{z^n}{(w-z)w^n}$$

valid for arbitrary z and w with $z \neq w$ and $w \neq 0$, we can rewrite $f(z)$ as

$$f(z) = \frac{1}{2\pi i} \int_\gamma f(w)\left(\sum_{k=0}^{n-1} \frac{(z-z_0)^k}{(w-z_0)^{k+1}} \right.$$

$$\left. + \frac{(z-z_0)^n}{[(w-z_0)-(z-z_0)](w-z_0)^n} \right)\mathrm{d}w$$

$$= \sum_{k=0}^{n-1} \frac{(z-z_0)^k}{2\pi i} \int_\gamma \frac{f(w)}{(w-z_0)^{k+1}}\mathrm{d}w$$

$$+ \frac{(z-z_0)^n}{2\pi i} \int_\gamma \frac{f(w)}{[(w-z_0)-(z-z_0)](w-z_0)^n}\mathrm{d}w$$

$$= \sum_{k=0}^{n-1} \frac{f^{(k)}(z_0)}{k!}(z-z_0)^k + \rho_n(z), \tag{8.2}$$

having used the Cauchy integral formula

$$f^{(k)}(z_0) = \frac{k!}{2\pi i} \int_\gamma \frac{f(w)}{(w-z_0)^{k+1}}\mathrm{d}w$$

and set

$$\rho_n(z) = \frac{(z-z_0)^n}{2\pi i} \int_\gamma \frac{f(w)}{[(w-z_0)-(z-z_0)](w-z_0)^n}\mathrm{d}w.$$

To prove (8.1) it is now sufficient to take the limit $n \to \infty$ of (8.2) showing that $\lim_{n\to\infty} \rho_n(z) = 0$. Let $M = \sup_{w \in \{\gamma\}} |f(w)|$, by Darboux's inequality we get

$$|\rho_n(z)| \leq \frac{|z-z_0|^n}{2\pi} 2\pi r \frac{M}{(r-|z-z_0|)r^n} = \frac{M(|z-z_0|/r)^n}{1-|z-z_0|/r} \xrightarrow{n\to\infty} 0$$

as $|z-z_0| < r$. $\qquad\square$

In the particular case $z_0 = 0$, the Taylor series expansion

$$f(z) = \sum_{n=0}^{\infty} \frac{f^{(n)}(0)}{n!} z^n \tag{8.3}$$

is often referred to as the Maclaurin series expansion of f.

Theorem 8.2 (Integral of a Power Series) *Let* $\sum_{n=0}^{\infty} a_n (z - z_0)^n = f(z)$ *within the circle of convergence with center z_0 and radius R. If γ is a piecewise regular curve contained in $B(z_0, R)$ and g is a continuous function on $\{\gamma\}$, then*

$$\sum_{n=0}^{\infty} a_n \int_{\gamma} g(z)(z - z_0)^n \mathrm{d}z = \int_{\gamma} g(z) f(z) \mathrm{d}z. \tag{8.4}$$

Proof Note that g and f are continuous functions on $\{\gamma\}$. The function g is continuous by hypothesis. As for f, observe that since γ is contained in $B(z_0, R)$ $\exists r < R$ such that $\{\gamma\} \subset \overline{B}(z_0, r)$. The power series $\sum_{n=0}^{\infty} a_n (z - z_0)^n$ converges uniformly to $f(z)$ in $\overline{B}(z_0, r)$, it then follows that f is continuous in $\overline{B}(z_0, r)$ and therefore on $\{\gamma\}$. The integral along γ of the product gf therefore exists. Furthermore, $\forall z \in B(z_0, R)$ we have

$$g(z) f(z) = \sum_{k=0}^{n-1} a_k g(z)(z - z_0)^k + g(z) \rho_n(z), \tag{8.5}$$

where $\rho_n(z)$ is the n-th remainder of the power series convergent to f

$$\rho_n(z) = \sum_{k=n}^{\infty} a_k (z - z_0)^k.$$

From (8.5) it follows that $g\rho_n$, as difference of continuous functions on $\{\gamma\}$, is a continuous function on $\{\gamma\}$ and hence integrable along γ. We can then write

$$\int_{\gamma} g(z) f(z) \mathrm{d}z = \sum_{k=0}^{n-1} a_k \int_{\gamma} g(z)(z - z_0)^k \mathrm{d}z + \int_{\gamma} g(z) \rho_n(z) \mathrm{d}z.$$

Let M be the maximum value of $|g(z)|$ on $\{\gamma\}$ and L be the length of γ. Since the power series considered converges uniformly to f in $\overline{B}(z_0, r) \supset \{\gamma\}$, $\forall \varepsilon > 0 \, \exists N(\varepsilon)$ such that $|\rho_n(z)| < \varepsilon \, \forall z \in \{\gamma\}$ and $\forall n > N$. By Darboux's inequality we get

$$\left| \int_{\gamma} g(z) \rho_n(z) \mathrm{d}z \right| < L M \varepsilon, \qquad \forall n > N(\varepsilon).$$

The above inequality holds for arbitrarily small ε provided $N(\varepsilon)$ is sufficiently large, it then follows that

$$\lim_{n \to \infty} \int_\gamma g(z)\rho_n(z)\mathrm{d}z = 0$$

and hence the statement. $\qquad\square$

Corollary 8.3 (Analyticity of a Power Series) *Let $\sum_{n=0}^{\infty} a_n(z - z_0)^n = f(z)$ within the circle of convergence with center z_0 and radius R, then f is analytic in $B(z_0, R)$.*

Proof From Theorem 8.2 with the choice $g(z) = 1$ and γ arbitrary piecewise regular closed curve contained in $B(z_0, R)$, we have

$$\int_\gamma f(z)\mathrm{d}z = \sum_{n=0}^{\infty} a_n \int_\gamma (z - z_0)^n \mathrm{d}z = 0.$$

By Morera's Theorem (7.37) we conclude that f is analytic in $B(z_0, R)$. $\qquad\square$

Corollary 8.4 (Derivative of a Power Series) *Let $\sum_{n=0}^{\infty} a_n(z - z_0)^n = f(z)$ within the circle of convergence with center z_0 and radius R, then $\forall z \in B(z_0, R)$ we have*

$$f'(z) = \sum_{n=1}^{\infty} a_n n (z - z_0)^{n-1}. \tag{8.6}$$

Proof Arbitrarily fix $z \in B(z_0, R)$, from Theorem 8.2 with the choice $g(w) = 1/(2\pi \mathrm{i}(w - z)^2)$ and γ piecewise regular simple closed curve, positively oriented, such that $\{\gamma\} \subset B(z_0, R)$ and $z \in \mathrm{Int}(\gamma)$, we have

$$\frac{1}{2\pi \mathrm{i}} \int_\gamma \frac{f(w)}{(w - z)^2}\mathrm{d}w = \sum_{n=0}^{\infty} a_n \frac{1}{2\pi \mathrm{i}} \int_\gamma \frac{(w - z_0)^n}{(w - z)^2}\mathrm{d}w. \tag{8.7}$$

By Corollary 8.3 the function $f(w)$ is analytic in $B(z_0, R)$ and also analytic within this ball are the powers $(w - z_0)^n$, with $n = 0, 1, 2, \ldots$. Since $\{\gamma\} \subset B(z_0, R)$ and $z \in \mathrm{Int}(\gamma)$, we can then apply the Cauchy integral formula (Theorem 7.35) and obtain

$$\frac{1}{2\pi \mathrm{i}} \int_\gamma \frac{f(w)}{(w - z)^2}\mathrm{d}w = f'(z),$$

$$\frac{1}{2\pi \mathrm{i}} \int_\gamma \frac{(w - z_0)^n}{(w - z)^2}\mathrm{d}w = n(z - z_0)^{n-1}.$$

Plugging these results into (8.7) the statement is proven. $\qquad\square$

Corollary 8.5 (Uniqueness of Taylor Series Expansion) *Let $\sum_{n=0}^{\infty} a_n (z-z_0)^n = f(z)$ within the circle of convergence with center z_0 and radius R, then this is the Taylor series expansion of f around z_0, that is, $a_n = f^{(n)}(z_0)/n!$, $n = 0, 1, 2, \ldots$.*

Proof Fixed an arbitrary index $n = 0, 1, 2, \ldots$, by Theorem 8.2 with the choice $g(z) = 1/(2\pi i(z - z_0)^{n+1})$ and γ piecewise regular simple closed curve, positively oriented, such that $\{\gamma\} \subset B(z_0, R)$ and $z_0 \in \mathrm{Int}(\gamma)$, we have

$$\frac{1}{2\pi i} \int_\gamma \frac{f(z)}{(z - z_0)^{n+1}} dz = \sum_{k=0}^{\infty} a_k \frac{1}{2\pi i} \int_\gamma (z - z_0)^{k-n-1} dz. \tag{8.8}$$

By Corollary 8.3 the function f is analytic in $B(z_0, R)$ and since $\{\gamma\} \subset B(z_0, R)$ and $z_0 \in \mathrm{Int}(\gamma)$, for the Cauchy integral formula (Theorem 7.35) we have

$$\frac{1}{2\pi i} \int_\gamma \frac{f(z)}{(z - z_0)^{n+1}} dz = \frac{f^{(n)}(z_0)}{n!}.$$

The integrals appearing in the right-hand side of (8.8) can evaluated giving

$$\frac{1}{2\pi i} \int_\gamma (z - z_0)^{k-n-1} dz = \delta_{k-n-1,-1} = \delta_{k,n}.$$

Combining the last two results we find $f^{(n)}(z_0)/n! = a_n$. $\qquad\qquad\square$

Example 8.6 Find the Maclaurin series expansion of $f(z) = e^z$. The function is entire and for every $n \in \mathbb{N}$ we have $f^{(n)}(z) = e^z$, that is, $f^{(n)}(0) = 1$. Therefore

$$e^z = \sum_{n=0}^{\infty} \frac{1}{n!} z^n$$

$$= 1 + z + \frac{z^2}{2} + \frac{z^3}{6} + \frac{z^4}{24} + \cdots, \qquad |z| < \infty.$$

Example 8.7 Determine the Maclaurin series expansion of $f(z) = \sin z$. The function is entire and remembering that $\sin z = (e^{iz} - e^{-iz})/2i$ from the previous result we have

$$\sin z = \frac{1}{2i} \left(\sum_{n=0}^{\infty} \frac{(iz)^n}{n!} - \sum_{n=0}^{\infty} \frac{(-iz)^n}{n!} \right)$$

$$= \frac{1}{2i} \sum_{n=0}^{\infty} (1 - (-1)^n) \frac{i^n z^n}{n!}$$

$$= \sum_{k=0}^{\infty} \frac{(-1)^k}{(2k+1)!} z^{2k+1}$$

$$= z - \frac{z^3}{6} + \frac{z^5}{5!} - \frac{z^7}{7!} + \cdots, \qquad |z| < \infty.$$

Example 8.8 Determine the Maclaurin series expansion of $f(z) = \cos z$. The function is entire and recalling that $\cos z = d \sin z / dz$ from the previous result we have

$$\cos z = \frac{d}{dz} \sum_{k=0}^{\infty} \frac{(-1)^k}{(2k+1)!} z^{2k+1}$$

$$= \sum_{k=0}^{\infty} \frac{(-1)^k}{(2k)!} z^{2k}$$

$$= 1 - \frac{z^2}{2} + \frac{z^4}{24} - \frac{z^6}{6!} + \cdots, \qquad |z| < \infty.$$

Example 8.9 Determine the Maclaurin series expansion of $f(z) = \cosh z$. The function is entire and recalling that $\cosh z = \cos(iz)$ from the previous result we have

$$\cosh z = \sum_{k=0}^{\infty} \frac{(-1)^k}{(2k)!} (iz)^{2k}$$

$$= \sum_{k=0}^{\infty} \frac{1}{(2k)!} z^{2k}$$

$$= 1 + \frac{z^2}{2} + \frac{z^4}{24} + \frac{z^6}{6!} + \cdots, \qquad |z| < \infty.$$

Example 8.10 Determine the Maclaurin series expansion of $f(z) = \sinh z$. The function is entire and remembering that $\sinh z = d \cosh z / dz$ from the previous result we have

$$\sinh z = \frac{d}{dz} \sum_{k=0}^{\infty} \frac{1}{(2k)!} z^{2k}$$

$$= \sum_{k=1}^{\infty} \frac{1}{(2k-1)!} z^{2k-1}$$

$$= \sum_{n=0}^{\infty} \frac{1}{(2k+1)!} z^{2k+1}$$

$$= z + \frac{z^3}{6} + \frac{z^5}{5!} + \frac{z^7}{7!} + \ldots, \qquad |z| < \infty.$$

Example 8.11 We have already studied the convergence of the geometric series

$$\sum_{n=0}^{\infty} z^n = \frac{1}{1-z}, \qquad |z| < 1. \tag{8.9}$$

Let $f(z) = 1/(1-z)$, by the uniqueness of the Taylor series expansion from (8.9) we obtain $f^{(n)}(0) = n!$. With the transformations $z \to -z$ and $z \to 1 - z$, from (8.9) we also obtain

$$\frac{1}{1+z} = \sum_{n=0}^{\infty} (-1)^n z^n, \qquad |z| < 1,$$

$$\frac{1}{z} = \sum_{n=0}^{\infty} (-1)^n (z-1)^n, \qquad |z-1| < 1.$$

Example 8.12 Determine the Taylor series expansion of $f(z) = 1/z^2$ around the point $z_0 \neq 0$. Observing that

$$\frac{1}{z} = \frac{1}{z_0} \frac{1}{1 + (z - z_0)/z_0}$$

$$= \frac{1}{z_0} \sum_{n=0}^{\infty} (-1)^n \left(\frac{z - z_0}{z_0} \right)^n, \qquad \left| \frac{z - z_0}{z_0} \right| < 1,$$

we find

$$\frac{1}{z^2} = -\frac{d}{dz} \frac{1}{z}$$

$$= -\frac{1}{z_0} \sum_{n=1}^{\infty} (-1)^n \frac{n}{z_0} \left(\frac{z - z_0}{z_0} \right)^{n-1}$$

$$= \frac{1}{z_0^2} \sum_{k=0}^{\infty} (-1)^k (k+1) \left(\frac{z - z_0}{z_0} \right)^k, \qquad |z - z_0| < |z_0|.$$

Example 8.13 Determine the Taylor series expansion of the principal branch of $f(z) = \log z$ around $z_0 \neq -t$ with $0 \leq t < \infty$. The largest circle with center z_0 contained in the domain of analyticity of the principal branch of $\log z$ has radius $|\mathrm{Im} z_0|$ if $\mathrm{Re} z_0 < 0$ and $|z_0|$ if $\mathrm{Re} z_0 \geq 0$. Within this circle we have

$$\log z - \log z_0 = \int_{z_0}^{z} \frac{1}{w}\, dw$$

$$= \int_{z_0}^{z} \frac{1}{z_0} \sum_{n=0}^{\infty} (-1)^n \left(\frac{w - z_0}{z_0} \right)^n dw$$

$$= \sum_{n=0}^{\infty} \frac{(-1)^n}{n+1} \left(\frac{w - z_0}{z_0} \right)^{n+1} \Bigg|_{w=z_0}^{w=z}$$

$$= \sum_{n=0}^{\infty} \frac{(-1)^n}{n+1} \left(\frac{z - z_0}{z_0} \right)^{n+1}.$$

Therefore

$$\log z = \log z_0 + \sum_{k=1}^{\infty} \frac{(-1)^{k-1}}{k z_0^k} (z - z_0)^k$$

$$= \log z_0 + \frac{1}{z_0}(z - z_0) - \frac{1}{2 z_0^2}(z - z_0)^2 + \frac{1}{3 z_0^3}(z - z_0)^3 + \dots. \tag{8.10}$$

Now let us consider the power series appearing in (8.10) and ask ourselves which analytic function corresponds to its sum within the circle of convergence of the series. Let us first calculate the radius of convergence. Using the ratio criterion, we have

$$R = \lim_{k \to \infty} \left| \frac{(-1)^{k-1}}{k z_0^k} \right| \left| \frac{(k+1) z_0^{k+1}}{(-1)^k} \right| = |z_0|.$$

So the sum of the series in (8.10) is an analytic function in $B(z_0, |z_0|)$. Such a function cannot be the principal branch of $\log z$ in the case where $\mathrm{Re}\, z_0 < 0$. It will be one of the branches of $\log z$ whose branch line does not cross the ball $B(z_0, |z_0|)$. If instead $\mathrm{Re}\, z_0 \geq 0$ the principal branch of $\log z$ is one of the possible branches that represent the sum of the series (8.10). All the branches allowed to represent the sum of the series in (8.10) have the same value within the convergence circle (Fig. 8.1).

Example 8.14 Determine the Taylor series expansion of the principal branch of $f(z) = \sqrt{z}$ around $z_0 = 1$. The function considered has a branch line coinciding with the negative real semi-axis and is analytic in the circle $|z - 1| < 1$ where it can be expanded in the Taylor series. Observing that

$$f^{(0)}(z) = z^{1/2},$$

$$f^{(1)}(z) = \frac{1}{2} z^{1/2 - 1},$$

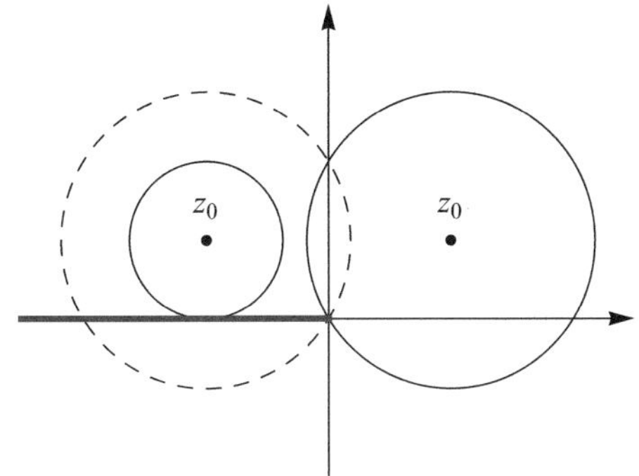

Fig. 8.1 Circle of convergence for the Taylor series expansion around z_0 of the principal branch of $\log z$ in the cases $\mathrm{Re}\, z_0 < 0$ and $\mathrm{Re}\, z_0 > 0$

$$f^{(2)}(z) = \frac{1}{2}\left(\frac{1}{2}-1\right) z^{1/2-2},$$

$$f^{(3)}(z) = \frac{1}{2}\left(\frac{1}{2}-1\right)\left(\frac{1}{2}-2\right) z^{1/2-3},$$

$$\vdots$$

$$f^{(n)}(z) = \frac{1}{2}\left(\frac{1}{2}-1\right)\left(\frac{1}{2}-2\right)\cdots\left(\frac{1}{2}-n+1\right) z^{1/2-n},$$

so that

$$f^{(n)}(1) = \frac{1}{2}\left(\frac{1}{2}-1\right)\left(\frac{1}{2}-2\right)\cdots\left(\frac{1}{2}-n+1\right) = \binom{\frac{1}{2}}{n} n!,$$

we conclude

$$\sqrt{z} = \sum_{n=0}^{\infty} \binom{\frac{1}{2}}{n}(z-1)^n$$

$$= 1 + \frac{1}{2}(z-1) - \frac{1}{8}(z-1)^2 + \frac{1}{16}(z-1)^3 - \frac{5}{128}(z-1)^4 + \dots.$$

The previous result extends immediately to the binomial series

$$(1+z)^c = \sum_{n=0}^{\infty}\binom{c}{n} z^n$$

$$= 1 + cz + \frac{c(c-1)}{2}z^2 + \frac{c(c-1)(c-2)}{6}z^3 + \dots,$$

convergent for $|z| < 1$ with $c \in \mathbb{C}$.

8.2 Laurent Series

Definition 8.15 (Annulus) Let $0 \le R_1 < R_2 \le \infty$ and $z_0 \in \mathbb{C}$, we call annulus with center z_0 and radii R_1, R_2 the set

$$A(z_0, R_1, R_2) = \{z \in \mathbb{C} : \ R_1 < |z - z_0| < R_2\}. \tag{8.11}$$

An annulus is a set that is not simply connected. For $R_1 = 0$ we also use the notation $\dot{B}(z_0, R_2) = A(z_0, 0, R_2)$, called pointed ball. The domain of analyticity of $1/(z - z_0)$ is the annulus $A(z_0, 0, \infty)$.

Theorem 8.16 (Laurent Series Expansion) *Let f be analytic in the annulus $A(z_0, R_1, R_2)$, then $\forall z \in A(z_0, R_1, R_2)$ the following Laurent series expansion holds*

$$f(z) = \sum_{n=0}^{\infty} a_n (z - z_0)^n + \sum_{n=1}^{\infty} \frac{b_n}{(z - z_0)^n}, \tag{8.12}$$

where

$$a_n = \frac{1}{2\pi i} \int_{\gamma} \frac{f(z)}{(z - z_0)^{n+1}} \mathrm{d}z, \quad n = 0, 1, 2, \ldots, \tag{8.13}$$

$$b_n = \frac{1}{2\pi i} \int_{\gamma} f(z)(z - z_0)^{n-1} \mathrm{d}z, \quad n = 1, 2, 3, \ldots, \tag{8.14}$$

or, equivalently,

$$f(z) = \sum_{n=-\infty}^{\infty} c_n (z - z_0)^n, \tag{8.15}$$

where

$$c_n = \frac{1}{2\pi i} \int_{\gamma} \frac{f(z)}{(z - z_0)^{n+1}} \mathrm{d}z, \quad n \in \mathbb{Z}, \tag{8.16}$$

with γ arbitrary piecewise regular simple closed curve, positively oriented, such that $\{\gamma\} \subset A(z_0, R_1, R_2)$ and $z_0 \in \mathrm{Int}(\gamma)$.

Proof Arbitrarily choose $z \in A(z_0, R_1, R_2)$ and γ piecewise regular simple closed curve, positively oriented, with $\{\gamma\} \subset A(z_0, R_1, R_2)$ and $z_0 \in \mathrm{Int}(\gamma)$, and consider two positively oriented circular paths γ_1 and γ_2 with center z_0 and, respectively, radii r_1 and r_2 such that $z \in A(z_0, r_1, r_2)$ and $\{\gamma\} \subset A(z_0, r_1, r_2)$. Finally, consider a positively oriented circular path γ_3 with center z and radius r sufficiently small so that $\{\gamma_3\} \subset A(z_0, r_1, r_2)$ (Fig. 8.2). Since $f(w)/(w - z)$ is a function of w analytic

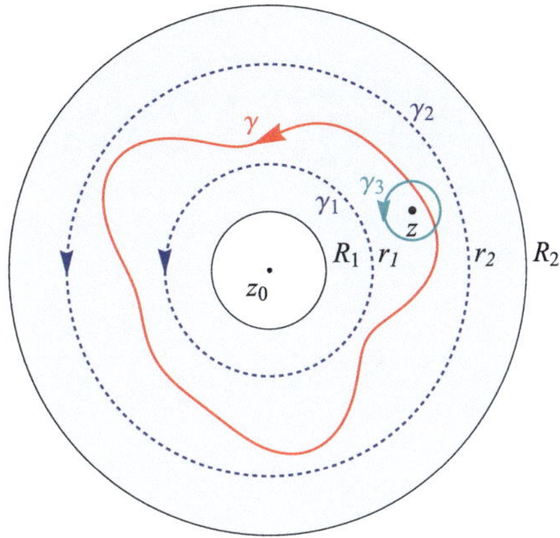

Fig. 8.2 Paths used in the proof of Theorem 8.16. Note that z can be internal to γ, as in the present example, external to γ, or even lie on γ

on γ_2, γ_1 and γ_3 and in the region between them, by the path deformation principle we have

$$\int_{\gamma_2} \frac{f(w)}{w-z}\,\mathrm{d}w = \int_{\gamma_1} \frac{f(w)}{w-z}\,\mathrm{d}w + \int_{\gamma_3} \frac{f(w)}{w-z}\,\mathrm{d}w.$$

By (7.20) the integral over γ_3 is $2\pi\mathrm{i}\,f(z)$ and therefore

$$f(z) = \frac{1}{2\pi\mathrm{i}} \int_{\gamma_2} \frac{f(w)}{w-z}\,\mathrm{d}w - \frac{1}{2\pi\mathrm{i}} \int_{\gamma_1} \frac{f(w)}{w-z}\,\mathrm{d}w$$

$$= \frac{1}{2\pi\mathrm{i}} \int_{\gamma_2} \frac{f(w)}{(w-z_0)-(z-z_0)}\,\mathrm{d}w + \frac{1}{2\pi\mathrm{i}} \int_{\gamma_1} \frac{f(w)}{(z-z_0)-(w-z_0)}\,\mathrm{d}w.$$

Using the identities

$$\frac{1}{w-z} = \frac{1}{w}\frac{1}{1-z/w}$$

$$= \frac{1}{w}\frac{1-(z/w)^n}{1-z/w} + \frac{1}{w}\frac{(z/w)^n}{1-z/w}$$

$$= \sum_{k=0}^{n-1} \frac{z^k}{w^{k+1}} + \frac{z^n}{(w-z)w^n},$$

$$\frac{1}{z-w} = \sum_{k=0}^{n-1} \frac{w^k}{z^{k+1}} + \frac{w^n}{(z-w)z^n}$$

$$= \sum_{j=1}^{n} \frac{w^{j-1}}{z^j} + \frac{w^n}{(z-w)z^n},$$

valid for arbitrary z and w with $z \neq w$, $z \neq 0$ and $w \neq 0$, we can rewrite $f(z)$ as

$$f(z) = \sum_{k=0}^{n-1} a_k (z - z_0)^k + \rho_n(z) + \sum_{j=1}^{n} \frac{b_j}{(z - z_0)^j} + \sigma_n(z), \qquad (8.17)$$

where

$$a_k = \frac{1}{2\pi i} \int_{\gamma_2} \frac{f(w)}{(w - z_0)^{k+1}} dw, \qquad k = 0, 1, \ldots, n-1,$$

$$\rho_n(z) = \frac{(z - z_0)^n}{2\pi i} \int_{\gamma_2} \frac{f(w)}{[(w - z_0) - (z - z_0)](w - z_0)^n} dw,$$

$$b_j = \frac{1}{2\pi i} \int_{\gamma_1} f(w)(w - z_0)^{j-1} dw, \qquad j = 1, 2, \ldots, n,$$

$$\sigma_n(z) = \frac{1}{2\pi i(z - z_0)^n} \int_{\gamma_1} \frac{f(w)(w - z_0)^n}{[(w - z_0) - (z - z_0)]} dw.$$

Note that, by the path deformation principle, the integrals over γ_2 and γ_1 that appear in the expressions of a_k and b_j can both be equivalently evaluated over γ. To conclude our proof it is then sufficient to show that $\lim_{n \to \infty} \rho_n(z) = 0$ and $\lim_{n \to \infty} \sigma_n(z) = 0$, in fact in this limit (8.17) reproduces the desired Laurent series expansion ((8.12)–(8.15)). Let $M_2 = \sup_{w \in \{\gamma_2\}} |f(w)|$, by Darboux's inequality we find

$$|\rho_n(z)| \leq \frac{|z - z_0|^n}{2\pi} 2\pi r_2 \frac{M_2}{(r_2 - |z - z_0|)r_2^n} = \frac{M_2(|z - z_0|/r_2)^n}{1 - |z - z_0|/r_2} \xrightarrow{n \to \infty} 0$$

as $|z - z_0| < r_2$. Similarly, let $M_1 = \sup_{w \in \{\gamma_1\}} |f(w)|$, we have

$$|\sigma_n(z)| \leq \frac{1}{2\pi |z - z_0|^n} 2\pi r_1 \frac{M_1 r_1^n}{|z - z_0| - r_1} = \frac{M_1(r_1/|z - z_0|)^n}{|z - z_0|/r_1 - 1} \xrightarrow{n \to \infty} 0$$

as $r_1 < |z - z_0|$. $\square$

The convergence of a series of the type

$$\sum_{n=-\infty}^{\infty} c_n (z - z_0)^n = f(z), \qquad (8.18)$$

which we will agree to call a generalized power series, is equivalent, by definition, to the simultaneous convergence of the two series

$$\sum_{n=0}^{\infty} a_n (z - z_0)^n = f_a(z), \qquad a_n = c_n, \quad n = 0, 1, 2, \ldots, \tag{8.19}$$

$$\sum_{n=1}^{\infty} \frac{b_n}{(z - z_0)^n} = f_b(z), \qquad b_n = c_{-n}, \quad n = 1, 2, 3, \ldots, \tag{8.20}$$

with $f_a(z) + f_b(z) = f(z)$. The power series (8.19) is convergent within the circle with center z_0 and radius R_2 given by

$$\frac{1}{R_2} = \limsup_{n \to \infty} |c_n|^{1/n} . \tag{8.21}$$

On the other hand, (8.20) with the position $w = 1/(z - z_0)$ becomes the power series $\sum_{n=1}^{\infty} b_n w^n$ convergent for $|w| < 1/\limsup_{n \to \infty} |c_{-n}|^{1/n}$. Therefore, the series (8.20) converges outside the circle of center z_0 and radius

$$R_1 = \limsup_{n \to \infty} |c_{-n}|^{1/n} . \tag{8.22}$$

In conclusion, we have the simultaneous convergence of (8.19) and (8.20), hence of the series (8.18), in the annulus $A(z_0, R_1, R_2)$.

Theorem 8.17 (Integral of a Generalized Power Series) *Let $\sum_{n=-\infty}^{\infty} c_n (z - z_0)^n = f(z)$ within the annulus with center z_0 and radii R_1, R_2. If γ is a piecewise smooth curve contained in $A(z_0, R_1, R_2)$ and g is a continuous function on $\{\gamma\}$, then*

$$\sum_{n=-\infty}^{\infty} c_n \int_\gamma g(z)(z - z_0)^n \mathrm{d}z = \int_\gamma g(z) f(z) \mathrm{d}z. \tag{8.23}$$

Proof For each of the two power series (8.19) and (8.20) the analogous Theorem 8.2 holds with the curve γ contained in the respective convergence domains and g continuous on $\{\gamma\}$. Therefore, if $\{\gamma\} \subset A(z_0, R_1, R_2)$ and g is continuous on $\{\gamma\}$, simultaneously we have

$$\sum_{n=0}^{\infty} a_n \int_\gamma g(z)(z - z_0)^n \mathrm{d}z = \int_\gamma g(z) f_a(z) \mathrm{d}z,$$

$$\sum_{n=1}^{\infty} b_n \int_\gamma \frac{g(z)}{(z - z_0)^n} \mathrm{d}z = \int_\gamma g(z) f_b(z) \mathrm{d}z,$$

and therefore the statement. $\square$

From Theorem 8.17 the following three corollaries follow similarly to what seen in the case of the Taylor series.

Corollary 8.18 (Analyticity of a Generalized Power Series) *Let* $\sum_{n=-\infty}^{\infty} c_n (z - z_0)^n = f(z)$ *within the annulus with center* z_0 *and radii* R_1, R_2, *then* f *is analytic in* $A(z_0, R_1, R_2)$.

Corollary 8.19 (Derivative of a Generalized Power Series) *Let* $\sum_{n=-\infty}^{\infty} c_n (z - z_0)^n = f(z)$ *within the annulus with center* z_0 *and radii* R_1, R_2, *then* $\forall z \in A(z_0, R_1, R_2)$ *we have* $f'(z) = \sum_{n=-\infty}^{\infty} c_n n (z - z_0)^{n-1}$.

Corollary 8.20 (Uniqueness of Laurent Series Expansion) *Let* $\sum_{n=-\infty}^{\infty} c_n (z - z_0)^n = f(z)$ *within the annulus with center* z_0 *and radii* R_1, R_2, *then this is the Laurent series expansion of* f *around* z_0, *that is,* $c_n = (1/2\pi i) \int_\gamma [f(z)/(z - z_0)^{n+1}] \mathrm{d}z$, $n \in \mathbb{Z}$, *with* γ *arbitrary piecewise regular simple closed curve, positively oriented, such that* $\{\gamma\} \subset A(z_0, R_1, R_2)$ *and* $z_0 \in \mathrm{Int}(\gamma)$.

Example 8.21 Determine the Laurent series expansion of $f(z) = \exp(1/z)$ in the annulus $0 < |z| < \infty$. Recalling the Maclaurin series expansion of $\exp(w)$, valid $\forall w \in \mathbb{C}$, that is, for $|w| < \infty$, with the change of variable $w = 1/z$, $z \neq 0$, we obtain

$$\exp(1/z) = \sum_{n=0}^{\infty} \frac{1}{n! z^n}, \qquad 0 < |z| < \infty.$$

Example 8.22 Determine the series expansion around $z_0 = 0$ of

$$f(z) = \frac{-1}{(z - 1)(z - 2)} = \frac{1}{z - 1} - \frac{1}{z - 2}.$$

The function f is analytic in $\mathbb{C}$ except at the points $z = 1$ and $z = 2$. Therefore, it is possible to expand f around $z_0 = 0$ in the Taylor series for $|z| < 1$ and in the Laurent series for $1 < |z| < 2$ and $2 < |z| < \infty$. If $|z| < 1$, observing that $|z/2| < 1$, we have

$$f(z) = -\frac{1}{1 - z} + \frac{1}{2} \frac{1}{1 - z/2}$$

$$= -\sum_{n=0}^{\infty} z^n + \frac{1}{2} \sum_{n=0}^{\infty} (z/2)^n$$

$$= \sum_{n=0}^{\infty} (2^{-n-1} - 1) z^n.$$

If $1 < |z| < 2$, observing that $|1/z| < 1$ and $|z/2| < 1$, we have

$$f(z) = \frac{1}{z}\frac{1}{1 - 1/z} + \frac{1}{2}\frac{1}{1 - z/2}$$

$$= \frac{1}{z}\sum_{n=0}^{\infty}(1/z)^n + \frac{1}{2}\sum_{n=0}^{\infty}(z/2)^n$$

$$= \sum_{k=0}^{\infty}\frac{z^k}{2^{k+1}} + \sum_{k=1}^{\infty}\frac{1}{z^k}.$$

If $2 < |z| < \infty$, observing that $|1/z| < 1$ and $|2/z| < 1$, we have

$$f(z) = \frac{1}{z}\frac{1}{1 - 1/z} - \frac{1}{z}\frac{1}{1 - 2/z}$$

$$= \frac{1}{z}\sum_{n=0}^{\infty}(1/z)^n - \frac{1}{z}\sum_{n=0}^{\infty}(2/z)^n$$

$$= \sum_{k=1}^{\infty}\frac{1 - 2^{k+1}}{z^k}.$$

8.3 Multiplication and Division of Two Power Series

Theorem 8.23 (Product of Two Power Series) *Let* $\sum_{n=0}^{\infty} a_n(z - z_0)^n = f(z)$ *for* $z \in B(z_0, R_1)$ *and* $\sum_{n=0}^{\infty} b_n(z - z_0)^n = g(z)$ *for* $z \in B(z_0, R_2)$. *Let* $R = \min(R_1, R_2)$, *then* $\forall z \in B(z_0, R)$ *we have*

$$f(z)g(z) = \sum_{n=0}^{\infty} c_n(z - z_0)^n, \qquad c_n = \sum_{k=0}^{n} a_k b_{n-k}. \tag{8.24}$$

Proof The functions f and g as sums of power series are analytic in $B(z_0, R_1)$ and $B(z_0, R_2)$ respectively. Furthermore due to the uniqueness of the Taylor series expansion we have $a_n = f^{(n)}(z_0)/n!$ and $b_n = g^{(n)}(z_0)/n!$ for $n = 0, 1, 2, \ldots$. Therefore, fg is analytic in $B(z_0, R)$, where $R = \min(R_1, R_2)$, and in this ball the following Taylor series expansion holds

$$f(z)g(z) = \sum_{n=0}^{\infty} c_n(z - z_0)^n,$$

with

$$c_n = \frac{1}{n!} \frac{d^n}{dz^n} f(z)g(z)\bigg|_{z=z_0}.$$

Using the Leibniz rule for the derivative of order n of the product of two functions, we conclude

$$c_n = \frac{1}{n!} \sum_{k=0}^{n} \binom{n}{k} f^{(k)}(z_0) g^{(n-k)}(z_0)$$

$$= \sum_{k=0}^{n} \frac{f^{(k)}(z_0)}{k!} \frac{g^{(n-k)}(z_0)}{(n-k)!}$$

$$= \sum_{k=0}^{n} a_k b_{n-k}.$$

$\square$

Theorem 8.24 (Ratio of Two Power Series) *Let* $\sum_{n=0}^{\infty} a_n (z - z_0)^n = f(z)$ *for* $z \in B(z_0, R_1)$ *and* $\sum_{n=0}^{\infty} b_n (z - z_0)^n = g(z) \neq 0$ *for* $z \in B(z_0, R_2)$. *Let* $R = \min(R_1, R_2)$, *then* $\forall z \in B(z_0, R)$ *we have*

$$f(z)/g(z) = \sum_{n=0}^{\infty} d_n (z - z_0)^n, \qquad d_n = \left(a_n - \sum_{k=0}^{n-1} d_k b_{n-k} \right) \frac{1}{b_0}. \tag{8.25}$$

Proof The functions f and g as sums of power series are analytic in $B(z_0, R_1)$ and $B(z_0, R_2)$ respectively. Moreover $g(z) \neq 0 \ \forall z \in B(z_0, R_2)$. Therefore f/g is analytic in $B(z_0, R)$, where $R = \min(R_1, R_2)$, and in this ball the following Taylor series expansion holds

$$f(z)/g(z) = \sum_{n=0}^{\infty} d_n (z - z_0)^n.$$

To determine the coefficients d_n observe that $(f/g)\, g = f$. According to Theorem 8.23 we then have

$$\sum_{k=0}^{n} d_k b_{n-k} = a_n \qquad n = 0, 1, 2, \ldots .$$

Solving for d_n we find

$$d_n = \left(a_n - \sum_{k=0}^{n-1} d_k b_{n-k} \right) \frac{1}{b_0}.$$

The coefficients d_n must be evaluated iteratively. For $n = 0$ we have $d_0 = a_0/b_0$. Once d_0 is known, for $n = 1$ we have $d_1 = (a_1 - d_0 b_1)/b_0$. Once d_0 and d_1 are known, for $n = 2$ we have $d_2 = (a_2 - d_1 b_1 - d_0 b_2)/b_0$, and so on. $\square$

Example 8.25 Determine the first terms of the Taylor series expansion around $z_0 = 0$ of $f(z) = e^z/(1 + z)$. The function f is analytic in $\mathbb{C}$ except at the point $z = -1$. Therefore, f can be expanded in the Maclaurin series for $|z| < 1$. Using the Maclaurin series expansion of e^z and the geometric series, both series being convergent for $|z| < 1$, in the same ball we have

$$\frac{e^z}{1+z} = \left(1 + z + \frac{z^2}{2} + \frac{z^3}{6} + \frac{z^4}{24} + \frac{z^5}{120} + \dots\right)$$

$$\times \left(1 - z + z^2 - z^3 + z^4 - z^5 + \dots\right)$$

$$= 1 + \frac{1}{2}z^2 - \frac{1}{3}z^3 + \frac{3}{8}z^4 - \frac{11}{30}z^5 + \dots$$

Example 8.26 Determine the first terms of the Laurent series expansion around $z_0 = 0$ of $f(z) = 1/\sin z$. The function f is analytic in $\mathbb{C}$ except at the points $z = k\pi$, $k \in \mathbb{Z}$. Therefore, f can be expanded in the Laurent series around $z_0 = 0$ in the annulus $0 < |z| < \pi$. Using the Maclaurin series expansion of $\sin z$, we can write

$$\frac{1}{\sin z} = \frac{1}{z - \frac{z^3}{3!} + \frac{z^5}{5!} - \frac{z^7}{7!} + \dots}$$

$$= \frac{1}{z} \frac{1}{1 - \left(\frac{z^2}{3!} - \frac{z^4}{5!} + \frac{z^6}{7!} + \dots\right)}$$

$$= \frac{1}{z}\left[1 + \left(\frac{z^2}{3!} - \frac{z^4}{5!} + \frac{z^6}{7!} + \dots\right) + \left(\frac{z^2}{3!} - \frac{z^4}{5!} + \frac{z^6}{7!} + \dots\right)^2\right.$$

$$\left. + \left(\frac{z^2}{3!} - \frac{z^4}{5!} + \frac{z^6}{7!} + \dots\right)^3 + \dots\right]$$

$$= \frac{1}{z}\left[1 + \frac{z^2}{3!} - \frac{z^4}{5!} + \frac{z^4}{3!3!} + \frac{z^6}{7!} - \frac{2z^6}{3!5!} + \frac{z^6}{3!3!} + \dots\right]$$

$$= \frac{1}{z} + \frac{1}{6}z + \frac{7}{360}z^3 + \frac{31}{15120}z^5 + \dots,$$

where we used

$$\frac{1}{1 - h(z)} = 1 + h(z) + h(z)^2 + h(z)^3 + \dots,$$

valid for $|h(z)| < 1$. This condition is certainly satisfied for $|z|$ sufficiently small if, as in our case, $\lim_{z \to 0} h(z) = 0$. By the uniqueness theorem, the series just found is the Laurent series expansion of $f(z)$ in the annulus $0 < |z| < \pi$. The same result is reached by applying Theorem 8.24. Writing

$$\frac{1}{\sin z} = \frac{1}{z} \frac{1}{(\sin z)/z},$$

since $(\sin z)/z$ vanishes for $z = k\pi$, $k = \pm 1, \pm 2, \ldots$, but not for $z = 0$ where by continuity the function is assumed to be 1, in the ball $|z| < \pi$ the ratio $1/((\sin z)/z)$ can be evaluated as the ratio of two series, the one in the numerator having coefficients $a_0 = 1$ and $a_k = 0$ for $k \geq 1$, and the one in the denominator with coefficients $b_0 = 1$, $b_2 = -1/6$, $b_4 = 1/120$, $\ldots$, and $b_k = 0$ for k odd. By using formula (8.25) we find

$$d_0 = a_0/b_0 = 1,$$

$$d_1 = (a_1 - d_0 b_1)/b_0 = 0,$$

$$d_2 = (a_2 - d_0 b_2 - d_1 b_1)/b_0 = \frac{1}{6},$$

$$d_3 = (a_3 - d_0 b_3 - d_1 b_2 - d_2 b_1)/b_0 = 0,$$

$$d_4 = (a_4 - d_0 b_4 - d_1 b_3 - d_2 b_2 - d_3 b_1)/b_0 = \frac{7}{360},$$

$$\vdots$$

and therefore

$$\frac{1}{\sin z} = \frac{1}{z} + \frac{1}{6} z + \frac{7}{360} z^3 + \cdots.$$

Exercises

8.1 Expand in the Maclaurin series the function

$$f(z) = \frac{1}{az + b}, \qquad a, b \neq 0,$$

and determine the radius of convergence of the power series thus obtained.

8.2 Expand in the Taylor series around $z_0 = 0$ the function

$$f(z) = \int_0^z e^{w^2} \, dw$$

and determine the radius of convergence of the power series thus obtained.

8.3 Assuming the principal branch for the logarithm, determine the domain of analyticity of the function

$$f(z) = \log \frac{1+z}{1-z}$$

and then expand it in the Taylor series around $z = 0$.

8.4 Expand in the Taylor series with center $z_0 = 3$ the principal branch of $\log z$ and determine the radius of convergence of the resulting power series.

8.5 Assuming the principal branch for the logarithm, determine the region of analyticity of the function

$$f(z) = (z+1) \log \left(1 + z^2 \right)$$

and then expand it in the Taylor series around $z = 0$. Determine the radius of convergence of the power series thus obtained.

8.6 Expand in the Taylor series with center $z_0 = -1$ the principal branch of $\log(z^2)$ and determine the radius of convergence of the resulting power series.

8.7 Expand in the Taylor series around the point $z_0 = 0$ the function

$$f(z) = e^z \sin z.$$

8.8 Determine the domain of analyticity of the principal branch of $\log(z^2)$ and its expansion in the Taylor series around the point $z_0 = 1$.

8.9 Expand in the Taylor series around the point $z_0 = 0$ the function $e^{iz} \sinh z$.

8.10 Expand in the Taylor series around the point $z_0 = 0$ the function $\arctan z$ and determine the radius of convergence of the power series thus obtained.

8.11 Consider the power series

$$\sum_{n=0}^{\infty} a_n (z - z_0)^n$$

convergent to $f(z) = \sin(z)/(z^3 + 1)$ for $|z - z_0| < R$ with $z_0 = 2$. Determine the radius of convergence R of the series.

8.12 Determine the nature of the isolated singularity at $z = 0$ of the function

$$f(z) = \left(\frac{1}{\tan z} - \frac{1}{\sin z} \right) \frac{1}{z^2}$$

and calculate its corresponding residue (coefficient c_{-1} of the Laurent series expansion in $A(0, 0, R)$).

8.13 Determine the residue at $z = 0$ (coefficient c_{-1} of the Laurent series expansion in $A(0, 0, R)$) of the function

$$f(z) = \int_2^3 \frac{t^2 \sin t}{t - 1/z} \, dt.$$

8.14 Determine the nature of the singularity at $z = 0$ of the function

$$f(z) = \left(\frac{\exp(\cos z)}{\sin z} \right)^3$$

and calculate its corresponding residue (coefficient c_{-1} of the Laurent series expansion in $A(0, 0, R)$). Also calculate the residue at $z = 0$ of the functions $z f(z)$ and $z^2 f(z)$.

8.15 Expand in the Laurent series around $z_0 = i$ the function

$$f(z) = \frac{1}{(z^2 + 1)^2}$$

specifying the annular domain of validity of the generalized power series thus obtained.

8.16 Expand in the Laurent series around $z_0 = 0$, in both annular regions $0 < |z| < 1$ and $1 < |z| < \infty$, the function

$$f(z) = \frac{1}{(z(1 - z))^2}.$$

8.17 Expand in the Laurent series around $z_0 = 0$, in both annular regions $0 < |z| < 2$ and $2 < |z| < \infty$, the function

$$f(z) = \frac{1}{z^3 + 2z^2}.$$

8.18 Expand in the Laurent series around $z_0 = 0$, in both annular regions $0 < |z| < \sqrt{2}$ and $\sqrt{2} < |z| < \infty$, the function

$$f(z) = \frac{1}{z^3 + 2z}.$$

8.19 Expand in the Laurent series around $z_0 = i$, in both the annular regions $0 < |z - i| < 2$ and $2 < |z - i| < \infty$, the function

$$f(z) = \frac{1}{z^2 + 1}.$$

8.20 Expand in the Laurent series around $z_0 = 0$ in the annular region $0 < |z| < 1$ the function

$$f(z) = \frac{1}{z + z^3}.$$

8.21 Expand in the Laurent series around $z_0 = 0$, in both annular regions $0 < |z| < 2$ and $2 < |z| < \infty$, the function

$$f(z) = \frac{3 + z}{z^3 + 2z^2}.$$

8.22 Expand in the Laurent series in the annulus $0 < |z - 1| < 1$ la function

$$f(z) = \frac{1}{(z - 1)^2(z - 2)^2}.$$

8.23 Expand in the Laurent series around $z = 0$, in both annular regions $0 < |z| < 1$ and $1 < |z| < \infty$, the function

$$f(z) = \frac{2z}{z^2(1 + z^2)^2}.$$

8.24 Expand in the Laurent series around $z = i$, in both annular regions $0 < |z - i| < 1$ and $2 < |z - i| < \infty$, the function

$$f(z) = \frac{1}{z + z^3}.$$

8.25 Determine the nature of all isolated singularities of the function

$$f(z) = \frac{z^5 + 1}{(z^7 + 1)z}$$

and expand it in the Laurent series with center $z = 0$ in the region $|z| > 1$. Finally, determine the residue of $f(z)$ at $z = 0$ (coefficient c_{-1} of the Laurent series expansion in $A(0, 0, R)$).

8.26 Assuming the principal branch for multi-valued functions, determine the value of the following limit justifying the answer

$$\lim_{z \to 0} (\cos z)^{1/z^2}.$$

8.27 Determine the first three non-zero terms of the series expansion of $\tan z$ in the two cases (a) around $z_0 = 0$ and (b) around $z_0 = \pi/2$.

8.28 Expand in the Laurent series around $z = 0$ up to order $O(z^1)$ included the function

$$f(z) = \frac{1}{z^2 \sinh z}.$$

8.29 Find the fourth derivative at $z = 0$ of the function $f(z) = 1/\cos z$.

8.30 Determine, at each order, the power series defined by the ratio

$$\left(\sum_{n=0}^{\infty} z^n \right) \left(\sum_{n=0}^{\infty} 2^n z^n \right)^{-1}$$

and calculate its radius of convergence.

8.31 The Legendre polynomials $P_n(x)$, with $x \in [-1, 1]$, can be defined as the coefficients of the series expansion

$$(1 - 2xz + z^2)^{-1/2} = 1 + \sum_{k=1}^{\infty} P_k(x)z^k, \qquad z \in \mathbb{C}.$$

Determine P_1, P_2 and P_3.

8.32 Find the singular part of the Laurent series expansion around $z_0 = 0$ of

$$f(z) = \frac{\cos z}{\sin(z^2)(e^z - 1)}.$$

8.33 Expand in the Taylor series around $z_0 = 0$ the Fresnel function

$$S(z) = \int_0^z \sin(\pi w^2/2)\,dw.$$

Determine the Laurent series expansion of $1/S(z)$ in the annulus $0 < |z| < \infty$ up to order $O(z^0)$ included.

8.34 Determine up to order $O(z^6)$ included the Taylor series expansion around $z_0 = 0$ of the principal branch of the function

$$f(z) = \frac{\sin z}{\sqrt{\cos z}}.$$

Chapter 9
Residues

Abstract Singularities and isolated singularities. Residue of a function at an isolated singularity. Residue theorem. Residue theorem with the residue at infinity. Classification of isolated singularities: removable singularities, poles of order m, essential singularities. Necessary and sufficient condition for an isolated singular point of an analytic function to be a pole of order m and formula for the corresponding residue. Zeros of order m of analytic functions. Necessary and sufficient condition for an analytic function to have a zero of order m. Identity theorem. The zeros of non-constant analytic functions are isolated and of finite order. Sufficient condition for a function of the type $f(z) = p(z)/q(z)$ to have a pole of order m and formula for the corresponding residue. Behavior of an analytic function in the vicinity of isolated singularities. Riemann lemma. Casorati-Weierstrass theorem.

9.1 Isolated Singular Points: Residues

Throughout this chapter, unless otherwise specified, we will understand f as a function $f : G \mapsto \mathbb{C}$ with $G \subset \mathbb{C}$ open.

Definition 9.1 (Singular Point) A point z_0 is called a singular point of f if f is not analytic at z_0 but $\forall \varepsilon > 0 \; \exists z \in B(z_0, \varepsilon)$ such that f is analytic at z.

Definition 9.2 (Isolated Singular Point) A singular point z_0 of f is called isolated if $\exists r > 0$ such that f is analytic in $A(z_0, 0, r)$. In other words, z_0 is an isolated singular point of f if $\exists r > 0$ such that f is analytic in $A(z_0, 0, r)$ but not in $B(z_0, r)$.

Definition 9.3 (Residue) Let z_0 be an isolated singular point of f, then $\exists r > 0$ such that $\forall z \in A(z_0, 0, r)$ the Laurent series expansion holds

$$f(z) = \sum_{n=0}^{\infty} a_n (z - z_0)^n + \sum_{n=1}^{\infty} \frac{b_n}{(z - z_0)^n}. \tag{9.1}$$

C. Presilla, *Complex Functions of a Variable*, La Matematica per il 3+2 179,
https://doi.org/10.1007/978-3-032-12494-4_9

The coefficient b_1 is called the residue of $f(z)$ at z_0 and we write

$$\operatorname*{Res}_{z=z_0} f(z) = b_1. \tag{9.2}$$

Theorem 9.4 (Of Residues) *Let γ be a simple closed piecewise regular positively oriented curve. If f is an analytic function on and within γ except for a finite number of singular (isolated) points $z_k \in \operatorname{Int}(\gamma)$, $k = 1, \ldots, n$, then*

$$\int_\gamma f(z)\mathrm{d}z = 2\pi\mathrm{i} \sum_{k=1}^{n} \operatorname*{Res}_{z=z_k} f(z). \tag{9.3}$$

Proof Let γ_k, $k = 1, \ldots, n$, be n positively oriented circles with center z_k and radius sufficiently small so that they are all within γ and disjoint pairwise. Since f is analytic on γ, on the γ_k circles and in the multiply connected region between these curves, by the path deformation principle we have

$$\int_\gamma f(z)\mathrm{d}z = \sum_{k=1}^{n} \int_{\gamma_k} f(z)\mathrm{d}z.$$

On the other hand, from the definition of residue it follows

$$\int_{\gamma_k} f(z)\mathrm{d}z = 2\pi\mathrm{i} \operatorname*{Res}_{z=z_k} f(z),$$

and hence the statement. $\square$

Theorem 9.5 (Of the Residue at Infinity) *Let γ be a simple closed piecewise regular positively oriented curve. If f is an analytic function on and outside γ, then*

$$\int_\gamma f(z)\mathrm{d}z = 2\pi\mathrm{i} \operatorname*{Res}_{z=0} \frac{1}{z^2} f\left(\frac{1}{z}\right). \tag{9.4}$$

Proof By the Jordan curve theorem, $\exists R < \infty$ such that $\{\gamma\} \subset B(0, R)$. The function f is analytic in $A(0, R, \infty)$ and therefore in this annular region the Laurent series expansion holds

$$f(z) = \sum_{n=-\infty}^{\infty} c_n z^n, \qquad c_n = \frac{1}{2\pi\mathrm{i}} \int_{\gamma_0} \frac{f(z)}{z^{n+1}}\mathrm{d}z, \qquad n \in \mathbb{Z}. \tag{9.5}$$

The path γ_0 is an arbitrary simple closed piecewise regular positively oriented path such that $\{\gamma_0\} \subset A(0, R, \infty)$ and $0 \in \operatorname{Int}(\gamma_0)$. For γ_0 we can then choose the circle

$|z| = R_0$ with $R_0 > R$. Since f is analytic on γ, on γ_0 and in the region between γ and γ_0, by the path deformation principle we have

$$\int_\gamma f(z)\mathrm{d}z = \int_{\gamma_0} f(z)\mathrm{d}z = 2\pi\mathrm{i}\, c_{-1}. \tag{9.6}$$

Let us now establish the role of the coefficient c_{-1}. Since f admits the expansion (9.5) for $R < |z| < \infty$, setting $z = 1/w$ for $0 < |w| < 1/R$ we get

$$\frac{1}{w^2} f\left(\frac{1}{w}\right) = \frac{1}{w^2} \sum_{n=-\infty}^{\infty} c_n \left(\frac{1}{w}\right)^n = \sum_{n=-\infty}^{\infty} c_n \frac{1}{w^{n+2}} = \sum_{k=-\infty}^{\infty} c_{-k-2} w^k.$$

From this follows

$$\operatorname*{Res}_{w=0} \frac{1}{w^2} f\left(\frac{1}{w}\right) = c_{-1}. \tag{9.7}$$

Therefore, the coefficient c_{-1} is the residue at $z = 0$ of the function $f(1/z)/z^2$, certainly not of $f(z)$ which could also be analytic at $z = 0$ or have a non-isolated singularity at that point. $\qquad\square$

Example 9.6 Find the value of the integral of $\sin(z^{-1})$ along the circle $\gamma(t) = \mathrm{e}^{\mathrm{i}t}$, $0 \le t \le 2\pi$. The function $\sin(z^{-1})$ is analytic on and within γ except for the isolated singularity at $z = 0$. For $0 < |z| < \infty$ the following Laurent series expansion holds

$$\sin(z^{-1}) = \sum_{k=0}^{\infty} \frac{(-1)^k}{(2k+1)!} z^{-(2k+1)}$$

$$= z^{-1} - \frac{1}{3!} z^{-3} + \frac{1}{5!} z^{-5} - \frac{1}{7!} z^{-7} + \cdots,$$

from which we find

$$\operatorname*{Res}_{z=0} \sin(z^{-1}) = 1$$

and therefore by Theorem 9.3 we conclude

$$\int_\gamma \sin(z^{-1})\mathrm{d}z = 2\pi\mathrm{i}\operatorname*{Res}_{z=0} \sin(z^{-1}) = 2\pi\mathrm{i}.$$

Example 9.7 Determine the value of the integral of $1/\sin(z^{-1})$ along the circle $\gamma(t) = \mathrm{e}^{\mathrm{i}t}$, $0 \le t \le 2\pi$. The function $1/\sin(z^{-1})$ is analytic everywhere except for the isolated singularities at the points $z_n = (n\pi)^{-1}$, $n = \pm 1, \pm 2, \ldots$, and the

non-isolated singularity at $z = 0$. Therefore, the function $1/\sin(z^{-1})$ is analytic on
the closed path γ and on its outside, and by Theorem 9.5 we have

$$\int_\gamma \frac{1}{\sin(z^{-1})}\,dz = 2\pi i \operatorname*{Res}_{z=0} \frac{1}{z^2} \frac{1}{\sin(z)}.$$

This residue can be calculated considering that for $0 < |z| < \infty$ the following
Laurent series expansion holds

$$\frac{1}{z^2}\frac{1}{\sin(z)} = \frac{1}{z^2}\frac{1}{z - \frac{z^3}{3!} + \frac{z^5}{5!} - \frac{z^7}{7!} + \dots}$$

$$= \frac{1}{z^3}\frac{1}{1 - \left(\frac{z^2}{3!} - \frac{z^4}{5!} + \dots\right)}$$

$$= \frac{1}{z^3}\left[1 + \left(\frac{z^2}{3!} - \frac{z^4}{5!} + \dots\right) + \left(\frac{z^2}{3!} - \frac{z^4}{5!} + \dots\right)^2 + \dots\right]$$

$$= z^{-3} + \frac{1}{3!}z^{-1} + \left(-\frac{1}{5!} + \frac{1}{(3!)^2}\right) z + \dots.$$

In conclusion

$$\int_\gamma \frac{1}{\sin(z^{-1})}\,dz = 2\pi i \frac{1}{3!} = \frac{\pi}{3}i.$$

9.2   Classification of Isolated Singularities

Definition 9.8 (Pole of Order m, Removable Singularity, Essential Singularity)
Let z_0 be an isolated singular point of f, then $\exists r > 0$ such that $\forall z \in A(z_0, 0, r)$ the
following Laurent series expansion holds

$$f(z) = \sum_{k=0}^{\infty} a_k (z - z_0)^k + \sum_{k=1}^{\infty} \frac{b_k}{(z - z_0)^k}. \tag{9.8}$$

Depending on the number of terms actually present in the series with negative
powers (often referred to as principal part or singular part of f), z_0 is said to be:

(a) a pole of order m if $b_m \neq 0$ and $b_k = 0\ \forall k > m$;
(b) a removable singularity if $b_k = 0\ \forall k \geq 1$ (pole of order 0);
(c) an essential singularity if $b_k \neq 0$ for infinite values of the index k (pole of
     infinite order).

Example 9.9 The function $z^{-4} \sinh z$ has a pole of order 3 at the isolated singular point $z_0 = 0$. In fact for $z \in A(0, 0, \infty)$ we have

$$\frac{\sinh z}{z^4} = \frac{1}{z^4} \sum_{k=0}^{\infty} \frac{z^{2k+1}}{(2k+1)!} = \frac{1}{z^3} + \frac{1}{3!z} + \frac{z}{5!} + \frac{z^3}{7!} + \dots.$$

Example 9.10 The function $(1 - \cos z)/z^2$ has an isolated singularity at $z_0 = 0$ that is removable. In fact for $z \in A(0, 0, \infty)$ we have

$$\frac{1 - \cos z}{z^2} = \frac{1}{z^2} \left(1 - \sum_{k=0}^{\infty} (-1)^{2k} \frac{z^{2k}}{(2k)!} \right) = \frac{1}{2!} - \frac{z^2}{4!} + \frac{z^4}{6!} + \dots.$$

The name, removable singularity, follows from the fact that in fact it is possible to define a function $f(z)$ associated with the given one

$$f(z) = \begin{cases} (1 - \cos z)/z^2 & z \neq 0 \\ 1/2 & z = 0 \end{cases}$$

which is analytic also at z_0.

Example 9.11 The function $\exp(1/z)$ has an essential singularity at the isolated singular point $z_0 = 0$. In fact for $z \in A(0, 0, \infty)$ we have

$$e^{1/z} = \sum_{k=0}^{\infty} \frac{1}{k!} \frac{1}{z^k}.$$

Theorem 9.12 *Let z_0 be an isolated singular point of f. The point z_0 is a pole of order $m > 0$ if and only if $f(z) = g(z)/(z - z_0)^m$ with g analytic at z_0 and $g(z_0) \neq 0$. Furthermore*

$$\operatorname*{Res}_{z=z_0} f(z) = \frac{g^{(m-1)}(z_0)}{(m-1)!}.$$

Proof Let $f(z) = g(z)/(z - z_0)^m$ with g analytic at z_0 and $g(z_0) \neq 0$. Since g is analytic at z_0, $\exists r > 0$ such that g is analytic in $B(z_0, r)$ and in this ball the following Taylor series expansion holds

$$g(z) = \sum_{k=0}^{\infty} \frac{g^{(k)}(z_0)}{k!} (z - z_0)^k.$$

Therefore, $\forall z \in A(z_0, 0, r)$ we have

$$f(z) = \frac{g(z)}{(z - z_0)^m} = \sum_{k=0}^{\infty} \frac{g^{(k)}(z_0)}{k!} (z - z_0)^{k-m}.$$

From this expression, since $g(z_0) \neq 0$, we conclude that z_0 is a pole of order m and also $\mathrm{Res}_{z=z_0} f(z) = g^{(m-1)}(z_0)/(m-1)!$.

Conversely, suppose that z_0 is a pole of order m for f. Then $\exists r > 0$ such that f is analytic in $A(z_0, 0, r)$ and in this annulus the following Laurent series expansion holds

$$f(z) = \sum_{k=0}^{\infty} a_k (z - z_0)^k + \sum_{k=1}^{m} \frac{b_k}{(z - z_0)^k},$$

with $b_m \neq 0$. Set

$$g(z) = \begin{cases} (z - z_0)^m f(z) & z \in A(z_0, 0, r) \\ b_m & z = z_0 \end{cases},$$

$\forall z \in B(z_0, r)$ we have

$$g(z) = b_m + b_{m-1}(z - z_0) + \cdots + b_1 (z - z_0)^{m-1} + \sum_{k=0}^{\infty} a_k (z - z_0)^{k+m}.$$

Being the sum of a convergent power series, g is analytic in $B(z_0, r)$ and therefore analytic at z_0. Furthermore, $g(z_0) = b_m \neq 0$. Finally, by the uniqueness of the Taylor series expansion we have $b_1 = g^{(m-1)}(z_0)/(m-1)!$. $\qquad\qquad\square$

9.3 Zeros of Analytic Functions

Definition 9.13 (Zero of Order m) Let f be analytic at z_0. The point z_0 is called zero of order m of f if $f^{(k)}(z_0) = 0$ for $k = 0, 1, 2, \ldots, m-1$ and $f^{(m)}(z_0) \neq 0$.

Theorem 9.14 *Let f be analytic at z_0. The point z_0 is a zero of order m of f if and only if $f(z) = g(z)(z - z_0)^m$ with g analytic in z_0 and $g(z_0) \neq 0$.*

Proof Let $f(z) = (z - z_0)^m g(z)$ with $g(z)$ analytic and nonzero in z_0. Since g is analytic at z_0, $\exists r > 0$ such that g is analytic in $B(z_0, r)$ and in this ball the following Taylor series expansion holds

$$g(z) = \sum_{k=0}^{\infty} \frac{g^{(k)}(z_0)}{k!} (z - z_0)^k.$$

Then $\forall z \in B(z_0, r)$ we have

$$f(z) = \sum_{k=0}^{\infty} \frac{g^{(k)}(z_0)}{k!}(z - z_0)^{k+m}.$$

From the uniqueness of the Taylor series expansion of f it follows that

$$f^{(k)}(z_0) = 0, \quad k = 0, 1, \ldots, m - 1, \qquad f^{(m)}(z_0) = m!\, g(z_0) \neq 0,$$

that is, z_0 is a zero of order m of $f(z)$.

Conversely, suppose that z_0 is a zero of order m of f, that is,

$$f^{(k)}(z_0) = 0, \quad k = 0, 1, \ldots, m - 1, \qquad f^{(m)}(z_0) \neq 0.$$

Since f is analytic at z_0, $\exists r > 0$ such that f is analytic in $B(z_0, r)$ and in this ball the following Taylor series expansion holds

$$f(z) = \sum_{k=m}^{\infty} \frac{f^{(k)}(z_0)}{k!}(z - z_0)^k = (z - z_0)^m g(z),$$

where

$$g(z) = \sum_{k=m}^{\infty} \frac{f^{(k)}(z_0)}{k!}(z - z_0)^{k-m}.$$

The function g, being the sum of a convergent power series in $B(z_0, r)$, is analytic at z_0. Furthermore $g(z_0) - f^{(m)}(z_0)/m! \neq 0$. $\qquad\square$

Theorem 9.15 *Let f be analytic in $D \subset \mathbb{C}$ open and connected. Then the following properties are equivalent:*

(a) *f is identically zero in D;*
(b) *$\exists z_0 \in D$ such that $f^{(k)}(z_0) = 0 \;\forall k \geq 0$;*
(c) *the set $Z = \{z \in D : f(z) = 0\}$ has a limit point in D.*

Proof The implications (a) $\Rightarrow$ (b) and (b) $\Rightarrow$ (c) are obvious. We show the inverse implications.

(c) $\Rightarrow$ (b) Let $z_0 \in D$ be a limit point of Z. Since f is continuous, we must have $f(z_0) = 0$. If z_0 is a zero of f of finite order m, then $f(z) = g(z)(z - z_0)^m$ with g analytic at z_0 and $g(z_0) \neq 0$. On the other hand, since g is continuous and non-zero at z_0, by Theorem 3.28 $\exists \delta > 0$ such that $g(z) \neq 0 \;\forall z \in B(z_0, \delta)$. Therefore, $f(z) \neq 0$ for $0 < |z - z_0| < \delta$ which contradicts the fact that z_0 is a limit point of Z. We must therefore conclude that z_0 is a zero of order m infinite, that is, $f^{(k)}(z_0) = 0$ $\forall k \geq 0$.

(b) $\Rightarrow$ (a) Let $A = \{z \in D : \; f^{(k)}(z) = 0 \; \forall k \geq 0\}$. By hypothesis $A \neq \emptyset$. It is then sufficient to show that A is simultaneously open and closed in D, to conclude, since D is connected, that $A = D$ and therefore f is identically zero in D. We show that $\overline{A} \subset A$ and therefore that A is closed. Consider an arbitrary $z \in \overline{A}$ and let (z_j) be a sequence of points $z_j \in A$ convergent to z. Since $f^{(k)}$ is, for every k, a continuous function, it follows that $f^{(k)}(z) = \lim_{j \to \infty} f^{(k)}(z_j) = 0$. Therefore, $z \in A$. To show that A is open in D, consider an arbitrary point $z \in A$. Since $z \in D$ and D is open, $\exists r > 0$ such that $B(z, r) \subset D$. Then $\forall w \in B(z, r)$ we have $f(w) = \sum_{n=0}^{\infty} a_n(w - z)^n = 0$ since for every n we have $a_n = f^{(n)}(z)/n! = 0$. Differentiating this series k times, with k arbitrary, it follows that $f^{(k)}(w) = 0$, that is, $w \in A$. We conclude that $B(z, r) \subset A$ and therefore A is open. $\qquad\square$

Corollary 9.16 (Identity Theorem) *Let f and g be analytic in $D \subset \mathbb{C}$ open and connected. Then $f = g$ in D if and only if the set $\{z \in D : \; f(z) = g(z)\}$ has a limit point in D.*

Proof It follows from Theorem 9.15 applied to the function $h = f - g$. $\qquad\square$

Corollary 9.17 (Zeroes of Non-constant Analytic Functions) *Let f be analytic and non-constant in $D \subset \mathbb{C}$ open and connected. Then every zero of f in D is isolated and of finite order.*

Proof Let $z_0 \in D$ be such that $f(z_0) = 0$. Since f is not constant in D, it is not identically zero in D. By Theorem 9.15 it follows that the set Z of zeros of f in D has no limit points, so z_0 is isolated, namely, $\exists \delta > 0$ such that $f(z) \neq 0$ for $0 < |z - z_0| < \delta$. Furthermore, by the same theorem there must exist an integer $m \geq 1$ such that $f^{(k)}(z_0) = 0$ for $0 \leq k < m$ while $f^{(m)}(z_0) \neq 0$. As a consequence z_0 is a zero of order m. $\qquad\square$

Example 9.18 Consider the functions $f(z) = \log(z^2)$ and $g(z) = 2\log(z)$, in both cases meaning by log the principal branch of the logarithm. These functions are analytic in $D_f = \{z \in \mathbb{C} : \; z \neq it, \; t \in (-\infty, +\infty)\}$ and $D_g = \{z \in \mathbb{C} : \; z \neq t, \; t \in (-\infty, 0]\}$, respectively. Let $D = \{z \in \mathbb{C} : \; \mathrm{Re}\, z > 0\} \subset D_f \cap D_g$. Note that D is open and connected and f and g are analytic in D. At any point $x > 0$ of the positive real semi-axis we have $\log(x^2) = \ln(x^2) = 2\ln(x) = 2\log(x)$. By the Identity Theorem 9.16 we conclude that

$$\log(z^2) = 2\log(z), \qquad \mathrm{Re}\, z > 0.$$

This result does not hold in the third quadrant ($\mathrm{Re}\, z < 0, \mathrm{Im}\, z > 0$) and fourth quadrant ($\mathrm{Re}\, z < 0, \mathrm{Im}\, z < 0$) where it is respectively $\log(z^2) - 2\log(z) = \mp 2\pi i$.

9.4   Zeros and Poles

Theorem 9.19 *Let* $f(z) = p(z)/q(z)$ *with* p *and* q *analytic functions at* z_0. *If* $p(z_0) \neq 0$ *and* q *has a zero of order* m *at* z_0, *then* f *has a pole of order* m *at* z_0.

Proof Since q has a zero of order m at z_0 and zeros of finite order are isolated, $\exists \delta > 0$ such that $q(z) \neq 0 \ \forall z \in A(z_0, 0, \delta)$. Therefore, $f(z) = p(z)/q(z)$ has an isolated singularity at z_0. Furthermore, since $q(z) = g(z)(z - z_0)^m$ with g analytic at z_0 and $g(z_0) \neq 0$, we have

$$f(z) = \frac{p(z)}{g(z)(z - z_0)^m} = \frac{h(z)}{(z - z_0)^m},$$

where $h(z) = p(z)/g(z)$ is analytic and non-zero at z_0. By Theorem 9.12 we conclude that z_0 is a pole of order m for f. $\qquad\square$

From the result of the previous theorem, one could, in principle, also calculate the residue of f at z_0 using the formula

$$\operatorname*{Res}_{z=z_0} f(z) = \frac{h^{(m-1)}(z_0)}{(m - 1)!}.$$

However, this is only possible if one knows the function g that defines $h = p/g$. In the particular case $m = 1$ one has the following simple result.

Theorem 9.20 *Let* $f(z) = p(z)/q(z)$ *with* p *and* q *analytic functions at* z_0. *If* $p(z_0) \neq 0$, $q(z_0) = 0$ *and* $q'(z_0) \neq 0$, *then* f *has a simple pole at* z_0 *and*

$$\operatorname*{Res}_{z=z_0} f(z) = \frac{p(z_0)}{q'(z_0)}. \tag{9.9}$$

Proof By hypothesis q has a simple zero at z_0 and so $q(z) = g(z)(z - z_0)$ with g analytic at z_0 and $g(z_0) \neq 0$. By Theorem 9.19 f has a simple pole at z_0 and therefore $f(z) = h(z)/(z - z_0)$ with h analytic at z_0 and $h(z_0) \neq 0$. From this last expression and observing that $h(z) = p(z)/g(z)$, it follows

$$\operatorname*{Res}_{z=z_0} f(z) = h(z_0) = \frac{p(z_0)}{g(z_0)}.$$

Differentiating $q(z) = g(z)(z - z_0)$ we obtain $q'(z) = g'(z)(z - z_0) + g(z)$, which evaluated at z_0 gives $q'(z_0) = g(z_0)$ and hence the statement. $\qquad\square$

The formula for the residue of f at z_0 in the case $q(z)$ has a double zero at z_0 and therefore f a double pole is illustrated in Exercise 9.30.

Example 9.21 Calculate the residue of $f(z) = z/(z^4 + 4)$ at $z_0 = 1 + i$. Given $f(z) = p(z)/q(z)$ with $p(z) = z$ and $q(z) = z^4 + 4$, we have that p and q are analytic at z_0 with $p(z_0) = z_0$, $q(z_0) = 0$ and $q'(z_0) = 4z_0^3$. By Theorem 9.20, we have therefore

$$\operatorname*{Res}_{z=z_0} \frac{z}{z^4 + 4} = \frac{z_0}{4z_0^3} = \frac{1}{4z_0^2} = \frac{1}{8e^{i\pi/2}} = -\frac{i}{8}.$$

9.5   Behavior Near Isolated Singularities

The behavior of an analytic function near an isolated singular point is established by the following theorems.

Theorem 9.22 *Let z_0 be a pole of f, then*

$$\lim_{z \to z_0} f(z) = \infty. \tag{9.10}$$

Proof Assuming that the pole is of order m, by Theorem 9.12 we have $f(z) = g(z)/(z - z_0)^m$ with g analytic at z_0 and $g(z_0) \neq 0$. Therefore

$$\lim_{z \to z_0} \frac{1}{f(z)} = \lim_{z \to z_0} \frac{(z - z_0)^m}{g(z)} = \frac{0}{g(z_0)} = 0,$$

from which, using Theorem 3.9, the statement follows. $\square$

Theorem 9.23 *Let z_0 be a removable singularity of f, then $\exists \varepsilon > 0$ such that f is analytic and bounded in $A(z_0, 0, \varepsilon)$.*

Proof Since z_0 is a removable singularity of f, $\exists r > 0$ such that f is analytic in $A(z_0, 0, r)$ and in such domain the following Laurent series expansion holds

$$f(z) = \sum_{n=0}^{\infty} a_n (z - z_0)^n.$$

Define $g(z) = \sum_{n=0}^{\infty} a_n (z - z_0)^n$ for $z \in B(z_0, r)$. We have $g(z) = f(z)$ for $z \in B(z_0, r) \setminus \{z_0\}$. Choosing $0 < \varepsilon < r$, the function g is analytic, hence continuous, in the compact $\overline{B}(z_0, \varepsilon)$. It follows that g is bounded in $\overline{B}(z_0, \varepsilon)$ and therefore f is bounded in $A(z_0, 0, \varepsilon)$. $\square$

Lemma 9.24 (Riemann) *Let f be analytic and bounded in $A(z_0, 0, \varepsilon)$ with $\varepsilon > 0$. If f is non-analytic at z_0 then it has a removable singularity in z_0.*

Proof Suppose that f is non-analytic at z_0. Since f is analytic in $A(z_0, 0, \varepsilon)$, in this domain the following Laurent series expansion holds

$$f(z) = \sum_{n=0}^{\infty} a_n(z - z_0)^n + \sum_{n=1}^{\infty} \frac{b_n}{(z - z_0)^n}.$$

The coefficients of the main part of this series can be calculated with the formula

$$b_n = \frac{1}{2\pi i} \int_{\gamma} f(z)(z - z_0)^{n-1} dz, \qquad n = 1, 2, 3, \ldots,$$

where γ is the positively oriented circle $|z - z_0| = \rho$ with arbitrary $\rho < \varepsilon$. Since f is bounded in $A(z_0, 0, \varepsilon)$, that is, $\exists M > 0$ such that $|f(z)| \leq M$ for $0 < |z - z_0| < \varepsilon$, we have

$$|b_n| \leq \frac{1}{2\pi} M\rho^{n-1} 2\pi\rho = M\rho^n, \qquad n = 1, 2, 3, \ldots.$$

Due to the arbitrariness of ρ it must be $b_n = 0 \ \forall n \geq 1$. We conclude that z_0 is a removable singularity of f. $\qquad\qquad\Box$

Theorem 9.25 (Casorati-Weierstrass) *Let z_0 be an essential singularity of f, then $\forall \delta > 0$ the set $f(A(z_0, 0, \delta))$ is dense in $\mathbb{C}$.*

Proof Let f be analytic in $A(z_0, 0, r)$ with $r > 0$. We need to show that $\forall \delta > 0$, provided that $\delta < r$, then $\overline{f(A(z_0, 0, \delta))} = \mathbb{C}$. In other words, given an arbitrary $w \in \mathbb{C}$ and an arbitrary $\varepsilon > 0$, we need to show that for an arbitrary δ, with $r > \delta > 0$, it is possible to find a point $z \in A(z_0, 0, \delta)$ such that $|f(z) - w| < \varepsilon$. Let us argue by contradiction and suppose that this is not possible. We then assume that there exist $w \in \mathbb{C}$ and $\varepsilon > 0$ such that $|f(z) - w| \geq \varepsilon \ \forall z \in A(z_0, 0, \delta)$ for some δ, with $r > \delta > 0$. Let $g(z) = 1/(f(z) - w)$, then g is analytic and bounded in $A(z_0, 0, \delta)$. By Lemma 9.24, z_0 is a removable singularity of g. Let us then define $g(z_0)$ such that g is analytic at z_0. If $g(z_0) \neq 0$, it follows that $f(z) = w + 1/g(z)$ becomes analytic at z_0 if we define $f(z_0) = w + 1/g(z_0)$. In other words, z_0 is a removable singularity of f. If $g(z_0) = 0$, it follows that g, not being identically zero for $|z - z_0| < \delta$, has at z_0 a zero of order $m < \infty$. As a consequence, $f(z) = w - 1/g(z)$ has at z_0 a pole of order m. In both cases, there is a contradiction with the hypothesis that z_0 is an essential singularity of f. $\qquad\qquad\Box$

Exercises

9.1 Determine the value of the integral

$$\int_{\gamma} \frac{1}{\sin(z^{-1})} dz,$$

where γ is the positively oriented circle with center 0 and radius $1/4$.

Fig. 9.1 Non-analyticity
regions of function f and
integration paths

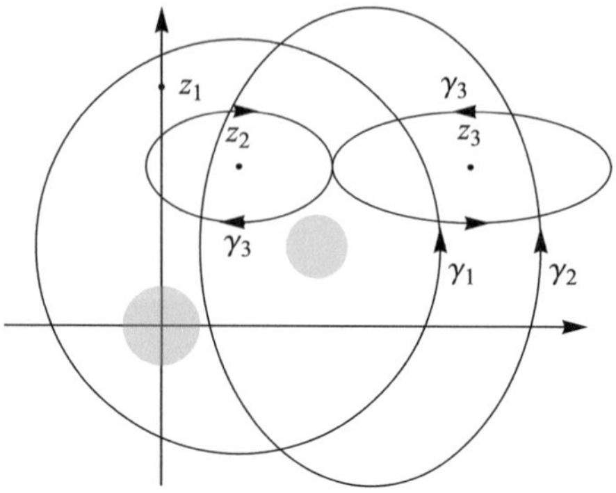

9.2 A function f is analytic everywhere in the complex plane with the exception of
the isolated singularities at z_1, z_2 and z_3 and the shaded disks indicated in Fig. 9.1,
within which there is no information about the behavior of f. Express the integral
of f along the curves γ_1, γ_2 and γ_3 in terms of suitable residues. If the information
is not sufficient to evaluate the integral, answer that the integral is not computable.

9.3 Calculate the integral

$$\int_\gamma \frac{1}{\sin(z^{-2})}\,dz,$$

where γ is the circle of unit radius centered at the origin and traveled counterclockwise.

9.4 Calculate the integral

$$\int_\gamma \frac{1}{\sin(z^{-1})}\,dz,$$

where γ is the circle of radius $1/5$ centered at the origin and traveled counterclockwise.

9.5 Calculate the integral

$$\int_\gamma \frac{z\,dz}{(z-2)(z^2+1)}, \qquad \gamma(t) = 1 + i + 2e^{it}, \quad 0 \le t \le 2\pi.$$

9.6 Calculate the integral

$$\int_\gamma \frac{z^2 dz}{e^{\pi z}+1}, \qquad \gamma(t) = 1 + 2i + 2e^{it}, \quad 0 \le t \le 2\pi.$$

9.7 Let γ_n be the boundary of the square with vertices $(-1-\mathrm{i})a_n$, $(1-\mathrm{i})a_n$, $(1+\mathrm{i})a_n$ and $(-1+\mathrm{i})a_n$, with $a_n = (n+1/2)\pi$ and $n \in \mathbb{N}$. Show that

$$\lim_{n\to\infty} \int_{\gamma_n} \frac{1}{z^2 \sin z}\,\mathrm{d}z = 0.$$

Then calculate the same integral for generic n in terms of suitable residues. Finally, consider the series obtained for $n \to \infty$ and determine its sum.

9.8 Classify all the isolated singularities of the following functions and calculate their corresponding residues:

$$\text{(a)}\ \ \cos\left(\frac{1}{z-2}\right), \qquad \text{(b)}\ \ z^3 \cos\left(\frac{1}{z-2}\right).$$

9.9 Determine the nature of the isolated singularity at $z = 0$ of the following function and calculate its corresponding residue

$$f(z) = \frac{1}{(e^z - 1)\sin z}.$$

9.10 Assuming the principal branch for multi-valued functions, determine, up to order z^4 included, the expansion in power series around $z = 0$ of the function

$$f(z) = \frac{1}{z^2}\log(\cos z).$$

Classify the nature of the singularity of $f(z)$ at $z = 0$.

9.11 Each of the following functions (consider the principal branch for the multi-valued ones) has a singularity at $z = 0$. Classify its nature and, if applicable, calculate the corresponding residue:

$$\text{(a)}\ \ \frac{\sqrt{z}}{z}, \qquad \text{(b)}\ \ \frac{z}{1-\cos z}, \qquad \text{(c)}\ \ (z^3 + 3)\exp(z^{-1}).$$

9.12 Determine the nature of the isolated singularity at $z = 0$ of the following function and calculate its corresponding residue

$$f(z) = \frac{e^z \sin z}{z(1 - \cos z)}.$$

9.13 Determine the nature of the isolated singularity at $z = 3$ of the following functions and calculate the corresponding residue:

$$\text{(a)}\ \ z \sin\left(\frac{1}{z-3}\right), \qquad \text{(b)}\ \ \frac{z}{(z-3)\sin(z-3)}.$$

9.14 Each of the following functions (consider the principal branch for the multi-valued ones) has a singularity at $z = 0$. Classify its nature and, if applicable, calculate the corresponding residue:

$$\text{(a)} \ \log(z^2), \qquad \text{(b)} \ z^3 \cos\left(\frac{1}{z^2}\right), \qquad \text{(c)} \ \frac{1}{z^4 \sinh z}.$$

9.15 Each of the following functions (consider the principal branch for the multi-valued ones) has a singularity at $z = 0$. Classify its nature and, if applicable, calculate the corresponding residue:

$$\text{(a)} \ \frac{1}{1 - \cos z}, \qquad \text{(b)} \ \frac{\sqrt{z}}{z^2}, \qquad \text{(c)} \ z \exp(z^{-2}).$$

9.16 Each of the following functions (consider the principal branch for the multi-valued ones) has a singularity at $z = 0$. Classify their nature and, if applicable, calculate the corresponding residue:

$$\text{(a)} \ \frac{3+z}{z^3 + 2z^2}, \qquad \text{(b)} \ z \log z, \qquad \text{(c)} \ \frac{1}{\log(1+z)}.$$

9.17 Classify all the isolated singularities of the following functions and calculate the corresponding residues:

$$\text{(a)} \ z \cos(z^{-1}), \qquad \text{(b)} \ z^{-1} \cos(z).$$

9.18 Each of the following functions (consider the principal branch for the multi-valued ones) has a singularity at $z = 0$. Classify its nature and, if applicable, calculate the corresponding residue:

$$\text{(a)} \ \tan\frac{1}{z}, \qquad \text{(b)} \ z^n \sin\frac{1}{z}, \quad n \in \mathbb{N}, \qquad \text{(c)} \ \sqrt{1 + \frac{1}{z}}.$$

9.19 Determine the nature of the singularity at $z = 0$ of the following functions (consider the principal branch for the multi-valued ones) and, if applicable, the value of the corresponding residue:

$$\text{(a)} \ \frac{\exp(\sin z)}{z^2}, \qquad \text{(b)} \ \frac{\sin z}{\log(\cos z)}.$$

9.20 Suppose that $f(z)$ has a simple pole at z_0 and $g(z)$ is analytic at z_0. Prove that

$$\operatorname*{Res}_{z=z_0} f(z)g(z) = g(z_0) \operatorname*{Res}_{z=z_0} f(z).$$

9.21 Determine the nature of all the isolated singularities of the following function and calculate their corresponding residue

$$f(z) = \frac{e^z}{z^2 \sin z}.$$

9.22 Determine the nature of all the isolated singularities of the following function and calculate their corresponding residue

$$f(z) = z^3 \cos\left(\frac{1}{z-2}\right).$$

9.23 Determine the nature of all the isolated singularities of the following function and calculate their corresponding residue

$$f(z) = \frac{\log(1+z)}{z \sin z}.$$

Assume the principal branch for the logarithm.

9.24 Calculate the residue at $z = 0$ of the functions

$$f_n(z) = \frac{z^{-n}}{(z-1)^2(z-2)}, \qquad n = 1, 2, 3, 4.$$

Determine a function $g(z)$, multi-valued but analytic at $z = 0$, such that $f_2(z)g(z)$ has a double pole at $z = 0$ with zero residue.

9.25 Find the residue at $z = -2$ of the function

$$f(z) = \frac{(z+2)^4}{z-3} \sin\left(\frac{1}{z+2}\right).$$

9.26 Assuming that $f(z)$ and $p(z)$ are analytic at z_0 and that $f(z)$ has a zero of order m at z_0, determine

$$\operatorname*{Res}_{z=z_0}\left(\frac{f'(z)}{f(z)}p(z)\right).$$

9.27 Assuming that $f(z) = 1/q(z)^2$ with $q(z)$ analytic at z_0 and with $q(z_0) = 0$ and $q'(z_0) \neq 0$, prove that f has a pole of order 2 at z_0 and that

$$\operatorname*{Res}_{z=z_0} f(z) = -\frac{q''(z_0)}{q'(z_0)^3}.$$

9.28 Suppose that the function $f(z)$ has a pole of order m in z_0. Prove that

$$\operatorname*{Res}_{z=z_0} f(z) = \lim_{z \to z_0} \frac{1}{(m-1)!} \frac{\mathrm{d}^{m-1}}{\mathrm{d}z^{m-1}} \left((z - z_0)^m f(z) \right).$$

9.29 Suppose the identity $\sin^2 z + \cos^2 z = 1$ is proven $\forall z \in \mathbb{R}$. Prove that the identity is valid $\forall z \in \mathbb{C}$.

9.30 Let $f(z) = p(z)/q(z)$ with p and q analytic functions at z_0. Prove that if p is nonzero at z_0 and q has a double zero at z_0, then f has a double pole at z_0 and we have

$$\operatorname*{Res}_{z=z_0} f(z) = \frac{p'(z_0)q''(z_0)/2 - p(z_0)q'''(z_0)/6}{(q''(z_0)/2)^2}.$$

9.31 Let $f(z)$ be an analytic function at $|z| < 2$ and such that

$$\int_{|z|=1} \frac{f(z)}{nz - 1} \mathrm{d}z = 0, \qquad \forall n \in \mathbb{N}.$$

Prove that $f(z) = 0 \; \forall z \in B(0, 2)$.

9.32 Assuming the principal branch for the logarithm, expand in the Taylor series around $z = 0$ the function

$$f(z) = \log \frac{1 + z}{1 - z}.$$

Chapter 10
Applications of Residues

Abstract Improper integrals: convergence and Cauchy principal value. Integrals of trigonometric functions. Integrals of rational functions. Integrals of rational functions multiplied by a trigonometric function. Jordan's lemma. Rectangular integration paths. Indented paths around a simple pole. Indented paths around a branch point. Paths along a branch line. Other paths. Bromwich integrals: inverse Laplace transform.

10.1 Improper Integrals: Convergence and Cauchy Principal Value

The theory of residues has many applications, a classic example is the calculation of real integrals, proper or improper, by using Theorem 9.4. In the following we will illustrate different solution schemes depending on the different classes of integrals considered. We begin by recalling the following definitions valid for functions of real variable with real or complex values.

Definition 10.1 (Improper Integral) An improper integral is defined as an integral in which either the integration interval diverges or the integrand function has a divergence. In these cases the integral is defined by a limit as specified below:

$$\int_a^\infty f(x)\mathrm{d}x = \lim_{R\to\infty} \int_a^R f(x)\mathrm{d}x, \tag{10.1}$$

$$\int_\infty^b f(x)\mathrm{d}x = \lim_{R\to\infty} \int_R^b f(x)\mathrm{d}x, \tag{10.2}$$

$$\int_{-\infty}^{+\infty} f(x)\mathrm{d}x = \lim_{R_1,R_2\to\infty} \int_{-R_1}^{R_2} f(x)\mathrm{d}x, \tag{10.3}$$

C. Presilla, *Complex Functions of a Variable*, La Matematica per il 3+2 179,
https://doi.org/10.1007/978-3-032-12494-4_10

$$\int_a^b f(x)\mathrm{d}x = \lim_{r\to 0^+} \int_a^{b-r} f(x)\mathrm{d}x, \qquad \lim_{x\to b^-} |f(x)| = \infty, \tag{10.4}$$

$$\int_a^b f(x)\mathrm{d}x = \lim_{r\to 0^+} \int_{a+r}^{b} f(x)\mathrm{d}x, \qquad \lim_{x\to a^+} |f(x)| = \infty, \tag{10.5}$$

$$\int_a^b f(x)\mathrm{d}x = \lim_{r_1,r_2\to 0^+} \left(\int_a^{c-r_1} f(x)\mathrm{d}x + \int_{c+r_2}^{b} f(x)\mathrm{d}x \right), \qquad \lim_{x\to c} |f(x)| = \infty, \tag{10.6}$$

where $a < c < b$. An improper integral is said to converge if the limit that defines it exists and the result is a finite number (real or complex).

Definition 10.2 (Cauchy Principal Value) For the improper integrals (10.3) and (10.6), we define the Cauchy principal value as the limit

$$\mathrm{PV} \int_{-\infty}^{+\infty} f(x)\mathrm{d}x = \lim_{R\to\infty} \int_{-R}^{R} f(x)\mathrm{d}x,$$

$$\mathrm{PV} \int_a^b f(x)\mathrm{d}x = \lim_{r\to 0^+} \left(\int_a^{c-r} f(x)\mathrm{d}x + \int_{c+r}^{b} f(x)\mathrm{d}x \right), \qquad \lim_{x\to c} |f(x)| = \infty.$$

An improper integral is said to converge according to the Cauchy principal value if the limit specified above exists and is finite.

Obviously if an improper integral is convergent then it is also convergent according to the Cauchy principal value. The vice versa is generally false. Consider, for example,

$$\mathrm{PV} \int_{-\infty}^{+\infty} \frac{x}{x^2+1}\mathrm{d}x = 0, \qquad \mathrm{PV} \int_{-1}^{1} \frac{1}{x}\mathrm{d}x = 0,$$

whose corresponding improper integrals are not convergent.

10.2   Integrals of Trigonometric Functions

Theorem 10.3 *Let $f(\cos\theta, \sin\theta)$ be a rational function of $\cos\theta$ and $\sin\theta$ integrable for $0 \le \theta \le 2\pi$. Then the integral of f between 0 and 2π is equivalent to the complex integral*

$$\int_0^{2\pi} f(\cos\theta, \sin\theta)\mathrm{d}\theta = \int_\gamma f\left(\frac{z+z^{-1}}{2}, \frac{z-z^{-1}}{2i} \right) \frac{\mathrm{d}z}{iz}, \tag{10.7}$$

where γ is the circle centered at the origin, of unit radius and positively oriented.

Proof Choosing for the circle γ the parametric representation $\gamma(\theta) = e^{i\theta}$, $0 \le \theta \le 2\pi$, and using Euler's formulas $\cos\theta = (e^{i\theta} + e^{-i\theta})/2$ and $\sin\theta = (e^{i\theta} - e^{-i\theta})/2i$, we have

$$
\int_0^{2\pi} f(\cos\theta, \sin\theta)\, d\theta
$$

$$
= \int_0^{2\pi} f\left(\frac{e^{i\theta} + e^{-i\theta}}{2}, \frac{e^{i\theta} - e^{-i\theta}}{2i}\right) \frac{1}{ie^{i\theta}} ie^{i\theta}\, d\theta
$$

$$
= \int_0^{2\pi} f\left(\frac{\gamma(\theta) + \gamma(\theta)^{-1}}{2}, \frac{\gamma(\theta) - \gamma(\theta)^{-1}}{2i}\right) \frac{1}{i\gamma(\theta)} \gamma'(\theta)\, d\theta
$$

$$
= \int_\gamma f\left(\frac{z + z^{-1}}{2}, \frac{z - z^{-1}}{2i}\right) \frac{dz}{iz},
$$

which is formula (10.7). $\square$

Example 10.4 Show that

$$
\int_0^{2\pi} \frac{d\theta}{1 + a\cos\theta} = \frac{2\pi}{\sqrt{1 - a^2}}, \qquad -1 < a < 1.
$$

The integral is equivalent to

$$
\int_0^{2\pi} \frac{d\theta}{1 + a\cos\theta} = \int_\gamma \frac{1}{1 + a\frac{z + z^{-1}}{2}} \frac{dz}{iz} = \frac{2}{ia} \int_\gamma \frac{1}{z^2 + 2z/a + 1}\, dz,
$$

where $\gamma(\theta) = e^{i\theta}$, $0 \le \theta \le 2\pi$. For $|a| < 1$ the equation $z^2 + 2z/a + 1 = 0$ has two distinct real solutions $z_\pm = (-1 \pm \sqrt{1 - a^2})/a$ one of which lies outside the circle γ, $|z_-| > 1$, and the other within it, $|z_+| < 1$, as can be seen by observing that $z_+ z_- = 1$. Therefore, the complex function to be integrated along γ is analytic on γ and within it except for the simple pole at z_+. By the residue theorem

$$
\frac{2}{ia} \int_\gamma \frac{1}{z^2 + \frac{2}{a}z + 1}\, dz = \frac{2}{ia} 2\pi i \operatorname*{Res}_{z=z_+} \frac{1}{z^2 + \frac{2}{a}z + 1}
$$

$$
= \frac{4\pi}{a} \operatorname*{Res}_{z=z_+} \frac{1/(z - z_-)}{z - z_+}
$$

$$
= \frac{4\pi}{a} \frac{1}{z_+ - z_-}
$$

$$
= \frac{2\pi}{\sqrt{1 - a^2}}.
$$

10.3   Integrals of Rational Functions

Theorem 10.5 *Let $P_n(x)$ and $Q_m(x)$ be two real polynomials of degree n and m. If $m \geq n + 2$ and $Q_m(x) \neq 0 \; \forall x \in \mathbb{R}$, then*

$$\int_{-\infty}^{+\infty} \frac{P_n(x)}{Q_m(x)} dx = 2\pi i \sum_k \operatorname*{Res}_{z=z_k} \frac{P_n(z)}{Q_m(z)}, \tag{10.8}$$

where z_k are the zeros of $Q_m(z)$ contained in the complex half-plane $\operatorname{Im} z > 0$.

Proof The improper integral in question exists and therefore coincides with the corresponding principal value

$$\int_{-\infty}^{+\infty} \frac{P_n(x)}{Q_m(x)} dx = \mathrm{PV} \int_{-\infty}^{+\infty} \frac{P_n(x)}{Q_m(x)} dx = \lim_{R \to \infty} \int_{-R}^{R} \frac{P_n(x)}{Q_m(x)} dx.$$

The latter can be evaluated by integrating the complex rational function $P_n(z)/Q_m(z)$ along the positively oriented simple closed curve $\gamma = \lambda + \gamma_R$ represented by the paths

$$\lambda(x) = x, \qquad -R \leq x \leq R,$$

$$\gamma_R(t) = Re^{it}, \qquad 0 \leq t \leq \pi.$$

Assuming R sufficiently large so that all zeros of $Q_m(z)$ having positive imaginary part are contained within γ, by the residue theorem we have

$$\int_{\lambda+\gamma_R} \frac{P_n(z)}{Q_m(z)} dz = \int_{\lambda} \frac{P_n(z)}{Q_m(z)} dz + \int_{\gamma_R} \frac{P_n(z)}{Q_m(z)} dz = 2\pi i \sum_k \operatorname*{Res}_{z=z_k} \frac{P_n(z)}{Q_m(z)}.$$

The statement follows immediately by taking the limit $R \to \infty$ of this expression considering that

$$\int_{\lambda} \frac{P_n(z)}{Q_m(z)} dz = \int_{-R}^{R} \frac{P_n(\lambda(x))}{Q_m(\lambda(x))} \lambda'(x) dx = \int_{-R}^{R} \frac{P_n(x)}{Q_m(x)} dx,$$

while

$$\lim_{R \to \infty} \int_{\gamma_R} \frac{P_n(z)}{Q_m(z)} dz = 0.$$

To prove this last limit we can reason as follows. Let $P_n(z) = p_n z^n + p_{n-1} z^{n-1} + \cdots + p_0$ and $Q_m(z) = q_m z^m + q_{m-1} z^{m-1} + \cdots + q_0$ with $p_n \neq 0$ and $q_m \neq 0$, then

$$\lim_{z \to \infty} \frac{z P_n(z)}{Q_m(z)} = \lim_{z \to \infty} \frac{z z^n}{z^m} \frac{p_n + p_{n-1} z^{-1} + \cdots + p_0 z^{-n}}{q_m + q_{m-1} z^{-1} + \cdots + q_0 z^{-m}}$$

$$= \lim_{z \to \infty} \frac{z^{n+1}}{z^m} \lim_{z \to \infty} \frac{p_n + p_{n-1} z^{-1} + \cdots + p_0 z^{-n}}{q_m + q_{m-1} z^{-1} + \cdots + q_0 z^{-m}}$$

$$= 0 \, \frac{p_n}{q_m}$$

$$= 0,$$

since $m \geq n + 2$. Therefore, $\forall \varepsilon > 0$ it is possible to find a sufficiently large $R > 0$ such that $|z P_n(z)/Q_m(z)| < \varepsilon/\pi$ if $|z| \geq R$ and hence, in particular, $|P_n(z)/Q_m(z)| < \varepsilon/(\pi R)$ if $|z| = R$. Choosing the radius R of γ_R in this way, from Darboux's inequality we have

$$\left| \int_{\gamma_R} \frac{P_n(z)}{Q_m(z)} dz \right| \leq \pi R \sup_{z \in \{\gamma_R\}} \left| \frac{P_n(z)}{Q_m(z)} \right| < \pi R \frac{\varepsilon}{\pi R} = \varepsilon.$$

From the arbitrariness of ε it follows that $\int_{\gamma_R} P_n(z)/Q_m(z) dz \xrightarrow{R \to \infty} 0.$ $\qquad \square$

Note that, since the coefficients of the polynomial Q_m are real, we have $Q_m(\bar{z}) = \overline{Q_m(z)}$ and therefore the zeros of $Q_m(z)$ are complex conjugate pairs z_k and $\bar{z}_k$. The condition $Q_m(x) \neq 0 \; \forall x \in \mathbb{R}$ cannot therefore be satisfied if m is odd: in this case, in fact, at least one zero of $Q_m(z)$ must coincide with its complex conjugate, that is, it must be real. Obviously, also in the case m even $Q_m(z)$ may have real zeros, necessarily of even order.

In the above proof the integral of $P_n(z)/Q_m(z)$ can also be evaluated along the closed path $\gamma = \lambda + \gamma_R$, where $\gamma_R(t) = R e^{-it}$ with $0 \leq t \leq \pi$. Note that this path is negatively oriented and therefore the application of the residue theorem provides

$$\int_{-\infty}^{+\infty} \frac{P_n(x)}{Q_m(x)} dx = -2\pi i \sum_k \operatorname*{Res}_{z=\bar{z}_k} \frac{P_n(z)}{Q_m(z)}. \tag{10.9}$$

This result, however, coincides with the formula (10.8). In fact, we have

$$\operatorname*{Res}_{z=\bar{z}_k} \frac{P_n(z)}{Q_m(z)} = \overline{\operatorname*{Res}_{z=z_k} \frac{P_n(z)}{Q_m(z)}}$$

and therefore, observing that the value of the improper integral is real, we conclude

$$-2\pi i \sum_k \operatorname*{Res}_{z=\overline{z_k}} \frac{P_n(z)}{Q_m(z)} = 2\pi \sum_k \operatorname{Im} \operatorname*{Res}_{z=\overline{z_k}} \frac{P_n(z)}{Q_m(z)}$$

$$= -2\pi \sum_k \operatorname{Im} \operatorname*{Res}_{z=z_k} \frac{P_n(z)}{Q_m(z)}$$

$$= 2\pi i \sum_k \operatorname*{Res}_{z=z_k} \frac{P_n(z)}{Q_m(z)}.$$

In the case $m = n + 1$ and always assuming $Q_m(x) \neq 0 \; \forall x \in \mathbb{R}$, the integral of $P_n(x)/Q_m(x)$ over $\mathbb{R}$ exists in the sense of Cauchy principal value and can be evaluated by the formula

$$\mathrm{PV} \int_{-\infty}^{+\infty} \frac{P_n(x)}{Q_m(x)}\,dx = \lim_{R\to\infty} \int_{-R}^{R} \frac{P_n(x)}{Q_m(x)}\,dx$$

$$= \lim_{R\to\infty} \left(\lim_{a\to 0} \int_{-R}^{R} \frac{P_n(x)}{Q_m(x)} \cos(ax)\,dx \right)$$

$$= \lim_{a\to 0} \left(\lim_{R\to\infty} \int_{-R}^{R} \frac{P_n(x)}{Q_m(x)} \cos(ax)\,dx \right)$$

$$= \lim_{a\to 0} \int_{-\infty}^{+\infty} \frac{P_n(x)}{Q_m(x)} \cos(ax)\,dx. \tag{10.10}$$

The convergence of the improper integral in the last line and the computation of its value are discussed in the next section.

Example 10.6 Show that

$$\int_{-\infty}^{+\infty} \frac{x}{(x^2 + 1)(x^2 + 2x + 2)}\,dx = -\frac{\pi}{5}.$$

Consider the complex function $f(z) = z/((z^2 + 1)(z^2 + 2z + 2))$ (Fig. 10.1). It has simple poles at $z = \pm i$ and $z = -1 + \sqrt{1 - 2} = -1 \pm i$ with residues

$$\operatorname*{Res}_{z=\pm i} f(z) = \left. \frac{z}{(z - (\mp i))(z^2 + 2z + 2)} \right|_{z=\pm i} = \frac{\pm i}{\pm 2i(-1 \pm 2i + 2)} = \frac{1 \mp 2i}{10},$$

$$\operatorname*{Res}_{z=-1\pm i} f(z) = \left. \frac{z}{(z^2 + 1)(z - (-1 \mp i))} \right|_{z=-1\pm i} = \frac{-1 \pm i}{(1 \mp 2i)(\pm 2i)} = \frac{-1 \pm 3i}{10}.$$

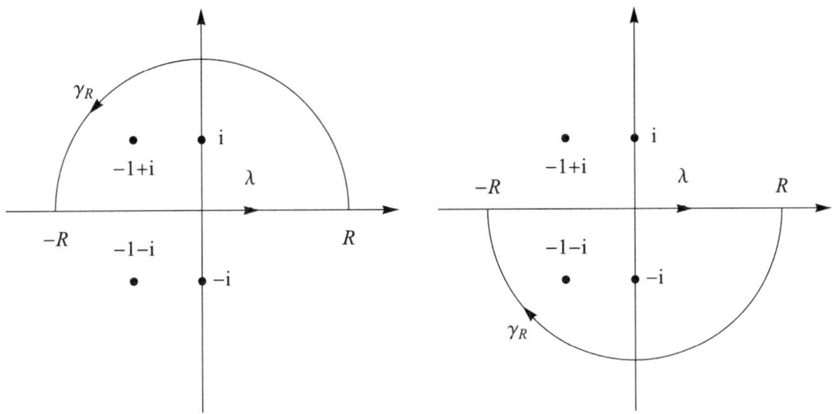

Fig. 10.1 Different closed paths $\gamma = \lambda + \gamma_R$ used to integrate the rational function $f(z) = z/((z^2 + 1)(z^2 + 2z + 1))$

Applying formula (10.8) or (10.9) we obtain

$$
\int_{-\infty}^{+\infty} \frac{x}{(x^2 + 1)(x^2 + 2x + 2)}\,\mathrm{d}x = \pm 2\pi\mathrm{i}\left(\frac{1 \mp 2\mathrm{i}}{10} + \frac{-1 \pm 3\mathrm{i}}{10}\right)
$$

$$
= \pm 2\pi\mathrm{i}\frac{\pm\mathrm{i}}{10}
$$

$$
= -\frac{\pi}{5}.
$$

10.4 Integrals of Rational and Trigonometric Functions

In this section we will consider improper integrals of the type

$$
\int_{-\infty}^{+\infty} R(x)\mathrm{e}^{\mathrm{i}ax}\,\mathrm{d}x = \int_{-\infty}^{+\infty} R(x)\cos(ax)\,\mathrm{d}x + \mathrm{i}\int_{-\infty}^{+\infty} R(x)\sin(ax)\,\mathrm{d}x,
$$

where $R(x)$ is a rational function and a is a real constant.

Lemma 10.7 (Jordan Lemma) *Let f be analytic in $A = \{z \in \mathbb{C} : |z| > R_0,\ \mathrm{Im}\,z \geq 0\}$ and consider the semicircular path $\gamma_R(\theta) = R\mathrm{e}^{\mathrm{i}\theta}, 0 \leq \theta \leq \pi$, with $R > R_0$ so that $\{\gamma_R\} \subset A$. If $\exists M_R > 0$ such that $|f(z)| \leq M_R\ \forall z \in \{\gamma_R\}$ and furthermore $\lim_{R\to\infty} M_R = 0$, then*

$$
\lim_{R\to\infty} \int_{\gamma_R} f(z)\mathrm{e}^{\mathrm{i}az}\,\mathrm{d}z = 0, \qquad a > 0. \tag{10.11}
$$

Proof The modulus of the integral over γ_R can be upper bounded as follows

$$\left| \int_{\gamma_R} f(z)\mathrm{e}^{iaz}\mathrm{d}z \right| = \left| \int_0^\pi f(R\mathrm{e}^{i\theta})\mathrm{e}^{iaR\mathrm{e}^{i\theta}} iR\mathrm{e}^{i\theta}\mathrm{d}\theta \right|$$

$$\leq \int_0^\pi \left| f(R\mathrm{e}^{i\theta})\mathrm{e}^{iaR\mathrm{e}^{i\theta}} iR\mathrm{e}^{i\theta} \right| \mathrm{d}\theta$$

$$\leq M_R R \int_0^\pi \mathrm{e}^{-aR\sin\theta}\mathrm{d}\theta$$

$$= 2M_R R \int_0^{\pi/2} \mathrm{e}^{-aR\sin\theta}\mathrm{d}\theta. \qquad (10.12)$$

Note that for $\theta \in [0, \pi/2]$ we have $\sin\theta \geq 2\theta/\pi$. Therefore, since $aR > 0$, the following Jordan's inequality holds

$$\int_0^{\pi/2} \mathrm{e}^{-aR\sin\theta}\mathrm{d}\theta \leq \int_0^{\pi/2} \mathrm{e}^{-aR2\theta/\pi}\mathrm{d}\theta$$

$$= \frac{\mathrm{e}^{-2aR\theta/\pi}}{-2aR/\pi}\Bigg|_0^{\pi/2}$$

$$= \frac{\pi}{2aR}\left(1 - \mathrm{e}^{-aR}\right). \qquad (10.13)$$

Inserting the inequality (10.13) into the upper bound (10.12) we conclude

$$\left| \int_{\gamma_R} f(z)\mathrm{e}^{iaz}\mathrm{d}z \right| \leq \frac{\pi M_R}{a}\left(1 - \mathrm{e}^{-aR}\right) \xrightarrow{R\to\infty} 0,$$

hence the statement. $\qquad\qquad\qquad\qquad\qquad\qquad\qquad\qquad\qquad\qquad\qquad\qquad\qquad$ $\square$

Theorem 10.8 *Let $P_n(x)$ and $Q_m(x)$ be two real polynomials of degree n and m. If $m \geq n + 1$ and $Q_m(x) \neq 0\ \forall x \in \mathbb{R}$, then $\forall a > 0$ we have*

$$\mathrm{PV}\int_{-\infty}^{+\infty} \frac{P_n(x)}{Q_m(x)}\mathrm{e}^{iax}\mathrm{d}x = 2\pi i \sum_k \operatorname*{Res}_{z=z_k} \frac{P_n(z)}{Q_m(z)}\mathrm{e}^{iaz}, \qquad (10.14)$$

where z_k are the zeros of $Q_m(z)$ contained in the complex half-plane $\operatorname{Im} z > 0$.

Proof Reason as in the case of the formula (10.8). The only difference is that the integral over γ_R of $P_n(z)\mathrm{e}^{iaz}/Q_m(z)$ vanishes in the limit $R \to \infty$ by virtue of Jordan's lemma. $\qquad\qquad\qquad\qquad\qquad\qquad\qquad\qquad\qquad\qquad\qquad$ $\square$

Using a different integration path in the complex plane, it is possible to prove formula (10.14) together with the convergence of the improper integral in the ordinary sense.

Theorem 10.9 *Let $P_n(x)$ and $Q_m(x)$ be two real polynomials of degree n and m. If $m \geq n + 1$ and $Q_m(x) \neq 0 \; \forall x \in \mathbb{R}$, then $\forall a > 0$ we have*

$$\int_{-\infty}^{+\infty} \frac{P_n(x)}{Q_m(x)} e^{iax} dx = 2\pi i \sum_k \operatorname*{Res}_{z=z_k} \frac{P_n(z)}{Q_m(z)} e^{iaz}, \tag{10.15}$$

where z_k are the zeros of $Q_m(z)$ contained in the complex half-plane $\operatorname{Im} z > 0$.

Proof Integrate the function $P_n(z)e^{iaz}/Q_m(z)$ along the boundary of the square $\gamma = \lambda_1 + \lambda_2 + \lambda_3 + \lambda_4$ represented by the four straight paths

$$\lambda_1(x) = x, \qquad -R_1 \leq x \leq R_2,$$

$$\lambda_2(y) = R_2 + iy, \qquad 0 \leq y \leq R_1 + R_2,$$

$$\lambda_3(x) = x + i(R_1 + R_2), \qquad R_2 \geq x \geq -R_1,$$

$$\lambda_4(y) = -R_1 + iy, \qquad R_1 + R_2 \geq y \geq 0.$$

Assuming $R_1, R_2 > 0$ sufficiently large so that all zeros of $Q_m(z)$ having positive imaginary part are contained within γ, the residue theorem gives

$$\int_\gamma \frac{P_n(z)}{Q_m(z)} e^{iaz} dz = \sum_{j=1}^{4} \int_{\lambda_j} \frac{P_n(z)}{Q_m(z)} e^{iaz} dz = 2\pi i \sum_k \operatorname*{Res}_{z=z_k} \frac{P_n(z)}{Q_m(z)} e^{iaz}.$$

For the integrals along the individual paths forming γ we have

$$\int_{\lambda_1} \frac{P_n(z)}{Q_m(z)} e^{iaz} dz = \int_{-R_1}^{R_2} \frac{P_n(\lambda_1(x))}{Q_m(\lambda_1(x))} e^{ia\lambda_1(x)} \lambda_1'(x) dx$$

$$= \int_{-R_1}^{R_2} \frac{P_n(x)}{Q_m(x)} e^{iax} dx,$$

$$\int_{\lambda_2} \frac{P_n(z)}{Q_m(z)} e^{iaz} dz = \int_{0}^{R_1+R_2} \frac{P_n(\lambda_2(y))}{Q_m(\lambda_2(y))} e^{ia\lambda_2(y)} \lambda_2'(y) dy$$

$$= \int_{0}^{R_1+R_2} \frac{P_n(R_2 + iy)}{Q_m(R_2 + iy)} e^{iaR_2} e^{-ay} i \, dy,$$

$$\int_{\lambda_3} \frac{P_n(z)}{Q_m(z)} e^{iaz} dz = \int_{R_2}^{-R_1} \frac{P_n(\lambda_3(x))}{Q_m(\lambda_3(x))} e^{ia\lambda_3(x)} \lambda_3'(x) dx$$

$$= \int_{R_2}^{-R_1} \frac{P_n(x + i(R_1 + R_2))}{Q_m(x + i(R_1 + R_2))} e^{iax} e^{-a(R_1+R_2)} dx,$$

$$\int_{\lambda_4} \frac{P_n(z)}{Q_m(z)} e^{iaz} dz = \int_{R_1+R_2}^{0} \frac{P_n(\lambda_4(y))}{Q_m(\lambda_4(y))} e^{ia\lambda_4(y)} \lambda_4'(y) dy$$

$$= \int_{R_1+R_2}^{0} \frac{P_n(-R_1 + iy)}{Q_m(-R_1 + iy)} e^{-iaR_1} e^{-ay} i\, dy.$$

Since $m \geq n + 1$, it turns out that $P_n(z)/Q_m(z) \xrightarrow{z \to \infty} 0$ and therefore $\forall \varepsilon > 0$ $\exists R(\varepsilon) > 0$ such that $|P_n(z)/Q_m(z)| < \varepsilon a$ if $|z| \geq R$. Then, choosing $R_1, R_2 > R$, the moduli of the integrals on λ_j, with $j = 2, 3, 4$, can be upper bounded as follows:

$$\left| \int_{\lambda_2} \frac{P_n(z)}{Q_m(z)} e^{iaz} dz \right| \leq \int_{0}^{R_1+R_2} \left| \frac{P_n(R_2 + iy)}{Q_m(R_2 + iy)} e^{iaR_2} e^{-ay} i \right| dy$$

$$< \int_{0}^{R_1+R_2} \varepsilon a e^{-ay} dy$$

$$= \varepsilon \left(1 - e^{-a(R_1+R_2)} \right),$$

$$\left| \int_{\lambda_3} \frac{P_n(z)}{Q_m(z)} e^{iaz} dz \right| \leq \int_{-R_1}^{R_2} \left| \frac{P_n(x + i(R_1 + R_2))}{Q_m(x + i(R_1 + R_2))} e^{iax} e^{-a(R_1+R_2)} \right| dx$$

$$< \int_{-R_1}^{R_2} \varepsilon a e^{-a(R_1+R_2)} dx$$

$$= \varepsilon a (R_1 + R_2) e^{-a(R_1+R_2)},$$

$$\left| \int_{\lambda_4} \frac{P_n(z)}{Q_m(z)} e^{iaz} dz \right| \leq \int_{0}^{R_1+R_2} \left| \frac{P_n(-R_1 + iy)}{Q_m(-R_1 + iy)} e^{-iaR_1} e^{-ay} i \right| dy$$

$$< \int_{0}^{R_1+R_2} \varepsilon a e^{-ay} dy$$

$$= \varepsilon \left(1 - e^{-a(R_1+R_2)} \right).$$

From the arbitrariness of ε it follows that these integrals converge to 0 for $R_1, R_2 \to \infty$ and therefore exists the limit

$$\lim_{R_1, R_2 \to \infty} \int_{-R_1}^{R_2} \frac{P_n(x)}{Q_m(x)} e^{iax} dx = 2\pi i \sum_{k} \operatorname*{Res}_{z=z_k} \frac{P_n(z)}{Q_m(z)} e^{iaz}.$$

$\square$

Changing a to $-a$ gives

$$\int_{-\infty}^{+\infty} \frac{P_n(x)}{Q_m(x)} e^{-iax}\,dx = 2\pi i \sum_k \operatorname*{Res}_{z=z_k} \overline{\frac{P_n(z)}{Q_m(z)} e^{iaz}}$$

$$= -2\pi i \sum_k \operatorname*{Res}_{z=z_k} \overline{\frac{P_n(z)}{Q_m(z)}} e^{iaz}$$

$$= -2\pi i \operatorname*{Res}_{z=\overline{z_k}} \frac{P_n(z)}{Q_m(z)} e^{-iaz}. \tag{10.16}$$

This formula can also be obtained directly by integrating the complex function $f(z) = P_n(z)e^{-iaz}/Q_m(z)$, with $a > 0$, along the boundary of the square with vertices $-R_1$, R_2, $R_2 - i(R_1 + R_2)$, $-R_1 - i(R_1 + R_2)$, which is negatively oriented and contains the zeros $\overline{z_k}$ of $Q_m(z)$ lying in the lower complex half-plane.

Example 10.10 Show that

$$\int_{-\infty}^{+\infty} \frac{(x+1)\cos x}{x^2 + 4x + 5}\,dx = \frac{\pi}{e}(\sin 2 - \cos 2),$$

$$\int_{-\infty}^{+\infty} \frac{(x+1)\sin x}{x^2 + 4x + 5}\,dx = \frac{\pi}{e}(\sin 2 + \cos 2).$$

Consider the function $f(z) = (z+1)e^{iz}/((z^2 + 4z + 5)$ (Fig. 10.2). This has simple poles at $z = -2 + \sqrt{4-5} = -2 \pm i$ and the residue in the pole contained in the half-plane $\operatorname{Im} z > 0$ is

$$\operatorname*{Res}_{z=-2+i} f(z) = \left.\frac{(z+1)e^{iz}}{z - (-2 - i)}\right|_{z=-2+i} = \frac{(-1+i)e^{-2i-1}}{2i} = \frac{(1+i)e^{-2i}}{2e}.$$

Taking the real part of the formula (10.15) we get

$$\int_{-\infty}^{+\infty} \frac{(x+1)\cos x}{x^2 + 4x + 5}\,dx = -2\pi \operatorname{Im} \frac{(1+i)(\cos 2 - i\sin 2)}{2e}$$

$$= \frac{\pi}{e}(\sin 2 - \cos 2).$$

Taking the imaginary part of the same formula we get

$$\int_{-\infty}^{+\infty} \frac{(x+1)\sin x}{x^2 + 4x + 5}\,dx = 2\pi \operatorname{Re} \frac{(1+i)(\cos 2 - i\sin 2)}{2e}$$

$$= \frac{\pi}{e}(\sin 2 + \cos 2).$$

Fig. 10.2 Closed square path $\gamma = \lambda_1 + \lambda_2 + \lambda_3 + \lambda_4$ used for integrate the function $f(z) = (z+1)\mathrm{e}^{\mathrm{i}z}/((z^2 + 4z + 5)$

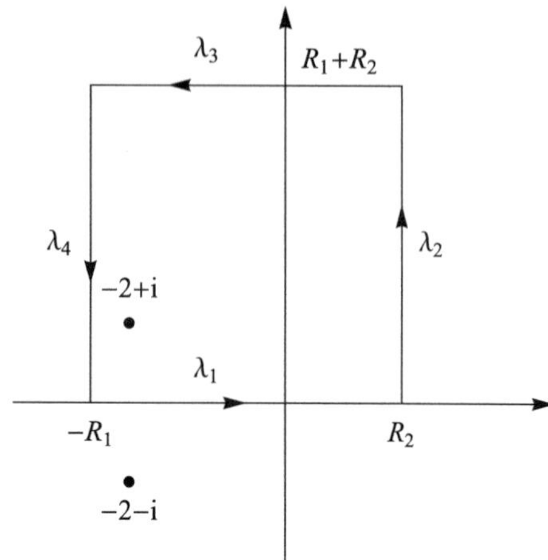

Example 10.11 Show that

$$\mathrm{PV} \int_{-\infty}^{+\infty} \frac{(x+1)}{x^2 + 4x + 5}\,\mathrm{d}x = -\pi.$$

Consider the function $f(z) = (z+1)\mathrm{e}^{\mathrm{i}az}/((z^2 + 4z + 5)$ with $a > 0$. From the previous example it immediately follows

$$\int_{-\infty}^{+\infty} \frac{(x+1)\cos(ax)}{x^2 + 4x + 5}\,\mathrm{d}x = -2\pi\,\mathrm{Im}\,\frac{(1+\mathrm{i})(\cos(2a) - \mathrm{i}\sin(2a))}{2\mathrm{e}^a}$$

$$= \frac{\pi}{\mathrm{e}^a}(\sin(2a) - \cos(2a))$$

and then

$$\mathrm{PV} \int_{-\infty}^{+\infty} \frac{(x+1)}{x^2 + 4x + 5}\,\mathrm{d}x = \lim_{a \to 0} \int_{-\infty}^{+\infty} \frac{(x+1)\cos(ax)}{x^2 + 4x + 5}\,\mathrm{d}x = -\pi.$$

10.5 Indented Paths Around a Simple Pole

Lemma 10.12 *If f has a simple pole at z_0 and $\gamma_r^{\pm}$ are the two semicircles defined by $\gamma_r^{\pm}(\theta) = z_0 + r\mathrm{e}^{\pm\mathrm{i}\theta}$ with $-\pi \leq \theta \leq 0$, then*

$$\lim_{r \to 0} \int_{\gamma_r^{\pm}} f(z)\,\mathrm{d}z = \pm\pi\mathrm{i}\,\mathop{\mathrm{Res}}_{z=z_0} f(z). \tag{10.17}$$

Proof Since f has a simple pole at z_0, then $\exists R > 0$ such that, for $0 < |z - z_0| < R$, we can write

$$f(z) = \frac{b_1}{z - z_0} + g(z),$$

with g analytic in the ball $B(z_0, R)$. For each $r < R$ we then have

$$\int_{\gamma_r^\pm} f(z)\mathrm{d}z = b_1 \int_{\gamma_r^\pm} \frac{\mathrm{d}z}{z - z_0} + \int_{\gamma_r^\pm} g(z)\mathrm{d}z = \pm\pi i b_1 + \int_{\gamma_r^\pm} g(z)\mathrm{d}z.$$

The first term independent of r is precisely $\pm\,\pi i \operatorname*{Res}_{z=z_0} f(z)$, while the second one vanishes in the limit $r \to 0$. In fact, choose $r < r_0 < R$, the function $|g|$ is continues in $\overline{B}(z_0, r_0)$ and therefore $\exists M > 0$ such that $|g(z)| \le M \ \forall z \in \overline{B}(z_0, r_0)$. It follows that $\forall r \le r_0$ we have the inequality

$$\left| \int_{\gamma_r^\pm} g(z)\mathrm{d}z \right| \le \pi r M$$

and therefore $\int_{\gamma_r^\pm} g(z)\mathrm{d}z \xrightarrow{r \to 0} 0.$ $\qquad\square$

The previous lemma allows us to extend formulas (10.8) and (10.15) to the case in which the polynomial in denominator $Q_m(x)$ has simple real zeros. It is evident that the real integrals converge in this case only in the sense of Cauchy principal value.

Theorem 10.13 *Let $P_n(x)$ and $Q_m(x)$ be two real polynomials of degree n and m. If $m \ge n + 2$ and $Q_m(x)$ has real zeros x_j only of order 1, then*

$$\mathrm{PV} \int_{-\infty}^{+\infty} \frac{P_n(x)}{Q_m(x)}\mathrm{d}x = 2\pi i \sum_k \operatorname*{Res}_{z=z_k} \frac{P_n(z)}{Q_m(z)} + \pi i \sum_j \operatorname*{Res}_{z=x_j} \frac{P_n(z)}{Q_m(z)}, \qquad (10.18)$$

where z_k are the zeros of $Q_m(z)$ contained in the complex half-plane $\operatorname{Im} z > 0$.

Proof Let $\gamma = \lambda + \gamma_R$ be the path used in the proof of formula (10.8) and deform it by going around each of the simple real poles x_j with semicircles $\gamma_{r_j}^\pm$, the signs $\pm$ indicating the positive or negative direction of travel. Finally, take the limits $r_j \to 0$ and $R \to \infty$. $\qquad\square$

Theorem 10.14 *Let $P_n(x)$ and $Q_m(x)$ be two real polynomials of degree n and m. If $m \geq n + 1$ and $Q_m(x)$ has real zeros x_j only of order 1, then $\forall a > 0$ we have*

$$\mathrm{PV} \int_{-\infty}^{+\infty} \frac{P_n(x)}{Q_m(x)} \mathrm{e}^{\mathrm{i}ax} \, \mathrm{d}x = 2\pi \mathrm{i} \sum_k \operatorname*{Res}_{z=z_k} \frac{P_n(z)}{Q_m(z)} \mathrm{e}^{\mathrm{i}az}$$

$$+ \pi \mathrm{i} \sum_j \operatorname*{Res}_{z=x_j} \frac{P_n(z)}{Q_m(z)} \mathrm{e}^{\mathrm{i}az}, \tag{10.19}$$

where z_k are the zeros of $Q_m(z)$ contained in the complex half-plane $\operatorname{Im} z > 0$.

Proof Deform the path $\gamma = \lambda_1 + \lambda_2 + \lambda_3 + \lambda_4$ used in the proof of formula (10.15) by going around each of the simple real poles x_j with semicircles $\gamma_{r_j}^{\pm}$. Finally, take the limits $r_j \to 0$ and $R_1, R_2 \to \infty$. $\qquad\qquad\square$

Example 10.15 Show that

$$\mathrm{PV} \int_{-\infty}^{+\infty} \frac{x}{x^3 - 1} \, \mathrm{d}x = \frac{\pi}{\sqrt{3}}.$$

The rational function $f(z) = p(z)/q(z)$, with $p(z) = z$ and $q(z) = z^3 - 1$, has simple poles at the points $z_k = \sqrt[3]{1} = \mathrm{e}^{\mathrm{i}2\pi k/3}$, $k = 0, 1, 2$, one of which, $z_0 = 1$, is real (Fig. 10.3). The residues of f at the poles lying in the upper complex half-plane, real axis included, are

$$\operatorname*{Res}_{z=z_0} f(z) = \frac{p(z_0)}{q'(z_0)} = \frac{1}{3},$$

$$\operatorname*{Res}_{z=z_1} f(z) = \frac{p(z_1)}{q'(z_1)} = \frac{\mathrm{e}^{\mathrm{i}2\pi/3}}{3\mathrm{e}^{\mathrm{i}4\pi/3}} = \frac{1}{3\left(-\frac{1}{2} + \mathrm{i}\frac{\sqrt{3}}{2}\right)} = -\frac{1}{6} - \mathrm{i}\frac{\sqrt{3}}{6},$$

therefore

$$\mathrm{PV} \int_{-\infty}^{+\infty} \frac{x}{x^3 - 1} \, \mathrm{d}x = 2\pi \mathrm{i} \operatorname*{Res}_{z=z_1} f(z) + \pi \mathrm{i} \operatorname*{Res}_{z=z_0} f(z)$$

$$= 2\pi \mathrm{i} \left(-\frac{1}{6} - \mathrm{i}\frac{\sqrt{3}}{6}\right) + \pi \mathrm{i}\frac{1}{3}$$

$$= \frac{\pi\sqrt{3}}{3}.$$

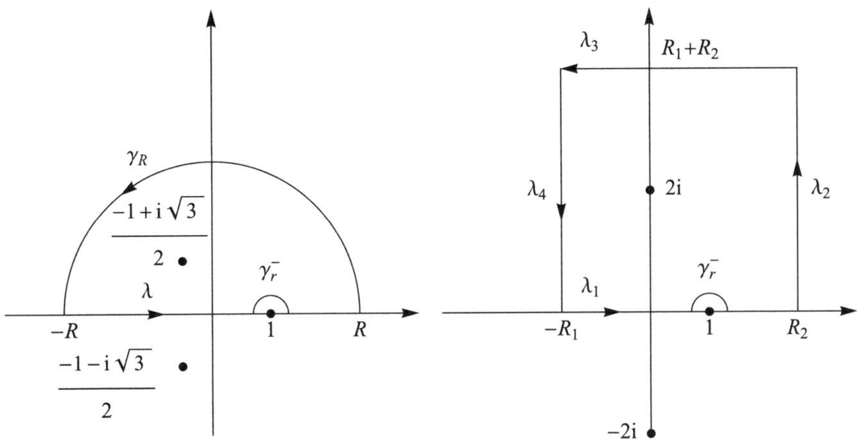

Fig. 10.3 Indentation of closed paths $\gamma = \lambda + \gamma_R$ (left) and $\gamma = \lambda_1 + \lambda_2 + \lambda_3 + \lambda_4$ (right) used to integrate the functions, respectively, $f(z) = z/(z^3 - 1)$ and $f(z) = e^{iz}/((z-1)(z^2+4))$

Example 10.16 Show that

$$\mathrm{PV} \int_{-\infty}^{+\infty} \frac{\sin x}{(x-1)(x^2+4)}\,dx = \frac{\pi}{5}\left(\cos 1 - \frac{1}{e^2}\right).$$

The function $f(z) = e^{iz}/q(z)$, with $q(z) = (z-1)(z^2+4)$, has simple poles at the points 1 and $\pm 2i$ (Fig. 10.3). The residues of f at the poles lying in the upper complex half-plane, including the real axis, are

$$\operatorname*{Res}_{z=1} f(z) = \left.\frac{e^{iz}}{z^2+4}\right|_{z=1} = \frac{\cos 1 + i\sin 1}{5},$$

$$\operatorname*{Res}_{z=2i} f(z) = \left.\frac{e^{iz}}{(z-1)(z+2i)}\right|_{z=2i} = \frac{e^{-2}}{-8-4i} = \frac{-2+i}{20e^2},$$

therefore

$$\mathrm{PV} \int_{-\infty}^{+\infty} \frac{\sin x}{(x-1)(x^2+4)}\,dx = \operatorname{Im}\left(2\pi i \operatorname*{Res}_{z=2i} f(z) + \pi i \operatorname*{Res}_{z=1} f(z)\right)$$

$$= 2\pi \frac{-2}{20e^2} + \pi \frac{\cos 1}{5}$$

$$= \frac{\pi}{5}\left(\cos 1 - \frac{1}{e^2}\right).$$

Example 10.17 Show that

$$\int_0^\infty \frac{\sin x}{x}\,dx = \frac{\pi}{2}.$$

First note that due to the parity of the integrand function and since $\lim_{x\to 0}\sin x/x = 1$, we have

$$
\begin{aligned}
\int_0^\infty \frac{\sin x}{x}\,dx &= \frac{1}{2}\int_{-\infty}^{+\infty}\frac{\sin x}{x}\,dx \\
&= \frac{1}{2}\lim_{r\to 0}\left(\int_{-\infty}^{-r}\frac{\sin x}{x}\,dx + \int_r^\infty \frac{\sin x}{x}\,dx\right) \\
&= \frac{1}{2}\lim_{r\to 0}\operatorname{Im}\left(\int_{-\infty}^{-r}\frac{e^{ix}}{x}\,dx + \int_r^\infty \frac{e^{ix}}{x}\,dx\right) \\
&= \frac{1}{2}\operatorname{Im}\left(\mathrm{PV}\int_{-\infty}^{+\infty}\frac{e^{ix}}{x}\,dx\right).
\end{aligned}
$$

According to formula (10.19) there exists the Cauchy principal value of the integral of e^{ix}/x between $-\infty$ and $+\infty$, therefore according to the above equality there exists the integral of $\sin x/x$ between 0 and ∞. To find its value, observe that the complex function $f(z) = e^{iz}/z$ is analytic everywhere except for the simple pole at $z = 0$, where it has residue

$$\operatorname*{Res}_{z=0} f(z) = e^{iz}\Big|_{z=0} = 1.$$

Therefore

$$\int_0^\infty \frac{\sin x}{x}\,dx = \frac{1}{2}\operatorname{Im}\left(\pi i\operatorname*{Res}_{z=0} f(z)\right) = \frac{\pi}{2}.$$

10.6　Indented Paths Around a Branch Point

Example 10.18 Show that

$$\int_0^\infty \frac{\ln x}{x^2 + a^2}\,dx = \frac{\pi \ln a}{2a}, \qquad a > 0.$$

Set $f(z) = (\log z)/(z^2 + a^2)$ assuming for the logarithm that branch whose branch line coincides with the negative imaginary semi-axis (Fig. 10.4),

$$\log z = \ln r + i\theta, \qquad z = re^{i\theta}, \quad r > 0, \quad -\frac{\pi}{2} < \theta < \frac{3\pi}{2},$$

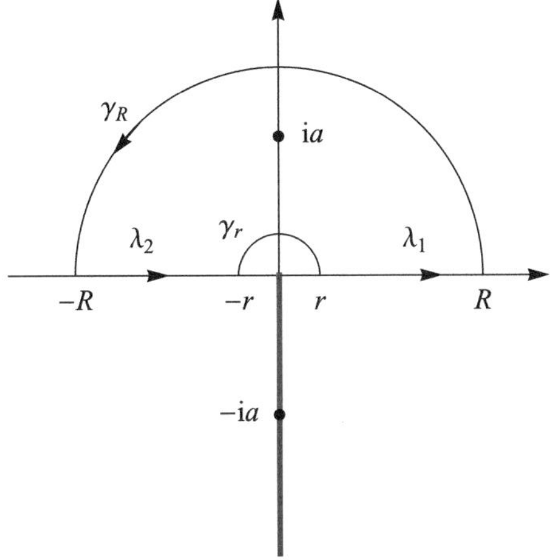

Fig. 10.4 Indented closed path $\gamma = \lambda_1 + \gamma_R + \lambda_2 + \gamma_r$ used to integrate the function $f(z) = \log z/(z^2 + a^2)$ with $a > 0$

and integrate f along the positively oriented closed path $\gamma = \lambda_1 + \gamma_R + \lambda_2 + \gamma_r$, where

$$\lambda_1(x) = x, \qquad r \le x \le R,$$

$$\gamma_R(\theta) = Re^{i\theta}, \qquad 0 \le \theta \le \pi,$$

$$\lambda_2(x) = xe^{i\pi}, \qquad R \ge x \ge r,$$

$$\gamma_r(\theta) = re^{i\theta}, \qquad \pi \ge \theta \ge 0.$$

Since f is analytic on and within γ except for the simple pole at $z = ia$, for $r < a < R$ we have

$$\int_\gamma f(z)\mathrm{d}z = \int_{\lambda_1} f(z)\mathrm{d}z + \int_{\gamma_R} f(z)\mathrm{d}z + \int_{\lambda_2} f(z)\mathrm{d}z + \int_{\gamma_r} f(z)\mathrm{d}z$$

$$= 2\pi i \operatorname*{Res}_{z=ia} f(z)$$

$$= 2\pi i \frac{\log\left(ae^{i\pi/2}\right)}{ia + ia}$$

$$= \frac{\pi}{a}\left(\ln a + i\frac{\pi}{2}\right).$$

For the integrals along the individual paths that compose γ we find

$$\int_{\lambda_1} f(z)\mathrm{d}z = \int_r^R \frac{\ln x}{x^2 + a^2}\mathrm{d}x,$$

$$\int_{\gamma_R} f(z)\mathrm{d}z = \int_0^\pi \frac{\ln R + i\theta}{R^2 e^{2i\theta} + a^2}\, iRe^{i\theta}\,\mathrm{d}\theta,$$

$$\int_{\lambda_2} f(z)\mathrm{d}z = \int_R^r \frac{\ln x + i\pi}{x^2 e^{i2\pi} + a^2}\, e^{i\pi}\, \mathrm{d}x$$

$$= \int_r^R \frac{\ln x}{x^2 + a^2}\mathrm{d}x + i\pi \int_r^R \frac{1}{x^2 + a^2}\mathrm{d}x,$$

$$\int_{\gamma_r} f(z)\mathrm{d}z = \int_\pi^0 \frac{\ln r + i\theta}{r^2 e^{2i\theta} + a^2}\, ire^{i\theta}\mathrm{d}\theta.$$

The integrals over γ_R and γ_r tend to 0 respectively for $R \to \infty$ and for $r \to 0$. In fact, using Darboux's inequality, we have

$$\left| \int_{\gamma_R} f(z)\mathrm{d}z \right| \le \pi R \frac{\ln R + \pi}{R^2 - a^2} \xrightarrow{R \to \infty} 0,$$

$$\left| \int_{\gamma_r} f(z)\mathrm{d}z \right| \le \pi r \frac{|\ln r| + \pi}{a^2 - r^2} \xrightarrow{r \to 0} 0.$$

Therefore, in the limit $r \to 0$ and $R \to \infty$ we get

$$2 \int_0^\infty \frac{\ln x}{x^2 + a^2}\mathrm{d}x + i\pi \int_0^\infty \frac{1}{x^2 + a^2}\mathrm{d}x = \frac{\pi \ln a}{a} + i\frac{\pi^2}{2a},$$

from which, taking the real and imaginary parts, we conclude

$$\int_0^\infty \frac{\ln x}{x^2 + a^2}\mathrm{d}x = \frac{\pi \ln a}{2a}, \qquad \int_0^\infty \frac{1}{x^2 + a^2}\mathrm{d}x = \frac{\pi}{2a}.$$

10.7   Paths Coinciding with a Branch Line

Example 10.19 Show that

$$\int_0^\infty \frac{x^a}{x^2 + 1}\mathrm{d}x = \frac{\pi/2}{\cos(\pi a/2)}, \qquad -1 < a < 1.$$

Let $f(z) = z^a/(z^2 + 1) = e^{a \log z}/(z^2 + 1)$, assuming for the logarithm that branch whose branch line coincides with the positive real semi-axis (Fig. 10.5)

$$\log z = \ln r + i\theta, \qquad z = re^{i\theta}, \quad r > 0, \quad 0 < \theta < 2\pi.$$

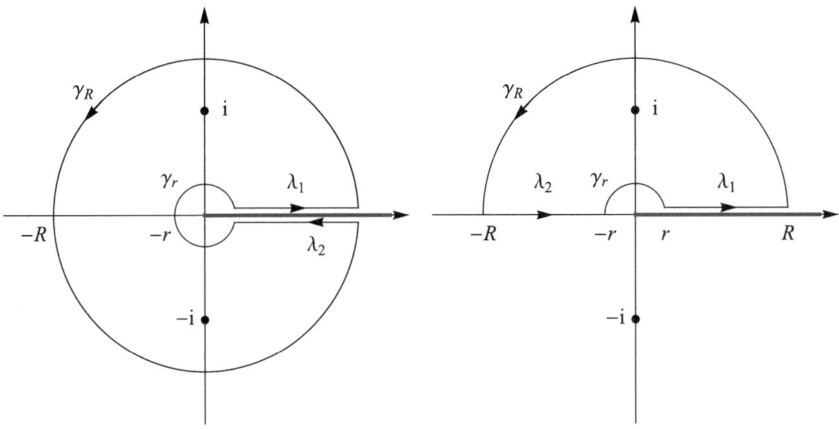

Fig. 10.5 Different closed paths $\gamma = \lambda_1 + \gamma_R + \lambda_2 + \gamma_r$ used to integrate the multi-valued function $f(z) = z^a/(z^2 + 1)$ with $-1 < a < 1$

The function f is analytic everywhere except on the positive real semi-axis and the two simple poles at $z = \pm i$ where it has residues

$$\operatorname*{Res}_{z=i} f(z) = \left. \frac{e^{a \log z}}{z + i} \right|_{z=i} = \frac{e^{a(\ln 1 + i\pi/2)}}{i + i} = \frac{e^{i\pi a/2}}{2i},$$

$$\operatorname*{Res}_{z=-i} f(z) = \left. \frac{e^{a \log z}}{z - i} \right|_{z=-i} = \frac{e^{a(\ln 1 + i3\pi/2)}}{-i - i} = \frac{e^{i3\pi a/2}}{-2i}.$$

Integrate f along the positively oriented closed path $\gamma = \lambda_1 + \gamma_R + \lambda_2 + \gamma_r$, where

$$\lambda_1(x) = x + i0, \qquad r \le x \le R,$$

$$\gamma_R(\theta) = Re^{i\theta}, \qquad 0 \le \theta \le 2\pi,$$

$$\lambda_2(x) = x - i0, \qquad R \ge x \ge r,$$

$$\gamma_r(\theta) = re^{i\theta}, \qquad 2\pi \ge \theta \ge 0.$$

As usual, $x \pm i0$ stands for $x \pm i\varepsilon$ with ε positive infinitesimal, which, in polar coordinates with $0 < \theta < 2\pi$, implies $x + i0 = xe^{i\varepsilon}$ and $x - i0 = xe^{i(2\pi - \varepsilon)}$. For $r < 1 < R$, we have

$$\int_\gamma f(z)\mathrm{d}z = \int_{\lambda_1} f(z)\mathrm{d}z + \int_{\gamma_R} f(z)\mathrm{d}z + \int_{\lambda_2} f(z)\mathrm{d}z + \int_{\gamma_r} f(z)\mathrm{d}z$$

$$= 2\pi \mathrm{i} \left(\operatorname*{Res}_{z=\mathrm{i}} f(z) + \operatorname*{Res}_{z=-\mathrm{i}} f(z) \right)$$

$$= \pi \left(\mathrm{e}^{\mathrm{i}\pi a/2} - \mathrm{e}^{\mathrm{i}3\pi a/2} \right).$$

For the integrals along the individual paths that compose γ we find

$$\int_{\lambda_1} f(z)\mathrm{d}z = \int_r^R \frac{\mathrm{e}^{a(\ln x + \mathrm{i}0)}}{(x\mathrm{e}^{\mathrm{i}0})^2 + 1}\mathrm{e}^{\mathrm{i}0}\mathrm{d}x = \int_r^R \frac{\mathrm{e}^{a \ln x}}{x^2 + 1}\mathrm{d}x,$$

$$\int_{\lambda_2} f(z)\mathrm{d}z = \int_R^r \frac{\mathrm{e}^{a(\ln x + \mathrm{i}2\pi)}}{(x\mathrm{e}^{\mathrm{i}2\pi})^2 + 1}\mathrm{e}^{\mathrm{i}2\pi}\mathrm{d}x = -\mathrm{e}^{\mathrm{i}2\pi a}\int_r^R \frac{\mathrm{e}^{a \ln x}}{x^2 + 1}\mathrm{d}x,$$

$$\left| \int_{\gamma_R} f(z)\mathrm{d}z \right| \le \frac{\mathrm{e}^{a \ln R}}{R^2 - 1} 2\pi R = \frac{2\pi R^{a+1}}{R^2 - 1} \xrightarrow{R \to \infty} 0, \qquad a + 1 < 2,$$

$$\left| \int_{\gamma_r} f(z)\mathrm{d}z \right| \le \frac{\mathrm{e}^{a \ln r}}{1 - r^2} 2\pi r = \frac{2\pi r^{a+1}}{1 - r^2} \xrightarrow{r \to 0} 0, \qquad a + 1 > 0.$$

For $-1 < a < 1$ in the limit $r \to 0$ and $R \to \infty$ we thus conclude

$$\int_0^\infty \frac{x^a}{x^2 + 1}\mathrm{d}x = \frac{\pi \left(\mathrm{e}^{\mathrm{i}\pi a/2} - \mathrm{e}^{\mathrm{i}3\pi a/2} \right)}{1 - \mathrm{e}^{\mathrm{i}2\pi a}}$$

$$= \frac{\pi \mathrm{e}^{\mathrm{i}\pi a} \left(\mathrm{e}^{-\mathrm{i}\pi a/2} - \mathrm{e}^{\mathrm{i}\pi a/2} \right)}{\mathrm{e}^{\mathrm{i}\pi a} \left(\mathrm{e}^{-\mathrm{i}\pi a} - \mathrm{e}^{\mathrm{i}\pi a} \right)}$$

$$= \frac{\pi \sin(\pi a/2)}{\sin(\pi a)}$$

$$= \frac{\pi/2}{\cos(\pi a/2)}.$$

The same result, but avoiding the computation of one of the two residues, is obtained by integrating f along the closed path $\gamma = \lambda_1 + \gamma_R + \lambda_2 + \gamma_r$, where

$$\lambda_1(x) = x + \mathrm{i}0, \qquad r \le x \le R,$$

$$\gamma_R(\theta) = R\mathrm{e}^{\mathrm{i}\theta}, \qquad 0 \le \theta \le \pi,$$

$$\lambda_2(x) = x\mathrm{e}^{\mathrm{i}\pi}, \qquad R \ge x \ge r,$$

$$\gamma_r(\theta) = r\mathrm{e}^{\mathrm{i}\theta}, \qquad \pi \ge \theta \ge 0.$$

In this case, always assuming $r < 1 < R$, we have

$$\int_\gamma f(z)\,dz = \int_{\lambda_1} f(z)\,dz + \int_{\gamma_R} f(z)\,dz + \int_{\lambda_2} f(z)\,dz + \int_{\gamma_r} f(z)\,dz$$

$$= 2\pi i \operatorname*{Res}_{z=i} f(z)$$

$$= \pi\, e^{i\pi a/2},$$

while for the integrals along the individual paths we find

$$\int_{\lambda_1} f(z)\,dz = \int_r^R \frac{e^{a(\ln x + i0)}}{(xe^{i0})^2 + 1}\, e^{i0}\,dx = \int_r^R \frac{e^{a\ln x}}{x^2 + 1}\,dx,$$

$$\int_{\lambda_2} f(z)\,dz = \int_R^r \frac{e^{a(\ln x + i\pi)}}{(xe^{i\pi})^2 + 1}\, e^{i\pi}\,dx = e^{i\pi a} \int_r^R \frac{e^{a\ln x}}{x^2 + 1}\,dx,$$

$$\left| \int_{\gamma_R} f(z)\,dz \right| \le \frac{e^{a\ln R}}{R^2 - 1}\, \pi R = \frac{\pi R^{a+1}}{R^2 - 1} \xrightarrow{R\to\infty} 0, \qquad a + 1 < 2,$$

$$\left| \int_{\gamma_r} f(z)\,dz \right| \le \frac{e^{a\ln r}}{1 - r^2}\, \pi r = \frac{\pi r^{a+1}}{1 - r^2} \xrightarrow{r\to 0} 0, \qquad a + 1 > 0.$$

For $-1 < a < 1$, in the limit $r \to 0$ and $R \to \infty$ we still conclude

$$\int_0^\infty \frac{x^a}{x^2 + 1}\,dx = \frac{\pi\, e^{i\pi a/2}}{1 + e^{i\pi a}}$$

$$= \frac{\pi/2}{\cos(\pi a/2)}.$$

Note that this second type of integration path can be used by virtue of the parity of the function in the denominator, that assumes the same value, $x^2 + 1$, both on $\lambda_2(x)$ and on $\lambda_1(x)$.

10.8 Miscellaneous Paths

Other real integrals, not belonging to the classes described so far, can be evaluated by integrating a suitable complex function along a suitable path in $\mathbb{C}$ and again by exploiting the residue theorem. It is not always possible to theorize in a general way how these suitable functions and, above all, these suitable paths should be chosen. Some notable examples are given below.

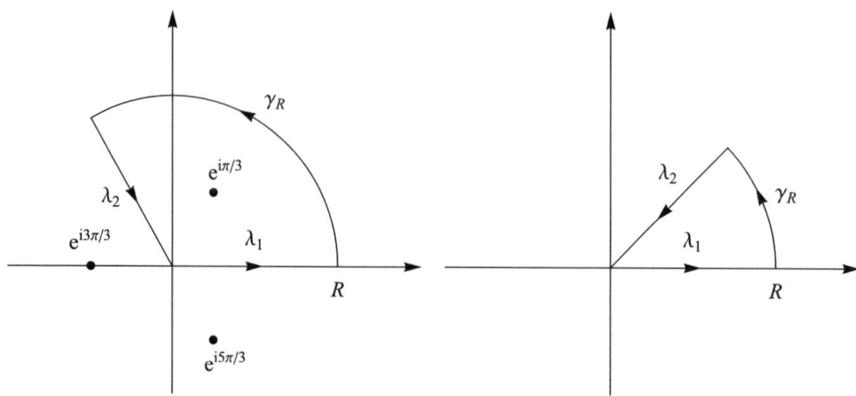

Fig. 10.6 Closed paths $\gamma = \lambda_1 + \gamma_R + \lambda_2$ used to integrate the functions respectively $f(z) = z/(z^3 + 1)$ (left) and $f(z) = e^{iz^2}$ (right)

Example 10.20 Show that

$$\int_0^\infty \frac{x}{x^3 + 1}\,\mathrm{d}x = \frac{2\pi}{3\sqrt{3}}.$$

The function $f(z) = z/(z^3 + 1)$ has simple poles at $z_k = e^{i(\pi + 2\pi k)/3}$, $k = 0, 1, 2$ (Fig. 10.6). Let $\gamma = \lambda_1 + \gamma_R + \lambda_2$ be the closed integration path defined by

$$\lambda_1(x) = x, \qquad 0 \le x \le R,$$

$$\gamma_R(\theta) = Re^{i\theta}, \qquad 0 \le \theta \le 2\pi/3,$$

$$\lambda_2(x) = xe^{i\frac{2}{3}\pi}, \qquad R \ge x \ge 0,$$

since, for $R > 1$, f is analytic on and within γ except for the simple pole at $z = z_0$, we have

$$\int_\gamma f(z)\mathrm{d}z = \int_{\lambda_1} f(z)\mathrm{d}z + \int_{\gamma_R} f(z)\mathrm{d}z + \int_{\lambda_2} f(z)\mathrm{d}z$$

$$= 2\pi i \operatorname*{Res}_{z=z_0} f(z)$$

$$= 2\pi i \,\frac{z_0}{3z_0^2}$$

$$= \frac{2\pi i}{3} e^{-i\frac{\pi}{3}}.$$

The integrals along the individual paths that compose γ give

$$\int_{\lambda_1} f(z)\,dz = \int_0^R \frac{x}{x^3+1}\,dx,$$

$$\int_{\gamma_R} f(z)\,dz = \int_0^{\frac{2}{3}\pi} \frac{Re^{i\theta}}{R^3 e^{3i\theta}+1}\, iRe^{i\theta}\,d\theta \xrightarrow{R\to\infty} 0,$$

$$\int_{\lambda_2} f(z)\,dz = \int_R^0 \frac{xe^{i\frac{2}{3}\pi}}{x^3 e^{i2\pi}+1}\, e^{i\frac{2}{3}\pi}\,dx = -e^{i\frac{4}{3}\pi}\int_0^R \frac{x}{x^3+1}\,dx.$$

Therefore, in the limit $R\to\infty$ we have

$$\left(1-e^{i\frac{4}{3}\pi}\right)\int_0^\infty \frac{x}{x^3+1}\,dx = \frac{2\pi i}{3}e^{-i\frac{\pi}{3}}$$

and, in conclusion,

$$\int_0^\infty \frac{x}{x^3+1}\,dx = \frac{2\pi i}{3}\frac{e^{-i\frac{\pi}{3}}}{1-e^{i\frac{4}{3}\pi}}$$

$$= \frac{2\pi i}{3}\frac{e^{-i\frac{\pi}{3}}e^{-i\frac{2}{3}\pi}}{e^{-i\frac{2}{3}\pi}-e^{i\frac{2}{3}\pi}}$$

$$= \frac{\pi/3}{\sin(2\pi/3)}$$

$$= \frac{2\pi}{3\sqrt{3}}.$$

Example 10.21 (Fresnel Integrals) Show that

$$\int_0^\infty \cos(x^2)\,dx = \int_0^\infty \sin(x^2)\,dx = \frac{1}{2}\sqrt{\frac{\pi}{2}}.$$

Integrate the function $f(z) = e^{iz^2}$, analytic throughout $\mathbb{C}$, along the positively oriented simple closed path $\gamma = \lambda_1 + \gamma_R + \lambda_2$, where (Fig. 10.6)

$$\lambda_1(x) = x, \qquad 0 \le x \le R,$$

$$\gamma_R(\theta) = Re^{i\theta}, \qquad 0 \le \theta \le \pi/4,$$

$$\lambda_2(r) = re^{i\pi/4}, \qquad R \ge r \ge 0.$$

Since f is analytic on and within γ, we have

$$\int_\gamma f(z)\mathrm{d}z = \int_{\lambda_1} f(z)\mathrm{d}z + \int_{\gamma_R} f(z)\mathrm{d}z + \int_{\lambda_2} f(z)\mathrm{d}z = 0.$$

The integrals along the paths λ_1 and λ_2 give

$$\int_{\lambda_1} f(z)\mathrm{d}z = \int_0^R e^{\mathrm{i}x^2}\mathrm{d}x,$$

$$\int_{\lambda_2} f(z)\mathrm{d}z = \int_R^0 e^{\mathrm{i}r^2(\cos(\pi/2)+\mathrm{i}\sin(\pi/2))}e^{\mathrm{i}\pi/4}\mathrm{d}r = -e^{\mathrm{i}\pi/4}\int_0^R e^{-r^2}\mathrm{d}r.$$

For the integral along γ_R, using Jordan's inequality we find

$$\left|\int_{\gamma_R} f(z)\mathrm{d}z\right| \le \int_0^{\pi/4}\left|e^{\mathrm{i}R^2(\cos(2\theta)+\mathrm{i}\sin(2\theta))}\,\mathrm{i}Re^{\mathrm{i}\theta}\right|\mathrm{d}\theta$$

$$= \frac{R}{2}\int_0^{\pi/2}e^{-R^2\sin\varphi}\mathrm{d}\varphi$$

$$\le \frac{R}{2}\int_0^{\pi/2}e^{-R^2 2\varphi/\pi}\mathrm{d}\varphi$$

$$= \frac{\pi}{4}\frac{1-e^{-R^2}}{R}\xrightarrow{R\to\infty} 0.$$

Therefore, in the limit $R\to\infty$ we conclude

$$\int_0^\infty e^{\mathrm{i}x^2}\mathrm{d}x = \int_0^\infty \cos(x^2)\mathrm{d}x + \mathrm{i}\int_0^\infty \sin(x^2)\mathrm{d}x$$

$$= e^{\mathrm{i}\pi/4}\int_0^\infty e^{-r^2}\mathrm{d}r$$

$$= \left(\frac{1}{\sqrt{2}} + \frac{\mathrm{i}}{\sqrt{2}}\right)\frac{\sqrt{\pi}}{2}.$$

Example 10.22 Show that

$$\int_0^\infty x\cos(x^4)\mathrm{d}x = \int_0^\infty x\sin(x^4)\mathrm{d}x = \frac{1}{4}\sqrt{\frac{\pi}{2}}.$$

Integrate the function $f(z) = z e^{iz^4}$, analytic throughout $\mathbb{C}$, along the positively oriented simple closed path $\gamma = \lambda_1 + \gamma_R + \lambda_2$, where

$$\lambda_1(x) = x, \qquad 0 \le x \le R,$$

$$\gamma_R(\theta) = R e^{i\theta}, \qquad 0 \le \theta \le \pi/8,$$

$$\lambda_2(r) = r e^{i\pi/8}, \qquad R \ge r \ge 0.$$

Since f is analytic on and within γ, we have

$$\int_\gamma f(z)\mathrm{d}z = \int_{\lambda_1} f(z)\mathrm{d}z + \int_{\gamma_R} f(z)\mathrm{d}z + \int_{\lambda_2} f(z)\mathrm{d}z = 0.$$

The integrals along the paths λ_1 and λ_2 give

$$\int_{\lambda_1} f(z)\mathrm{d}z = \int_0^R x e^{ix^4}\,\mathrm{d}x,$$

$$\int_{\lambda_2} f(z)\mathrm{d}z = \int_R^0 r e^{i\pi/8} e^{ir^4(\cos(\pi/2)+i\sin(\pi/2))} e^{i\pi/8}\,\mathrm{d}r$$

$$= -e^{i\pi/4} \int_0^R r e^{-r^4}\,\mathrm{d}r$$

$$= -\frac{e^{i\pi/4}}{2} \int_0^R e^{-u^2}\,\mathrm{d}u.$$

For the integral along γ_R, using Jordan's inequality we find

$$\left| \int_{\gamma_R} f(z)\mathrm{d}z \right| \le \int_0^{\pi/8} \left| R e^{i\theta} e^{iR^4(\cos(4\theta)+i\sin(4\theta))}\, iR e^{i\theta} \right| \mathrm{d}\theta,$$

$$= \frac{R^2}{4} \int_0^{\pi/2} e^{-R^4 \sin\varphi}\,\mathrm{d}\varphi$$

$$\le \frac{R^2}{4} \int_0^{\pi/2} e^{-R^4 2\varphi/\pi}\,\mathrm{d}\varphi$$

$$= \frac{\pi}{8} \frac{1-e^{-R^4}}{R^2} \xrightarrow{R\to\infty} 0.$$

Therefore, in the limit $R \to \infty$ we conclude

$$\int_0^\infty x e^{ix^4} dx = \int_0^\infty x \cos(x^4) dx + i \int_0^\infty x \sin(x^4) dx$$

$$= \frac{e^{i\pi/4}}{2} \int_0^\infty e^{-u^2} du$$

$$= \left(\frac{1}{\sqrt{2}} + \frac{i}{\sqrt{2}} \right) \frac{\sqrt{\pi}}{4}.$$

Obviously the same result could be obtained by making the change of variable $x^2 = u$ in the starting integral and using the result of the previous example.

10.9   Bromwich Integrals: Inverse Laplace Transform

Definition 10.23 Let $g : \mathbb{C} \mapsto \mathbb{C}$ be analytic in $\mathbb{C}$ except for a finite set of isolated singular points z_k, $k = 1, \ldots, N$ and let $x_0 \in \mathbb{R}$ be such that $\max_k \operatorname{Re} z_k < x_0$. Set $\lambda_{y_0}(y) = x_0 + iy$ with $-y_0 \le y \le y_0$ and $y_0 > 0$, we call Bromwich integral, assuming the indicated limit exists,

$$f(t) = \frac{1}{2\pi i} \lim_{y_0 \to \infty} \int_{\lambda_{y_0}} e^{zt} g(z) dz, \qquad t > 0. \tag{10.20}$$

It can be shown that, under suitable regularity conditions, $g(z)$ is the Laplace transform of $f(t) : (0, \infty) \mapsto \mathbb{C}$

$$g(z) = \int_0^\infty e^{-zt} f(t) dt. \tag{10.21}$$

In other words, the Bromwich integral (10.20) represents an inversion formula for $g(z)$.

The Bromwich integral can often be evaluated by means of the residue theorem. For example, consider the positively oriented closed path $\eta_{y_0} = \lambda_{y_0} + \gamma_{y_0}$, where $\gamma_{y_0}(\theta) = x_0 + y_0 e^{i\theta}$ with $\pi/2 \le \theta \le 3\pi/2$ (Fig. 10.7). Chosen y_0 large enough such that all the singularities of g lie within the closed path η_{y_0}, we have

$$\int_{\lambda_{y_0}} e^{zt} g(z) dz = 2\pi i \sum_{k=1}^{N} \operatorname*{Res}_{z=z_k} \left(e^{zt} g(z) \right) - \int_{\gamma_{y_0}} e^{zt} g(z) dz.$$

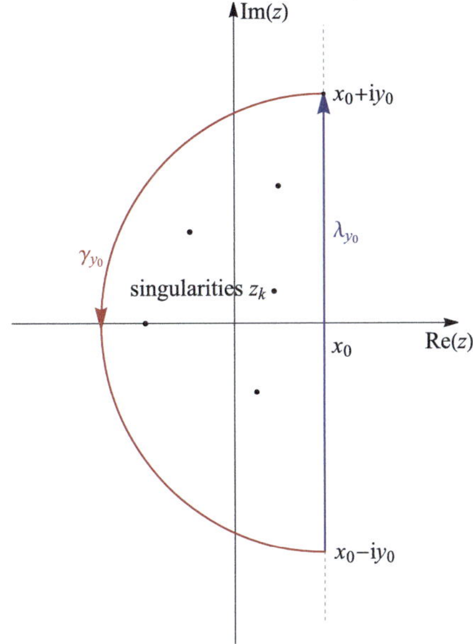

Fig. 10.7 Closed path $\eta_{y_0} = \lambda_{y_0} + \gamma_{y_0}$ used to evaluate the Bromwich integral. The singularities z_k of $g(z)$ are represented by dots which lie to the left of $x_0 > \max_k \mathrm{Re}\, z_k$

Using the inequality (7.3), the modulus of the integral on γ_{y_0} can be upper bounded as follows

$$\left| \int_{\gamma_{y_0}} e^{zt} g(z)\mathrm{d}z \right| \le \int_{\pi/2}^{3\pi/2} \left| e^{x_0 t + y_0 t \cos\theta + i y_0 t \sin\theta} g(x_0 + y_0 e^{i\theta}) i y_0 e^{i\theta} \right| \mathrm{d}\theta$$

$$\le e^{x_0 t} M_{y_0} y_0 \int_{\pi/2}^{3\pi/2} e^{y_0 t \cos\theta} \mathrm{d}\theta$$

$$= e^{x_0 t} M_{y_0} y_0 \int_{0}^{\pi} e^{-y_0 t \sin\varphi} \mathrm{d}\varphi$$

$$\le e^{x_0 t} M_{y_0} y_0 \frac{\pi}{y_0 t},$$

where we set $M_{y_0} = \sup_{z \in \{\gamma_{y_0}\}} |g(z)|$ and Jordan's inequality was used. If g is such that $\lim_{y_0 \to \infty} M_{y_0} = 0$, then we get the formula

$$f(t) = \sum_{k=1}^{N} \operatorname*{Res}_{z=z_k} \left(e^{zt} g(z) \right), \qquad t > 0. \tag{10.22}$$

In the case the singularities of g are countably infinite, the sum in (10.22) is replaced by the corresponding series. However, this expression is, in general, only formal, in the sense that it is necessary to verify that the function $f(t)$ thus found admits as a Laplace transform precisely the starting function $g(z)$.

The evaluation of the formula (10.22) is facilitated by the following observation. If $g(z)$ has at z_k a pole of order m, then $\exists \varepsilon > 0$ such that $\forall z \in A(z_k, 0, \varepsilon)$ we have

$$g(z) = \sum_{j=0}^{\infty} a_j (z - z_k)^j + \frac{b_1}{z - z_k} + \frac{b_2}{(z - z_k)^2} + \cdots + \frac{b_m}{(z - z_k)^m}.$$

On the other hand the function e^{zt} is entire and $\forall z \in \mathbb{C}$

$$e^{zt} = e^{z_k t} e^{(z - z_k)t} = e^{z_k t} \sum_{j=0}^{\infty} \frac{t^j}{j!} (z - z_k)^j.$$

Multiplying the two series, from the expression of the coefficient of the term $(z - z_k)^{-1}$ we obtain

$$\operatorname*{Res}_{z=z_k} \left(e^{zt} g(z) \right) = e^{z_k t} \left(b_1 + \frac{b_2 t}{1!} + \frac{b_3 t^2}{2!} + \cdots + \frac{b_m t^{m-1}}{(m-1)!} \right). \tag{10.23}$$

Example 10.24 Determine the function $f(t)$ whose Laplace transform is $g(z) = z/(z - z_1)^2$. Choose $x_0 > \operatorname{Re} z_1$ and define $\gamma_{y_0}(\theta) = x_0 + y_0 e^{i\theta}$ with $\pi/2 \leq \theta \leq 3\pi/2$ and $y_0 > 0$. For $z \in \{\gamma_{y_0}\}$ we have

$$|g(z)| = \frac{|z|}{|z - z_1|^2} \leq \frac{|x_0| + y_0}{||z| - |z_1||^2} \leq \frac{|x_0| + y_0}{(y_0 - |x_0| - |z_1|)^2} \xrightarrow{y_0 \to \infty} 0.$$

It is then possible to apply the formula (10.22), which in this case provides

$$f(t) = \operatorname*{Res}_{z=z_1} \left(e^{zt} g(z) \right), \qquad t > 0.$$

The function $g(z)$ has a double pole at z_1 and in the annulus $0 < |z - z_1| < \infty$ the following Laurent series expansion holds

$$g(z) = \frac{z - z_1 + z_1}{(z - z_1)^2} = \frac{1}{z - z_1} + \frac{z_1}{(z - z_1)^2}.$$

Using (10.23) we conclude

$$f(t) = e^{z_1 t}(1 + z_1 t), \qquad t > 0.$$

It's easy to verify that

$$\int_0^{\infty} e^{-zt} e^{z_1 t}(1 + z_1 t)\, dt = \frac{z}{(z - z_1)^2}, \qquad \operatorname{Re} z > \operatorname{Re} z_1.$$

Exercises

10.1 Calculate the real integral

$$\int_0^{2\pi} \frac{d\theta}{(a + b\cos\theta)^2}, \qquad a > b > 0.$$

10.2 Calculate the real integral

$$\int_0^{2\pi} \cos^2\theta \, d\theta.$$

10.3 Calculate the real integral

$$\int_0^{\pi} \frac{d\theta}{a + \cos\theta}, \qquad a > 1.$$

10.4 Calculate the real integral

$$\int_0^{2\pi} (\sin\theta)^{2n} d\theta, \qquad n \in \mathbb{N}.$$

10.5 Calculate the real integral

$$\int_0^{\pi} \frac{d\theta}{(a + \cos\theta)^2}, \qquad a > 1.$$

10.6 Calculate the real integral

$$\int_0^{2\pi} \frac{d\theta}{2 + \cos\theta}.$$

10.7 Calculate the integral

$$\int_0^{2\pi} \sin(\exp(\cos\theta + i\sin\theta)) d\theta.$$

10.8 Calculate the integral

$$\int_0^{2\pi} e^{e^{i\theta}} d\theta.$$

10.9 Calculate the real integral

$$\int_0^{\infty} \frac{1}{(x^2 + 1)^2} dx.$$

10.10 Calculate the real integral

$$\int_{-\infty}^{+\infty} \frac{x}{(x^2 + 4x + 13)^2}\,dx.$$

10.11 Calculate the real integral

$$\int_{-\infty}^{+\infty} \frac{x^2}{(x^2 + 1)(x^2 + 9)}\,dx.$$

10.12 Calculate the real integral

$$\int_{0}^{\infty} \frac{x^2}{(x^2 + 16)^2}\,dx.$$

10.13 Calculate the Cauchy principal value of the real integral

$$\mathrm{PV} \int_{-\infty}^{+\infty} \frac{x \sin(ax)}{x^2 + b^2}\,dx, \qquad a > 0, \quad b > 0.$$

10.14 Calculate the Cauchy principal value of the real integral

$$\mathrm{PV} \int_{-\infty}^{+\infty} \frac{x \sin(ax)}{(x - b)^2 + c^2}\,dx, \qquad a > 0, \quad b > 0, \quad c > 0.$$

10.15 Calculate the Cauchy principal value of the real integral

$$\mathrm{PV} \int_{-\infty}^{+\infty} \frac{x^2 \sin x}{(x - 1)(x^2 + 1)}\,dx$$

defined as the limit

$$\lim_{R_1,R_2 \to \infty} \lim_{r \to 0^+} \left(\int_{-R_1}^{1-r} \frac{x^2 \sin x}{(x - 1)(x^2 + 1)}\,dx + \int_{1+r}^{R_2} \frac{x^2 \sin x}{(x - 1)(x^2 + 1)}\,dx \right).$$

10.16 Calculate the Cauchy principal value of the improper integral

$$\mathrm{PV} \int_{-\infty}^{+\infty} \frac{e^{ix}}{x}\,dx = \lim_{R_1,R_2 \to \infty} \lim_{r \to 0} \left(\int_{-R_1}^{-r} \frac{e^{ix}}{x}\,dx + \int_{r}^{R_2} \frac{e^{ix}}{x}\,dx \right).$$

Draw the integration path and justify each single step.

10.17 Calculate the real integral

$$\int_{-\infty}^{+\infty} \frac{x \sin(ax)}{(x-b)^2 + c^2}\, dx, \qquad a > 0, \quad b > 0, \quad c > 0.$$

10.18 Calculate the real integral

$$\int_0^\infty \frac{\sin x}{x(x^2+1)}\, dx.$$

10.19 Calculate the integral

$$\int_{-\infty}^{+\infty} \frac{e^{-i\omega x}}{x^2 + a^2}\, dx, \qquad a > 0, \quad \omega \in \mathbb{R}.$$

Hint: Consider separately the cases $\omega > 0$ and $\omega < 0$.

10.20 Calculate the real integral

$$\int_{-\infty}^{+\infty} \frac{x \sin(\pi x)}{x^2 + 2x + 5}\, dx.$$

10.21 Calculate the real integral

$$\int_{-\infty}^{+\infty} \frac{\sin x}{x^2 - 6x + 10}\, dx.$$

10.22 Calculate the real integral

$$\int_0^\infty \frac{x \sin x}{x^4 + 1}\, dx.$$

10.23 Calculate the real integral

$$\int_{-\infty}^{+\infty} \frac{x \sin(x) + 3}{x^2 + 1}\, dx.$$

10.24 Calculate the real integral

$$\int_{-\infty}^{+\infty} \frac{x \sin(x) + 7}{(x^2 + 1)^2}\, dx.$$

10.25 Calculate the integral

$$\int_{-\infty}^{+\infty} \frac{\cos(px) - \cos(qx)}{x^2}\, dx,$$

with p, q relative integers. Draw the integration path and justify each single step.

10.26 Calculate the real integral

$$\int_0^\infty \frac{\ln x}{x^2 + 1}\, dx.$$

10.27 Calculate the real integral

$$\int_0^\infty \frac{\sqrt{x}}{x^2 + 1}\, dx.$$

10.28 Calculate the real integral

$$\int_0^\infty \frac{\ln x}{x^2 + 4}\, dx.$$

10.29 Calculate the real integral

$$\int_0^\infty \frac{1}{\sqrt[3]{x}(x^2 + 1)}\, dx.$$

10.30 Calculate the real integral

$$\int_0^\infty \frac{\ln x}{x^2 - 2x + 2}\, dx.$$

10.31 Calculate the real integral

$$\int_0^\infty \frac{\ln x}{x^2 - 2x + 2}\, dx.$$

10.32 Calculate the real integral

$$\int_0^\infty \frac{\log(1 + x^2)}{x^{a+1}}\, dx, \qquad 0 < a < 2.$$

Hint: First, perform an integration by parts.

10.33 Calculate the real integral

$$\int_0^\infty \frac{\sqrt[3]{x^2}}{x^2 + 2\sqrt{3}x + 4}\,\mathrm{d}x.$$

10.34 Calculate the real integral

$$\int_0^\infty \frac{x + 1}{\sqrt{x}(x^2 - 2x + 2)}\,\mathrm{d}x.$$

10.35 Calculate the real integral

$$\int_0^\infty \frac{x^{1-a}}{(1 + x^2)^2}\,\mathrm{d}x, \qquad -2 < a < 2.$$

10.36 Calculate the real integral

$$\int_0^\infty \frac{x^{1-a}}{1 + x^3}\,\mathrm{d}x, \qquad a \in \mathbb{R},$$

specifying for which values of a the integral is convergent.

10.37 Calculate the real integral

$$\int_0^\infty \frac{x^{1-a}}{(4 + x^2)^2}\,\mathrm{d}x, \qquad a \in \mathbb{R},$$

specifying for which values of a the integral is convergent.

10.38 Calculate the real integral

$$\int_0^\infty \frac{x^a}{x^2 + 2x + 1}\,\mathrm{d}x, \qquad a \in \mathbb{R},$$

specifying for which values of a the integral is convergent.

10.39 Calculate the real integral

$$\int_{-\infty}^{+\infty} \frac{e^{ax}}{\cosh x}\,\mathrm{d}x, \qquad -1 < a < 1.$$

Hint: consider the rectangular integration path of vertices $\pm R$ and $\pm R + i\pi$.

10.40 Calculate the real integral

$$\int_0^\infty \frac{1}{1 + x^{2n+1}}\,\mathrm{d}x, \qquad n - 1, 2, 3, \ldots$$

10.41 Calculate the real integral

$$\int_{-\infty}^{+\infty} \frac{\cos x}{e^x + e^{-x}}\,dx.$$

Hint: consider the rectangular integration path of vertices $\pm R$ and $\pm R + i\pi$.

10.42 Prove the convergence and calculate the value of the improper integral

$$\int_0^\infty e^{-wx^2}\,dx, \qquad w \in \mathbb{C}, \quad w \neq 0, \quad \operatorname{Re} w \geq 0.$$

Recall that $\int_0^\infty e^{-x^2}\,dx = \sqrt{\pi}/2$.

10.43 Calculate the real integral

$$\int_0^{2\pi} \log\left(a + a^{-1} - 2\cos\theta\right)d\theta, \qquad a > 1.$$

10.44 Show that the following improper integral is convergent and calculate its corresponding value:

$$\int_{-\infty}^{+\infty} \frac{x \sin(3x) + 3}{x^2 - 2x + 2}\,dx.$$

10.45 Show that the following improper integral is convergent and calculate its corresponding value

$$\int_{-\infty}^{+\infty} \frac{x \sin(3x) + 3}{x^2 + 1}\,dx.$$

10.46 Calculate, using the residue theorem, the integral

$$\int_{-\infty}^{+\infty} e^{-x^2/\sigma^2} e^{-i\lambda x}\,dx,$$

with $\sigma, \lambda \in \mathbb{R}$.

10.47 Calculate the Bromwich integral

$$\frac{1}{2\pi i} \lim_{R \to \infty} \int_{\lambda_R} e^{zt} \frac{1}{(z - z_0)(z - z_1)}\,dz, \qquad z_0 \neq z_1 \in \mathbb{C}, \quad t > 0,$$

where $\lambda_R(y) = a + iy$, $-R \leq y \leq R$, with $a > \max(\operatorname{Re} z_0, \operatorname{Re} z_1)$.

10.48 Determine the function $f(t) : (0, \infty) \mapsto \mathbb{R}$ such that

$$\int_0^\infty e^{-zt} f(t)\,dt = \frac{1}{(z-5)^3}, \qquad z \in \mathbb{C}, \quad \mathrm{Re}\, z > \mathrm{Re}\, -5.$$

10.49 Determine the function $f(t) : (0, \infty) \mapsto \mathbb{R}$ such that

$$\int_0^\infty e^{-zt} f(t)\,dt = \frac{z}{(z-3)^2}, \qquad z \in \mathbb{C}, \quad \mathrm{Re}\, z > \mathrm{Re}\, 3.$$

Chapter 11
Further Properties of Analytic Functions

Abstract Analytic continuation. Borel and Weierstrass continuations. Schwarz reflection principle. Continuation along a path, monodromy theorem. Principle of maximum (minimum) modulus. Meromorphic functions. The number of zeros and poles of a meromorphic function within a bounded region is finite. Integral along a simple closed curve of a meromorphic function multiplied by an analytic one: generalized principle of the argument. Index of a closed curve. Principle of the argument. Formula of the inverse function. Abel-Plana formula. Rouché theorem. Determination of the number of zeros of an analytic function within a given region. Harmonic and conjugate harmonic functions. The function $f = u + iv$ is analytic if and only if v is harmonic conjugate to u. Existence of the conjugate harmonic function and its determination. Laplace's and saddle point methods.

11.1 Analytic Continuation

Let D_1 and D_2 with $D_1 \cap D_2 \neq \emptyset$ be two open and connected subsets of $\mathbb{C}$. Let f_1 be an analytic function in D_1. If there exists an analytic function f_2 in D_2 such that $f_2(z) = f_1(z) \; \forall z \in D_1 \cap D_2$, then f_2 is called the analytic continuation of f_1 from D_1 to D_2. This analytic continuation may not exist, however, if it exists, by the Identity Theorem 9.16 it is unique.

Example 11.1 Consider the analytic functions $f_1 : D_1 \mapsto \mathbb{C}$ and $f_2 : D_2 \mapsto \mathbb{C}$ defined by

$$f_1(z) = \sum_{k=0}^{\infty} z^k, \qquad D_1 = B(0, 1),$$

$$f_2(z) = \frac{1}{1 - z}, \qquad D_2 = \{z \in \mathbb{C} : \operatorname{Re} z < 1\}.$$

C. Presilla, *Complex Functions of a Variable*, La Matematica per il 3+2 179,
https://doi.org/10.1007/978-3-032-12494-4_11

As we know $f_2(z) = f_1(z) \; \forall z \in D_1 \cap D_2$, therefore f_2 is the analytic continuation of f_1 from D_1 to D_2. Let us observe that

$$f_2(z) = \int_0^\infty e^{-t} \sum_{k=0}^\infty \frac{1}{k!} (zt)^k \, dt = \int_0^\infty e^{-(1-z)t} \, dt,$$

the integral being convergent for $\operatorname{Re} z < 1$.

Example 11.2 (Borel Continuation) The previous example admits the following generalization. Let $f_1 : D_1 \mapsto \mathbb{C}$ be the analytic function

$$f_1(z) = \sum_{k=0}^\infty a_k z^k, \qquad D_1 = B(0, R), \qquad \frac{1}{R} = \limsup_{k \to \infty} |a_k|^{1/k}.$$

Then set

$$f_2(z) = \int_0^\infty e^{-t} h(zt) \, dt, \qquad h(zt) = \sum_{k=0}^\infty \frac{a_k}{k!} (zt)^k.$$

Note that, in the non-trivial case $R > 0$, the radius of convergence of the series having $h(zt)$ as its sum is infinite. For $|z| < R$ we have $f_2(z) = f_1(z)$ since

$$\int_0^\infty e^{-t} \sum_{k=0}^\infty \frac{a_k}{k!} (zt)^k \, dt = \sum_{k=0}^\infty \frac{a_k}{k!} \int_0^\infty e^{-t} (zt)^k \, dt = \sum_{k=0}^\infty a_k z^k,$$

therefore f_2 is the analytic continuation of f_1 from D_1 to that region, to be determined given the sequence (a_k), in which the integral that defines f_2 is convergent.

Example 11.3 Consider again the analytic function $f_1 : D_1 \mapsto \mathbb{C}$

$$f_1(z) = \sum_{k=0}^\infty z^k, \qquad D_1 = B(0, 1).$$

Having chosen $z_1 \in D_1$, with $z_1 \neq 0$, we define a new analytic function $f_2 : D_2 \mapsto \mathbb{C}$ as the sum of the power series of center z_1

$$f_2(z) = \sum_{n=0}^\infty c_n (z - z_1)^n, \qquad D_2 = B(z_1, R), \qquad \frac{1}{R} = \limsup_{n \to \infty} |c_n|^{1/n}.$$

If we impose that $f_2(z) = f_1(z) \; \forall z \in D_1 \cap D_2$, since the function f_1 is analytic at z_1, by the Uniqueness Theorem 8.5 the series just written is the Taylor series expansion of f_1 with center z_1. Therefore, the coefficients c_n can be calculated as

$$
\begin{aligned}
c_n &= \frac{1}{n!} \frac{d^n}{dz^n} f_1(z) \bigg|_{z=z_1} \\[2mm]
&= \sum_{k=n}^{\infty} \frac{1}{n!} k(k-1) \ldots (k-n+1) z_1^{k-n} \\[2mm]
&= \sum_{j=0}^{\infty} \binom{n+j}{n} z_1^j.
\end{aligned}
$$

Using the relationship

$$
\binom{n+j}{n} = \binom{n+j}{j} = (-1)^j \binom{-n-1}{j}
$$

and the notable expansion

$$
(1+w)^a = \sum_{j=0}^{\infty} \binom{a}{j} w^j, \qquad a \in \mathbb{C}, \quad |w| < 1,
$$

we get $c_n = (1 - z_1)^{-n-1}$. In conclusion, it turns out

$$
f_2(z) = \sum_{n=0}^{\infty} \frac{1}{(1-z_1)^{n+1}} (z - z_1)^n,
$$

which has radius of convergence $R = |1 - z_1|$. Obviously, this result could be reached directly by assuming that we know the sum of the series that defines f_1

$$
f_1(z) = \frac{1}{1-z} = \frac{1}{1 - z_1 - (z - z_1)} = \frac{1}{1 - z_1} \frac{1}{1 - \frac{z - z_1}{1 - z_1}}.
$$

Note that, except for the case in which z_1 belongs to the real interval $(0, 1)$, the domain D_2 always represents an continuation of D_1 in the direction of z_1. We conclude that f_2 is the analytic continuation of f_1 from D_1 to D_2.

Example 11.4 (Weierstrass Continuation) Let f be an analytic function in D open and connected, and suppose we know the Taylor series expansion of such a function in a neighborhood of a point $z_0 \in D$

$$
f(z) = \sum_{n=0}^{\infty} c_n^{(0)} (z - z_0)^n, \qquad z \in B(z_0, r_0).
$$

The value of $f(z)$ at a generic point $z \in D$ can be obtained with a procedure, known as Weierstrass analytic continuation, which generalizes the previous example. Join z_0 to z by means of a polygonal P. Let $\delta > 0$ be the minimum distance of P from the boundary of D, identify a finite sequence of points $z_0, z_1, \ldots, z_m$, with $z_m = z$, belonging to P and such that $|z_k - z_{k-1}| < \delta$, $k = 1, \ldots, m$. In the ball $B(z_k, r_k)$, where r_k is the distance of z_k from ∂D, the function f can be expanded in the Taylor series

$$f(z) = \sum_{n=0}^{\infty} c_n^{(k)} (z - z_k)^n, \qquad z \in B(z_k, r_k),$$

and, since $z_k \in B(z_{k-1}, r_{k-1})$, the coefficients $c_n^{(k)}$ can be calculated as

$$c_n^{(k)} = \frac{1}{n!} \frac{d^n}{dz^n} \sum_{j=0}^{\infty} c_j^{(k-1)} (z - z_{k-1})^j \Bigg|_{z=z_k}$$

$$= \sum_{j=n}^{\infty} \binom{j}{n} c_j^{(k-1)} (z_k - z_{k-1})^{j-n}.$$

By iterating the process from $k = 1$ up to $k = m$, it is possible, in principle, to determine $f(z) = c_0^{(m)}$.

The methods discussed in the previous examples allow us to perform an analytic continuation between two open and connected domains, D_1 and D_2, having non-zero intersection. It is also possible to consider the analytical extension between two open and connected domains, D_1 and D_2, that are adjacent, that is, such that $D_1 \cap D_2 = \emptyset$ but with the Jordan curves ∂D_1 and ∂D_2 having as intersection some open curve. Let γ be such a curve deprived of its extreme points. In the case of adjacent domains we say that (f_1, D_1) can be analytically continued to (f_2, D_2) if there exists an analytic function $f : D \mapsto \mathbb{C}$, with $D = D_1 \cup D_2 \cup \{\gamma\}$ open and connected, such that $f(z) = f_1(z) \ \forall z \in D_1$ and $f(z) = f_2(z) \ \forall z \in D_2$. It is evident that, given two domains D_1 and D_2 with non-zero intersection, we can always introduce two adjacent domains $\tilde{D}_1$ and $\tilde{D}_2$ whose boundaries have in common an open curve deprived of the endpoints γ such that $D_1 \cup D_2 = \tilde{D}_1 \cup \tilde{D}_2 \cup \{\gamma\}$. The vice versa is also true. It follows then that the analytic continuation through domains with non-zero intersection is equivalent to that through adjacent domains.

To perform an analytic continuation in the case of adjacent domains the Schwarz reflection method may be useful.

Theorem 11.5 (Schwarz Reflection Principle) *Let $D \subset \mathbb{C}$ be open, connected and symmetric with respect to the real axis, namely, coincident with $D^* = \{z \in \mathbb{C} : \bar{z} \in D\}$. Set $D_\pm = \{z \in D : \mathrm{Im}\, z \gtrless 0\}$ and $D_0 = \{z \in D : \mathrm{Im}\, z = 0\}$, let $f : D_+ \cup D_0 \mapsto \mathbb{C}$ be continuous in $D_+ \cup D_0$, analytic in D_+ and real on D_0.*

Then, the function

$$g(z) = \begin{cases} f(z) & z \in D_+ \cup D_0 \\[2mm] \overline{f(\overline{z})} & z \in D_- \end{cases}$$

is analytic in D, that is, g is the analytic continuation of f from D_+ to D.

Proof First note that, since D is open, connected and symmetric with respect to the real axis, D_0 must contain at least one open interval of the real axis. It is easy to verify that g is continuous in D_0 and therefore in D. We then show that the integral of g along an arbitrary piecewise regular closed curve contained in D is zero and, by Morera's theorem, we can conclude that g is analytic in D. Let $\gamma : [a, b] \mapsto \mathbb{C}$ be a piecewise regular closed path such that $\{\gamma\} \subset D$. Various cases are possible. If $\{\gamma\} \subset D_+$, the integral of g along γ is zero since $g(z) = f(z) \; \forall z \in D_+$ and f is analytic in D_+. If $\{\gamma\} \subset D_-$, the integral of g along γ can be reduced to the integral of f along the path $\overline{\gamma(t)}, a \leq t \leq b$, contained in D_+

$$\int_\gamma g(z)\mathrm{d}z = \int_a^b \overline{f\left(\overline{\gamma(t)}\right)}\gamma'(t)\mathrm{d}t = \overline{\int_a^b f\left(\overline{\gamma(t)}\right)\overline{\gamma'(t)}\mathrm{d}t} = \overline{\int_{\overline{\gamma}} f(z)\mathrm{d}z},$$

so also this integral is zero. The cases in which $\{\gamma\} \subset D_+ \cup D_0$ or $\{\gamma\} \subset D_- \cup D_0$ are treated in an analogous way considering a limit procedure starting from a path γ_ε homotopic to γ, with $\{\gamma_\varepsilon\} \subset D_+$ or $\{\gamma_\varepsilon\} \subset D_-$ and such that $\lim_{\varepsilon \to 0} \gamma_\varepsilon = \gamma$. It remains to consider the situation in which $\{\gamma\}$ has non-empty intersection with both D_+ and D_-. In this case we can write

$$\int_\gamma g(z)\mathrm{d}z = \int_{\gamma_+} g(z)\mathrm{d}z + \int_{\gamma_-} g(z)\mathrm{d}z,$$

where $\gamma_\pm$ are two closed paths formed by joining the portions of γ contained in $D_\pm$ with one or more segments of D_0. Since $\{\gamma_\pm\} \subset D_\pm \cup D_0$, by what was said before, each of the two integrals on $\gamma_\pm$ is zero and therefore the integral of g along γ is zero. $\qquad\square$

In Theorem 11.5 it was assumed that the domain D in the plane z has an axis of symmetry λ coinciding with the real axis. It was also assumed that for $z \in \{\lambda\}$ the function f assumes values belonging to an axis Λ of the plane $w = f(z)$ coinciding with the real axis. The function g was then constructed by assigning to the point z^* symmetric of z with respect to λ the value w^* symmetric of $w = f(z)$ with respect to Λ. By applying the linear transformations $z' = az + b$ and $w' = cw + d$, with $a, b, c, d \in \mathbb{C}$, it is then possible to extend the Schwarz reflection principle to completely generic axes of symmetry λ and Λ. The case where λ and Λ are arcs of circles can be treated similarly, for details see [7, 10].

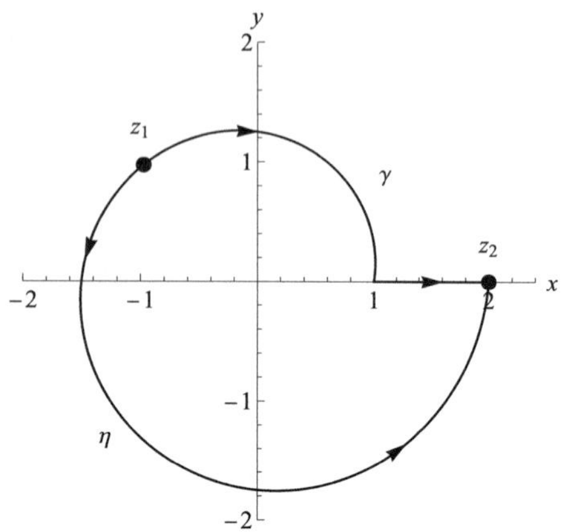

Fig. 11.1 Different paths γ and η having the same end points, namely, $z_1 = (1 + t_1)e^{i2\pi t_1}$, with $t_1 = 3/8$, and $z_2 = 2$, used to analytically continue the principal branch of $\log z$ from z_1 to z_2

Finally, let us return to the Weierstrass analytic continuation. Without going into details, it is clear that it is possible to define an analytic continuation along a path γ that is not necessarily a polygon. Let γ be a path with extreme points z_1 and z_2 and suppose that f is analytic in a neighborhood of z_1. We have no theorems that establish under which conditions f can be continued along γ up to z_2. However, if this is possible, the Identity Theorem 9.16 assures us that the f thus obtained is unique. What happens if we analytically continue f from z_1 to z_2, assuming that it is possible, along a different path η? In general, the result $f(z_2)$ will depend on the path chosen to perform the continuation.

Example 11.6 Continue the main branch of $\log z$ from $z_1 = (1 + t_1)e^{i2\pi t_1}$, with $t_1 = 3/8$, to $z_2 = 2$ along the γ and η paths shown in Fig. 11.1. The principal branch of $\log z$ is the analytic function $f_1 : D_1 \mapsto \mathbb{C}$, with $D_1 = \{z \in \mathbb{C} : z \neq -u,\ 0 \leq u < \infty\}$, defined by

$$f_1(z) = \ln r + i\theta, \qquad z = re^{i\theta}, \quad r > 0, \quad -\pi < \theta < \pi.$$

The path γ is internal to the domain D_1 so the analytic continuation of $\log z$ along γ gives $\log 2 = f_1(2) = \ln 2$. Now consider

$$f_2(z) = \ln r + i\theta, \qquad z = re^{i\theta}, \quad r > 0, \quad \pi/2 < \theta < 5\pi/2.$$

The function $f_2 : D_2 \mapsto \mathbb{C}$, analytic in $D_2 = \{z \in \mathbb{C} : z \neq iu,\ 0 \leq u < \infty\}$, has the property $f_2(z) = f_1(z)$ for every z such that $\operatorname{Re} z < 0$ and $\operatorname{Im} z > 0$. It is therefore the analytic continuation of $\log z$ from D_1 to D_2 along η. In this case we have $\log 2 = f_2(2) = \ln 2 + i2\pi$.

In the example just illustrated, the principal branch of the logarithm is continued to the non-simply connected domain $G = \mathbb{C} \setminus \{0\}$ within which the non-equivalent paths γ and η are non-homotopic. Sufficient conditions for the analytic continuation to be single-value (or monodromic) are identified by the following theorem.

Theorem 11.7 (Monodromy Theorem) *Let $f : B(z_1, r) \mapsto \mathbb{C}, r > 0$, be an analytic function in a neighborhood of z_1. If G is a simply connected open set containing $B(z_1, r)$ and it is possible to analytically continue f along every path connecting z_1 to a generic point $z_2 \in G$, then there exists an analytic function $g : G \mapsto \mathbb{C}$ such that $g(z) = f(z) \ \forall z \in B(z_1, r)$.*

For a precise definition of analytic continuation along a path and the proof of the monodromy theorem see [5, 6, 8].

11.2 Principle of Maximum Modulus

Theorem 11.8 (Principle of Maximum Modulus) *Let f be analytic and non-constant in $D \subset \mathbb{C}$ open and connected. Then $|f|$ has no maximum in D, that is, there does not exist a point $z_0 \in D$ such that $|f(z_0)| \geq |f(z)| \ \forall z \in D$.*

Proof Argue by contradiction and suppose that $\exists z_0 \in D$ such that $|f(z_0)| \geq |f(z)|$ $\forall z \in D$. Since D is open, $\exists r > 0$ such that $B(z_0, r) \subset D$. Let $\gamma_\rho(t) = z_0 + \rho e^{it}$ with $0 \leq t \leq 2\pi$ be a positively oriented circular path with center z_0 and radius $\rho < r$. For the Cauchy integral formula we have

$$f(z_0) = \frac{1}{2\pi i} \int_{\gamma_\rho} \frac{f(z)}{z - z_0} dz = \frac{1}{2\pi} \int_0^{2\pi} f(z_0 + \rho e^{it}) dt. \tag{11.1}$$

Formula (11.1) is known as the Gaussian mean value theorem. Using the inequality (7.3) and the assumed hypothesis, we get

$$|f(z_0)| \leq \frac{1}{2\pi} \int_0^{2\pi} \left| f(z_0 + \rho e^{it}) \right| dt$$

$$\leq \frac{1}{2\pi} 2\pi \sup_{t \in [0, 2\pi]} \left| f(z_0 + \rho e^{it}) \right|$$

$$\leq |f(z_0)|.$$

Therefore

$$\int_0^{2\pi} \left(|f(z_0)| - \left| f(z_0 + \rho e^{it}) \right| \right) dt = 0.$$

Since, by hypothesis, the integrand function is non-negative, the cancellation of the integral implies the cancellation of the integrand function, that is, $\left| f(z_0 + \rho e^{it}) \right| = |f(z_0)| \ \forall t \in [0, 2\pi]$. On the other hand $\rho < r$ is arbitrary and therefore $|f(z)| = |f(z_0)| \ \forall z \in B(z_0, r)$. By Theorem 5.20 we conclude that $f(z) = f(z_0) \ \forall z \in$

$B(z_0, r)$ and by the Identity Theorem 9.16 we arrive at the contradiction that $f(z) = f(z_0) \ \forall z \in D$. $\qquad\qquad\square$

Corollary 11.9 *Let f be continuous in $D \subset \mathbb{C}$ closed and bounded, and also analytic and non-constant in D°. Then the maximum of $|f|$ in D is assumed on ∂D.*

Proof First, since D is compact, $|f|$ has a maximum in D, that is, $\exists z_0 \in D$ such that $|f(z_0)| \geq |f(z)| \ \forall z \in D$. If $z_0 \in D^\circ$, since $\exists r > 0$ such that $B(z_0, r) \subset D^\circ$, there would be a contradiction of Theorem 11.8, that is, $|f(z)|$ would have a maximum in the connected open set $B(z_0, r)$ where f is analytic and non-constant. We must therefore conclude that $z_0 \in \partial D$. $\qquad\qquad\square$

Theorem 11.10 (Principle of Minimum Modulus) *Let f be analytic and non-constant in $D \subset \mathbb{C}$ open and connected. Then $|f|$ has no minimum in $D \setminus D_0$, where $D_0 = \{z \in D : \ f(z) = 0\}$.*

Proof Note that $D \setminus D_0$ is still open and connected. The function $g(z) = 1/f(z)$ is analytic and non-constant in $D \setminus D_0$ and therefore, by Theorem 11.8, $|g|$ has no maximum in $D \setminus D_0$. This is equivalent to saying that $|f|$ has no minimum in $D \setminus D_0$. $\qquad\qquad\square$

Corollary 11.11 *Let f be continuous in $D \subset \mathbb{C}$ closed and bounded, and also analytic, non-constant and never zero in D°. Then the minimum of $|f|$ in D is assumed on ∂D.*

Corollary 11.12 *Let f be continuous in $D \subset \mathbb{C}$ closed and bounded, and also analytic and non-constant in D°. Then the maximum and minimum of $\mathrm{Re}\, f$ and $\mathrm{Im}\, f$ in D are assumed on ∂D.*

Proof Let $g(z) = e^{f(z)}$. Since g is continuous in D, analytic and non-constant in D° and never zero, by the Corollaries 11.9 and 11.11 the maximum and minimum in D of $|g(z)| = e^{\mathrm{Re}\, f(z)}$ are assumed on ∂D. By the monotonicity of the real exponential, the same is true for the function $\mathrm{Re}\, f$. In the case of $\mathrm{Im}\, f$ an identical proof is valid by setting $g(z) = e^{-if(z)}$. $\qquad\qquad\square$

Example 11.13 Find the maximum of $|\sin z|$ for $z \in D = \{z = x + iy, \ -1 \leq x \leq 1, \ -1 \leq y \leq 1\}$. First, observe that $|\sin z|$ is continuous and D is compact, so $|\sin z|$ has a maximum in D. Since $\sin z$ is analytic and not constant in D, by the maximum modulus theorem $|\sin z|$ cannot assume the maximum in D°. The maximum of $|\sin z|$ must therefore be found on ∂D. We have

$$
\begin{aligned}
|\sin z|^2 &= \frac{e^{i(x+iy)} - e^{-i(x+iy)}}{2i} \, \frac{e^{-i(x-iy)} - e^{+i(x-iy)}}{-2i} \\[2mm]
&= \frac{e^{-2y} - e^{i2x} - e^{-i2x} + e^{2y}}{4} \\[2mm]
&= \frac{-e^{i2x} - e^{-i2x} + 2}{4} + \frac{e^{-2y} + e^{2y} - 2}{4}
\end{aligned}
$$

$$= \left(\frac{e^{ix} - e^{-ix}}{2i} \right)^2 + \left(\frac{e^{y} - e^{-y}}{2} \right)^2$$

$$= \sin^2 x + \sinh^2 y.$$

On the sides $y = \pm 1$ the maximum of $|\sin z|$ is assumed for $x = \pm 1$ and it is worth $(\sin^2 1 + \sinh^2 1)^{1/2}$. On the sides $x = \pm 1$ the maximum of $|\sin z|$ is assumed for $y = \pm 1$ and is $(\sin^2 1 + \sinh^2 1)^{1/2}$. In conclusion, $|\sin z|$ takes the maximum value $(\sin^2 1 + \sinh^2 1)^{1/2}$ at the four vertices $z = \pm 1 \pm i$.

11.3 Meromorphic Functions

Definition 11.14 (Meromorphic Function) Let $f : D \mapsto \mathbb{C}$ with $D \subset \mathbb{C}$ open and connected. The function f is said to be meromorphic in D if it is analytic in D except for poles (of finite order).

Theorem 11.15 *Let f be meromorphic within γ, a simple closed curve, and let f be analytic and non-zero on γ. Then the number of poles and zeros of f within γ is finite.*

Proof By the Jordan curve theorem, $\mathrm{Int}(\gamma)$ is a bounded set. Suppose, by contradiction, that the number of poles (zeros) of f internal to γ, and therefore contained in $\overline{\mathrm{Int}(\gamma)}$, is infinite. Let $(z_k)_{k=1}^{\infty}$ be the sequence of these poles (zeros). Since $\overline{\mathrm{Int}(\gamma)}$ is closed and bounded in $\mathbb{C}$, by the Heine-Borel theorem this set is compact, that is, sequentially compact, and therefore it is possible to extract from (z_k) a subsequence which is convergent in $\overline{\mathrm{Int}(\gamma)}$. This is in contradiction with the fact that the poles (the zeros) are isolated points. $\qquad\square$

Theorem 11.16 (Generalized Argument Principle) *Let f be meromorphic within γ, a simple closed piecewise regular positively oriented curve, and let f be analytic and non-zero on γ. If g is analytic on and within γ, then*

$$\int_{\gamma} g(z) \frac{f'(z)}{f(z)} \, dz = 2\pi i \left(\sum_{k=1}^{m} \alpha_k g(a_k) - \sum_{k=1}^{n} \beta_k g(b_k) \right), \tag{11.2}$$

where a_k, $k = 1, \ldots, m$, and b_k, $k = 1, \ldots, n$, are, respectively, the zeros and poles of f within γ of order α_k and β_k.

Proof By the residue theorem, we have

$$\int_{\gamma} g(z) \frac{f'(z)}{f(z)} \, dz = 2\pi i \sum_{k} \operatorname*{Res}_{z=z_k} g(z) \frac{f'(z)}{f(z)},$$

where z_k are the isolated singular points of gf'/f within γ. We show that such points are simple poles corresponding to the zeros and poles of f. If f has at a a zero of order α, then $f(z) = h(z)(z - a)^\alpha$ with h analytic and non-zero at a. Therefore

$$
\begin{aligned}
g(z)\frac{f'(z)}{f(z)} &= g(z)\frac{h'(z)(z - a)^\alpha + h(z)\alpha(z - a)^{\alpha-1}}{h(z)(z - a)^\alpha} \\
&= \frac{g(z)h'(z)}{h(z)} + \frac{\alpha}{z - a}g(z).
\end{aligned}
$$

Since g and gh'/h are analytic at a, the function gf'/f has a simple pole at a with

$$
\operatorname*{Res}_{z=a} g(z)\frac{f'(z)}{f(z)} = \alpha g(a).
$$

If f has a pole of order β at b, then $f(z) = h(z)(z - b)^{-\beta}$ with h analytic and non-zero at b. Therefore

$$
\begin{aligned}
g(z)\frac{f'(z)}{f(z)} &= g(z)\frac{h'(z)(z - b)^{-\beta} - h(z)\beta(z - b)^{-\beta-1}}{h(z)(z - b)^{-\beta}} \\
&= \frac{g(z)h'(z)}{h(z)} - \frac{\beta}{z - b}g(z).
\end{aligned}
$$

Since g and gh'/h are analytic at b, the function gf'/f has a simple pole at b with

$$
\operatorname*{Res}_{z=b} g(z)\frac{f'(z)}{f(z)} = -\beta g(b).
$$

$\square$

Definition 11.17 (Index of a Closed Curve) We call index, or number of turns, of a closed curve γ with respect to a point $z_0 \notin \{\gamma\}$ the number

$$
n(\gamma, z_0) = \frac{1}{2\pi i} \int_\gamma \frac{dz}{z - z_0}. \tag{11.3}
$$

Consider a simple closed γ curve. If $z_0 \in \mathrm{Ext}(\gamma)$ we have $n(\gamma, z_0) = 0$. If $z_0 \in \mathrm{Int}(\gamma)$, then $n(\gamma, z_0) = \pm 1$ depending on whether γ is oriented positively $(+1)$ or negatively (-1). In general, the following result holds.

Theorem 11.18 *The index of a closed curve is a relative integer.*

Proof Let γ be a generic closed curve and z_0 be a point not belonging to its trace. Let $\gamma : [a, b] \mapsto \mathbb{C}$ be a path that represents such a curve (let's assume that γ is piecewise regular so that the integral (11.3) is well defined), we introduce $\eta : [a, b] \mapsto \mathbb{C}$

$$\eta(t) = (\gamma(t) - z_0) \exp\left(-\int_a^t \frac{\gamma'(u)}{\gamma(u) - z_0} du\right). \tag{11.4}$$

Differentiating with respect to the parameter t, we have

$$\eta'(t) = \gamma'(t) \exp\left(-\int_a^t \frac{\gamma'(u)}{\gamma(u) - z_0} du\right)$$

$$- (\gamma(t) - z_0) \frac{\gamma'(t)}{\gamma(t) - z_0} \exp\left(-\int_a^t \frac{\gamma'(u)}{\gamma(u) - z_0} du\right)$$

$$= 0.$$

Therefore $\eta(b) = \eta(a)$, that is,

$$(\gamma(b) - z_0) \exp\left(-\int_a^b \frac{\gamma'(u)}{\gamma(u) - z_0} du\right) = (\gamma(a) - z_0).$$

Since $\gamma(a) = \gamma(b)$ and $z_0 \notin \{\gamma\}$, it must be $\gamma(a) - z_0 = \gamma(b) - z_0 \neq 0$. Therefore

$$\exp\left(-\int_a^b \frac{\gamma'(u)}{\gamma(u) - z_0} du\right) = \exp\left(-n(\gamma, z_0) 2\pi i\right) = 1, \tag{11.5}$$

which implies $n(\gamma, z_0) \in \mathbb{Z}$. $\qquad\square$

Theorem 11.19 (Argument Principle) *Let f be meromorphic within γ, a positively oriented piecewise regular simple closed curve, and let f be analytic and non-zero on γ. Then*

$$n(f \circ \gamma, 0) = Z - P, \tag{11.6}$$

where Z and P are the number of zeros and poles of f, counting their multiplicity, internal to γ.

Proof From Theorem 11.16 with the choice $g(z) = 1$, we have

$$\frac{1}{2\pi i} \int_\gamma \frac{f'(z)}{f(z)} dz = Z - P. \tag{11.7}$$

On the other hand, if $\gamma : [a, b] \mapsto \mathbb{C}$ is a path that represents the curve γ, we have

$$\frac{1}{2\pi i} \int_\gamma \frac{f'(z)}{f(z)} dz = \frac{1}{2\pi i} \int_a^b \frac{f'(\gamma(t))}{f(\gamma(t))} \gamma'(t) dt$$

$$= \frac{1}{2\pi i} \int_a^b \frac{1}{(f \circ \gamma)(t)} (f \circ \gamma)'(t) dt$$

$$= \frac{1}{2\pi i} \int_{f \circ \gamma} \frac{dz}{z - 0}$$

$$= n(f \circ \gamma, 0).$$

Equating the two previous expressions, the statement follows. $\qquad\square$

Example 11.20 Consider $f(z) = z^2$ and let $\gamma(t) = Re^{it}$, $0 \le t \le 2\pi$. Within the circle γ the function f has $Z = 2$ zeros and $P = 0$ poles. On the other hand $(f \circ \gamma)(t) = R^2 e^{2it}$ and so

$$n(f \circ \gamma, 0) = \frac{1}{2\pi i} \int_{f \circ \gamma} \frac{dz}{z - 0}$$

$$= \frac{1}{2\pi i} \int_0^{2\pi} \frac{1}{(f \circ \gamma)(t)} (f \circ \gamma)'(t) dt$$

$$= \frac{1}{2\pi i} \int_0^{2\pi} \frac{2i R^2 e^{2it}}{R^2 e^{2it}} dt$$

$$= 2.$$

Example 11.21 Consider $f(z) = z^{-2}$ and let $\gamma(t) = Re^{it}$, $0 \le t \le 2\pi$. Within the circle γ the function f has $Z = 0$ zeros and $P = 2$ poles. On the other hand $(f \circ \gamma)(t) = R^{-2} e^{-2it}$ and therefore

$$n(f \circ \gamma, 0) = \frac{1}{2\pi i} \int_{f \circ \gamma} \frac{dz}{z - 0}$$

$$= \frac{1}{2\pi i} \int_0^{2\pi} \frac{1}{(f \circ \gamma)(t)} (f \circ \gamma)'(t) dt$$

$$= \frac{1}{2\pi i} \int_0^{2\pi} \frac{-2i R^{-2} e^{-2it}}{R^{-2} e^{-2it}} dt$$

$$= -2.$$

Theorem 11.22 (Inverse Function Formula) *Let f be analytic in $\overline{B}(z_0, R)$ and bijective in $B(z_0, R)$. Then, called γ the positively oriented circle $|z - z_0| = R$, $f^{-1}(w)$ is defined $\forall w \in f(B(z_0, R))$ by the formula*

$$f^{-1}(w) = \frac{1}{2\pi i} \int_\gamma \frac{z f'(z)}{f(z) - w} dz. \tag{11.8}$$

Proof It follows from Theorem 11.16 with the choice $g(z) = z$. In fact $\forall w \in f(B(z_0, R))$ the function $f(z) - w$ is analytic on γ and within it, it has a single simple zero $z = f^{-1}(w) \in \mathrm{Int}(\gamma)$ and does not vanish for $z \in \{\gamma\}$. Since $g(z) = z$

is also analytic on and within γ, we can apply the formula (11.2) and obtain

$$\int_\gamma z \frac{f'(z)}{f(z) - w}\,\mathrm{d}z = 2\pi \mathrm{i} f^{-1}(w).$$

$\square$

Theorem 11.23 (Abel-Plana Formula) *Let $f(z)$ be analytic $\forall z \in \mathbb{C}$, then*

$$\sum_{n=0}^{\infty} f(n) - \int_0^\infty f(x)\,\mathrm{d}x = \frac{1}{2} f(0) + \mathrm{i} \int_0^\infty \frac{f(\mathrm{i}t) - f(-\mathrm{i}t)}{\mathrm{e}^{2\pi t} - 1}\,\mathrm{d}t. \tag{11.9}$$

Proof It follows from Theorem 11.16 with an appropriate choice of $g(z)$ and of the path γ.
$\square$

Theorem 11.24 (Rouché) *Let f and g be analytic in $\overline{\mathrm{Int}(\gamma)}$ with γ piecewise regular simple closed curve and also let $|f(z)| > |g(z)|\ \forall z \in \{\gamma\}$. Then f and $f + g$ have the same number of zeros, counting their multiplicity, within γ.*

Proof Let $h = f + g$. The functions f and h never cancel on γ, in fact $\forall z \in \{\gamma\}$ we have $|f(z)| > |g(z)| \geq 0$ and $|h(z)| \geq ||f(z)| - |g(z)|| > 0$. Moreover, f and h are analytic in $\overline{\mathrm{Int}(\gamma)}$. By the principle of the argument, the number of zeros, counting their multiplicity, of f and h within γ is therefore equal to

$$Z_f = n(f \circ \gamma, 0) = \frac{1}{2\pi \mathrm{i}} \int_\gamma \frac{f'(z)}{f(z)}\,\mathrm{d}z, \tag{11.10}$$

$$Z_h = n(h \circ \gamma, 0) = \frac{1}{2\pi \mathrm{i}} \int_\gamma \frac{h'(z)}{h(z)}\,\mathrm{d}z. \tag{11.11}$$

Observing that

$$\frac{h'}{h} = \frac{(f(1 + g/f))'}{f(1 + g/f)} = \frac{f'}{f} + \frac{(1 + g/f)'}{1 + g/f},$$

from (11.11) we have $Z_h = Z_f + n((1 + g/f) \circ \gamma, 0)$. Since, by hypothesis, we have $|(1 + g(z)/f(z)) - 1| < 1\ \forall z \in \{\gamma\}$, it turns out that the path $(1 + g/f) \circ \gamma$ is contained within the circle with center 1 and radius 1. It follows that $z = 0$ is a point external to the path $(1 + g/f) \circ \gamma$ and therefore $n((1 + g/f) \circ \gamma, 0) = 0$. We conclude that $Z_f = Z_h$.
$\square$

Example 11.25 Determine the number of solutions, by counting their multiplicity, of the equation $2z^5 - 6z^2 + z + 1 = 0$ within the region $1 \leq |z| < 2$. Let us first find the number of zeros of the equation in the region $|z| < 2$. Given $2z^5 - 6z^2 + z + 1 = f + g$, choose f and g such that $|f(z)| > |g(z)|$ for $|z| = 2$. One possible choice is $f(z) = 2z^5$ and $g(z) = -6z^2 + z + 1$; in fact for $|z| = 2$ we have $|f(z)| = 2^6 = 64$

and $|g(z)| \leq 6\, 2^2 + 2 + 1 = 27$. The functions f and g are entire, so Rouché theorem is applicable and gives $Z_{f+g} = Z_f = 5$ in $|z| < 2$, since in that region f has a single zero at $z = 0$ with multiplicity 5. Now consider the region $|z| < 1$. The decomposition $2z^5 - 6z^2 + z + 1 = f + g$ must be done in such a way that $|f(z)| > |g(z)|$ for $|z| = 1$. Setting $f(z) = 2z^5 - 6z^2$ and $g(z) = z+1$, for $|z| = 1$ we have $|f(z)| \geq |2\,|z|^5 - 6\,|z|^2\,| = 6 - 2 = 4$ and $|g(z)| \leq |z| + 1 = 2$. Since $f(z) = z^2(2z^3 - 6)$ has a double zero at $z = 0$ while the roots of $2z^3 - 6 = 0$ have modulus $3^{1/3} > 1$, we have $Z_{f+g} = Z_f = 2$ for $|z| < 1$. In conclusion, the equation $f + g = 0$ has $5 - 2 = 3$ solutions in the region $1 \leq |z| < 2$.

Example 11.26 Show that $az^n = e^z$, with $a \in \mathbb{C}$ and $|a| > e$, has n solutions in the region $|z| < 1$. Set $f(z) = az^n$ and $g(z) = -e^z$. The functions f and g are analytic for $|z| \leq 1$ and we have $|f(z)| > |g(z)|$ for $|z| = 1$. By Rouché theorem, $Z_{f+g} = Z_f$ in the region $|z| < 1$. Obviously $Z_f = n$.

Example 11.27 Show that $P_n(z) = a_0 + a_1 z + a_2 z^2 + \cdots + a_{n-1} z^{n-1} + a_n z^n$, with $a_k \in \mathbb{C}$, $k = 0, 1, \ldots, n$, and $a_n \neq 0$, has, counting their multiplicities, n zeros in $\mathbb{C}$. The result is already known from the fundamental theorem of algebra, we can prove it again using Rouché theorem. Let $f(z) = a_n z^n$ and $g(z) = a_0 + a_1 z + \cdots + a_{n-1} z^{n-1}$. Since f and g are entire and for $|z| = R$ with R sufficiently large we have $|f(z)| > |g(z)|$, from Theorem 11.24 it follows that $Z_{f+g} = Z_f$ in the region $|z| < R$. Evidently $Z_f = n$, therefore P_n has n zeros in $\mathbb{C}$.

11.4 Harmonic Functions

Definition 11.28 (Harmonic Function) Let $h(x, y) : D \mapsto \mathbb{R}$ with $D \subset \mathbb{R}^2$ open. The function h is said to be harmonic in D if $\forall (x, y) \in D$ it has continuous second derivatives and satisfies the Laplace equation

$$h_{xx}(x, y) + h_{yy}(x, y) = 0. \tag{11.12}$$

Definition 11.29 (Conjugate Harmonic Function) Let $u(x, y) : D \mapsto \mathbb{R}$ and $v(x, y) : D \mapsto \mathbb{R}$ be two harmonic functions in the open subset $D \subset \mathbb{R}^2$. We say that v is harmonic conjugate to u in D if the Cauchy-Riemann conditions $u_x = v_y$ and $u_y = -v_x$ hold $\forall (x, y) \in D$.

Theorem 11.30 *Let $f(x + iy) = u(x, y) + iv(x, y)$ with $(x, y) \in D \subset \mathbb{R}^2$ open. The function f is analytic in D if and only if v is harmonic conjugate to u in D.*

Proof Let f be analytic in D. Then the component functions u and v are infinitely differentiable in D and, according to Theorem 5.8, in this domain the Cauchy-Riemann conditions $u_x = v_y$ and $u_y = -v_x$ hold. It follows that u is harmonic in D because in D we have $u_{xx} = \partial_x v_y = \partial_y v_x = -u_{yy}$. Similarly, $v_{xx} = \partial_x(-u_y) =$

$\partial_y(-u_x) = -v_{yy}$, namely, v is harmonic in D. We conclude that v is harmonic conjugate to u in D.

Conversely, let v be harmonic conjugate to u in D. Then u and v, being harmonic, have continuous second derivatives in D and furthermore in D they satisfy the Cauchy-Riemann conditions $u_x = v_y$ and $u_y = -v_x$. By Theorem 5.12 it follows that $f'(x + iy) = u_x(x, y) - iu_y(x, y)$ exists $\forall(x, y) \in D$, that is, f is analytic in D. $\qquad\square$

Theorem 11.31 (Existence of the Conjugate Harmonic Function) *Let $u : D \mapsto \mathbb{R}$ be harmonic in $D \subset \mathbb{R}^2$ open and simply connected. Then u admits in D one and only one conjugate harmonic defined, up to an arbitrary real constant, by*

$$v(x, y) = \int_{(x_0, y_0)}^{(x,y)} (u_s(s, t)\mathrm{d}t - u_t(s, t)\mathrm{d}s), \qquad (11.13)$$

where the line integral is evaluated along any path contained in D that goes from an arbitrary fixed point $(x_0, y_0) \in D$ to the generic point $(x, y) \in D$. Up to an arbitrary complex constant, the function

$$f(x + iy) = u(x, y) + iv(x, y) \qquad (11.14)$$

is the only analytic function on D whose real part is, up to an arbitrary real constant, $u(x, y)$.

Proof Let $g(x + iy) = u_x(x, y) - iu_y(x, y)$. Since u is harmonic in D, $\forall(x, y) \in D$ there exists $g'(x + iy) = u_{xx}(x, y) - iu_{xy}(x, y)$, that is, g is analytic in D. Furthermore D is open and simply connected, hence g admits primitive in D. Let $f(x + iy)$ be this primitive. Up to a complex constant $f(x + iy)$ is uniquely defined $\forall(x, y) \in D$ by the integral along an arbitrary path contained in D that connects a fixed point (x_0, y_0) of D to (x, y)

$$f(x + iy) = \int_{(x_0, y_0)}^{(x,y)} (u_s(s, t) - iu_t(s, t))(\mathrm{d}s + i\mathrm{d}t). \qquad (11.15)$$

The point (x_0, y_0) is arbitrary and its choice corresponds to the choice of the constant up to which $f(x + iy)$ is unique. Up to a real constant, the real part of $f(x + iy)$ is $u(x, y)$

$$\mathrm{Re}\, f(x + iy) = \int_{(x_0, y_0)}^{(x,y)} (u_s(s, t)\mathrm{d}s + u_t(s, t)\mathrm{d}t) = u(x, y) - u(x_0, y_0).$$

Always up to a real constant, the function $v(x, y)$, harmonic conjugate to $u(x, y)$ in D, is the imaginary part of f

$$\operatorname{Im} f(x + iy) = \int_{(x_0, y_0)}^{(x, y)} (u_s(s, t)dt - u_t(s, t)ds) = v(x, y) - v(x_0, y_0).$$

$\square$

In case the domain of the function u is $D = B(0, R)$ with $0 < R \le \infty$, therefore in particular in the case $D = \mathbb{C}$, we can choose as fixed point $(x_0, y_0) = (0, 0)$ and as integration path the union of the two segments joining $(0, 0)$ to $(x, 0)$ and $(x, 0)$ to (x, y), respectively. The expression (11.13) then simplifies to

$$v(x, y) = -\int_0^x u_y(s, 0)ds + \int_0^y u_x(x, t)dt. \tag{11.16}$$

In case D is multiply connected, the integral that appears in (11.15) generally assumes different values for different choices of the path from (x_0, y_0) to (x, y) and therefore $f(x + iy)$ as well as $v(x, y)$ are multi-valued functions.

For some examples of applications of the theory of harmonic functions see [3].

11.5   Saddle Point Method

Consider the function $I(\lambda) : \mathbb{R} \mapsto \mathbb{C}$ defined by the integral

$$I(\lambda) = \int_\gamma e^{\lambda f(z)} g(z)dz, \tag{11.17}$$

where $f, g : D \mapsto \mathbb{C}$ are analytic in $D \subset \mathbb{C}$ and γ is a piecewise smooth curve contained in D. Without loss of generality we can assume λ to be real and positive. Otherwise, just set $\exp(\lambda f(z)) = \exp(|\lambda| \tilde{f}(z))$ with $\tilde{f}(z) = f(z) \exp(i \arg \lambda)$. We are interested in evaluating the asymptotic behavior of $I(\lambda)$ for $\lambda \to \infty$.

To understand the idea underlying the method we are going to expose, let us start by considering the case in which f and g are infinitely differentiable real functions and the path γ is the real interval $[a, b]$

$$I(\lambda) = \int_a^b e^{\lambda f(t)} g(t)dt. \tag{11.18}$$

Suppose that $f(t)$ has a unique maximum at $t_0 \in (a, b)$. The Taylor series expansion of f around the point t_0 is then

$$f(t) = f(t_0) + \frac{1}{2} f''(t_0)(t - t_0)^2 + O((t - t_0)^3),$$

where $f''(t_0) < 0$. Inserting this expansion into (11.18) and also expanding $g(t)$ in the Taylor series around t_0, with the change of variable $\sqrt{\lambda\,|f''(t_0)|}\,(t - t_0) = u$ we obtain

$$I(\lambda) = \int_a^b e^{\lambda\left(f(t_0) - \frac{1}{2}|f''(t_0)|(t-t_0)^2 + O((t-t_0)^3)\right)} \left(g(t_0) + O(t - t_0)\right) \mathrm{d}t$$

$$= \frac{e^{\lambda f(t_0)} g(t_0)}{\sqrt{\lambda\,|f''(t_0)|}} \int_{-(t_0-a)\sqrt{\lambda|f''(t_0)|}}^{(b-t_0)\sqrt{\lambda|f''(t_0)|}} e^{-\frac{1}{2}u^2 + O((u/\sqrt{\lambda|f''(t_0)|})^3)}$$

$$\times \left(1 + O(u/\sqrt{\lambda\,|f''(t_0)|})\right) \mathrm{d}u.$$

For simplicity, we assume $g(t_0) \neq 0$. Approximating the integral appearing in this expression with $\sqrt{2\pi}$, namely, its asymptotic value for $\lambda \to \infty$, we get

$$I(\lambda) \simeq \sqrt{\frac{2\pi}{\lambda\,|f''(t_0)|}}\, e^{\lambda f(t_0)} g(t_0).$$

This is a heuristic estimate of $I(\lambda)$ where $\simeq$ is to emphasize that we have no control over the error committed.

A rigorous evaluation of (11.18) can be obtained by performing the change of variable $f(t_0) - f(t) = u$ separately in the two intervals $[a, t_0)$ and $(t_0, b]$ where, by hypothesis, $f(t)$ is strictly monotone, increasing and decreasing respectively,

$$I(\lambda) = \int_a^{t_0} e^{\lambda f(t)} g(t)\mathrm{d}t + \int_{t_0}^b e^{\lambda f(t)} g(t)\mathrm{d}t$$

$$= e^{\lambda f(t_0)} \left(\int_0^{f(t_0)-f(a)} e^{-\lambda u} G_-(u)\mathrm{d}u - \int_0^{f(t_0)-f(b)} e^{-\lambda u} G_+(u)\mathrm{d}u \right).$$

For convenience, we introduced two functions

$$G_\pm(u) = \left. \frac{g(t)}{f'(t)} \right|_{t = f_\pm^{-1}(f(t_0)-u)}$$

defined in terms of $f_\pm^{-1}$, the two monotone branches of the inverse function of f. We have $f_-^{-1}(f(t_0) - u) < t_0$ and $f_+^{-1}(f(t_0) - u) > t_0$, with $u > 0$ in both cases. The functions $G_\pm(u)$ have a divergence for $u \to 0^+$ whose behavior in terms of u can be estimated as follows. Expand the ratio $g(t)/f'(t)$ in powers of $t - t_0$

$$\frac{g(t)}{f'(t)} = \frac{g(t_0) + g'(t_0)(t - t_0) + \frac{1}{2}g''(t_0)(t - t_0)^2 + O((t - t_0)^3)}{f''(t_0)(t - t_0) + \frac{1}{2}f'''(t_0)(t - t_0)^2 + O((t - t_0)^3)}$$

$$= \frac{\frac{g(t_0)}{f''(t_0)}(t-t_0)^{-1} + \frac{g'(t_0)}{f''(t_0)} + \frac{g''(t_0)}{2f''(t_0)}(t-t_0) + O((t-t_0)^2)}{1 + \frac{f'''(t_0)}{2f''(t_0)}(t-t_0) + O((t-t_0)^2)}$$

$$= \frac{g(t_0)}{f''(t_0)}(t-t_0)^{-1} + \left(\frac{g'(t_0)}{f''(t_0)} - \frac{g(t_0)f'''(t_0)}{2(f''(t_0))^2} \right) + O(t-t_0).$$

Up to terms $O((t-t_0)^3)$ we have $u = f(t_0) - f(t) = \frac{1}{2}|f''(t_0)|(t-t_0)^2$ and therefore $t - t_0 = \pm\sqrt{2u/|f''(t_0)|}$, where the sign $\pm$ depends on which branch between $f_\pm^{-1}$ is used for the inverse function $t(u)$. Inserting this result into the previous expansion, we have

$$G_\pm(u) = \mp\frac{g(t_0)}{\sqrt{2|f''(t_0)|}}u^{-1/2} + \left(\frac{g'(t_0)}{f''(t_0)} - \frac{g(t_0)f'''(t_0)}{2(f''(t_0))^2} \right) + O(u^{1/2}).$$

By Watson's lemma [2], integrating term by term the expansion of $G_\pm(u)$, we obtain, for $\lambda \to \infty$, the expansion of the two integrals in which $I(\lambda)$ has been split

$$\int_0^{f(t_0)-f(a)} e^{-\lambda u} G_-(u)\,du$$

$$= +\frac{g(t_0)}{\sqrt{2|f''(t_0)|}}\sqrt{\frac{\pi}{\lambda}} + \left(\frac{g'(t_0)}{f''(t_0)} - \frac{g(t_0)f'''(t_0)}{2(f''(t_0))^2} \right)\frac{1}{\lambda} + O(\lambda^{-3/2}),$$

$$\int_0^{f(t_0)-f(b)} e^{-\lambda u} G_+(u)\,du$$

$$= -\frac{g(t_0)}{\sqrt{2|f''(t_0)|}}\sqrt{\frac{\pi}{\lambda}} + \left(\frac{g'(t_0)}{f''(t_0)} - \frac{g(t_0)f'''(t_0)}{2(f''(t_0))^2} \right)\frac{1}{\lambda} + O(\lambda^{-3/2}),$$

where it was used

$$\int_0^U e^{-\lambda u}(au^{-1/2} + b + cu^{1/2})\,du = \left(\frac{a\sqrt{\pi}}{\lambda^{1/2}} + \frac{b}{\lambda} + \frac{c\sqrt{\pi}}{2\lambda^{3/2}} + O(\lambda^{-5/2}) \right)$$

$$+ e^{-\lambda U}\left(-\frac{a + b\sqrt{U} + cU}{\lambda\sqrt{U}} + \frac{a - cU}{2\lambda^2 U^{3/2}} + O(\lambda^{-3}) \right)$$

$$= \frac{a\sqrt{\pi}}{\lambda^{1/2}} + \frac{b}{\lambda} + O(\lambda^{-3/2}).$$

In conclusion,

$$I(\lambda) = e^{\lambda f(t_0)} \left(\sqrt{\frac{2\pi}{\lambda \, |f''(t_0)|}} \, g(t_0) + O(\lambda^{-3/2}) \right). \tag{11.19}$$

The estimate just illustrated, known as Laplace's method, extends easily to functions f that have multiple maxima in (a, b). In this case, the asymptotic behavior for $\lambda \to \infty$ of $I(\lambda)$ is given by a sum of terms like (11.19) corresponding to the maxima of f. In the case of a maximum realized at a boundary point, $t_0 = a$ or $t_0 = b$, based on what we saw previously, the following slightly different formula is valid

$$I(\lambda) = e^{\lambda f(t_0)} \left(\sqrt{\frac{\pi}{2\lambda \, |f''(t_0)|}} \, g(t_0) + O(\lambda^{-1}) \right). \tag{11.20}$$

Note, in particular, the different order of the error which is now proportional to λ^{-1} instead of $\lambda^{-3/2}$.

Example 11.32 Derive the Stirling approximation of the factorial of n for n large

$$n! \simeq \sqrt{2\pi} \, n^{n+1/2} e^{-n}.$$

The factorial can be defined in terms of the real integral

$$n! = \int_0^\infty e^{-x} x^n \, dx.$$

With the change of variable $x = nt$, we have

$$n! = \int_0^\infty e^{-nt} (nt)^n n \, dt = n^{n+1} \int_0^\infty e^{n(\ln t - t)} \, dt.$$

Setting $\lambda = n$, $f(t) = \ln t - t$ and $g(t) = 1$, since $f'(t_0) = 0$ has solution $t_0 = 1 \in (0, \infty)$, for $n \to \infty$ we can apply the formula (11.19). Observing that $f(t_0) = -1$ and $f''(t_0) = -1$, we obtain

$$n! = n^{n+1} e^{-n} \left(\sqrt{\frac{2\pi}{n}} + O(n^{-3/2}) \right).$$

Let's go back to determining the asymptotic behavior for $\lambda \to \infty$ of the complex function (11.17). For reasons that will become clear in a moment, the method is now called the saddle point or steepest descent method. Let $\gamma(t) : [a, b] \mapsto \mathbb{C}$ be a path representing the curve γ, so that we can write

$$I(\lambda) = \int_{\gamma} e^{\lambda f(z)} g(z) dz$$

$$= \int_{a}^{b} e^{\lambda f(\gamma(t))} g(\gamma(t)) \gamma'(t) dt.$$

If $\operatorname{Re} f(\gamma(t))$ had a maximum at $t_0 \in (a, b)$ and, up to terms $O((t - t_0)^3)$, it were $\operatorname{Im} f(\gamma(t)) = \operatorname{Im} f(\gamma(t_0))$, we could estimate $I(\lambda)$ for large λ by applying the Laplace method. This possibility, however, cannot happen for a generic curve γ. Let $z_0 = \gamma(t_0)$ with z_0 a simple critical point of f, that is, let $f'(z_0) = 0$ and $f''(z_0) \neq 0$. These conditions must necessarily be satisfied if we want to have $(\operatorname{Re} f(\gamma(t_0)))' = (\operatorname{Im} f(\gamma(t_0)))' = 0$ and $(\operatorname{Re} f(\gamma(t_0)))'' < 0$. However, the point z_0 cannot be a local maximum for the function $\operatorname{Re} f(z)$. This statement follows from the fact that f is analytic at z_0, therefore $\exists \varepsilon > 0$ such that f is analytic in $B(z_0, \varepsilon)$. By the principle of maximum modulus, $\operatorname{Re} f$ takes maximum (and minimum) value on the boundary of this ball and not within it. The critical point z_0 is instead a saddle point for $\operatorname{Re} f(z)$ (Fig. 11.2). The proof is immediate. Since f can be expanded in Taylor series in $B(z_0, \varepsilon)$, setting $z = z_0 + re^{i\theta}$ and $f''(z_0) = |f''(z_0)| e^{i \operatorname{Arg} f''(z_0)}$, for $z \in B(z_0, \varepsilon)$ we have

$$f(z) = f(z_0) + \frac{1}{2} |f''(z_0)| r^2 \operatorname{cis}(2\theta + \operatorname{Arg} f''(z_0)) + O((z - z_0)^3).$$

When $2\theta + \operatorname{Arg} f''(z_0) = k\pi$, $k \in \mathbb{Z}$, up to terms $O((z - z_0)^3)$ the imaginary part of $f(z)$ is constant, while the real part has a second-order variation that is maximum

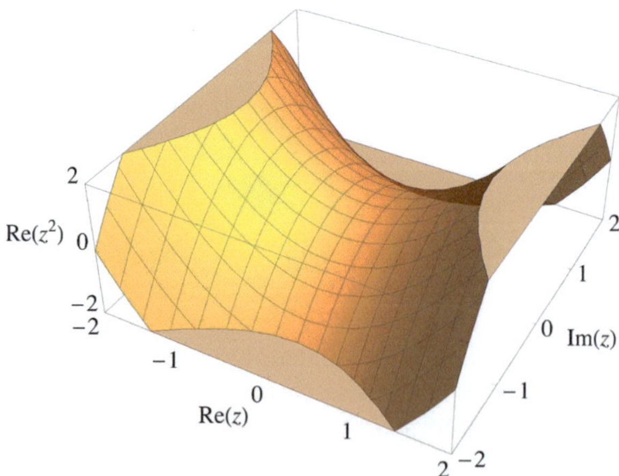

Fig. 11.2 Graph of the function $\operatorname{Re}(z^2)$ around the simple critical point $z = 0$

in modulus and of positive or negative sign

$$\mathrm{Re}\, f(z) = \mathrm{Re}\, f(z_0)) + \frac{1}{2}\left|f''(z_0)\right| r^2 \cos(k\pi) + O((z - z_0)^3)$$

$$= \mathrm{Re}\, f(z_0)) + \frac{1}{2}\left|f''(z_0)\right| r^2 (-1)^k + O((z - z_0)^3),$$

$$\mathrm{Im}\, f(z) = \mathrm{Im}\, f(z_0)) + \frac{1}{2}\left|f''(z_0)\right| r^2 \sin(k\pi) + O((z - z_0)^3)$$

$$= \mathrm{Im}\, f(z_0) + O((z - z_0)^3).$$

Let us then consider the two mutually orthogonal paths

$$\gamma_+(t) = z_0 + (t - t_0)e^{-i\frac{1}{2}\,\mathrm{Arg}\, f''(z_0)}, \qquad t_0 - \varepsilon < t < t_0 + \varepsilon,$$

$$\gamma_-(t) = z_0 + (t - t_0)e^{i(\frac{\pi}{2} - \frac{1}{2}\,\mathrm{Arg}\, f''(z_0))}, \qquad t_0 - \varepsilon < t < t_0 + \varepsilon,$$

corresponding, respectively, to the angles $\theta = -\frac{1}{2}\,\mathrm{Arg}\, f''(z_0) + \frac{\pi}{2}k$ obtained from the previous relation for $k = 0$ or $k = 1$ (k even or k odd, in general, for θ defined up to a multiple of 2π). These are the paths of steepest ascent (γ_+) and steepest descent (γ_-) for $\mathrm{Re}\, f(z)$ and, at the same time, of stationarity for $\mathrm{Im}\, f(z)$:

$$\mathrm{Re}\, f(\gamma_\pm(t)) = \mathrm{Re}\, f(z_0) \pm \frac{1}{2}\left|f''(z_0)\right| (t - t_0)^2 + O((t - t_0)^3),$$

$$\mathrm{Im}\, f(\gamma_\pm(t)) = \mathrm{Im}\, f(z_0) + O((t - t_0)^3).$$

The curve γ in general does not coincide in a neighborhood of z_0 with γ_-. However, thanks to the analyticity of f and g we can deform γ into a path η with this property without changing the value of the integral defining $I(\lambda)$. Let then $\eta(t) : [a, b] \mapsto \mathbb{C}$ be a piecewise regular path that coincides with $\gamma_-(t)$ for $t \in [t_0 - \varepsilon/2, t_0 + \varepsilon/2]$, connects with γ on the boundary of the ball $B(z_0, \varepsilon)$ and finally coincides with γ outside $B(z_0, \varepsilon)$, as shown in Fig. 11.3. For the path deformation principle, we obtain

$$I(\lambda) = \int_\eta e^{\lambda f(z)} g(z)\,dz$$

$$= \int_a^{t_0 - \varepsilon/2} e^{\lambda f(\eta(t))} g(\eta(t))\eta'(t)\,dt$$

$$+ \int_{t_0 - \varepsilon/2}^{t_0 + \varepsilon/2} e^{\lambda f(\gamma_-(t))} g(\gamma_-(t))\gamma_-'(t)\,dt$$

$$+ \int_{t_0 + \varepsilon/2}^{b} e^{\lambda f(\eta(t))} g(\eta(t))\eta'(t)\,dt.$$

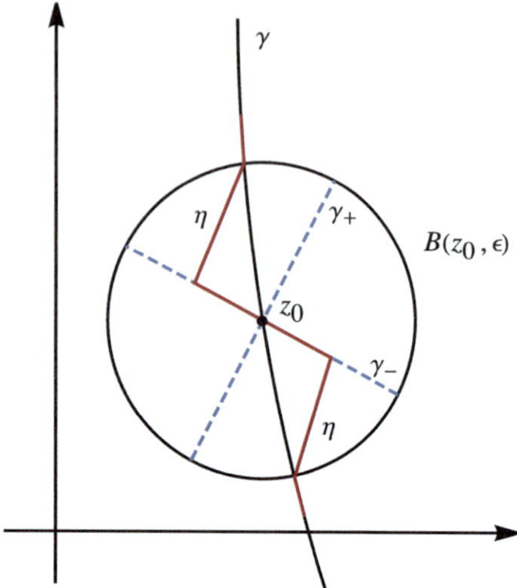

Fig. 11.3 Deformation of the path γ in the path η coinciding in a neighborhood of z_0, simple critical point of f, with the path γ_- along which $\operatorname{Re} f(z)$ realizes the steepest descent and $\operatorname{Im} f(z)$ is stationary

The Laplace method is now applicable. For large λ the dominant contribution to $I(\lambda)$ comes from the sole integral along the steepest descent path γ_-. Observing that $\gamma'_-(t_0) = \exp(\mathrm{i}(\frac{\pi}{2} - \frac{1}{2}\operatorname{Arg} f''(z_0)))$, we conclude that for $\lambda \to \infty$

$$I(\lambda) = \mathrm{e}^{\lambda f(z_0)}\left(\sqrt{\frac{2\pi}{\lambda\,|f''(z_0)|}}\, g(z_0)\mathrm{e}^{\mathrm{i}(\frac{\pi}{2} - \frac{1}{2}\operatorname{Arg} f''(z_0))} + O(\lambda^{-3/2}) \right). \qquad (11.21)$$

In case there are more than one simple critical point of f along the γ curve, the asymptotic behavior of $I(\lambda)$ for $\lambda \to \infty$ is given by a sum of terms like (11.21). The formula (11.21) can be extended, see [2], to critical points of order higher than one, namely, to points z_0 that are non-simple zeros of f'.

Exercises

11.1 Let $f : D \mapsto \mathbb{C}$ with $D = \{z \in \mathbb{C} : \operatorname{Re} z > 0\}$ defined by

$$f(z) = \int_0^\infty \mathrm{e}^{-zx^2}\,\mathrm{d}x$$

and let $g : E \mapsto \mathbb{C}$ with $E = \{z \in \mathbb{C} : z \neq -t,\ t \in [0, \infty)\}$ defined by

$$g(z) = \frac{\sqrt{\pi}}{2\sqrt{z}}, \qquad \sqrt{z} = |z|^{1/2}\,\mathrm{e}^{\frac{\mathrm{i}}{2}\operatorname{Arg} z}.$$

Show that g is the analytic continuation of f from D to E.

11.2 Determine the position and the value of the maximum of $|\sinh z|$ in the region $R = \{z = x + iy : \ 0 \le x \le 3, \ 0 \le y \le \pi\}$.

11.3 Determine the number of roots, counting their multiplicity, of the equation $z^4 - 5z + 1 = 0$ contained in the region $1 \le |z| < 2$.

11.4 Determine the number of solutions of the equation $az^n = \sin z$ contained in the internal part of the square $Q = \{z = x + iy : \ -1 \le x \le 1, \ -1 \le y \le 1\}$. Consider n integer and $a \in \mathbb{C}$ with $|a| > e$.

11.5 Find the number of solutions of the equation $z^8 + 4z^5 - z^2 + 1 = 0$ contained in the region $|z| < 1$.

11.6 Find the number of solutions of the equation $z^8 + 4z^3 - z^2 + 1 = 0$ contained in the region $1 \le |z| < \sqrt{2}$.

11.7 Determine the number of solutions of the equation $az^n = \cos z$ contained in the internal part of the square $Q = \{z = x + iy : \ -1 \le x \le 1, \ -1 \le y \le 1\}$. Consider n integer and $a \in \mathbb{C}$ with $|a| > e$.

11.8 Determine the harmonic function $v(x, y)$ conjugate to the function

$$u(x, y) = x^2 - y^2 + x$$

in the region $0 \le |x + iy| < \infty$.

11.9 Determine the harmonic function $v(x, y)$ conjugate to the function

$$u(x, y) = x + \sin(x^2 - y^2)\cosh(2xy)$$

throughout the plane xy.

11.10 Show that the function $u : \mathbb{R}^2 \mapsto \mathbb{R}$ defined by

$$u(x, y) = x^2 - y^2 - 2xy - 3x^2 y + y^3$$

is harmonic in $\mathbb{R}^2$. Determine the function $v(x, y) : \mathbb{R}^2 \mapsto \mathbb{R}$ harmonic conjugate to $u(x, y)$ in $\mathbb{R}^2$.

11.11 Show that the function $u : \mathbb{R}^2 \mapsto \mathbb{R}$ defined by

$$u(x, y) = x^2 - y^2 + e^{-y}\cos x$$

is harmonic in $\mathbb{R}^2$. Determine the function $v(x, y) : \mathbb{R}^2 \mapsto \mathbb{R}$ harmonic conjugate to $u(x, y)$ in $\mathbb{R}^2$.

11.12 The function $u : \mathbb{R}^2 \mapsto \mathbb{R}$ defined by

$$u(x, y) = e^{-2xy} \sin(x^2 - y^2)$$

is harmonic in $\mathbb{R}^2$. Determine the function $v(x, y) : \mathbb{R}^2 \mapsto \mathbb{R}$ harmonic conjugate to $u(x, y)$ in $\mathbb{R}^2$.

11.13 Determine the harmonic function $v(x, y)$ conjugate to the function

$$u(x, y) = \sin(x^2 - y^2) \cosh(2xy)$$

throughout the plane xy.

11.14 Let $u : \mathbb{C} \mapsto \mathbb{R}$ be a harmonic function such that $u(z) \geq 3$ for every $z \in \mathbb{C}$. Prove that u is constant.

11.15 Determine in which domain the function

$$u(x, y) = \ln((x - 1)^2 + y^2)$$

is harmonic and determine the function $v(x, y)$ harmonic conjugate to $u(x, y)$.

11.16 Determine the first term of the asymptotic expansion for $\lambda \to \infty$ of the integral

$$I(\lambda) = \int_\gamma \frac{1}{z} \exp\left[i\lambda \left(z^3/3 - z \right) \right] dz,$$

where γ is any straight line parallel to the real axis represented by the path $\gamma(t) = t + ib$ with $t \in (-\infty, +\infty)$ and $b > 0$.

11.17 Determine the first term of the asymptotic expansion for $\lambda \to \infty$ of the integral

$$I(\lambda) = \int_\gamma \frac{1}{\sqrt{z}} \exp\left[i\lambda \left(z^2/2 - z \right) \right] dz,$$

where γ is any straight line parallel to the real axis represented by the path $\gamma(t) = t + ib$ with $t \in (-\infty, +\infty)$ and $b > 0$.

11.18 Determine the first term of the asymptotic expansion for $\lambda \to \infty$ of the integral

$$I(\lambda) = \int_\gamma z^{3/2} \exp\left[i\lambda \left(z^2/2 - z \right) \right] dz,$$

where γ is a straight line parallel to the real axis at a non-zero distance from it.

11.19 A system of N spins 1/2, not mutually interacting, is characterized by microstates of energy $E_{\sigma_1,\sigma_2,\ldots,\sigma_N} = -J \sum_{j=1}^{N} \sigma_j$, where $\sigma_j = \pm 1$, $j = 1, \ldots, N$, and $J > 0$, whose energy density is

$$\Omega(E) = \sum_{\sigma_1=\pm 1} \sum_{\sigma_2=\pm 1} \cdots \sum_{\sigma_N=\pm 1} \delta(E - E_{\sigma_1,\sigma_2,\ldots,\sigma_N})$$

$$= \frac{2^N}{2\pi} \int_{-\infty}^{+\infty} dk \, e^{Nf(k)}, \qquad f(k) = ik\frac{E}{N} + \log(\cos(kJ)).$$

Evaluate the dominant term in $\Omega(E)$ in the thermodynamic limit $N \to \infty$ with E/N constant.

11.20 A system of N one-dimensional quantum harmonic oscillators, not mutually interacting, is characterized by microstates of energy $E_{n_1,n_2,\ldots,n_N} = \sum_{j=1}^{N} \hbar\omega(n_j + \frac{1}{2})$, where $n_j = 0, 1, 2, \ldots$, $j = 1, \ldots, N$, and $\hbar\omega > 0$, whose energy density is

$$\Omega(E) = \sum_{n_1=0}^{\infty} \sum_{n_2=0}^{\infty} \cdots \sum_{n_N=0}^{\infty} \delta(E - E_{n_1,n_2,\ldots,n_N})$$

$$= \frac{1}{2\pi} \int_{-\infty}^{+\infty} dk \, e^{Nf(k)}, \qquad f(k) = ik\frac{E}{N} - \log(2i\sin(k\hbar\omega/2)).$$

Evaluate the dominant term in $\Omega(E)$ in the thermodynamic limit $N \to \infty$ with E/N constant.

Chapter 12
Solutions to the Exercises

12.1 Exercises of Chap. 1

1.1

(a)

$$\mathrm{Im}\left(i^{37}\right) = \mathrm{Im}\left(i^{1+4\times9}\right) = \mathrm{Im}\left(i\,1^{9}\right) = 1.$$

(b)

$$\mathrm{Re}\left(\frac{2}{1-3i}\right) = \mathrm{Re}\left(\frac{2}{1-3i}\frac{1+3i}{1+3i}\right) = \mathrm{Re}\left(\frac{2+6i}{1+9}\right) = \frac{1}{5}.$$

(c)

$$\mathrm{Im}\left(\left(1+i\sqrt{3}\right)^{3}e^{i\frac{\pi}{4}}\right) = \mathrm{Im}\left(\left(2e^{i\frac{\pi}{3}}\right)^{3}e^{i\frac{\pi}{4}}\right) = -8\,\mathrm{Im}\left(e^{i\frac{\pi}{4}}\right) = -4\sqrt{2}.$$

1.2

(a)

$$\frac{i^{101}-3}{1-4i} = \frac{i^{4\,25+1}-3}{1-4i} = \frac{i-3}{1-4i} = \frac{(i-3)(1+4i)}{(1-4i)(1+4i)}$$

$$= \frac{i-4-3-12i}{1+16} = -\frac{7}{17} - \frac{11}{17}i.$$

C. Presilla, *Complex Functions of a Variable*, La Matematica per il 3+2 179,
https://doi.org/10.1007/978-3-032-12494-4_12

(b)

$$\text{Im}\,\frac{3i}{2+2i} = \text{Im}\,\frac{3i(2-2i)}{(2+2i)(2-2i)} = \text{Im}\,\frac{6i+6}{4+4} = \frac{6}{8} = \frac{3}{4}.$$

(c)

$$\text{Im}\left(e^{i\pi/4}\,\text{Re}\left(e^{i\pi/4}\right)\right) = \text{Im}\left(e^{i\pi/4}\,\text{Re}\left(\cos\frac{\pi}{4}+i\sin\frac{\pi}{4}\right)\right)$$

$$= \text{Im}\left(\left(\cos\frac{\pi}{4}+i\sin\frac{\pi}{4}\right)\cos\frac{\pi}{4}\right)$$

$$= \text{Im}\left(\cos^2\frac{\pi}{4}+i\sin\frac{\pi}{4}\cos\frac{\pi}{4}\right)$$

$$= \sin\frac{\pi}{4}\cos\frac{\pi}{4}$$

$$= \frac{1}{\sqrt{2}}\,\frac{1}{\sqrt{2}} = \frac{1}{2}.$$

1.3

(a) Since $17\pi/5 = 4\pi - 3\pi/5$ with $-3\pi/5 \in (-\pi, \pi]$, we have

$$\text{Arg}(\text{cis}(17\pi/5)) = -3\pi/5.$$

(b) Observing that $1 + z + z^2 + \cdots + z^n = (1 - z^{n+1})/(1-z)$ for $z \neq 1$, we have

$$\sum_{k=1}^{19}(1+i)^k = -1 + \sum_{k=0}^{19}(1+i)^k$$

$$= -1 + \frac{1-(1+i)^{20}}{1-(1+i)}$$

$$= -1 + \frac{1-(\sqrt{2}e^{i\pi/4})^{20}}{-i}$$

$$= -1 + \frac{1-2^{10}e^{i5\pi}}{-i}$$

$$= -1 + i(1+2^{10})$$

$$= -1 + i1025.$$

(c) By repeatedly applying $|zw| = |z|\,|w|$, we have

$$\left|\frac{ie^{-5i}(1+i)^3}{1+3i}\right| = \frac{|i|\,\left|e^{-5i}\right|\,|1+i|^3}{|1+3i|}$$

$$= \frac{2^{3/2}}{\sqrt{10}}$$

$$= \frac{2}{\sqrt{5}}.$$

1.4 Using the polar representation

$$z = re^{i\theta}, \qquad r \geq 0, \qquad -\pi < \theta \leq \pi,$$

we immediately obtain that the inequality $\mathrm{Re}(z^n) = \mathrm{Re}(r^n e^{in\theta}) \geq 0$ is equivalent to $\cos(n\theta) \geq 0$, that is

$$-\frac{\pi}{2} + 2\pi k \leq n\theta \leq \frac{\pi}{2} + 2\pi k, \qquad k \in \mathbb{Z},$$

which, dividing by n, gives

$$\frac{2\pi k - \pi/2}{n} \leq \theta \leq \frac{2\pi k + \pi/2}{n}, \qquad k \in \mathbb{Z}.$$

From the arbitrariness of n we get the result $\theta = 0$, namely, $z = r$ real non-negative.

1.5 Set $z = e^{i\theta}$ with $-\pi < \theta < \pi$ so that $|z| = 1$ and $z \neq -1$. The following identity applies

$$z = \frac{z(\bar{z} + 1)}{\bar{z} + 1} = \frac{1 + z}{1 + \bar{z}} = \frac{1 + \cos\theta + i\sin\theta}{1 + \cos\theta - i\sin\theta} = \frac{1 + i\sin\theta/(1 + \cos\theta)}{1 - i\sin\theta/(1 + \cos\theta)}.$$

Setting $t = \sin\theta/(1 + \cos\theta)$, as θ varies in $(-\pi, \pi)$ we have $t \in (-\infty, +\infty)$ which is the statement to prove.

1.6 Expand the product

$$(1 + z + z^2 + \cdots + z^n)(1 - z) = 1 + z + z^2 + \cdots + z^n$$
$$- (z + z^2 + z^3 + \cdots + z^{n+1})$$
$$= 1 - z^{n+1}$$

noting that all powers of z are cancelled except those of degree 0 and $n + 1$. Starting from this equality, valid for every integer n and for every complex number z, in the case $z \neq 1$ it is possible to divide both sides by $1 - z$ and obtain

$$\sum_{k=0}^{n} z^k = \frac{1 - z^{n+1}}{1 - z}.$$

1.7 Let $w = z^k$. If $k = np$ with $p = 1, 2, 3, \ldots$, then $w = 1$ and therefore

$$S_k = \sum_{j=0}^{n-1} w^j = \sum_{j=0}^{n-1} 1 = n.$$

If instead k is not an integer multiple of n then $w \neq 1$ and therefore

$$S_k = \sum_{j=0}^{n-1} w^j = \frac{1 - w^n}{1 - w} = \frac{1 - 1}{1 - w} = 0.$$

In conclusion

$$S_k = \sum_{j=0}^{n-1} z^{jk} = \begin{cases} n & k = np, \qquad p = 1, 2, 3, \ldots \\ 0 & \text{otherwise} \end{cases}.$$

1.8

$$\left| \operatorname{Im} \left(1 + 2\bar{z} + z^2 \right) \right| = \left| \operatorname{Im} \left(2\bar{z} + z^2 \right) \right|$$
$$\leq \left| 2\bar{z} + z^2 \right|$$
$$\leq |2\bar{z}| + \left| z^2 \right|$$
$$= 2 |z| + |z|^2$$
$$< 2 + 1 = 3.$$

1.9 It turns out that

$$\left| \operatorname{Re} \left(i + 2\bar{z} + z^2 \right) \right| = \left| \operatorname{Re} \left(2\bar{z} + z^2 \right) \right|$$
$$\leq \left| 2\bar{z} + z^2 \right|$$
$$\leq |2\bar{z}| + \left| z^2 \right|$$
$$= 2 |z| + |z|^2.$$

So, to have $\left| \operatorname{Re} \left(i + 2\bar{z} + z^2 \right) \right| < 3$ it is sufficient that

$$2 |z| + |z|^2 < 3.$$

This inequality is satisfied for $-3 < |z| < 1$, therefore $a = 1$.

1.10 For $n = 2$ the relation is the triangle inequality

$$|z_1 + z_2| \leq |z_1| + |z_2| \, .$$

Let us now reason by induction. Let us assume that the relation is valid for $n - 1$ terms

$$|z_1 + z_2 + \cdots + z_{n-1}| \leq |z_1| + |z_2| + \cdots + |z_{n-1}|$$

and prove that it is also valid by adding an additional term z_n. Setting $z_1 + z_2 + \cdots + z_{n-1} = w$ and using again the triangle inequality we have

$$
\begin{aligned}
|z_1 + z_2 + \cdots + z_{n-1} + z_n| &= |w + z_n| \\
&\leq |w| + |z_n| \\
&\leq |z_1| + |z_2| + \cdots + |z_{n-1}| + |z_n| \, .
\end{aligned}
$$

1.11 For every $z \in \mathbb{C}$ the following trivial inequality holds

$$(|\operatorname{Re} z| - |\operatorname{Im} z|)^2 \geq 0,$$

which can be rewritten

$$|\operatorname{Re} z|^2 + |\operatorname{Im} z|^2 \geq 2 \, |\operatorname{Re} z| \, |\operatorname{Im} z| \, .$$

By adding to both members of this inequality the quantity $|z|^2 = |\operatorname{Re} z|^2 + |\operatorname{Im} z|^2$ we arrive at

$$2 \, |z|^2 \geq |\operatorname{Re} z|^2 + |\operatorname{Im} z|^2 + 2 \, |\operatorname{Re} z| \, |\operatorname{Im} z| \, ,$$

from which, taking the square root, we conclude

$$|\operatorname{Re} z| + |\operatorname{Im} z| \leq \sqrt{2} \, |z| \, .$$

1.12

(a)

$$\sqrt[3]{-8} = \left(8 e^{i(\pi + 2n\pi)} \right)^{1/3} = \sqrt[3]{8} \exp\left(i\frac{\pi}{3} + i\frac{2n\pi}{3} \right), \qquad n = 0, 1, 2,$$

$$w_0 = 2 \left(\cos \frac{\pi}{3} + i \sin \frac{\pi}{3} \right) = 1 + i\sqrt{3},$$

$$w_1 = 2 \left(\cos \pi + i \sin \pi \right) = -2,$$

$$w_2 = 2\left(\cos\frac{5\pi}{3} + i\sin\frac{5\pi}{3}\right) = 1 - i\sqrt{3}.$$

(b)

$$\sqrt[2]{1-i} = \left(\sqrt{2}\,e^{i\left(-\frac{\pi}{4}+2n\pi\right)}\right)^{1/2} = \sqrt[4]{2}\exp\left(-i\frac{\pi}{8} + in\pi\right), \qquad n = 0, 1,$$

$$w_0 = \sqrt[4]{2}\left(\cos\frac{\pi}{8} - i\sin\frac{\pi}{8}\right),$$

$$w_1 = \sqrt[4]{2}\left(\cos\frac{7\pi}{8} + i\sin\frac{7\pi}{8}\right).$$

1.13

(a)

$$\sqrt[2]{i^3} = \sqrt[2]{-i} = \left(e^{i\left(\frac{3\pi}{2}+2\pi k\right)}\right)^{1/2} = e^{i\left(\frac{3\pi}{4}+\pi k\right)}, \qquad k = 0, 1,$$

$$w_0 = e^{i\frac{3\pi}{4}} = \cos\frac{3\pi}{4} + i\sin\frac{3\pi}{4} = -\frac{1}{\sqrt{2}} + i\frac{1}{\sqrt{2}},$$

$$w_1 = e^{i\frac{7\pi}{4}} = \cos\frac{7\pi}{4} + i\sin\frac{7\pi}{4} = \frac{1}{\sqrt{2}} - i\frac{1}{\sqrt{2}}.$$

(b)

$$\sqrt[3]{8+i} = \left(\sqrt[2]{64+1}\,e^{i(\theta+2\pi k)}\right)^{1/3} = \sqrt[6]{65}\,e^{i\left(\frac{\theta}{3}+\frac{2\pi k}{3}\right)}, \qquad k = 0, 1, 2,$$

$$w_0 = \sqrt[6]{65}\,e^{i\frac{\theta}{3}} = \sqrt[6]{65}\left(\cos\frac{\theta}{3} + i\sin\frac{\theta}{3}\right),$$

$$w_1 = \sqrt[6]{65}\,e^{i\left(\frac{\theta}{3}+\frac{2\pi}{3}\right)} = \sqrt[6]{65}\left(\cos\frac{\theta+2\pi}{3} + i\sin\frac{\theta+2\pi}{3}\right),$$

$$w_2 = \sqrt[6]{65}\,e^{i\left(\frac{\theta}{3}+\frac{4\pi}{3}\right)} = \sqrt[6]{65}\left(\cos\frac{\theta+4\pi}{3} + i\sin\frac{\theta+4\pi}{3}\right),$$

where $\theta = \arctan\frac{1}{8} \simeq \frac{1}{8} + O\left(\left(\frac{1}{8}\right)^3\right)$.

1.14

(a) Setting $w = z^2$, the equation to solve is equivalent to

$$w^2 + w + 1 = 0,$$

which has solutions

$$w_\pm = \frac{-1+\sqrt{1-4}}{2} = -\frac{1}{2}+\frac{\sqrt{-3}}{2} = -\frac{1}{2}\pm i\frac{\sqrt{3}}{2}.$$

Then take the two distinct roots of the equation $z = \sqrt{w_\pm}$

$$z_{k\pm} = \sqrt{-\frac{1}{2}\pm i\frac{\sqrt{3}}{2}} = \sqrt{e^{i(\pi\mp\pi/3)}} = e^{i(\pi\mp\pi/3+2\pi k)/2}, \qquad k=0,1.$$

In conclusion, the assigned equation has four distinct solutions:

$$z_{0\pm} = e^{i(\pi/2\mp\pi/6)} = i\left(\frac{\sqrt{3}}{2}\mp i\frac{1}{2}\right) = \pm\frac{1}{2}+i\frac{\sqrt{3}}{2},$$

$$z_{1\pm} = e^{i(\pi/2\mp\pi/6+\pi)} = -i\left(\frac{\sqrt{3}}{2}\mp i\frac{1}{2}\right) = \mp\frac{1}{2}-i\frac{\sqrt{3}}{2}.$$

(b) The distinct solutions of the given equation are

$$z_k = \sqrt[3]{i} = \left(e^{i(\pi/2)}\right)^{1/3} = e^{i(\pi/2+2\pi k)/3}, \qquad k=0,1,2,$$

that is, explicitly

$$z_0 = e^{i\pi/6} = \frac{\sqrt{3}}{2}+i\frac{1}{2},$$

$$z_1 = e^{i5\pi/6} = -\frac{\sqrt{3}}{2}+i\frac{1}{2},$$

$$z_2 = e^{i3\pi/2} = -i.$$

1.15 Rewrote the equation in the form

$$z(z^4 - (1+i)) = 0.$$

It is evident that this is solved by $z = 0$ and

$$z = (1+i)^{1/4} = \left(\sqrt{2}e^{i\pi/4}\right)^{1/4} = 2^{1/8}e^{i(\pi/4+2\pi k)/4}, \qquad k=0,1,2,3.$$

In conclusion we have the five distinct solutions

$$0, \quad 2^{1/8}e^{i\pi/16}, \quad 2^{1/8}e^{i9\pi/16}, \quad 2^{1/8}e^{i17\pi/16}, \quad 2^{1/8}e^{i25\pi/16}.$$

1.16

(a) The equation to be solved and its complex conjugate constitute a linear system in the unknowns z and $\bar{z}$:

$$\begin{cases} 3z + i\bar{z} = 4 \\ -iz + 3\bar{z} = 4 \end{cases}.$$

This system admits the unique solution

$$z = \frac{12 - 4i}{9 + i^2} = \frac{3}{2} - \frac{i}{2}.$$

(b) The assigned equation is equivalent to its complex conjugate

$$z^4 = 1 - i = \sqrt{2}\, e^{-i\pi/4},$$

which admits the four distinct solutions

$$z_k = \sqrt[8]{2}\, e^{i(-\pi/4 + 2\pi k)/4}, \qquad k = 0, 1, 2, 3.$$

1.17 For $z \neq 0$ the given equation is equivalent to

$$z^5 = 1 + i$$

whose distinct solutions are

$$z = (1 + i)^{1/5}$$
$$= \left(\sqrt{2}e^{i\pi/4}\right)^{1/5}$$
$$= 2^{1/10} e^{i(\pi/4 + 2\pi k)/5}, \qquad k = 0, 1, 2, 3, 4.$$

The starting equation, therefore, admits the five distinct solutions

$$\sqrt[10]{2}e^{i\pi/20}, \quad \sqrt[10]{2}e^{i9\pi/20}, \quad \sqrt[10]{2}e^{i17\pi/20}, \quad \sqrt[10]{2}e^{i25\pi/20}, \quad \sqrt[10]{2}e^{i33\pi/20}.$$

1.18

(a) The equation has two distinct solutions:

$$z = \frac{-i + \sqrt{-1 - 8}}{2} = \frac{-i \pm 3i}{2} = -2i,\ i.$$

(b) Setting $w = \bar{z}^2$, the equation to solve is equivalent to

$$w^2 + w + 1 = 0,$$

which has solutions

$$w_\pm = \frac{-1 + \sqrt{1-4}}{2} = -\frac{1}{2} + \frac{\sqrt{-3}}{2} = -\frac{1}{2} \pm i\frac{\sqrt{3}}{2}.$$

Then take the two distinct roots of the equation $z = \overline{\sqrt{w_\pm}}$

$$z_{k\pm} = \overline{\left(-\frac{1}{2} \pm i\frac{\sqrt{3}}{2}\right)^{1/2}}$$

$$= \overline{\left(e^{\pm i2\pi/3}\right)^{1/2}} = \overline{e^{(\pm i2\pi/3 + i2\pi k)/2}} = e^{\mp i\pi/3 - i\pi k}, \qquad k = 0, 1.$$

In conclusion, the assigned equation has the four distinct solutions:

$$z_{0\pm} = e^{\mp i\pi/3} = \frac{1}{2} \mp i\frac{\sqrt{3}}{2},$$

$$z_{1\pm} = e^{\mp i\pi/3 - i\pi} = -\frac{1}{2} \pm i\frac{\sqrt{3}}{2}.$$

1.19 The proposed equation and its complex conjugate form the linear system of equations

$$\begin{cases} az + b\bar{z} = -c \\ \bar{b}z + \bar{a}\bar{z} = -\bar{c} \end{cases}$$

in the unknowns z and $\bar{z}$. The system admits one and only one solution if the associated determinant is non-zero:

$$\begin{vmatrix} a & b \\ \bar{b} & \bar{a} \end{vmatrix} = |a|^2 - |b|^2 \neq 0.$$

In this case, using Cramer's rule the solution is

$$z = \frac{\begin{vmatrix} -c & b \\ -\bar{c} & \bar{a} \end{vmatrix}}{\begin{vmatrix} a & b \\ \bar{b} & \bar{a} \end{vmatrix}} = \frac{b\bar{c} - \bar{a}c}{|a|^2 - |b|^2}.$$

1.20 Setting $\theta = \arccos(1/7)$, from the de Moivre's formula we have

$$(\cos\theta + i\sin\theta)^3 = \cos 3\theta + i\sin 3\theta$$

that is

$$\cos(3\theta) = \mathrm{Re}\left((\cos\theta + i\sin\theta)^3\right) = \cos^3\theta - 3\cos\theta\sin^2\theta.$$

Using $\sin^2\theta = 1 - \cos^2\theta$ and $\cos\theta = 1/7$ we obtain

$$\cos(3\theta) = 4\cos^3\theta - 3\cos\theta = 4\frac{1}{7^3} - 3\frac{1}{7} = -\frac{143}{343}.$$

1.21 Setting $\theta = \arctan(3)$, from the de Moivre's formula we have

$$(\cos\theta + i\sin\theta)^4 = \cos(4\theta) + i\sin(4\theta)$$

that is

$$\cos(4\theta) = \mathrm{Re}\left((\cos\theta + i\sin\theta)^4\right) = \cos^4\theta - 6\cos^2\theta\sin^2\theta + \sin^4\theta.$$

Using $\cos^2\theta = 1 - \tan^2\theta\cos^2\theta$, for $\tan\theta = 3$ we find

$$\cos^2\theta = \frac{1}{1 + \tan^2\theta} = \frac{1}{10}, \qquad \sin^2\theta = 1 - \cos^2\theta = \frac{9}{10}.$$

In conclusion

$$\cos(4\arctan(3)) = \frac{1}{100} - 6\frac{1}{10}\frac{9}{10} + \frac{81}{100} = \frac{7}{25}.$$

1.22 Setting $\theta = \arccos(1/\sqrt{17})$, we have

$$e^{i5\theta} = \cos(5\theta) + i\sin(5\theta) = (\cos\theta + i\sin\theta)^5.$$

Therefore

$$\sin(5\theta) = \mathrm{Im}(\cos\theta + i\sin\theta)^5$$

$$= \mathrm{Im}\left(\cos^5\theta + 5\cos^4\theta(i\sin\theta) + 10\cos^3\theta(i\sin\theta)^2\right.$$

$$\left. + 10\cos^2\theta(i\sin\theta)^3 + 5\cos\theta(i\sin\theta)^4 + (i\sin\theta)^5\right)$$

$$= 5\cos^4\theta\sin\theta - 10\cos^2\theta\sin^3\theta + \sin^5\theta$$

$$= \sin\theta\left(5\cos^4\theta - 10\cos^2\theta\sin^2\theta + \sin^4\theta\right).$$

We know that $\cos\theta = 1/\sqrt{17}$ and therefore $\sin\theta = \pm\sqrt{1 - 1/17}$. We can conclude that

$$\sin(5\theta) = \pm\sqrt{\frac{16}{17}}\left(5\frac{1}{17^2} - 10\frac{1}{17}\frac{16}{17} + \frac{16^2}{17^2}\right)$$

$$= \mp\sqrt{\frac{1}{17}\frac{404}{289}}.$$

1.23 Let's set

$$z = e^{i\frac{\pi}{11}} = \cos\frac{\pi}{11} + i\sin\frac{\pi}{11},$$

using the de Moivre's formula $(\cos\theta + i\sin\theta)^n = \cos(n\theta) + i\sin(n\theta)$, we find

$$S = \operatorname{Re}\sum_{k=0}^{4} z^{2k+1} = \operatorname{Re}\left(z\sum_{k=0}^{4}(z^2)^k\right) = \operatorname{Re}\left(z\frac{1 - (z^2)^5}{1 - z^2}\right) = \operatorname{Re}\frac{z - z^{11}}{1 - z^2}.$$

Observing that $z^{11} = -1$ and $|z| = 1$, we obtain

$$S = \operatorname{Re}\frac{1 + z}{1 - z^2} = \operatorname{Re}\frac{1}{1 - z} = \operatorname{Re}\frac{1 - \bar{z}}{(1 - z)(1 - \bar{z})} = \frac{1 - \operatorname{Re}z}{1 + |z|^2 - 2\operatorname{Re}z}$$

$$= \frac{1 - \operatorname{Re}z}{2(1 - \operatorname{Re}z)} = \frac{1}{2}.$$

1.24 Let's set

$$z = e^{i\frac{\pi}{9}} = \cos\frac{\pi}{9} + i\sin\frac{\pi}{9},$$

using the de Moivre's formula $(\cos\theta + i\sin\theta)^n = \cos(n\theta) + i\sin(n\theta)$, we find

$$S = \operatorname{Im}\sum_{k=0}^{3} z^{2k+1} = \operatorname{Im}\left(z\sum_{k=0}^{3}(z^2)^k\right) = \operatorname{Im}\left(z\frac{1 - (z^2)^4}{1 - z^2}\right) = \operatorname{Im}\frac{z - z^9}{1 - z^2}.$$

Observing that $z^9 = -1$ and $|z| = 1$, we obtain

$$S = \operatorname{Im}\frac{1}{1 - z} = \operatorname{Im}\frac{1 - \bar{z}}{(1 - z)(1 - \bar{z})} = \frac{\operatorname{Im}(1 - \bar{z})}{1 + |z|^2 - 2dz} = \frac{\operatorname{Im}z}{2(1 - dz)}$$

$$= \frac{1}{2}\frac{\sin\frac{\pi}{9}}{1 - \cos\frac{\pi}{9}}.$$

1.25 Since $|w| = 1$, for $|z| < R$ we have $|w^m z| < R$ and therefore we can write

$$
\frac{1}{n} \sum_{m=0}^{n-1} w^{-jm} f(w^m z) = \frac{1}{n} \sum_{m=0}^{n-1} w^{-jm} \left(\sum_{k=0}^{\infty} a_k w^{mk} z^k \right)
$$

$$
= \sum_{k=0}^{\infty} a_k z^k \frac{1}{n} \sum_{m=0}^{n-1} w^{m(k-j)}
$$

$$
= \sum_{k=0}^{\infty} a_k z^k \frac{1}{n} \sum_{m=0}^{n-1} \left(e^{i2\pi (k-j)/n} \right)^m
$$

$$
= \sum_{k=0}^{\infty} a_k z^k \begin{cases} 1 & (k-j)/n \in \mathbb{Z} \\ 0 & \text{otherwise} \end{cases},
$$

where for $(k-j)/n \in \mathbb{Z}$ we used the fact that

$$
\left(e^{i2\pi (k-j)/n} \right)^m = 1^m = 1,
$$

while for $(k-j)/n \notin \mathbb{Z}$ we used the formula

$$
\sum_{m=0}^{n-1} \left(e^{i2\pi (k-j)/n} \right)^m = \frac{1 - \left(e^{i2\pi (k-j)/n} \right)^n}{1 - e^{i2\pi (k-j)/n}} = \frac{1 - e^{i2\pi (k-j)}}{1 - e^{i2\pi (k-j)/n}} = 0.
$$

Observing that k is an integer, including 0, while $0 \le j/n < 1$, we have that $(k-j)/n \in \mathbb{Z}$ if and only if $k - j = qn$, with $q = 0, 1, 2, \ldots$. In conclusion,

$$
\frac{1}{n} \sum_{m=0}^{n-1} w^{-jm} f(w^m z) = \sum_{k=0}^{\infty} a_k z^k \sum_{q=0}^{\infty} \delta_{k-j,qn}
$$

$$
= \sum_{q=0}^{\infty} a_{j+qn} z^{j+qn}.
$$

1.26

(a) The set of points whose distance from $z_0 = -3 + i$ is less than or equal to 7 and greater than 5, that is, the annular region of internal radius 5 and external radius 7 centered at z_0, including the external boundary, excluding the internal one.

(b) The set of points whose distance from $z_1 = i$ is equal to the distance from $z_2 = -1$, that is, the straight line bisecting the second and fourth quadrants.

1.27 The proposed inequality is equivalent to

$$
|z - 3| < 2\,|z + 3|,
$$

or, setting $z = x + iy$, with $x, y \in \mathbb{R}$,

$$\sqrt{(x-3)^2 + y^2} < 2\sqrt{(x+3)^2 + y^2}.$$

Squaring and simplifying we have

$$x^2 + y^2 + 10x + 9 > 0,$$

or $(x+5)^2 + y^2 > 16$, which rewritten as $|z + 5| > 4$ represents the set of points external to the circle of radius 4 centered at $z = -5$.

1.28 Let $w = (z - 1)^2$, in the plane $(\operatorname{Re} w, \operatorname{Im} w)$ the region $0 \le \operatorname{Arg} w \le \pi/2$ coincides with the first quadrant, that is, it is the set of points of the type $w = ue^{i\varphi}$ with $u \ge 0$ and $0 \le \varphi \le \pi$. Since

$$z = 1 + \sqrt{w}$$
$$= 1 + \sqrt{u}e^{i(\varphi + 2\pi k)/2}, \qquad k = 0, 1,$$

the region in which $0 \le \operatorname{Arg}((z-1)^2) \le \pi/2$ is

$$A = \left\{ z \in \mathbb{C} : z = 1 + re^{i\theta},\ r \ge 0,\ 0 \le \theta \le \pi/4,\ -\pi \le \theta \le -3\pi/4 \right\}.$$

12.2   Exercises of Chap. 2

2.1 It is necessary to show that d satisfies all the properties of a distance. The properties

$$d(x, y) \ge 0,$$
$$d(x, y) = 0 \text{ if and only if } x = y,$$
$$d(x, y) = d(y, x),$$

with arbitrary $x, y \in S$, are verified by definition. To verify the triangle inequality

$$d(x, y) \le d(x, z) + d(z, y),$$

with arbitrary $x, y, z \in S$, just analyze the following five possible cases:

$$x = y,\ x = z,\ y = z: \qquad 0 \le 0 + 0 = 0,$$
$$x = y,\ x \ne z,\ y \ne z: \qquad 0 \le 1 + 1 = 2,$$

$$x \neq y, \ x = z, \ y \neq z: \qquad 1 \leq 0 + 1 = 1,$$

$$x \neq y, \ x \neq z, \ y = z: \qquad 1 \leq 1 + 0 = 1,$$

$$x \neq y, \ x \neq z, \ y \neq z: \qquad 1 \leq 1 + 1 = 2.$$

2.2 It is necessary to verify that d satisfies all the properties of a distance. From the properties of the modulus it immediately follows that $\forall x, y \in S$

$$d(x, y) \geq 0,$$

$$d(x, y) = 0 \text{ if and only if } x = y,$$

$$d(x, y) = d(y, x).$$

To verify the triangle property, observe that the function $g(t) = t/(1 + t)$ is monotonically increasing for $t \in [0, \infty)$, in fact in this interval we have $g'(t) = 1/(1 + t)^2 > 0$. On the other hand by the triangle inequality of the module we have $|x - y| \leq |x - z| + |z - y|$. Therefore, $\forall x, y, z \in S$ it turns out

$$
\begin{aligned}
d(x, y) &= g(|x - y|) \\
&\leq g(|x - z| + |z - y|) \\
&= \frac{|x - z| + |z - y|}{1 + |x - z| + |z - y|} \\
&= \frac{|x - z|}{1 + |x - z| + |z - y|} + \frac{|z - y|}{1 + |x - z| + |z - y|} \\
&\leq \frac{|x - z|}{1 + |x - z|} + \frac{|z - y|}{1 + |z - y|} \\
&= d(x, z) + d(z, y).
\end{aligned}
$$

With this distance the diameter of $S = \mathbb{R}$ is finite, in fact

$$\operatorname{diam} S = \sup_{x, y \in S} \frac{|x - y|}{1 + |x - y|} = \lim_{t \to \infty} \frac{t}{1 + t} = 1.$$

2.3 To prove that $(\mathbb{C}, d)$ is a metric space we need to show that d is a distance, that is, that $\forall z_1, z_2, z_3 \in \mathbb{C}$ the following properties are satisfied:

(a) $d(z_1, z_2) \geq 0$;
(b) $d(z_1, z_2) = 0$ if and only if $z_1 = z_2$;
(c) $d(z_1, z_2) = d(z_2, z_1)$;
(d) $d(z_1, z_2) \leq d(z_1, z_3) + d(z_3, z_2)$.

The first property is true because the absolute value of a real number is always non-negative. The second follows from the fact that $d(z_1, z_2) = 0$ if and only if $|\mathrm{Re}(z_1 - z_2)| = |\mathrm{Im}(z_1 - z_2)| = 0$ which in turn is true if and only if $z_1 = z_2$. The third follows from the fact that $\forall x_1, x_2 \in \mathbb{R}$ we have $|x_1 - x_2| = |x_2 - x_1|$. For the triangle property we get

$$
\begin{aligned}
d(z_1, z_2) &= |\mathrm{Re}(z_1 - z_2)| + |\mathrm{Im}(z_1 - z_2)| \\
&\leq |\mathrm{d}(z_1 - z_3)| + |\mathrm{d}(z_3 - z_2)| + |\mathrm{Im}(z_1 - z_3)| + |\mathrm{Im}(z_3 - z_2)| \\
&= (|\mathrm{Re}(z_1 - z_3)| + |\mathrm{Im}(z_1 - z_3)|) + (|\mathrm{Re}(z_3 - z_2)| + |\mathrm{Im}(z_3 - z_2)|) \\
&= d(z_1, z_3) + d(z_3, z_2),
\end{aligned}
$$

having exploited the fact that $\forall x_1, x_2, x_3 \in \mathbb{R}$ we have $|x_1 - x_2| \leq |x_1 - x_3| + |x_3 - x_2|$.

By definition we have

$$
\begin{aligned}
B(1 + i3, 1) &= \{z \in \mathbb{C} : d(z, 1 + i3) < 1\} \\
&= \{z = x + iy \in \mathbb{C} : |x - 1| + |y - 3| < 1\}.
\end{aligned}
$$

The above inequality is satisfied by the points $x, y \in \mathbb{R}$ such that

$$
-1 + |x - 1| < y - 3 < 1 - |x - 1|.
$$

For $x > 1$ it must be $x - 2 < y - 3 < 2 - x$, that is, $1 + x < y < 5 - x$. For $x < 1$ it must be $-x < y - 3 < x$, that is, $3 - x < y < 3 + x$. It is easy to verify that these points correspond to the square, edge excluded, of vertices $(1, 2)$, $(2, 3)$, $(1, 4)$ and $(0, 3)$.

2.4 To prove that $(\mathbb{C}, d)$ is a metric space, we need to show that d is a distance, that is, that $\forall z_1, z_2, z_3 \in \mathbb{C}$ the following properties are satisfied

(a) $d(z_1, z_2) \geq 0$;
(b) $d(z_1, z_2) = 0$ if and only if $z_1 = z_2$;
(c) $d(z_1, z_2) = d(z_2, z_1)$;
(d) $d(z_1, z_2) \leq d(z_1, z_3) + d(z_3, z_2)$.

The first property is true because the absolute value of a real number is always non-negative. The second follows from the fact that $d(z_1, z_2) = 0$ if and only if $|\mathrm{Re}(z_1 - z_2)| = |\mathrm{Im}(z_1 - z_2)| = 0$ which in turn is true if and only if $z_1 = z_2$. The third follows from the fact that $\forall x_1, x_2 \in \mathbb{R}$ we have $|x_1 - x_2| = |x_2 - x_1|$. For the triangle property we get

$$
\begin{aligned}
d(z_1, z_2) &= \max\left(|\mathrm{Re}(z_1 - z_2)|, |\mathrm{Im}(z_1 - z_2)|\right) \\
&\leq \max\left(|\mathrm{d}(z_1 - z_3)| + |\mathrm{d}(z_3 - z_2)|, |\mathrm{Im}(z_1 - z_3)| + |\mathrm{Im}(z_3 - z_2)|\right)
\end{aligned}
$$

$$\leq \max\left(|\mathrm{Re}(z_1 - z_3)|, |\mathrm{Im}(z_1 - z_3)|\right)$$

$$+ \max\left(|\mathrm{Re}(z_3 - z_2)|, |\mathrm{Im}(z_3 - z_2)|\right)$$

$$= d(z_1, z_3) + d(z_3, z_2),$$

having exploited the fact that $\forall x_1, x_2, x_3 \in \mathbb{R}$ we have $|x_1 - x_2| \leq |x_1 - x_3| + |x_3 - x_2|$.

By definition, we have

$$\overline{B}(3 + i, 2) = \{z \in \mathbb{C} : d(z, 3 + i) \leq 2\}$$

$$= \{z = x + iy \in \mathbb{C} : \max\left(|x - 3|, |y - 1|\right) \leq 2\}.$$

The previous inequality is satisfied by the points $x, y \in \mathbb{R}$ such that $|y - 1| \leq |x - 3| \leq 2$ or $|x - 3| \leq |y - 1| \leq 2$. It is easy to verify that such points correspond to the square, including the edge, obtained from the intersection of the strips $|x - 3| \leq 2$ and $|y - 1| \leq 2$, that is, the square of vertices $(1, -1)$, $(5, -1)$, $(5, 3)$ and $(1, 3)$.

2.5

(a) True. Let $\{A_k\}_{k=1}^{\infty}$ be a countable infinity of open sets and consider a generic point $x \in \bigcup_{k=1}^{\infty} A_k$. Then there must exist at least one k_0 such that $x \in A_{k_0}$. Since A_{k_0} is open by definition, $\exists\, \varepsilon > 0$ such that $B(x, \varepsilon) \subset A_{k_0} \subset \bigcup_{k=1}^{\infty} A_k$ from which the statement follows.

(b) False. Consider in $\mathbb{C}$ the countable infinity of open sets $\{A_k\}_{k=1}^{\infty}$ where $A_k = B(0, 1 + 1/k)$. Evidently we have $\bigcap_{k=1}^{\infty} A_k = \overline{B}(0, 1)$ which is a closed set.

(c) False. Consider in $\mathbb{C}$ the countable infinity of closed sets $\{A_k\}_{k=1}^{\infty}$ where $A_k = \overline{B}(0, 1 - 1/k)$. Evidently we have $\bigcup_{k=1}^{\infty} A_k = B(0, 1)$ which is an open set.

(d) True. Let $\{A_k\}_{k=1}^{\infty}$ be a countable infinity of closed sets. By de Morgan's laws

$$\left(\bigcap_{k=1}^{\infty} A_k\right)^c = \bigcup_{k=1}^{\infty} (A_k)^c.$$

By virtue of point 1 the set in right hand side, namely, the union of a countable infinity of open sets, is open and therefore the intersection of the closed sets A_k is also a closed set.

2.6

(a) False. Consider in $\mathbb{R}$ the two closed sets $A = [0, 1]$ and $B = [1, 2]$. The internal part of their union is $(A \cup B)^{\circ} = (0, 2)$. On the other hand we have $A^{\circ} = (0, 1)$ and $B^{\circ} = (1, 2)$ and therefore $A^{\circ} \cup B^{\circ} = (0, 2) \setminus \{1\}$.

(b) True. We prove separately the two inclusions $(A \cap B)^{\circ} \subset A^{\circ} \cap B^{\circ}$ and $A^{\circ} \cap B^{\circ} \subset (A \cap B)^{\circ}$. Since $A \cap B \subset A$ it follows $(A \cap B)^{\circ} \subset A^{\circ}$. Similarly $(A \cap B)^{\circ} \subset B^{\circ}$ and therefore $(A \cap B)^{\circ} \subset A^{\circ} \cap B^{\circ}$. On the other hand $A^{\circ} \cap B^{\circ}$ is an open set with the property $A^\circ \cap B^\circ \subset A \cap B$. Since $(A \cap B)^\circ$ represents the union of all open sets contained in $A \cap B$, it must be $A^\circ \cap B^\circ \subset (A \cap B)^\circ$.

(c) True. In fact, using the identity $\overline{A} = ((A^c)^\circ)^c$ and de Morgan's laws we have

$$\overline{A \cup B} = (((A \cup B)^c)^\circ)^c = ((A^c \cap B^c)^\circ)^c = ((A^c)^\circ \cap (B^c)^\circ)^c$$

$$= ((A^c)^\circ)^c \cup ((B^c)^\circ)^c = \overline{A} \cup \overline{B}.$$

(d) False. Consider in $\mathbb{R}$ the two open sets $A = (0, 1)$ and $B = (1, 2)$. The closure of their intersection is $\overline{A \cap B} = \emptyset$. On the other hand, we have $\overline{A} = [0, 1]$ and $\overline{B} = [1, 2]$ and therefore $\overline{A} \cap \overline{B} = \{1\}$.

2.7   The statement is true. To prove it, let's show that $\operatorname{diam} A \leq \operatorname{diam} \overline{A}$ and $\operatorname{diam} A \geq \operatorname{diam} \overline{A}$. The first inequality follows trivially from the definition of diameter and from the fact that $A \subset \overline{A}$. For the second, consider two arbitrary points $x, y \in \overline{A}$. Since $\overline{A}$ coincides with the set of limit points of A, there exist two sequences (x_n) and (y_n) in A converging to x and y respectively. It follows that $\forall \varepsilon > 0 \ \exists N(\varepsilon)$ integer such that $d(x_n, x) < \varepsilon$ and $d(y_n, y) < \varepsilon \ \forall n \geq N$. Let $n \geq N$, for the triangle inequality of the distance we have

$$d(x, y) \leq d(x, x_n) + d(x_n, y_n) + d(y_n, y) < \varepsilon + d(x_n, y_n) + \varepsilon,$$

and so

$$\operatorname{diam} A = \sup_{w, z \in A} d(w, z) \geq d(x_n, y_n) > d(x, y) - 2\varepsilon.$$

From the arbitrariness of $x, y \in \overline{A}$ it follows that $\operatorname{diam} A \geq \operatorname{diam} \overline{A} - 2\varepsilon$ and from the arbitrariness of ε it follows that $\operatorname{diam} A \geq \operatorname{diam} \overline{A}$.

2.8   Let us show that $\overline{A} \subset \overline{B}$. By definition

$$\overline{A} = \bigcap_{\text{chiusi } F \supset A} F, \qquad \overline{B} = \bigcap_{\text{chiusi } F \supset B} F.$$

So $\overline{B}$ is a closed set such that $\overline{B} \supset B \supset A$. In other words, $\overline{B}$ is one of the closed sets F containing A whose intersection forms $\overline{A}$. Necessarily then $\overline{B} \supset \overline{A}$.

Let us show that $A^\circ \subset B^\circ$. By definition

$$A^\circ = \bigcup_{\text{open } G \subset A} G, \qquad B^\circ = \bigcup_{\text{open } G \subset B} G.$$

So A° is an open set such that $A^\circ \subset A \subset B$. In other words, A° is one of the open sets G contained in B whose union forms B°. Necessarily then $A^\circ \subset B^\circ$.

2.9 Since $A^\circ \subset A$, it is always true that $\overline{A^\circ} \subset \overline{A}$. The reversed inclusion is in general false. Consider the metric space $(\mathbb{R}, d)$ with d Euclidean metric and take $A = \{1\}$. We have

$$\overline{A} = A, \qquad A^\circ = \left(\overline{A^c}\right)^c = \emptyset,$$

so $\overline{A} = \{1\}$ is not contained in $A^\circ = \emptyset$.

Since $A \subset \overline{A}$, it is always true that $A^\circ \subset (\overline{A})^\circ$. The reversed inclusion is in general false. Consider the metric space $(\mathbb{R}, d)$ with d Euclidean metric and take $A = \mathbb{R} \setminus \{1\}$. We have

$$A^\circ = A, \qquad \overline{A} = \left((A^c)^\circ\right)^c = \mathbb{R},$$

so $(\overline{A})^\circ = \mathbb{R}$ is not contained in $A^\circ = \mathbb{R} \setminus \{1\}$.

2.10

(a) A is open in $(\mathbb{Q}, d)$. In fact, $\forall x \in A$ put $r = \max(|x - \sqrt{2}|, |x - \sqrt{3}|)$ and consider the open ball $B(x, r/2) = \{y \in \mathbb{Q} : d(x, y) < r/2\}$. It is clear that $B(x, r/2) \subset A$ and for the arbitrariness x we get the result.

(b) A is closed in $(\mathbb{Q}, d)$. In fact, since $\sqrt{2}, \sqrt{3} \notin \mathbb{Q}$ the complement of A is $A^c = \{x \in \mathbb{Q} : x < \sqrt{2}\} \cup \{x \in \mathbb{Q} : x > \sqrt{3}\}$. The set A^c is open in $(\mathbb{Q}, d)$ as it is the union of two sets both open in $(\mathbb{Q}, d)$. It follows that A is closed in $(\mathbb{Q}, d)$.

(c) A is totally bounded. In fact, consider the metric space (A, d) and let $\varepsilon > 0$ be an arbitrary real number. Then consider the n real intervals $[y_k, y_{k+1}]$ where $y_k = \sqrt{2} + k\varepsilon$ with $k = 1, 2, \ldots, n$ and $n = \lfloor(\sqrt{3} - \sqrt{2})/\varepsilon\rfloor$. Let x_k be any rational number such that $x_k \in (y_k, y_{k+1})$ and put $B(x_k, \varepsilon) = \{x \in A : d(x, x_k) < \varepsilon\}$, we have $A = \cup_{k=1}^{n} B(x_k, \varepsilon)$.

(d) A is not compact. This follows immediately from the fact that A is not complete. Consider in fact the so-called Babylonian method for computing square roots using a sequence of rational numbers. Explicitly consider the monotonically decreasing sequence $(x_k)_{k=0}^{\infty}$ defined recursively by $x_{k+1} = (x_k + 2/x_k)/2$ with $x_0 = 1$. Such a sequence is a Cauchy sequence in A that converges to $\sqrt{2} \notin A$.

12.3 Exercises of Chap. 3

3.1

(a) We have

$$\lim_{z \to \infty} \frac{6(z^3 - 2)^2}{(z - 1)^4} = \infty,$$

because

$$\lim_{z \to 0} \frac{1}{\dfrac{6\left(\left(\frac{1}{z}\right)^3 - 2\right)^2}{\left(\frac{1}{z} - 1\right)^4}} = \lim_{z \to 0} \frac{\left(\frac{1}{z} - 1\right)^4}{6\left(\left(\frac{1}{z}\right)^3 - 2\right)^2} = \lim_{z \to 0} \frac{z^2 (1 - z)^4}{6\left(1 - 2z^3\right)^2} = 0.$$

(b) Exploiting the identity $1 + z + z^2 + \cdots + z^n = (1 - z^{n+1})/(1 - z)$ with $z \neq 1$, we have

$$\lim_{z \to 1} \frac{z^{101} - 1}{z - 1} = \lim_{z \to 1} (1 + z + z^2 + \cdots + z^{100}) = 101.$$

(c) Observing that $z^2 + 1 = (z - i)(z + i)$ we have

$$\lim_{z \to i} \frac{(z^2 + 1)^2}{(z - i)^2} = \lim_{z \to i} \frac{(z - i)^2 (z + i)^2}{(z - i)^2} = \lim_{z \to i}(z + i)^2 = -4.$$

3.2

(a) We have

$$\lim_{z \to i} \frac{z}{z^2 + 1} = \infty.$$

In fact, for $f(z) = z/(z^2 + 1)$, by the continuity of the polynomials, we have

$$\lim_{z \to i} \frac{1}{f(z)} = \lim_{z \to i} \frac{z^2 + 1}{z} = \frac{i^2 + 1}{i} = 0.$$

(b) By the continuity of the function $\sqrt{z}$ when $-\pi < \operatorname{Arg} z < \pi$ and of the identity function in $\mathbb{C}$, it follows

$$\lim_{z \to i} \frac{\sqrt{z}}{z} = \frac{\sqrt{i}}{i} = \frac{e^{i\pi/4}}{i} = \frac{1}{\sqrt{2}} - i\frac{1}{\sqrt{2}}.$$

(c) By the continuity of polynomials we have

$$\lim_{z \to i} \frac{z^4 - z^3 - z^2 + z}{z + 1} = \frac{1 - (-i) - (-1) + i}{i + 1} = 2.$$

3.3

(a) By the continuity of rational functions and since the limit of the real part is equal to the real part of the limit, we have

$$\lim_{z \to 1} \operatorname{Re}\left(\frac{1+z}{z-3i}\right) = \operatorname{Re}\left(\frac{1+1}{1-3i}\right) = \operatorname{Re}\left(\frac{2(1+3i)}{1+9}\right) = \frac{1}{5}.$$

(b) The limit does not exist. In fact, put $z = 3e^{i\theta}$ and consider that $z = -3$ can be obtained from $3e^{i\theta}$ for $\theta \to \pi$ and $\theta \to -\pi$. In the two cases we have

$$\lim_{z \to -3} |z|^{1/2} e^{\frac{i}{2} \operatorname{Arg} z} = \lim_{\theta \to \pi} 3^{1/2} e^{i\theta/2} = i\sqrt{3},$$

$$\lim_{z \to -3} |z|^{1/2} e^{\frac{i}{2} \operatorname{Arg} z} = \lim_{\theta \to -\pi} 3^{1/2} e^{i\theta/2} = -i\sqrt{3}.$$

(c) The limit does not exist. In fact, choose $z = x$, with $x \in \mathbb{R}$, then

$$\lim_{z \to 0} \left(\frac{z}{\bar{z}}\right)^3 = \lim_{x \to 0} \left(\frac{x}{x}\right)^3 = 1.$$

Instead, choosing $z = x + ix$, with $x \in \mathbb{R}$, we have

$$\lim_{z \to 0} \left(\frac{z}{\bar{z}}\right)^3 = \lim_{x \to 0} \left(\frac{x+ix}{x-ix}\right)^3 = \lim_{x \to 0} \left(\frac{(x+ix)^2}{2x^2}\right)^3 = \lim_{x \to 0} \left(\frac{i}{x}\right)^3 = \infty.$$

3.4

(a) The limit does not exist. In fact, if we set $z = x + ix$ we have

$$\lim_{z \to 0} \frac{z^2}{|z|^2} = \lim_{x \to 0} \frac{x^2 - x^2 + 2ix^2}{x^2 + x^2} = \lim_{x \to 0} \frac{2i}{2} = i,$$

while for $z = x - ix$ we have

$$\lim_{z \to 0} \frac{z^2}{|z|^2} = \lim_{x \to 0} \frac{x^2 - x^2 - 2ix^2}{x^2 + x^2} = \lim_{x \to 0} \frac{-2i}{2} = -i.$$

(b) Multiplying and dividing by $\sqrt{z+i} + \sqrt{z-i}$ we have

$$\lim_{z \to \infty} \left(\sqrt{z+i} - \sqrt{z-i}\right)\sqrt{z} = \lim_{z \to \infty} \frac{2i\sqrt{z}}{\sqrt{z+i} + \sqrt{z-i}}$$

$$= \lim_{z \to \infty} \frac{2i}{\sqrt{1 + i/z} + \sqrt{1 - i/z}}$$

$$= i,$$

having used $\lim_{z \to \infty} c/z = \lim_{z \to 0} cz = 0$ and the continuity of rational functions and of $\sqrt{z}$.

(c) We have

$$\lim_{z \to \infty} \frac{1 - |z|^2}{z^2 + i} = \lim_{w \to 0} \frac{1 - |w^{-1}|^2}{w^{-2} + i} = \lim_{w \to 0} \frac{w^2 - (w/|w|)^2}{1 + iw^2}.$$

This limit does not exist because, as demonstrated at point (a), it does not exist the limit of $(w/|w|)^2$ for $w \to 0$.

3.5

(a) Observing that $z^2 - 2i = 0$ has solutions $z = \sqrt{2}e^{i\pi/4} = 1 + i$ and $z = \sqrt{2}e^{i5\pi/4} = -1 - i$, while $z^2 - 2z + 2 = 0$ has solutions $z = 1 + i$ and $z = 1 - i$, we have

$$\lim_{z \to 1+i} \frac{z^2 - 2i}{z^2 - 2z + 2} = \lim_{z \to 1+i} \frac{(z - 1 - i)(z + 1 + i)}{(z - 1 - i)(z - 1 + i)}$$

$$= \lim_{z \to 1+i} \frac{z + 1 + i}{z - 1 + i}$$

$$= \frac{1 + i + 1 + i}{1 + i - 1 + i}$$

$$= 1 - i.$$

(b) The limit does not exist. In fact for $z = 3e^{i(\pi - \varepsilon)}$ we have

$$\lim_{\varepsilon \to 0} \frac{\sqrt{3e^{i(\pi - \varepsilon)}}}{3e^{i(\pi - \varepsilon)}} = \lim_{\varepsilon \to 0} \frac{\sqrt{3}e^{i(\pi - \varepsilon)/2}}{3e^{i(\pi - \varepsilon)}} = \frac{1}{\sqrt{3}}e^{-i\pi/2} = -\frac{i}{\sqrt{3}},$$

instead for $z = 3e^{-i(\pi - \varepsilon)}$ we have

$$\lim_{\varepsilon \to 0} \frac{\sqrt{3e^{-i(\pi - \varepsilon)}}}{3e^{-i(\pi - \varepsilon)}} = \lim_{\varepsilon \to 0} \frac{\sqrt{3}e^{-i(\pi - \varepsilon)/2}}{3e^{-i(\pi - \varepsilon)}} = \frac{1}{\sqrt{3}}e^{i\pi/2} = \frac{i}{\sqrt{3}}.$$

(c) Multiplying and dividing by $\sqrt{z + 2i} + \sqrt{z + i}$ we have

$$\lim_{z \to \infty} (\sqrt{z + 2i} - \sqrt{z + i}) = \lim_{z \to \infty} \frac{(z + 2i) - (z + i)}{\sqrt{z + 2i} + \sqrt{z + i}}$$

$$= \lim_{z \to \infty} \frac{i}{\sqrt{z + 2i} + \sqrt{z + i}}$$

$$= 0.$$

3.6 If $\lim_{z \to z_0} f(z) = w_0$, then $\forall \varepsilon > 0 \; \exists \delta(\varepsilon) > 0$ such that $|f(z) - w_0| < \varepsilon$ for $0 < |z - z_0| < \delta$. From the triangle inequality we have

$$\big||f(z)| - |w_0|\big| \le |f(z) - w_0|$$

for every value of z. It follows immediately that $\lim_{z \to z_0} |f(z)| = |w_0|$.

To show that the reversed implication is false, consider for example

$$f(z) = \sqrt{|z|}\, e^{i \, \mathrm{Arg}(z)/2}.$$

For every $z_0 = -t$ with $t > 0$ we have $\lim_{z \to -t} |f(z)| = \sqrt{t}$ while $\lim_{z \to -t} f(z)$ does not exist. Note that the example does not work for $z_0 = 0$, namely, $w_0 = 0$. In fact, for a generic function f, the property $\lim_{z \to z_0} |f(z)| = 0$, by the relation $\big||f(z)| - 0\big| = |f(z) - 0|$, implies $\lim_{z \to z_0} f(z) = 0$.

3.7

(a) Since $|z_n| = 1/n$ we find

$$\lim_{n \to \infty} \frac{i^n}{n} = 0.$$

(b) Since $1 + i = \sqrt{2}\, e^{i\pi/4}$ we have

$$|z_n| = \left|\frac{(1+i)^n}{n}\right| = \left|\frac{2^{n/2}}{n}\, e^{in\pi/4}\right| = \frac{2^{n/2}}{n}.$$

Note that $n/2^{n/2} \xrightarrow{n \to \infty} 0$, so that

$$\lim_{n \to \infty} \frac{1}{z_n} = 0,$$

and therefore $z_n \xrightarrow{n \to \infty} \infty$.

(c) Multiplying and dividing z_n by $\sqrt{n + 2i} + \sqrt{n + i}$ we have

$$\lim_{n \to \infty} \sqrt{n}\left(\sqrt{n + 2i} - \sqrt{n + i}\right) = \lim_{n \to \infty} \frac{i\sqrt{n}}{\sqrt{n + 2i} + \sqrt{n + i}}$$

$$= \lim_{n \to \infty} \frac{i}{\sqrt{1 + 2i/n} + \sqrt{1 + i/n}}$$

$$= \frac{i}{2}.$$

3.8

(a) Let $z = x + iy$, the function $\operatorname{Arg} z$ is defined by the equation (1.17) at the points $(x, y) \in \mathbb{R}^2$ except at the point $(x, y) = (0, 0)$. This function presents a discontinuity of amplitude 2π at the points $(x, 0)$ with $x < 0$. In fact for $x < 0$ we have $\lim_{\varepsilon \to 0+} \operatorname{Arg}(x + i\varepsilon) = \pi$ while $\lim_{\varepsilon \to 0+} \operatorname{Arg}(x - i\varepsilon) = -\pi$. At all points outside the negative real semi-axis, $\operatorname{Arg} z$ is continuous being a composition of continuous functions.

(b) As a rational function, $f(z)$ is continuous everywhere in $\mathbb{C}$ except at the zeros of the polynomial in the denominator. These are the solutions of $z^3 + 27 = 0$, that is, the numbers

$$z_k = (-27)^{1/3} = \left(27 e^{i\pi}\right)^{1/3} = 3 e^{i(\pi + 2\pi k)/3}, \qquad k = 0, 1, 2.$$

Note that $z_0 = -3$, $z_1 = 3/2 - i3\sqrt{3}/2$ and $z_2 = 3/2 + i3\sqrt{3}/2$.

(c) This function is not continuous at the points where $\operatorname{Arg} z = \pi$, that is, along the negative real axis excluding the origin. In fact, if $z = -r$ with $r > 0$, consider the circle $\gamma(t) = r e^{it}$, $-\pi \leq t \leq \pi$ passing through the point $z = \gamma(\pi) = \gamma(-\pi)$. We have $f(\gamma(t)) = \sqrt{r} e^{it/2}$ and therefore

$$\lim_{t \to \pi} f(\gamma(t)) = \sqrt{r} e^{i\pi/2} = i\sqrt{r},$$

$$\lim_{t \to -\pi} f(\gamma(t)) = \sqrt{r} e^{-i\pi/2} = -i\sqrt{r}.$$

This is enough to conclude that $\lim_{w \to z} f(w)$ does not exist and therefore f is not continuous at z. At the point $z = 0$, where it is $f(0) = 0$, the function is continuous because

$$|f(w) - f(0)| = \sqrt{|w|} \xrightarrow{w \to 0} 0.$$

At all other points of $\mathbb{C}$ f is continuous as composition of continuous functions.

3.9 If $f(z)$ were uniformly continuous in A, $\forall \varepsilon > 0$ there would exist $\delta(\varepsilon) > 0$ such that $\forall z, w \in A$ with $|z - w| < \delta$ then $|f(z) - f(w)| < \varepsilon$. We show that there exist pairs of points $z, w \in A$ for which this property is false. Fix $\varepsilon > 0$ and choose real positive z and w in the form $z = \delta$ and $w = \delta/(1 + 2\varepsilon)$ with $0 < \delta \leq 2$ arbitrary. It turns out

$$|z - w| = \delta - \frac{\delta}{1 + 2\varepsilon} = \delta \frac{2\varepsilon}{1 + 2\varepsilon} < \delta,$$

$$\left| \frac{1}{z} - \frac{1}{w} \right| = \frac{1 + 2\varepsilon}{\delta} - \frac{1}{\delta} = \frac{2\varepsilon}{\delta} \geq \varepsilon.$$

The statement follows from the arbitrariness of δ.

3.10 As a rational function with a never zero denominator, f is continuous in D. To demonstrate that f is not uniformly continuous in D, we show that, choosing $\varepsilon > 0$, there exist pairs of points $z, w \in D$ that are arbitrarily small apart and whose images $f(z)$, $f(w)$ have a distance greater than or equal to ε. Consider two points of the type $z = x$ and $w = x + \delta$ with $x \in (0, 3]$ and $\delta > 0$. Determine $\delta(x, \varepsilon)$ such that

$$|f(z) - f(w)| = \frac{1}{x^2} - \frac{1}{(x + \delta)^2} = \frac{\delta^2 + 2x\delta}{x^2(\delta^2 + 2x\delta + x^2)} = \varepsilon.$$

This condition is equivalent to

$$(1 - \varepsilon x^2)\delta^2 + 2x(1 - \varepsilon x^2)\delta - \varepsilon x^4 = 0,$$

whose positive solution is

$$\delta(x, \varepsilon) = x \left(\sqrt{1 + \frac{\varepsilon x^2}{1 - \varepsilon x^2}} - 1 \right).$$

In deriving this solution we have assumed that ε is sufficiently small so that $1 - \varepsilon x^2 > 0 \ \forall x \in (0, 3]$, that is we have assumed $\varepsilon < 1/9$. Since

$$\inf_{x \in (0,3]} \delta(x, \varepsilon) = 0,$$

it follows that $\forall \delta > 0 \ \exists z, w \in D$ with $|z - w| < \delta$ and $|f(z) - f(w)| \geq \varepsilon$.

Consider the domain $D_r = \{z \in \mathbb{C} : r \leq |z| \leq 3\}$ with $0 < r < 3$. The function f is continuous in D_r and D_r, being a closed and bounded subset of $\mathbb{C}$, is compact. It follows that f is uniformly continuous in D_r.

3.11 By hypothesis, arbitrarily chosen $\varepsilon > 0$ there exist $\delta_f(\varepsilon) > 0$ and $\delta_g(\varepsilon) > 0$ such that

$$|f(x) - f(y)| < \frac{\varepsilon}{2}, \qquad \forall x, y \in S, \quad d(x, y) \leq \delta_f,$$

$$|g(x) - g(y)| < \frac{\varepsilon}{2}, \qquad \forall x, y \in S, \quad d(x, y) \leq \delta_g.$$

From these inequalities and using the triangle inequality, for the function $f + g : (S, d) \mapsto \mathbb{C}$ we obtain

$$|(f + g)(x) - (f + g)(y)| = |f(x) + g(x) - f(y) - g(y)|$$

$$= |(f(x) - f(y)) + (g(y) - g(y))|$$

$$\leq |f(x) - f(y)| + |g(x) - g(y)|$$

$$< \frac{\varepsilon}{2} + \frac{\varepsilon}{2} = \varepsilon, \qquad \forall x, y \in S, \quad d(x, y) \leq \delta,$$

where $\delta(\varepsilon) = \min(\delta_f(\varepsilon), \delta_g(\varepsilon))$. We conclude that $f + g$ is uniformly continuous in S.

3.12 By hypothesis there exist two positive real numbers M_f and M_g such that

$$|f(x) - f(y)| \leq M_f d(x, y), \qquad \forall x, y \in S,$$

$$|g(x) - g(y)| \leq M_g d(x, y), \qquad \forall x, y \in S.$$

From these and using the triangle inequality, for the function $f + g : (S, d) \mapsto \mathbb{C}$ we obtain

$$
\begin{aligned}
|(f + g)(x) - (f + g)(y)| &= |f(x) + g(x) - f(y) - g(y)| \\
&= |(f(x) - f(y)) + (g(y) - g(y))| \\
&\leq |f(x) - f(y)| + |g(x) - g(y)| \\
&= (M_f + M_g)d(x, y), \qquad \forall x, y \in S,
\end{aligned}
$$

that is, $f + g$ is Lipschitz continuous in S.

12.4 Exercises of Chap. 4

4.1 For each $z \in \overline{B}(0, r)$ we have

$$|f_n(z) - f(z)| = \left| z^n - 0 \right| = |z|^n \leq r^n.$$

Since $0 < r < 1, \forall \varepsilon > 0$ it is possible to find an integer $N(\varepsilon)$, for example

$$N(\varepsilon) = \left\lceil \frac{\ln \varepsilon}{\ln r} \right\rceil,$$

such that $\forall n > N(\varepsilon)$ we have $r^n < \varepsilon$ and therefore $|f_n(z) - f(z)| < \varepsilon \ \forall z \in \overline{B}(0, r)$, which is the definition of uniform convergence of (f_n) to f in $\overline{B}(0, r)$.

On the other hand (f_n) does not converge uniformly to f in $B(0, 1)$ since, choosing $0 < \varepsilon < 1$, for every integer n it is possible to find a point $z \in B(0, 1)$, just take $\varepsilon^{1/n} < |z| < 1$, such that $|f_n(z) - f(z)| > \varepsilon$.

4.2 By hypothesis

$$\lim_{n\to\infty} \sum_{k=0}^{n} a_k = A, \qquad \lim_{n\to\infty} \sum_{k=0}^{n} b_k = B.$$

So $\forall \varepsilon > 0$ there exist integers $N_a(\varepsilon)$ and $N_b(\varepsilon)$ such that

$$\left| \sum_{k=0}^{n} a_k - A \right| < \frac{\varepsilon}{2}, \qquad \forall n > N_a,$$

$$\left| \sum_{k=0}^{n} b_k - B \right| < \frac{\varepsilon}{2}, \qquad \forall n > N_b.$$

Then setting $N = \max(N_a, N_b)$, $\forall n > N$ we have

$$\left| \sum_{k=0}^{k} (a_k + b_k) - (A + B) \right| \le \left| \sum_{k=0}^{n} a_k - A \right| + \left| \sum_{k=0}^{n} b_k - B \right| < \varepsilon.$$

This implies, as required, that

$$\lim_{n\to\infty} \sum_{k=0}^{n} (a_k + b_k) = A + B.$$

4.3

(a) The n-th coefficient of the series is

$$a_n = \cosh n = \frac{e^n + e^{-n}}{2}$$

and we have

$$\frac{a_n}{a_{n+1}} = \frac{e^{-n} + e^n}{e^{-(n+1)} + e^{n+1}} = \frac{e^{-2n} + 1}{e^{-2n-1} + e} \xrightarrow{n\to\infty} \frac{1}{e}.$$

The radius of convergence of the series is therefore $R = 1/e$.

(b) The n-th coefficient of the series is $a_n = n + c^n$, Furthermore

$$\frac{|a_n|}{|a_{n+1}|} = \frac{|n + c^n|}{\left|n + 1 + c^{n+1}\right|} \xrightarrow{n\to\infty} \begin{cases} 1 & |c| \le 1 \\ |c|^{-1} & |c| > 1 \end{cases}.$$

Therefore, we have

$$R = \lim_{n\to\infty} \frac{|a_n|}{|an + 1|} = \begin{cases} 1 & |c| \le 1 \\ |c|^{-1} & |c| > 1 \end{cases}.$$

(c) The n-th coefficient of the series is

$$
a_n = \begin{cases} 3 & n = \lfloor k \ln k \rfloor, \quad k = 1, 2, 3, \ldots \\ 0 & \text{otherwise} \end{cases}
$$

and so

$$
\begin{aligned}
R^{-1} &= \limsup_{m \to \infty} |a_m|^{1/m} \\
&= \lim_{m \to \infty} \sup_{n \geq m} \left\{ |a_n|^{1/n} \right\} \\
&= \lim_{m \to \infty} \sup_{k:\ \lfloor k \ln k \rfloor \geq m} \left\{ 3^{1/\lfloor k \ln k \rfloor} \right\} \\
&= \lim_{m \to \infty} 3^{1/\lfloor k_m \ln k_m \rfloor} \\
&= 1,
\end{aligned}
$$

where k_m is the smallest integer k such that $\lfloor k \ln k \rfloor \geq m$. In conclusion, $R = 1$.

4.4

(a) The n-th coefficient of the series is $a_n = 2^n/n^3$ and we have

$$
\frac{|a_n|}{|a_{n+1}|} = \frac{2^n}{n^3} \frac{(n+1)^3}{2^{n+1}} = \frac{1}{2} \left(1 + \frac{1}{n} \right)^3 \xrightarrow{n \to \infty} \frac{1}{2}.
$$

The radius of convergence of the series is therefore $R = 1/2$.

(b) The n-th coefficient of the series is

$$
a_n = i \sinh n = \frac{e^{-n} - e^n}{2i}
$$

and we have

$$
\frac{|a_n|}{|a_{n+1}|} = \frac{|e^{-n} - e^n|}{|e^{-(n+1)} - e^{n+1}|} = \frac{|e^{-2n} - 1|}{|e^{-2n-1} - e|} \xrightarrow{n \to \infty} \frac{1}{e}.
$$

The radius of convergence of the series is therefore $R = 1/e$.

(c) The n-th coefficient of the series is $a_n = n^3 4^n$ and we have

$$
\frac{a_n}{a_{n+1}} = \frac{n^3 4^n}{(n+1)^3 4^{n+1}} = \frac{1}{4(1 + 1/n)^3} \xrightarrow{n \to \infty} \frac{1}{4}.
$$

The radius of convergence of the series is therefore $R = 1/4$.

4.5

(a) The n-th coefficient of the series is $a_n = ne^n$ and we have

$$\left|\frac{a_n}{a_{n+1}}\right| = \frac{n}{(n+1)e} \xrightarrow{n\to\infty} \frac{1}{e}.$$

So

$$R = \lim_{n\to\infty} \left|\frac{a_n}{a_{n+1}}\right| = \frac{1}{e}.$$

(b) The n-th coefficient of the series rewritten in the form $\sum_{n=0}^{\infty} a_n z^n$ is

$$a_n = \begin{cases} 1 & n = k! + k, \quad k = 0, 1, 2, \ldots \\ 0 & \text{otherwise} \end{cases}$$

and so

$$R^{-1} = \limsup_{m\to\infty} |a_m|^{1/m}$$

$$= \lim_{m\to\infty} \sup_{n\geq m} \left\{ |a_n|^{1/n} \right\}$$

$$= \lim_{m\to\infty} 1$$

$$= 1,$$

that is $R = 1$.

(c) The n-th coefficient of the series is

$$a_n = n^n \left(\frac{1+i}{2}\right)^{n^2}$$

and we have

$$|a_n|^{1/n} = n \left|\frac{1+i}{2}\right|^n = \frac{n}{2^{n/2}} \xrightarrow{n\to\infty} 0.$$

So $R = \infty$.

4.6

(a) The n-th coefficient of the series is

$$a_n = \cosh(n^2) = \frac{e^{-n^2} + e^{n^2}}{2}$$

and we have

$$\frac{|a_n|}{|a_{n+1}|} = \frac{e^{-n^2} + e^{n^2}}{e^{-(n+1)^2} + e^{(n+1)^2}}$$

$$= \frac{e^{-2n^2} + 1}{e^{-2n^2-2n-1} + e^{2n+1}} \xrightarrow{n\to\infty} 0.$$

The radius of convergence of the series is therefore $R = 0$.

(b) The n-th coefficient of the series is

$$a_n = \frac{n^{3n}}{(3n)!}$$

and we have

$$\frac{|a_n|}{|a_{n+1}|} = \frac{n^{3n}}{(3n)!} \frac{(3(n+1))!}{(n+1)^{3(n+1)}}$$

$$= \frac{(3n+3)(3n+2)(3n+1)}{(n+1)^3} \frac{n^{3n}}{(n+1)^{3n}}$$

$$= \frac{27(n+1)\left(n+\frac{2}{3}\right)\left(n+\frac{1}{3}\right)}{(n+1)^3} \frac{1}{\left[\left(1+\frac{1}{n}\right)^n\right]^3} \xrightarrow{n\to\infty} \frac{27}{e^3}.$$

The radius of convergence of the series is therefore $R = 27/e^3$.

4.7

(a) The n-th coefficient of the series is $a_n = 3^n/(n+1)^3$. Furthermore

$$\left|\frac{a_n}{a_{n+1}}\right| = \frac{3^n}{(n+1)^3} \frac{(n+2)^3}{3^{n+1}} = \frac{1}{3}\left(\frac{n+2}{n+1}\right)^3 \xrightarrow{n\to\infty} \frac{1}{3}.$$

So we have

$$R = \lim_{n\to\infty} \left|\frac{a_n}{a_{n+1}}\right| = \frac{1}{3}.$$

(b) The n-th coefficient of the series rewritten in the form $\sum_{n=0}^{\infty} a_n z^n$ is

$$a_n = \begin{cases} 1 + n^3 & n = k^2, \quad k = 1, 2, 3, \ldots \\ 0 & \text{otherwise} \end{cases}$$

and so

$$R^{-1} = \limsup_{m \to \infty} |a_m|^{1/m}$$

$$= \lim_{m \to \infty} \sup_{n \geq m} \left\{ |a_n|^{1/n} \right\}$$

$$= \lim_{m \to \infty} \sup_{k:\, k^2 \geq m} \left\{ (1 + k^6)^{1/k^2} \right\}$$

$$= \lim_{m \to \infty} (1 + k_m^6)^{1/k_m^2}$$

$$= 1,$$

where k_m is the smallest integer k such that $k^2 \geq m$. In conclusion, $R = 1$.

4.8

(a) The n-th coefficient of the series is $a_n = n^3 e^{3n}$. Furthermore

$$\left| \frac{a_n}{a_{n+1}} \right| = \left(\frac{n}{n+1} \right)^3 \frac{e^{3n}}{e^{3(n+1)}} = \left(\frac{n}{n+1} \right)^3 \frac{1}{e^3} \xrightarrow{n \to \infty} \frac{1}{e^3}.$$

So we have

$$R = \lim_{n \to \infty} \left| \frac{a_n}{a_{n+1}} \right| = \frac{1}{e^3}.$$

(b) The n-th coefficient of the series is

$$a_n = \left(\frac{\ln n}{(1 + i)^n} \right)^n.$$

Furthermore

$$|a_n|^{1/n} = \frac{\ln n}{|1 + i|^n} = \frac{\ln n}{2^{n/2}} \xrightarrow{n \to \infty} 0$$

and so

$$R = \infty.$$

(c) The n-th coefficient of the series rewritten in the form $\sum_{n=0}^{\infty} a_n z^n$ is

$$a_n = \begin{cases} (-1)^k & n = k^3, \quad k = 0, 1, 2, \dots \\ 0 & \text{otherwise} \end{cases}$$

and so

$$R^{-1} = \limsup_{m\to\infty} |a_m|^{1/m} = \lim_{m\to\infty} \sup_{n\geq m} \left\{ |a_n|^{1/n} \right\} = \lim_{m\to\infty} 1 = 1,$$

that is $R = 1$.

4.9

(a) The n-th coefficient of the series is

$$a_n = \frac{(3n)! + 4^n}{(3n+1)!}.$$

Furthermore

$$\left| \frac{a_n}{a_{n+1}} \right| = \frac{(3n)! + 4^n}{(3n+1)!} \frac{(3n+4)!}{(3n+3)! + 4^{n+1}}$$

$$= \frac{(3n+4)(3n+3)(3n+2)[1 + 4^n/(3n)!]}{(3n+3)(3n+2)(3n+1) + 4^{n+1}/(3n)!} \xrightarrow{n\to\infty} 1,$$

having used $\ln n! \simeq n \ln n - n$ valid for n large. So we have

$$R = \lim_{n\to\infty} \left| \frac{a_n}{a_{n+1}} \right| = 1.$$

(b) The n-th coefficient of the series is

$$a_n = i^{-in} = e^{-in \log i} = e^{-in(\ln 1 + i\pi/2)} = e^{n\pi/2}.$$

Observing that $|a_n|^{1/n} = e^{\pi/2}$, we conclude

$$R = e^{-\pi/2}.$$

(c) The n-th coefficient of the series rewritten in the form $\sum_{n=1}^{\infty} a_n z^n$ is

$$a_n = \begin{cases} \log(in/2) & n \text{ even} \\ 0 & n \text{ odd} \end{cases}.$$

Observing that $|\log(in/2)| = \sqrt{(\ln(n/2))^2 + \pi^2/4}$ and remembering that $(\ln k)^{1/k}$ decreases as $k \to \infty$ monotonically to 1, we have

$$R^{-1} = \limsup_{m\to\infty} |a_m|^{1/m}$$

$$= \lim_{m\to\infty} \sup_{n\geq m} \left\{ |a_n|^{1/n} \right\}$$

$$= \lim_{m \to \infty} \sup_{n \geq m, \, n \text{ even}} \left\{ \left(\sqrt{(\ln(n/2))^2 + \pi^2/4} \right)^{1/n} \right\}$$

$$= \lim_{m \to \infty} \left(\sqrt{(\ln(n_m/2))^2 + \pi^2/4} \right)^{1/n_m}$$

$$= 1,$$

where $n_m = m$ if m is even and $n_m = m + 1$ if m is odd. In conclusion, $R = 1$.

4.10

(a) The n-th coefficient of the series rewritten in the form $\sum_{n=0}^{\infty} a_n z^n$ is

$$a_n = \begin{cases} (-1)^k \ n = 1 + k + k^2 + k^3, & k = 0, 1, 2, \ldots \\ 0 & \text{otherwise} \end{cases}$$

and so

$$R^{-1} = \limsup_{m \to \infty} |a_m|^{1/m} = \lim_{m \to \infty} \sup_{n \geq m} \left\{ |a_n|^{1/n} \right\} = \lim_{m \to \infty} 1 = 1,$$

that is $R = 1$.

(b) The n-th coefficient of the series is $a_n = n^2 3^n$ and we have

$$\frac{a_n}{a_{n+1}} = \frac{n^2 3^n}{(n+1)^2 3^{n+1}} = \frac{1}{3} \left(\frac{n}{n+1} \right)^2 \xrightarrow{n \to \infty} \frac{1}{3}.$$

The radius of convergence of the series is $R = 1/3$.

4.11

(a) The n-th coefficient of the series is $a_n = n^3 2^n$ and we have

$$\frac{a_n}{a_{n+1}} = \frac{n^3 2^n}{(n+1)^3 2^{n+1}} = \frac{1}{2} \left(\frac{n}{n+1} \right)^3 \xrightarrow{n \to \infty} \frac{1}{2}.$$

The radius of convergence of the series is $R = 1/2$.

(b) The n-th coefficient of the series rewritten in the form $\sum_{n=0}^{\infty} a_n z^n$ is

$$a_n = \begin{cases} (-5)^{k^3} \ n = 1 + k^3, & k = 0, 1, 2, \ldots \\ 0 & \text{otherwise} \end{cases}$$

and so

$$R^{-1} = \limsup_{m \to \infty} |a_m|^{1/m} = \lim_{m \to \infty} \sup_{n \ge m} \left\{ |a_n|^{1/n} \right\}$$

$$= \lim_{m \to \infty} \sup_{k:\ 1+k^3 \ge m} \left\{ 5^{k^3/(1+k^3)} \right\} = \lim_{m \to \infty} 5 = 5,$$

that is $R = 1/5$.

4.12 The convergence radius R of the series (b) is

$$R^{-1} = \limsup_{n \to \infty} \left| a_n n^c \right|^{1/n}$$

$$= \limsup_{n \to \infty} \left(|a_n|^{1/n} \ |n^c|^{1/n} \right).$$

Since the limit

$$\lim_{n \to \infty} |n^c|^{1/n} = \lim_{n \to \infty} \left(\sqrt{n^c n^{\bar c}} \right)^{1/n} = \lim_{n \to \infty} n^{\operatorname{Re} c/n} = 1$$

exists and is positive, for the properties of $\limsup$ we have

$$R^{-1} = \limsup_{n \to \infty} |a_n|^{1/n} \ \lim_{n \to \infty} |n^c|^{1/n}$$

$$= \limsup_{n \to \infty} |a_n|^{1/n},$$

which is the inverse of the radius of convergence of the series (a). In the case $a_n = e^{-n}$

$$R^{-1} = \lim_{n \to \infty} e^{-n/n} = e^{-1},$$

that is $R = e$.

4.13

(a) The n-th coefficient of the series rewritten in the form $\sum_{n=0}^{\infty} a_n z^n$ is

$$a_n = \begin{cases} e^{k!} & n = k!, \quad k = 1, 2, \ldots \\ 0 & \text{otherwise} \end{cases}$$

and therefore

$$R^{-1} = \limsup_{m \to \infty} |a_m|^{1/m}$$

$$= \lim_{m \to \infty} \sup_{n \geq m} \left\{ |a_n|^{1/n} \right\}$$

$$= \lim_{m \to \infty} \sup_{k:\, k! \geq m} \left\{ \left| e^{k!} \right|^{1/k!} \right\}$$

$$= \lim_{m \to \infty} e$$

$$= e.$$

In conclusion, $R = 1/e$.

(b) The n-th coefficient of the series rewritten in the form $\sum_{n=0}^{\infty} a_n z^n$ is

$$a_n = \begin{cases} e^k & n = k!, \quad k = 1, 2, \ldots \\ 0 & \text{otherwise} \end{cases}$$

and so

$$R^{-1} = \limsup_{m \to \infty} |a_m|^{1/m}$$

$$= \lim_{m \to \infty} \sup_{n \geq m} \left\{ |a_n|^{1/n} \right\}$$

$$= \lim_{m \to \infty} \sup_{k:\, k! \geq m} \left\{ \left| e^k \right|^{1/k!} \right\}$$

$$= 1,$$

having used the fact that the sequence $(e^{k/k!})$ is monotonically decreasing for $k \geq 1$ and tends to 1 for $k \to \infty$. In conclusion, $R = 1$.

(c) The series is already written in canonical form and its n-th coefficient is

$$a_n = e^{n!}.$$

We thus have

$$|a_n|^{1/n} = e^{n!/n} = e^{(n-1)!} \xrightarrow{n \to \infty} \infty,$$

and therefore $R = 0$.

4.14

(a) The n-th coefficient of the series rewritten in the form $\sum_{n=0}^{\infty} a_n z^n$ is

$$a_n = \begin{cases} n^2 \pi^n & n \text{ odd} \\ 0 & n \text{ even} \end{cases}$$

and therefore

$$R^{-1} = \limsup_{m \to \infty} |a_m|^{1/m}$$

$$= \lim_{m \to \infty} \sup_{n \ge m} \left\{ |a_n|^{1/n} \right\}$$

$$= \lim_{m \to \infty} \sup_{n \ge m:\ n \text{ even}} \left\{ n^{2/n} \pi \right\}$$

$$= \pi \lim_{m \to \infty} n_m^{2/n_m}$$

$$= \pi,$$

where $n_m = m$ if m is odd and $n_m = m + 1$ if m is even. In conclusion, $R = 1/\pi$.

(b) The n-th coefficient of the series is $a_n = n^{2+i}\pi^n = n^2\pi^n e^{i \ln n}$ and we have

$$\left| \frac{a_n}{a_{n+1}} \right| = \frac{n^2\pi^n}{(n+1)^2\pi^{n+1}} = \frac{1}{\pi} \left(\frac{n}{n+1} \right)^2 \xrightarrow{n \to \infty} \frac{1}{\pi}.$$

The radius of convergence of the series is $R = 1/\pi$.

4.15

(a) The n-th coefficient of the series is $a_n = n^3 e^{-3n}$. Furthermore

$$\left| \frac{a_n}{a_{n+1}} \right| = \left(\frac{n}{n+1} \right)^3 \frac{e^{3(n+1)}}{e^{3n}} = \left(\frac{n}{n+1} \right)^3 e^3 \xrightarrow{n \to \infty} e^3.$$

So we have

$$R = \lim_{n \to \infty} \left| \frac{a_n}{a_{n+1}} \right| = e^3.$$

(b) The n-th coefficient of the series rewritten in the form $\sum_{n=0}^{\infty} a_n z^n$ is

$$a_n = \begin{cases} (-1)^k / k \ n = k(k+1), & k = 1, 2, 3, \ldots \\ 0 & \text{otherwise} \end{cases}$$

and so

$$R^{-1} = \limsup_{m \to \infty} |a_m|^{1/m}$$

$$= \lim_{m \to \infty} \sup_{n \ge m} \left\{ |a_n|^{1/n} \right\}$$

$$= \lim_{m \to \infty} \sup_{k:\, k(k+1) \geq m} \left\{ \left(\frac{1}{k} \right)^{1/(k(k+1))} \right\}$$

$$= \lim_{m \to \infty} 1$$

$$= 1.$$

In conclusion, $R = 1$.

4.16

(a) By hypothesis we know that $\limsup_{n \to \infty} |c_n|^{1/n} = 1/R_0$, therefore

$$\frac{1}{R} = \limsup_{n \to \infty} \left| c_n^3 \right|^{1/n}$$

$$= \lim_{n \to \infty} \sup_{k \geq n} \left\{ \left(|c_k|^{1/k} \right)^3 \right\}$$

$$= \lim_{n \to \infty} \left(\sup_{k \geq n} \left\{ |c_k|^{1/k} \right\} \right)^3$$

$$= \left(\lim_{n \to \infty} \sup_{k \geq n} \left\{ |c_k|^{1/k} \right\} \right)^3$$

$$= \left(\frac{1}{R_0} \right)^3 .$$

In conclusion, $R = R_0^3$.

(b) The n-th coefficient of the series rewritten in the form $\sum_{n=0}^{\infty} a_n z^n$ is

$$a_n = \begin{cases} c_k & n = 3k, \quad k = 0, 1, 2, \ldots \\ 0 & \text{otherwise} \end{cases}$$

and so

$$\frac{1}{R} = \limsup_{m \to \infty} |a_m|^{1/m}$$

$$= \lim_{m \to \infty} \sup_{n \geq m} \left\{ |a_n|^{1/n} \right\}$$

$$= \lim_{m \to \infty} \sup_{k:\, 3k \geq m} \left\{ |c_k|^{1/(3k)} \right\}$$

$$= \lim_{j \to \infty} \sup_{k \geq j} \left\{ \left(|c_k|^{1/k} \right)^{1/3} \right\}$$

$$= \left(\lim_{j \to \infty} \sup_{k \geq j} \left\{ |c_k|^{1/k} \right\} \right)^{1/3}$$

$$= \left(\frac{1}{R_0} \right)^{1/3}.$$

In conclusion, $R = R_0^{1/3}$. Alternatively, set $w = z^3$. The series $\sum_{n=0}^{\infty} c_n w^n$ has radius of convergence R_0 in the w plane and $R = R_0^{1/3}$ in the $z = w^{1/3}$ plane.

4.17

(a) The n-th coefficient of the assigned series is $a_n = n + c^n$ and we have

$$|a_n|^{1/n} = |n + c^n|^{1/n} = \begin{cases} n^{1/n} |1 + c^n/n|^{1/n} & |c| \leq 1 \\ |c| |1 + n/c^n|^{1/n} & |c| > 1 \end{cases}.$$

Observing that $\lim_{n \to \infty} n^{1/n} = 1$, we have

$$|a_n|^{1/n} \xrightarrow{n \to \infty} \begin{cases} 1 & |c| \leq 1 \\ |c| & |c| > 1 \end{cases}$$

and so the radius of convergence is

$$R = \begin{cases} 1 & |c| \leq 1 \\ 1/|c| & |c| > 1 \end{cases}.$$

(b) The n-th coefficient of the series rewritten in the form $\sum_{n=1}^{\infty} a_n z^n$ is

$$a_n = \begin{cases} 5^k & n = k!, \quad k = 1, 2, 3, \ldots \\ 0 & \text{otherwise} \end{cases}$$

and the Hadamard formula provides

$$\frac{1}{R} = \limsup_{m \to \infty} |a_m|^{1/m}$$

$$= \lim_{m \to \infty} \sup_{n \geq m} |a_n|^{1/n}$$

$$= \lim_{m \to \infty} \sup_{k:\, k! \geq m} 5^{k/k!}$$

$$= \lim_{m \to \infty} 5^{1/(k_m - 1)!}$$

$$= 5^0,$$

where k_m is the smallest integer k such that $k! \geq m$. In conclusion $R = 1$.

4.18

(a) $R = e^{-2}$. In fact, for $a_n = ne^{2n-\sqrt{n}}$ we have

$$\left|\frac{a_n}{a_{n+1}}\right| = \frac{ne^{2n-\sqrt{n}}}{(n+1)e^{2(n+1)-\sqrt{n+1}}}$$

$$= e^{-2}\frac{1}{1+1/n}e^{\sqrt{n}-\sqrt{n+1}} \xrightarrow{n\to\infty} e^{-2}.$$

(b) $R = 1$. In fact, for $a_n = [\ln(1+n)]^4$ we have

$$\left|\frac{a_n}{a_{n+1}}\right| = \left[\frac{\ln(n+1)}{\ln(n+2)}\right]^4 \xrightarrow{n\to\infty} 1.$$

(c) $R = 1/4$. In fact, for $a_n = n^2(2n)!/(n!)^2$ we have

$$\left|\frac{a_n}{a_{n+1}}\right| = \frac{n^2(2n)!}{n!n!}\frac{(n+1)!(n+1)!}{(n+1)^2(2n+2)!}$$

$$= \left(\frac{n}{n+1}\right)^2\frac{(n+1)(n+1)}{(2n+2)(2n+10)} \xrightarrow{n\to\infty} \frac{1}{4}.$$

4.19   Observing that $e^{nz} = (e^z)^n$ we recognize that the series considered is a geometric series of ratio e^z. The series therefore converges when $|e^z| < 1$. Given $z = x + iy$, the previous inequality is equivalent to $e^x < 1$, that is, $x < 0$. In conclusion, the series considered converges $\forall z \in \mathbb{C}$ such that $\operatorname{Re} z < 0$.

4.20

(a) The n-th coefficient of the series is $a_n = c^{n^2}$ and we have

$$|a_n|^{1/n} = \left|c^{n^2}\right|^{1/n} = |c|^n \xrightarrow{n\to\infty} \begin{cases} 0 & |c| < 1 \\ 1 & |c| = 1 \\ \infty & |c| > 1 \end{cases}.$$

Therefore, the radius of convergence is

$$R = \begin{cases} \infty & |c| < 1 \\ 1 & |c| = 1 \\ 0 & |c| > 1 \end{cases}.$$

(b) The n-th coefficient of the series rewritten in the form $\sum_{n=1}^{\infty} a_n z^n$ is

$$a_n = \begin{cases} c^k & n = k^2, \quad k = 1,2,3,\ldots \\ 0 & \text{otherwise} \end{cases}$$

and the Hadamard formula provides

$$
\frac{1}{R} = \limsup_{m \to \infty} |a_m|^{1/m}
$$

$$
= \lim_{m \to \infty} \sup_{n \geq m} |a_n|^{1/n}
$$

$$
= \lim_{m \to \infty} \sup_{k:\, k^2 \geq m} \left| c^k \right|^{1/k^2}
$$

$$
= \lim_{m \to \infty} \sup_{k:\, k^2 \geq m} |c|^{1/k}
$$

$$
= |c|^0 .
$$

Excluding the trivial case $c = 0$, we always have $R = 1$.

4.21 The convergence of the proposed series is equivalent to the simultaneous convergence of the two series

$$
\sum_{n=0}^{\infty} \frac{z^n}{n!}, \qquad \sum_{n=0}^{\infty} \frac{n^2}{z^n}.
$$

The first has sum $\exp(z)$ and radius of convergence infinite. The second, with the position $w = 1/z$, is the power series

$$
\sum_{n=0}^{\infty} n^2 w^n
$$

which has radius of convergence

$$
R = \lim_{n \to \infty} \frac{n^2}{(n + 1)^2} = 1.
$$

This series therefore converges for $|w| < 1$, namely, $|z| > 1$. We conclude that the proposed series converges for $1 < |z| < \infty$.

4.22 Since $a_n \neq 0$ for any integer n, we can write

$$
\frac{a_n}{a_{n+1}} = \frac{a_n}{a_n + a_{n+1} - a_n} = \frac{1}{1 + b_n},
$$

where

$$
b_n = \frac{a_{n+1} - a_n}{a_n}
$$

Note that b_n is a fraction whose numerator coincides with the $(n+1)$-th term of the series $\sum_{k=1}^{\infty}(a_k - a_{k-1})$. This series is absolutely convergent and therefore convergent. It follows that its n-th term, namely, $a_{n+1} - a_n$, tends to 0 for $n \to \infty$. The denominator of b_n,

$$a_n = a_0 + \sum_{k=1}^{n} (a_k - a_{k-1}),$$

admits limit for $n \to \infty$ always by virtue of the convergence of the series $\sum_{k=1}^{\infty}(a_k - a_{k-1})$. We can also assume that $\lim_{n\to\infty} a_n = a_0 \neq 0$, because, if this were not the case, it would be enough to consider a new sequence with only the term a_0 modified. We conclude that $b_n \xrightarrow{n\to\infty} 0$ and therefore there exists and is equal to 1 the limit

$$R = \lim_{n\to\infty} \left| \frac{a_n}{a_{n+1}} \right| = 1$$

which represents the radius of convergence of the power series $\sum_{k=0}^{\infty} a_k z^k$.

In the procedure just explained we made use of the property $a_n \neq 0$ for any integer n. It is possible to obtain the same result even in the weaker hypothesis that $a_n \neq 0$ for an infinite number of values of n. If $a_n = 0$ for a finite number of integers n nothing changes. If instead for an infinity of integers n we have $a_n = 0$ and for another infinity we have $a_n \neq 0$, let $m(n)$ be those integers for which $a_{m(n)} \neq 0$. We still have

$$\frac{1}{R} = \limsup_{n\to\infty} |a_n|^{1/n} = \lim_{n\to\infty} \left| a_{m(n)} \right|^{1/n} = 1,$$

since, for the hypotheses made, there exists in $\mathbb{C}$ the limit

$$\lim_{n\to\infty} a_{m(n)} = a_0 + \sum_{k=1}^{\infty} (a_k - a_{k-1}).$$

12.5   Exercises of Chap. 5

5.1 For the existence of the limit in hypothesis, setting $\Delta z = (\Delta x, 0)$ with $\Delta x \in \mathbb{R}$, there must exist

$$\lim_{\Delta x \to 0} \mathrm{Re} \left(\frac{u(x + \Delta x, y) + iv(x + \Delta x, y) - u(x, y) - iv(x, y)}{\Delta x} \right)$$

$$= \lim_{\Delta x \to 0} \frac{u(x + \Delta x, y) - u(x, y)}{\Delta x}$$

$$= u_x(x, y).$$

Similarly, setting $\Delta z = (0, \Delta y)$ with $\Delta y \in \mathbb{R}$, there must exist

$$\lim_{\Delta y \to 0} \mathrm{Re}\left(\frac{u(x, y + \Delta y) + iv(x, y + \Delta y) - u(x, y) - iv(x, y)}{i\Delta y}\right)$$

$$= \lim_{\Delta y \to 0} \frac{v(x, y + \Delta y) - v(x, y)}{\Delta y}$$

$$= v_y(x, y).$$

For the uniqueness of the limit it follows $u_x(x, y) = v_y(x, y)$.

5.2 For the existence of the limit in hypothesis, setting $\Delta z = (\Delta x, 0)$ with $\Delta x \in \mathbb{R}$, there must exist

$$\lim_{\Delta x \to 0} \mathrm{Im}\left(\frac{u(x + \Delta x, y) + iv(x + \Delta x, y) - u(x, y) - iv(x, y)}{\Delta x}\right)$$

$$= \lim_{\Delta x \to 0} \frac{v(x + \Delta x, y) - v(x, y)}{\Delta x}$$

$$= v_x(x, y).$$

Similarly, setting $\Delta z = (0, \Delta y)$ with $\Delta y \in \mathbb{R}$, there must exist

$$\lim_{\Delta y \to 0} \mathrm{Im}\left(\frac{u(x, y + \Delta y) + iv(x, y + \Delta y) - u(x, y) - iv(x, y)}{i\Delta y}\right)$$

$$= \lim_{\Delta y \to 0} \frac{-u(x, y + \Delta y) + u(x, y)}{\Delta y}$$

$$= -u_y(x, y).$$

For the uniqueness of the limit it follows $u_y(x, y) = -v_x(x, y)$.

5.3 Since $f(z_0) = g(z_0) = 0$, we can write

$$\lim_{z \to z_0} \frac{f(z)}{g(z)} = \lim_{z \to z_0} \frac{f(z) - f(z_0)}{z - z_0} \frac{z - z_0}{g(z) - g(z_0)}$$

$$= \lim_{z \to z_0} \frac{f(z) - f(z_0)}{z - z_0} \lim_{z \to z_0} \frac{z - z_0}{g(z) - g(z_0)}$$

$$= \frac{f'(z_0)}{g'(z_0)},$$

having used the properties

$$\exists \lim_{z \to z_0} \frac{f(z) - f(z_0)}{z - z_0} = f'(z_0),$$

$$\exists \lim_{z \to z_0} \frac{g(z) - g(z_0)}{z - z_0} = g'(z_0) \neq 0.$$

5.4 Setting $z = re^{i\theta}$ we find

$$\lim_{r \to 0} \frac{f(re^{i\theta}) - f(0)}{re^{i\theta} - 0} = \lim_{r \to 0} \frac{re^{i5\theta}}{re^{i\theta}} = e^{i4\theta}.$$

Since this limit takes on different values for different values of θ we conclude that $f'(0)$ does not exist. On the other hand, put $f(x + iy) = u(x, y) + iv(x, y)$, with $x, y \in \mathbb{R}$, we have

$$u_x(0, 0) = \lim_{x \to 0} \frac{u(x, 0) - u(0, 0)}{x - 0} = \lim_{x \to 0} \frac{\operatorname{Re} f(x)}{x} = \lim_{x \to 0} \frac{x^5 / |x|^4}{x} = 1,$$

$$u_y(0, 0) = \lim_{y \to 0} \frac{u(0, y) - u(0, 0)}{y - 0} = \lim_{y \to 0} \frac{\operatorname{Re} f(iy)}{y} = \lim_{y \to 0} \frac{0}{y} = 0,$$

$$v_x(0, 0) = \lim_{x \to 0} \frac{v(x, 0) - v(0, 0)}{x - 0} = \lim_{x \to 0} \frac{\operatorname{Im} f(x)}{x} = \lim_{x \to 0} \frac{0}{x} = 0,$$

$$v_y(0, 0) = \lim_{y \to 0} \frac{v(0, y) - v(0, 0)}{y - 0} = \lim_{y \to 0} \frac{\operatorname{Im} f(iy)}{y} = \lim_{y \to 0} \frac{y^5 / |y|^4}{y} = 1.$$

Therefore, $u_x(0, 0) = v_y(0, 0)$ and $u_y(0, 0) = -v_x(0, 0)$.

5.5 Let us prove that f^* is differentiable at every point z of G^*. Since G^* is open, $\forall z \in G^* \ \exists \delta(z) > 0$ such that $B(z, \delta) \subset G^*$. Then $\forall w \in B(z, \delta)$ we can write the incremental ratio of f^* relative to the point z as

$$\frac{f^*(w) - f^*(z)}{w - z} = \frac{\overline{f(\overline{w})} - \overline{f(\overline{z})}}{w - z} = \overline{\left(\frac{f(\overline{w}) - f(\overline{z})}{\overline{w} - \overline{z}} \right)}.$$

On the other hand, since $z \in G^*$ and $B(z, \delta) \subset G^*$, we have $\overline{z} \in G$ and $B(\overline{z}, \delta) \subset G$. For the analyticity of f in G there must exist the limit

$$\lim_{\overline{w} \to \overline{z}} \frac{f(\overline{w}) - f(\overline{z})}{\overline{w} - \overline{z}} = f'(\overline{z}).$$

We can therefore conclude that there exists the limit

$$\lim_{w \to z} \frac{f^*(w) - f^*(z)}{w - z} = \overline{f'(\overline{z})}.$$

5.6

(a) As the real part of a continuous function in $\mathbb{C}$, f is continuous in $\mathbb{C}$. To study the differentiability, observe that $\forall z \in \mathbb{C}$

$$\frac{f(z + \Delta z) - f(z)}{\Delta z} = \frac{\mathrm{Re}(z + \Delta z) - \mathrm{Re}\, z}{\Delta z} = \frac{\mathrm{Re}\, \Delta z}{\Delta z},$$

whose limit for $\Delta z \to 0$ does not exist, in fact it assumes different values when, for example, $\mathrm{Re}\, \Delta z = 0$ or $\mathrm{Im}\, \Delta z = 0$. Therefore, f is never differentiable and its domain of analyticity is empty.

(b) The function is continuous throughout $\mathbb{C}$ since its real and imaginary components,

$$u(x, y) = \mathrm{Re}\, f(x + \mathrm{i}y) = e^x \cos y,$$

$$v(x, y) = \mathrm{Im}\, f(x + \mathrm{i}y) = e^x \sin y,$$

are continuous in $\mathbb{R}^2$. These components are also differentiable with continuous first derivatives

$$u_x(x, y) = e^x \cos y, \qquad u_y(x, y) = -e^x \sin y,$$

$$v_x(x, y) = e^x \sin y, \qquad v_y(x, y) = e^x \cos y,$$

which satisfy the Cauchy-Riemann conditions

$$u_x(x, y) = v_y(x, y), \qquad u_y(x, y) = -v_x(x, y),$$

at every point $(x, y) \in \mathbb{R}^2$. Therefore, f is differentiable in $\mathbb{C}$ and therefore entire.

(c) The function is continuous throughout $\mathbb{C}$ since its real and imaginary components,

$$u(x, y) = \mathrm{Re}\, f(x + \mathrm{i}y) = x^2 - y^2,$$

$$v(x, y) = \mathrm{Im}\, f(x + \mathrm{i}y) = 3xy,$$

are continuous in $\mathbb{R}^2$. These components are also differentiable with continuous first derivatives

$$u_x(x, y) = 2x, \qquad u_y(x, y) = -2y,$$
$$v_x(x, y) = 3y, \qquad v_y(x, y) = 3x.$$

The Cauchy-Riemann conditions are satisfied only for $x = y = 0$. Therefore, $f(z)$ is differentiable only at the point $z = 0$. It follows that the domain of analyticity of f is empty.

5.7 In both cases (a) and (b) we have rational functions, therefore analytic in $\mathbb{C}$ except for the singularities corresponding to the zeros of the denominator.

(a) The denominator vanishes at $z = 0$ and when $z^2 + 1 = 0$. This equation has solutions

$$z = \sqrt{-1} = \pm i.$$

(b) The denominator vanishes at $z = 2$ and when $z^2 + 2z + 2 = 0$. This quadratic equation has solutions

$$z = -1 + \sqrt{1-2} = -1 \pm i.$$

5.8

(a) Let $f(z) = u(r, \theta) + iv(r, \theta)$, we have

$$u(r, \theta) = r^{-4}\cos(4\theta), \qquad v(r, \theta) = -r^{-4}\sin(4\theta).$$

First of all, observe that for $r = 0$, that is, $z = 0$, f is not defined nor could it be defined in such a way as to be continuous. Therefore, at this point f is not differentiable. For $r > 0$ the functions u and v are differentiable with continuous first derivatives

$$u_r(r, \theta) = -4r^{-5}\cos(4\theta), \qquad u_\theta(r, \theta) = -4r^{-4}\sin(4\theta),$$
$$v_r(r, \theta) = 4r^{-5}\sin(4\theta), \qquad v_\theta(r, \theta) = -4r^{-4}\cos(4\theta),$$

and satisfy the Cauchy-Riemann conditions $ru_r = v_\theta$, $u_\theta = -rv_r$. Therefore, $f(z)$ is differentiable for $z \neq 0$ and its derivative is

$$f'(z) = (u_r(r, \theta) + iv_r(r, \theta))\, e^{-i\theta}$$
$$= -4r^{-5}(\cos(-4\theta) + i\sin(-4\theta))\, e^{-i\theta}$$
$$= -4z^{-5}.$$

The analyticity domain of f is $D = \mathbb{C} \setminus \{0\}$.

(b) Let $f(z) = u(r, \theta) + iv(r, \theta)$, we have

$$u(r, \theta) = \cos(\ln r)e^{-\theta}, \qquad v(r, \theta) = \sin(\ln r)e^{-\theta}.$$

First of all, observe that for $r = 0$, that is, $z = 0$, f is not defined nor could it be defined in such a way as to be continuous. Therefore, at this point f is not differentiable. Note that for every $r > 0$ we have $u(r, \pi) \neq \lim_{\theta \to -\pi} u(r, \theta)$ as well as $v(r, \pi) \neq \lim_{\theta \to -\pi} v(r, \theta)$, therefore along the negative real semi-axis, excluding the origin, f is not continuous and therefore not differentiable. For $r > 0$ and $-\pi < \theta < \pi$ the functions u and v are differentiable with continuous first derivatives

$$u_r(r, \theta) = -\frac{1}{r}\sin(\ln r)e^{-\theta}, \qquad u_\theta(r, \theta) = -\cos(\ln r)e^{-\theta},$$

$$v_r(r, \theta) = \frac{1}{r}\cos(\ln r)e^{-\theta}, \qquad v_\theta(r, \theta) = -\sin(\ln r)e^{-\theta},$$

and satisfy the Cauchy-Riemann conditions $ru_r = v_\theta$, $u_\theta = -rv_r$. In the same domain therefore $f(z)$ results differentiable and its derivative is

$$f'(z) = (u_r(r, \theta) + iv_r(r, \theta))\, e^{-i\theta}$$

$$= \frac{1}{r}\left(-\sin(\ln r)e^{-\theta} + i\cos(\ln r)e^{-\theta}\right)e^{-i\theta}$$

$$= \frac{i}{z}f(z),$$

in agreement with the fact that $f(z) = e^{i\log z} = z^i$. The analyticity domain of f is $D = \{z \in \mathbb{C} : z \neq -t,\ t \in [0, \infty)\}$.

5.9

(a) Let $f(z) = u(r, \theta) + iv(r, \theta)$, we have

$$u(r, \theta) = \sqrt{r}\cos(\theta/2), \qquad v(r, \theta) = \sqrt{r}\sin(\theta/2).$$

Note that for every $r > 0$ we have $v(r, \pi) \neq \lim_{\theta \to -\pi} v(r, \theta)$, therefore along the negative real semi-axis, excluding the origin, f is not continuous and therefore not differentiable. At $z = 0$ f is continuous but not differentiable, as is easily seen from the definitions. For $r > 0$ and $-\pi < \theta < \pi$ the functions u and v are differentiable with continuous first derivatives

$$u_r(r, \theta) = \frac{1}{2\sqrt{r}}\cos(\theta/2), \qquad u_\theta(r, \theta) = -\frac{\sqrt{r}}{2}\sin(\theta/2),$$

$$v_r(r, \theta) = \frac{1}{2\sqrt{r}}\sin(\theta/2), \qquad v_\theta(r, \theta) = \frac{\sqrt{r}}{2}\cos(\theta/2),$$

and satisfy the Cauchy-Riemann conditions $ru_r = v_\theta$, $u_\theta = -rv_r$. In the same domain $f(z)$ turns out to be differentiable and its derivative is

$$f'(z) = (u_r(r, \theta) + iv_r(r, \theta))\, e^{-i\theta}$$

$$= \frac{1}{2\sqrt{r}}\,(\cos(\theta/2) + i\sin(\theta/2))\, e^{-i\theta}$$

$$= \frac{1}{2f(z)}.$$

The analyticity domain of f is $D = \{z \in \mathbb{C} : z \neq -t,\ t \in [0, \infty)\}$.

(b) Let $f(z) = u(r, \theta) + iv(r, \theta)$, we have

$$u(r, \theta) = (\ln r)^2 - \theta^2, \qquad v(r, \theta) = 2\theta \ln r.$$

First of all, note that for $r = 0$, that is, $z = 0$, f is not defined nor could it be defined in such a way as to be continuous. At this point, therefore, f is not differentiable. For every $r > 0$ we have $v(r, \pi) \neq \lim_{\theta \to -\pi} v(r, \theta)$, therefore along the negative real semi-axis, excluding the origin, f is not continuous and therefore not differentiable. For $r > 0$ and $-\pi < \theta < \pi$ the functions u and v are differentiable with continuous first derivatives

$$u_r(r, \theta) = \frac{2}{r}\ln r, \qquad u_\theta(r, \theta) = -2\theta,$$

$$v_r(r, \theta) = \frac{2\theta}{r}, \qquad v_\theta(r, \theta) = 2\ln r,$$

and satisfy the Cauchy-Riemann conditions $ru_r = v_\theta$, $u_\theta = -rv_r$. In the same domain $f(z)$ turns out to be differentiable and its derivative is

$$f'(z) = (u_r(r, \theta) + iv_r(r, \theta))\, e^{-i\theta}$$

$$= \left(\frac{2}{r}\ln r + i\frac{2\theta}{r}\right) e^{-i\theta}$$

$$= \frac{2}{z}g(z), \qquad g(z) = \ln r + i\theta.$$

in agreement with the fact that $f(z) = g(z)^2 = (\log z)^2$. The domain of analyticity of f is $D = \{z \in \mathbb{C} : z \neq -t,\ t \in [0, \infty)\}$.

5.10 As a composition of continuous functions in $\mathbb{C}$, f is continuous in $\mathbb{C}$. To study its differentiability, observe that, setting $z = x + iy$ with $x, y \in \mathbb{R}$, we have

$$f(x + iy) = x^3 + 3xy^2 + i(y^3 + 3x^2y),$$

therefore

$$u(x, y) = \operatorname{Re} f(x + iy) = x^3 + 3xy^2,$$

$$v(x, y) = \operatorname{Im} f(x + iy) = y^3 + 3x^2y.$$

The functions u and v are differentiable throughout $\mathbb{R}^2$ with derivatives

$$u_x(x, y) = 3x^2 + 3y^2, \qquad u_y(x, y) = 6xy,$$

$$v_x(x, y) = 6xy, \qquad v_y(x, y) = 3y^2 + 3x^2,$$

continuous throughout $\mathbb{R}^2$. The Cauchy-Riemann equations, $u_x = v_y$ and $u_y = -v_x$, therefore give

$$3x^2 + 3y^2 = 3x^2 + 3y^2,$$

$$6xy = -6xy.$$

The first equation is always satisfied, the second only if $x = 0$ or $y = 0$. It follows that f is differentiable only at the points of the coordinate axes. At none of these points, however, f is analytic. In fact $\forall \varepsilon > 0$ the ball $B(z, \varepsilon)$, with z real or pure imaginary, contains points w such that $\operatorname{Re} w \neq 0$ and $\operatorname{Im} w \neq 0$ where f is not differentiable.

5.11 Let $z = x + iy$ and $f(z) = u(x, y) + iv(x, y)$, since f is analytic in D, in D the Cauchy-Riemann conditions $u_x = v_y$ and $u_y = -v_x$ are satisfied. On the other hand place $\overline{f}(z) = U(x, y) + iV(x, y) = \overline{u(x, y) + iv(x, y)}$, we must have $U = u$ and $V = -v$. Therefore, $U_x = -V_y$ and $U_y = V_x \ \forall z \in D$, that is, $\overline{f}$ is irrotational and solenoidal in D.

5.12

(a) The function is continuous throughout $\mathbb{C}$ since its real and imaginary components,

$$u(x, y) = \operatorname{Re} f(x + iy) = y,$$

$$v(x, y) = \operatorname{Im} f(x + iy) = 0,$$

are continuous in $\mathbb{R}^2$. These components are also differentiable with continuous first derivatives

$$u_x(x, y) = 0, \qquad u_y(x, y) = 1,$$

$$v_x(x, y) = 0, \qquad v_y(x, y) = 0,$$

but they satisfy the Cauchy-Riemann conditions at no point $(x, y) \in \mathbb{R}^2$. Therefore, f is never differentiable and its domain of analyticity is empty.

(b) The function is continuous throughout $\mathbb{C}$ since its real and imaginary components,

$$u(x, y) = \operatorname{Re} f(x + iy) = e^x \cos y,$$

$$v(x, y) = \operatorname{Im} f(x + iy) = -e^x \sin y,$$

are continuous in $\mathbb{R}^2$. These components are also differentiable with continuous first derivatives

$$u_x(x, y) = e^x \cos y, \qquad u_y(x, y) = -e^x \sin y,$$

$$v_x(x, y) = -e^x \sin y, \qquad v_y(x, y) = -e^x \cos y,$$

but they satisfy the Cauchy-Riemann conditions at no point $(x, y) \in \mathbb{R}^2$. In fact, since $e^x \neq 0 \ \forall x \in \mathbb{R}$, it follows $u_x(x, y) = v_y(x, y)$ and $u_y(x, y) = -v_x(x, y)$ only if $\sin y = \cos y = 0$, which can never happen. Therefore, f is never differentiable and its domain of analyticity is empty.

(c) The function is continuous throughout $\mathbb{C}$ since its real and imaginary components,

$$u(x, y) = \operatorname{Re} f(x + iy) = x^2 - y^2,$$

$$v(x, y) = \operatorname{Im} f(x + iy) = 2xy,$$

are continuous in $\mathbb{R}^2$. These components are also differentiable with continuous first derivatives

$$u_x(x, y) = 2x, \qquad u_y(x, y) = -2y,$$

$$v_x(x, y) = 2y, \qquad v_y(x, y) = 2x.$$

The Cauchy-Riemann conditions are satisfied at every point $(x, y) \in \mathbb{R}^2$. Therefore, $f(z)$ is differentiable throughout $\mathbb{C}$ and $\mathbb{C}$ is its domain of analyticity (f is entire).

5.13 In polar representation $z = re^{i\theta}$, we have $D = \{(r, \theta) : r > 0, \ -\pi < \theta \leq \pi\}$. Since $a \in \mathbb{R}$, the real and imaginary components of $f(z)$ are

$$u(r, \theta) = \operatorname{Re} f(z) = \operatorname{Arg}(r^2 e^{i2\theta}),$$

$$v(r, \theta) = \operatorname{Im} f(z) = -2a \ln r.$$

Regardless of the value of a, the function $v(r, \theta)$ is continues in D while

$$u(r, \theta) = \begin{cases} 2\theta + 2\pi & -\pi < \theta \le -\pi/2 \\ 2\theta & -\pi/2 < \theta \le \pi/2 \\ 2\theta - 2\pi & \pi/2 < \theta \le \pi \end{cases}$$

is continuous in the subset $\tilde{D} = \{(r, \theta) : r > 0, \; -\pi < \theta \le \pi, \; \theta \ne \pm\pi/2\}$ which represents the complex plane with the exception of the imaginary axis. The same result can be reached even without explicitly calculating $u(r, \theta)$, noting that $\text{Arg}(z^2)$ is a continuous function in $\mathbb{C}$ with the exception of the points for which its argument is real negative, $z(t)^2 = -t, t \ge 0$, which yields $z(t) = \pm it, t \ge 0$. In conclusion, f is continuous in $\tilde{D}$ for every value of a.

Within the continuity domain $\tilde{D}$, the functions u and v are differentiable with continuous first derivatives

$$u_r(r, \theta) = 0, \qquad u_\theta(r, \theta) = 2,$$

$$v_r(r, \theta) = -2a/r, \qquad v_\theta(r, \theta) = 0.$$

The Cauchy-Riemann equations, $r u_r = v_\theta$ and $u_\theta = -r v_r$, are therefore

$$0 = 0,$$

$$2 = 2a.$$

For $a = 1$ both equations are satisfied at every point of $\tilde{D}$ and therefore f is differentiable in $\tilde{D}$. Since $\tilde{D}$ is an open set, it is also the domain of analyticity of f. This result could also be reached by observing that for $a = 1$ we have $f(z) = -i \log(z^2)$ with log principal branch of the logarithm. For $a \ne 1$ the Cauchy-Riemann equations are never satisfied, therefore f is never differentiable and its domain of analyticity is empty.

12.6 Exercises of Chap. 6

6.1

(a) Using the definition of $\cosh z$ we have

$$\cosh(2 + i) = \frac{e^{2+i} + e^{-2-i}}{2}$$

$$= \frac{e^2(\cos 1 + i \sin 1) + e^{-2}(\cos 1 - i \sin 1)}{2}$$

$$= \cosh 2 \cos 1 + i \sinh 2 \sin 1.$$

(b) Using $\cot z = \cos z / \sin z$ and the definition of $\cos z$ and $\sin z$, we have

$$\cot\left(\frac{\pi}{4} - \mathrm{i}\ln 2\right) = \frac{e^{\mathrm{i}\frac{\pi}{4}+\ln 2} + e^{-\mathrm{i}\frac{\pi}{4}-\ln 2}}{2} \cdot \frac{2\mathrm{i}}{e^{\mathrm{i}\frac{\pi}{4}+\ln 2} - e^{-\mathrm{i}\frac{\pi}{4}-\ln 2}}$$

$$= \mathrm{i}\,\frac{2\left(\frac{1}{\sqrt{2}} + \frac{\mathrm{i}}{\sqrt{2}}\right) + \frac{1}{2}\left(\frac{1}{\sqrt{2}} - \frac{\mathrm{i}}{\sqrt{2}}\right)}{2\left(\frac{1}{\sqrt{2}} + \frac{\mathrm{i}}{\sqrt{2}}\right) - \frac{1}{2}\left(\frac{1}{\sqrt{2}} - \frac{\mathrm{i}}{\sqrt{2}}\right)}$$

$$= \mathrm{i}\,\frac{\frac{5}{2\sqrt{2}} + \mathrm{i}\frac{3}{2\sqrt{2}}}{\frac{3}{2\sqrt{2}} + \mathrm{i}\frac{5}{2\sqrt{2}}}$$

$$= \frac{-3 + 5\mathrm{i}}{3 + 5\mathrm{i}}$$

$$= \frac{(-3 + 5\mathrm{i})(3 - 5\mathrm{i})}{9 + 25}$$

$$= \frac{8}{17} + \mathrm{i}\frac{15}{17}.$$

6.2 Setting $w = \arctan z$, we have $\tan w = z$ or

$$\mathrm{i}z = \frac{e^{\mathrm{i}w} - e^{-\mathrm{i}w}}{e^{\mathrm{i}w} + e^{-\mathrm{i}w}},$$

which when resolved with respect to $e^{2\mathrm{i}w}$ gives

$$e^{2\mathrm{i}w} = \frac{1 + \mathrm{i}z}{1 - \mathrm{i}z}$$

and so

$$w = \frac{1}{2\mathrm{i}} \log \frac{1 + \mathrm{i}z}{1 - \mathrm{i}z} = \frac{\mathrm{i}}{2} \log \frac{\mathrm{i} + z}{\mathrm{i} - z}.$$

Therefore

$$\arctan(2 + \mathrm{i}) = \frac{\mathrm{i}}{2} \log \frac{\mathrm{i} + 2 + \mathrm{i}}{\mathrm{i} - 2 - \mathrm{i}}$$

$$= \frac{\mathrm{i}}{2} \log(-1 - \mathrm{i})$$

$$= \frac{\mathrm{i}}{2} \log\left(\sqrt{2}\,e^{-\mathrm{i}3\pi/4}\right)$$

$$= \frac{i}{2} \left(\ln \sqrt{2} - i\frac{3}{4}\pi \right)$$

$$= \frac{3}{8}\pi + \frac{i}{4} \ln 2.$$

6.3 Place $w = \arctan z$, we have $\tan w = z$ i.e.

$$iz = \frac{e^{iw} - e^{-iw}}{e^{iw} + e^{-iw}},$$

which when resolved with respect to e^{2iw} gives

$$e^{2iw} = \frac{1 + iz}{1 - iz}$$

and so

$$w = \frac{1}{2i} \log \frac{1 + iz}{1 - iz}.$$

Therefore

$$\arctan(1 + 2i) = -\frac{i}{2} \log \frac{1 + i - 2}{1 - i + 2}$$

$$= -\frac{i}{2} \log \frac{-2 + i}{5}$$

$$= -\frac{i}{2} \log \left(\frac{1}{\sqrt{5}} e^{i\varphi} \right)$$

$$= -\frac{i}{2} \left(\ln \frac{1}{\sqrt{5}} + i\varphi \right)$$

$$= \frac{\varphi}{2} + \frac{i}{4} \ln 5,$$

where φ is defined by $\cos \varphi = -2/\sqrt{5}$ and $\sin \varphi = 1/\sqrt{5}$. In conclusion

$$\text{Im}\,(\arctan(1 + 2i)) = \frac{1}{4} \ln 5.$$

6.4 Using $z^w = \exp(w \, \text{Log}\, z)$ and $\text{Log}\, z = \ln |z| + i \, \text{Arg}\, z$ with $-\pi < \text{Arg}\, z \le \pi$, we have

$$(4 - 4i)^{1+i} = \exp \left[(1 + i) \, \text{Log}(4 - 4i) \right]$$

$$= \exp \left[(1 + i) \left(\ln |4 - 4i| + i \, \text{Arg}(4 - 4i) \right) \right]$$

$$= \exp\left[(1+i)\left(\ln\sqrt{32} - i\frac{\pi}{4}\right)\right]$$

$$= \exp\left[\left(\ln\sqrt{32} + \frac{\pi}{4}\right) + i\left(\ln\sqrt{32} - \frac{\pi}{4}\right)\right]$$

$$= \sqrt{32}\, e^{\frac{\pi}{4}}\left[\cos\left(\ln\sqrt{32} - \frac{\pi}{4}\right) + i\sin\left(\ln\sqrt{32} - \frac{\pi}{4}\right)\right].$$

6.5 Using $z^w = \exp(w\,\mathrm{Log}\,z)$ and $\mathrm{Log}\,z = \ln|z| + i\,\mathrm{Arg}\,z$ with $-\pi < \mathrm{Arg}\,z \le \pi$, we have

$$(1+i)^{2-i} = \exp\left[(2-i)\,\mathrm{Log}(1+i)\right]$$

$$= \exp\left[(2-i)\,\mathrm{Log}\left(\sqrt{2}e^{i\frac{\pi}{4}}\right)\right]$$

$$= \exp\left[(2-i)\left(\frac{1}{2}\log 2 + i\frac{\pi}{4}\right)\right]$$

$$= \exp\left[\ln 2 + \frac{\pi}{4} + i\left(\frac{\pi}{2} - \frac{1}{2}\ln 2\right)\right]$$

and then

$$\left|(1+i)^{2-i}\right| = \exp\left(\ln 2 + \frac{\pi}{4}\right) = 2e^{\frac{\pi}{4}}.$$

6.6 Using $z^w = \exp(w\,\mathrm{Log}\,z)$ with the principal value of the logarithm defined by $\mathrm{Log}\,z = \ln|z| + i\,\mathrm{Arg}\,z$, where $|z| > 0$ and $-\pi < \mathrm{Arg}\,z \le \pi$, we have

$$\left(\frac{e^\pi}{\sqrt{2}} + i\frac{e^\pi}{\sqrt{2}}\right)^{i+1} = \exp\left[(i+1)\,\mathrm{Log}\left(\frac{e^\pi}{\sqrt{2}} + i\frac{e^\pi}{\sqrt{2}}\right)\right]$$

$$= \exp\left[(i+1)\,\mathrm{Log}\left(e^\pi\, e^{i\frac{\pi}{4}}\right)\right]$$

$$= \exp\left[(i+1)\left(\ln e^\pi + i\frac{\pi}{4}\right)\right]$$

$$= \exp\left[(i+1)\left(\pi + i\frac{\pi}{4}\right)\right]$$

$$= \exp\left[\pi - \frac{\pi}{4} + i\left(\pi + \frac{\pi}{4}\right)\right]$$

$$= e^{\frac{3\pi}{4} + i\frac{5\pi}{4}}$$

and so

$$\mathrm{Arg}\left(\frac{e^\pi}{\sqrt{2}} + i\frac{e^\pi}{\sqrt{2}}\right)^{i+1} = -\frac{3\pi}{4}.$$

6.7

(a) Using $z^w = \exp(w \operatorname{Log} z)$ and $\operatorname{Log} z = \ln |z| + i \operatorname{Arg} z$ with $-\pi < \operatorname{Arg} z \le \pi$, we have

$$(1 - i)^{\sqrt{2}+i\pi} = \exp\left[\left(\sqrt{2} + i\pi\right)\left(\ln \sqrt{2} - i\frac{\pi}{4}\right)\right]$$

$$= \exp\left[\left(\frac{1}{\sqrt{2}} \ln 2 + \frac{\pi^2}{4}\right) + i\left(\frac{\pi}{2} \ln 2 - \frac{\pi}{4\sqrt{2}}\right)\right],$$

therefore

$$\ln \left|(1 - i)^{\sqrt{2}+i\pi}\right| = \frac{1}{\sqrt{2}} \ln 2 + \frac{\pi^2}{4}.$$

(b) Using $\arctan z = -(i/2)\log((i - z)/(i + z))$, we have

$$\operatorname{Im}\left(\arctan\left(\frac{i}{2}\right)\right) = \operatorname{Im}\left(-\frac{i}{2} \log\left(\frac{i/2}{3i/2}\right)\right)$$

$$= \operatorname{Im}\left(\frac{i}{2} \log(3)\right)$$

$$= \frac{1}{2} \ln 3.$$

6.8 Setting $z = x + iy$ with $x, y \in \mathbb{R}$, we have

$$\cos z = \frac{1}{2}\left(e^{ix-y} + e^{-ix+y}\right)$$

$$= \frac{1}{2}\left((\cos x + i \sin x)e^{-y} + (\cos x - i \sin x)e^{y}\right)$$

$$= \cos x \cosh y - i \sin x \sinh y.$$

We must therefore solve the system of equations

$$\begin{cases} \cos x \cosh y = a \\ \sin x \sinh y = 0 \end{cases}.$$

The solution $y = 0$ of the second equation is incompatible with the first equation. The solution $x = \pi k$, $k \in \mathbb{Z}$, of the second equation substituted in the first gives

$$(-1)^k \cosh y = a.$$

This equation has a solution only for k even and with y that satisfies $\cosh y = a$, that is

$$e^{2y} - 2ae^{y} + 1 = 0,$$

which gives

$$e^{y} = a \pm \sqrt{a^2 - 1}.$$

In conclusion, the solutions are

$$z_k = 2\pi k + i \ln\left(a \pm \sqrt{a^2 - 1}\right)$$
$$= 2\pi k \pm i \ln\left(a + \sqrt{a^2 - 1}\right), \qquad k \in \mathbb{Z}.$$

6.9 The equation to be solved is

$$\frac{e^z + e^{-z}}{2} = \frac{1}{2},$$

that is

$$\left(e^z\right)^2 - e^z + 1 = 0,$$

which gives

$$e^z = \frac{1 + \sqrt{1 - 4}}{2} = \frac{1}{2} \pm i\frac{\sqrt{3}}{2} = e^{\pm i\frac{\pi}{3}} = e^{\pm i\frac{\pi}{3} + i2\pi k}, \qquad k \in \mathbb{Z}.$$

In conclusion, the solutions are

$$z_k = i\pi\left(2k \pm \frac{1}{3}\right), \qquad k \in \mathbb{Z}.$$

6.10 Using $\sin z = (e^{iz} - e^{-iz})/2i$ and $\cos z = (e^{iz} + e^{-iz})/2$ and setting $z = x + iy$ with $x, y \in \mathbb{R}$, the equation to solve is

$$e^{i(2x+i2y)} - e^{-i(2x+i2y)} = 8i.$$

From this, taking the real and imaginary parts, we obtain the system

$$\begin{cases} \cos(2x)\sinh(2y) = 0 \\ \sin(2x)\cosh(2y) = 4 \end{cases}.$$

The solution $y = 0$ of the first equation is not compatible with the second equation. It must therefore be $\cos(2x) = 0$ or $2x = \pi/2 + \pi k$, $k \in \mathbb{Z}$. This solution is compatible with the second equation only for k even, when $\sin(2x) = 1$. In this case the second equation gives $\cosh(2y) = 4$ or

$$\left(e^{2y}\right)^2 - 8e^{2y} + 1 = 0,$$

which solved gives

$$e^{2y} = 4 \pm \sqrt{15}.$$

In conclusion

$$\begin{cases} x = \frac{\pi}{4} + \pi k, & k \in \mathbb{Z} \\ y = \frac{1}{2} \ln\left(4 \pm \sqrt{15}\right) = \pm\frac{1}{2} \ln\left(4 + \sqrt{15}\right) \end{cases}.$$

Alternatively, to solve $\sin(2z) = 4$ we set $w = e^{i2z}$ and obtain the equation

$$w^2 - 8iw - 1 = 0,$$

that provides

$$e^{i2z} = i\left(4 \pm \sqrt{15}\right).$$

Taking the logarithm, we obtain again

$$z = \frac{1}{2i}\left[\ln\left(4 \pm \sqrt{15}\right) + i\left(\frac{\pi}{2} + 2\pi k\right)\right]$$
$$= \left(\frac{\pi}{4} + \pi k\right) \pm \frac{i}{2} \ln\left(4 + \sqrt{15}\right), \qquad k \in \mathbb{Z}.$$

6.11 We have

$$\frac{e^z + e^{-z}}{2} + e^z = 2,$$

that is to say

$$3\left(e^z\right)^2 - 4e^z + 1 = 0,$$

which has solutions

$$e^z = \frac{2 + 1}{3}.$$

From the solution $e^z = 1$ we have

$$z = \log 1 = i2\pi k, \qquad k \in \mathbb{Z}.$$

From the solution $e^z = 1/3$ we obtain

$$z = \log \frac{1}{3} = -\ln 3 + i2\pi k, \qquad k \in \mathbb{Z}.$$

6.12 Let $w = x + iy$, with $x, y \in \mathbb{R}$, the solutions of the equation

$$\cos w = \frac{e^{iw} + e^{-iw}}{2} = 0$$

are the solutions of the system

$$\begin{cases} \cos x \cosh y = 0 \\ \sin x \sinh y = 0 \end{cases},$$

namely, $y = 0$ and $x = \pm(2k + 1)\pi/2$, with $k \in \mathbb{N}$. Therefore $\cos(3z^2) = 0$ has solutions

$$3z^2 = \pm\frac{\pi}{2}(2k + 1), \qquad k \in \mathbb{N},$$

that is

$$z = \sqrt{\pm\frac{\pi}{6}(2k + 1)} = \begin{cases} \pm\sqrt{(2k + 1)\pi/6} \\ \pm i\sqrt{(2k + 1)\pi/6} \end{cases}, \qquad k \in \mathbb{N}.$$

6.13 We have

$$\sinh z = \frac{e^z - e^{-z}}{2} = -i\pi,$$

that is to say

$$e^{2z} + 2i\pi e^z - 1 = 0,$$

whose solutions are

$$e^z = -i\left(\pi \pm \sqrt{\pi^2 - 1}\right).$$

Therefore, the solutions of the equation $\sinh z = -i\pi$ are

$$z = \log\left(-i\left(\pi \pm \sqrt{\pi^2 - 1}\right)\right)$$

$$= \ln\left(\pi \pm \sqrt{\pi^2 - 1}\right) + i\left(-\frac{\pi}{2} + 2\pi k\right), \qquad k \in \mathbb{Z}.$$

6.14 We have

$$\sin z = \frac{e^{iz} - e^{-iz}}{2i} = a,$$

or

$$e^{2iz} - 2iae^{iz} - 1 = 0,$$

whose solutions are

$$e^{iz} = ia \pm \sqrt{1 - a^2}.$$

Note that for $-1 \le a \le 1$ the radicand is non-negative. Therefore, the solutions are

$$iz = \log\left(ia \pm \sqrt{1 - a^2}\right)$$

$$= \ln\sqrt{a^2 + (1 - a^2)} + i\left(\arcsin\left(\frac{a}{\sqrt{a^2 + (1 - a^2)}}\right) + 2\pi k\right),$$

$$= \ln(1) + i\left(\arcsin(a) + 2\pi k\right), \qquad k \in \mathbb{Z},$$

which are the same solutions obtained in the real field

$$z = \arcsin a + 2\pi k, \qquad k \in \mathbb{Z}.$$

6.15 We have

$$\cosh z = \frac{e^z + e^{-z}}{2} = a,$$

that is to say

$$e^{2z} - 2ae^z + 1 = 0,$$

which provides

$$e^z = a \pm i\sqrt{1 - a^2}.$$

The solutions are therefore

$$z = \log\left(a + i\sqrt{1 - a^2}\right)$$

$$= \ln \sqrt{a^2 + (1 - a^2)} + i \left(\arccos a + 2\pi k\right), \qquad k \in \mathbb{Z},$$

where $\arccos a$ indicates the principal value between $-\pi/2$ and $\pi/2$. In conclusion we have

$$z = i \left(\arccos a + 2\pi k\right), \qquad k \in \mathbb{Z},$$

which for $a \in (0, 1)$ corresponds to $z = iy$ with $-\pi/2 + 2\pi k < y < \pi/2 + 2\pi k$, $k \in \mathbb{Z}$.

Alternatively, we set $z = x + iy$ with $x, y \in \mathbb{R}$. The equation to be solved is equivalent to the system

$$\begin{cases} \cosh x \cos y = a \\ \sinh x \sin y = 0 \end{cases}.$$

The solution $y = k\pi$, $k \in \mathbb{Z}$, of the second equation is incompatible with the first equation. The solution $x = 0$ of the second equation substituted in the first equation gives

$$\cos y = a,$$

which for $a \in (0, 1)$ has solution $y = \arccos a \in (-\pi/2 + 2\pi k, \pi/2 + 2\pi k), k \in \mathbb{Z}$.

6.16 The equation to solve is

$$\frac{e^z - e^{-z}}{2} = i,$$

that is to say

$$\left(e^z\right)^2 - 2ie^z - 1 = 0,$$

that provides

$$e^z = i + \sqrt{i^2 + 1} = i = e^{i\left(\frac{\pi}{2} + 2\pi k\right)}, \qquad k \in \mathbb{Z}.$$

In conclusion, the solutions are

$$z = i\pi \frac{4k + 1}{2}, \qquad k \in \mathbb{Z}.$$

6.17 The assigned equation

$$\left(\frac{e^{iz} - e^{-iz}}{2i}\right)^2 = -4$$

is equivalent to

$$e^{iz} - e^{-iz} = \sqrt{-4(2i)^2} = \pm 4,$$

that is to say

$$e^{2iz} \pm 4e^{iz} - 1 = 0,$$

whose solutions are $e^{iz} = \pm 2 \pm \sqrt{5}$. Of these four expressions, all real, two are positive

$$e^{iz} = \sqrt{5} \pm 2 = (\sqrt{5} \pm 2)e^{i0},$$

while the other two are negative

$$e^{iz} = -\sqrt{5} \mp 2 = (\sqrt{5} \pm 2)e^{i\pi}.$$

Taking the logarithm, in the two cases we have respectively

$$iz = \log(\sqrt{5} \pm 2) = \ln(\sqrt{5} \pm 2) + i2\pi k, \qquad k \in \mathbb{Z},$$

$$iz = \log(-\sqrt{5} \mp 2) = \ln(\sqrt{5} \pm 2) + i(\pi + 2\pi k), \qquad k \in \mathbb{Z}.$$

We conclude that the solutions of the given equation are

$$z = \pi k - i\ln(\sqrt{5} \pm 2), \qquad k \in \mathbb{Z}.$$

These are isolated points that lie on the lines $y = -\ln(\sqrt{5} + 2) < 0$ and $y = -\ln(\sqrt{5} - 2) > 0$ parallel to the real axis. Their only limit point is the point at infinity.

6.18 Taking the logarithm of both sides of the equation gives

$$z^2 = \log(-1) = \log(1\, e^{i\pi}) = \ln 1 + i(\pi + 2\pi k) = (2k + 1)i\pi, \qquad k \in \mathbb{Z}.$$

Taking the square root of this expression, for $k \geq 0$ we have

$$z = \sqrt{(2k + 1)\pi\, e^{i\frac{\pi}{2}}} = \sqrt{(2k + 1)\pi}\; e^{i\left(\frac{\pi}{4} + \frac{2\pi n}{2}\right)}, \qquad n = 0, 1,$$

while for $k < 0$

$$z = \sqrt{|2k + 1|\,\pi\, e^{i\frac{3\pi}{2}}} = \sqrt{|2k + 1|\,\pi}\; e^{i\left(\frac{3\pi}{4} + \frac{2\pi n}{2}\right)}, \qquad n = 0, 1.$$

Setting $k = -m - 1$, with $m = 0, 1, 2, \ldots$, it is possible to rewrite the second system of solutions as

$$z = \sqrt{(2m+1)\pi}\; e^{i\left(\frac{3\pi}{4} + \frac{2\pi n}{2}\right)}, \qquad n = 0, 1.$$

In conclusion, we have the following four sets of solutions

$$z = \pm(1+i)\sqrt{\frac{2k+1}{2}\,\pi}, \qquad z = \pm(1-i)\sqrt{\frac{2k+1}{2}\,\pi}, \qquad k \in \mathbb{N}.$$

6.19

(a) Using formula (1.20) we have

$$z = \sqrt[4]{i} = \left(e^{i\pi/2}\right)^{1/4} = e^{i(\pi/2 + 2\pi k)/4}, \qquad k = 0, 1, 2, 3,$$

which corresponds to the 4 solutions

$$z_0 = e^{i\pi/8},$$
$$z_1 = e^{i5\pi/8},$$
$$z_2 = e^{i9\pi/8} = e^{-i7\pi/8},$$
$$z_3 = e^{i13\pi/8} = e^{-i3\pi/8}.$$

(b) Using $z^w = \exp(w \log z)$, we have

$$z = 4^{1/i}$$
$$= 4^{-i}$$
$$= \exp\left[-i \log\left(4e^{i0}\right)\right]$$
$$= \exp\left[-i\left(\ln 4 + i2\pi k\right)\right]$$
$$= e^{2\pi k} e^{-i\ln 4}, \qquad k \in \mathbb{Z}.$$

6.20 Since $\cotan z = \cos z / \sin z$, the given equation is equivalent to

$$\cos(\sqrt{z}) = 0.$$

Setting $w = \sqrt{z}$, we have

$$\frac{e^{iw} + e^{-iw}}{2} = 0,$$

that is to say

$$\left(e^{iw}\right)^2 = -1,$$

that provides

$$iw = \log(\pm i) = \log\left(1e^{\pm i\pi/2}\right) = \ln 1 + i(\pm\pi/2 + 2\pi k), \qquad k \in \mathbb{Z}.$$

In conclusion, we have

$$\sqrt{z} = (2n+1)\pi/2, \qquad n \in \mathbb{Z},$$

therefore the solutions are

$$z = ((2n+1)\pi/2)^2, \qquad n \in \mathbb{N}.$$

6.21

(a) We have

$$z = \sqrt[3]{2i} = \left(2e^{i\pi/2}\right)^{1/3} = 2^{1/3}e^{i(\pi/2 + 2\pi k)/3}, \qquad k = 0, 1, 2,$$

which corresponds to the 3 solutions

$$z_0 = 2^{1/3}e^{i\pi/6},$$
$$z_1 = 2^{1/3}e^{i5\pi/6},$$
$$z_2 = 2^{1/3}e^{i9\pi/6} = -2^{1/3}i.$$

(b) Using $z^w = \exp(w \log z)$, we have

$$z = 3^{1/(2i)}$$
$$= 3^{-i/2}$$
$$= \exp\left[-i/2 \log\left(3e^{i0}\right)\right]$$
$$= \exp\left[-i/2 \left(\ln 3 + i2\pi k\right)\right]$$
$$= e^{\pi k}e^{-i(\ln 3)/2}, \qquad k \in \mathbb{Z}.$$

6.22

(a) The principal branch of $\log z$ is an analytic function everywhere except on
   the negative real semi-axis including the origin. Therefore, the function under
   consideration is analytic everywhere except at the points z that satisfy

$$(2 - z)^2 = -t, \qquad t \geq 0,$$

   which are the points $z(t) = 2 \pm i\sqrt{t}$, $t \geq 0$, namely, the axis $\operatorname{Re} z = 2$.
(b) The function z^z is defined by

$$z^z = \exp(z \log z).$$

   Since the exponential function is analytic everywhere, the domain of analyticity
   of the principal branch of z^z coincides with that of the principal branch of $\log z$,
   that is, the entire complex plane except the negative real semi-axis including the
   origin.

6.23

(a) The function $z^{\sinh z}$ is defined by

$$z^{\sinh z} = \exp(\sinh z \log z).$$

   Since the composition of analytic functions is analytic and the functions $\exp$
   and $\sinh$ are entire, the principal branch of $z^{\sinh z}$ is an analytic function in the
   domain in which the principal branch of $\log z$ is analytic, namely, in the entire
   complex plane except the negative real semi-axis including the origin.
(b) Reasoning as in point (a), the domain of analyticity of $\log(\cos z)$ coincides
   with the entire complex plane with the exception of the points z that satisfy

$$\cos z = -t, \qquad t \geq 0.$$

Let $z = x + iy$, with $x, y \in \mathbb{R}$, this condition is equivalent to

$$\begin{cases} \cos x \cosh y = -t, \qquad t \geq 0 \\ \sin x \sinh y = 0 \end{cases}.$$

The second equation of the system is satisfied when $\sin x = 0$ or $\sinh y = 0$.
In the first case $x = \pi k$, $k \in \mathbb{Z}$. The solutions with k even are incompatible
with the first equation while for k odd this equation reduces to $\cosh y = t$, with
$t \geq 0$. We therefore have

$$z = (2k + 1)\pi + iy, \qquad -\infty < y < \infty, \quad k \in \mathbb{Z},$$

which are the imaginary axes $\operatorname{Re} z = (2k+1)\pi$, $k \in \mathbb{Z}$. The other solution $\sinh y = 0$, that is, $y = 0$, replaced in the first equation gives $x = \arccos(-t)$, with $t \geq 0$. We therefore have

$$z = x, \qquad \frac{\pi}{2} + 2\pi k \leq x \leq \frac{3\pi}{2} + 2\pi k, \quad k \in \mathbb{Z}.$$

6.24

(a) The functions $\cosh z$, e^z and $\sinh z$ are entire and the composition of entire functions is entire. Therefore

$$\cosh\left(e^{\sinh z}\right)$$

is analytic throughout the complex plane.

(b) The principal branch of $\log z$ is an analytic function everywhere except on the negative real semi-axis including the origin. Therefore, the function under consideration is analytic everywhere except at the points z that satisfy

$$3 + z^2 = -t, \qquad t \geq 0,$$

namely, the points

$$z(t) = \sqrt{-3-t} = \pm i\sqrt{3+t}, \qquad t \geq 0.$$

6.25

(a) Observing that the function $\exp z$ is entire and the rational function $1/(1+z^2)$ is analytic everywhere except at the singular points $z = \pm i$, by the differentiability of composite functions, the function $\exp\left(1/(1+z^2)\right)$ is analytic everywhere except at the points $z = \pm i$.

(b) The principal branch of the function $\log z$ is analytic everywhere except on the negative real semi-axis, including the origin. Therefore, the principal branch of the function $\log(z^3)$ is analytic everywhere except at the points z such that

$$z^3 = -t, \qquad t \geq 0,$$

that is

$$z(t) = (-t)^{1/3} = \sqrt[3]{t}\, e^{i(\pi + 2\pi k)/3}, \qquad k = 0, 1, 2, \quad t \geq 0,$$

which represent the origin and the three semi-axes coming out from the origin at the angles $\pm \pi/3$ and π from the positive real semi-axis.

6.26

(a) Observing that

$$\left(1+z^2\right)^z = \exp(z\log(1+z^2)),$$

since the function $\exp z$ is entire and the principal branch of $\log z$ is analytic everywhere except on the negative real semi-axis, including the origin, the function $\left(1+z^2\right)^z$ is analytic everywhere except at the points z such that

$$1+z^2 = -t, \qquad t \geq 0,$$

that is,

$$z(t) = (-1-t)^{1/2} = \pm i\sqrt{1+t}, \qquad t \geq 0,$$

which are all the points of the imaginary axis except the interval $(-1, 1)$ of the same axis.

(b) Observing that the functions $\sinh z$ and $\sin z$ are entire and the rational function $1/(z^2+9)$ is analytic everywhere except at the singular points $z = \pm 3i$, by the differentiability of composite functions, the function $\sinh(\sin(z))/(z^2+9)$ is analytic everywhere except at the points $z = \pm 3i$.

6.27

(a) The principal branch of $\log z$ is a function analytic everywhere in the complex plane except on the negative real semi-axis, including the origin. Therefore, the function under consideration is analytic on $\mathbb{C}$ except at the points such that

$$z^4 = -t, \qquad t \in [0, \infty),$$

that is to say

$$z(t) = (te^{i\pi})^{1/4} = t^{1/4}e^{i(\pi+2\pi k)/4}, \qquad t \geq 0, \quad k = 0, 1, 2, 3,$$

which respectively represent the semi-axes bisecting of the quadrants $1, 2, 3, 4$. Within the domain of analyticity the derivative is

$$\frac{d}{dz}\log\left(z^4\right) = \frac{4}{z}.$$

(b) The function $\sin z$ is entire while the principal branch of $\sqrt{z} = \exp(\frac{1}{2}\log z)$ is a function analytic everywhere on the complex plane except on the negative real semi-axis, including the origin. Therefore, the function under consideration is analytic on $\mathbb{C}$ except on the negative real semi-axis, including the origin. Within the domain of analyticity the derivative is

$$\frac{d}{dz} \sin \sqrt{z} = \frac{\cos \sqrt{z}}{2\sqrt{z}}.$$

6.28

(a) The function $\sin(z^n)$ is entire since it is a composition of entire functions. At each point $z \in \mathbb{C}$ its derivative is

$$\frac{d}{dz} \sin(z^n) = \cos(z^n) n z^{n-1}.$$

(b) If n is a multiple of m we fall into the case (a). Otherwise the function $\sin(z^{n/m}) = \sin(\exp((n/m)\log z))$ is multi-valued with m distinct branches. The principal branch, corresponding to the principal branch of $\log z$, is a function analytic everywhere in the complex plane except on the negative real semi-axis, including the origin. Within the domain of analyticity the derivative is

$$\frac{d}{dz} \sin(z^{n/m}) = \cos(z^{n/m})(n/m) z^{n/m-1}.$$

(c) The function $\sin(z^c) = \sin(\exp(c \log z))$ is multi-valued with infinitely many distinct branches. The principal branch, corresponding to the principal branch of $\log z$, is a function analytic everywhere in the complex plane except on the negative real semi-axis, including the origin. Within the domain of analyticity the derivative is

$$\frac{d}{dz} \sin(z^c) = \cos(z^c) c z^{c-1}.$$

6.29 Let $\arccos z = w$, we have $z = \cos w = (e^{iw} + e^{-iw})/2$ that is to say

$$\left(e^{iw}\right)^2 - 2z e^{iw} + 1 = 0,$$

whose solution is conventionally written as $e^{iw} = z + i\sqrt{1 - z^2}$. With this choice in fact, assuming the principal branch for the root, this solution, and therefore the function $\arccos z$, can be expanded in the Taylor series around $z = 0$ (see Exercise 6.30). With the choice $e^{iw} = z + \sqrt{z^2 - 1}$, this would only be possible for the secondary branch of the root. In conclusion,

$$\arccos z = -i \log \left(z + i\sqrt{1 - z^2} \right).$$

The principal branch of such a function is defined by taking the principal branches of the root and the logarithm. The principal branch of $\sqrt{1-z^2} = \exp(\frac{1}{2}\log(1-z^2))$ is a function analytic everywhere in $\mathbb{C}$ except at the points z such that $1 - z^2 = -u$ with $u \in [0, \infty)$. Such points

$$z(u) = \pm\sqrt{1+u}, \qquad u \in [0, \infty),$$

represent the real semi-axes $(-\infty, -1]$ and $[1, +\infty)$. The principal branch of $\log(z + i\sqrt{1-z^2})$ is non-analytic at the points satisfying $z + i\sqrt{1-z^2} = -t$ with $t \in [0, \infty)$. The solution of this equation obtained by squaring the equivalent expression $i\sqrt{1-z^2} = -(t+z)$ gives

$$z(t) = -\frac{1+t^2}{2t}, \qquad t \in (0, \infty),$$

which represents the real semi-axis $(-\infty, -1]$ traveled twice, once for $t \in (0, 1]$ and once for $t \in [1, \infty)$. Note that only the points $z(t)$ obtained for $t \in [1, \infty)$ actually satisfy the starting equation in which the root is considered to be the principal branch. In fact $i\sqrt{1 - z(t)^2} = i\sqrt{-(t^2 - 1)^2/(4t^2)} = i^2\sqrt{(t^2-1)^2/(4t^2)} \leq 0$ $\forall t \in (0, \infty)$, while $-(t+z(t)) = (1-t^2)/2t \leq 0$ only for $t \in [1, \infty)$. In conclusion, the domain of analyticity of $\arccos z$ is the entire complex plane except for the real semi-axes $(-\infty, -1]$ and $[1, +\infty)$, see Fig. 12.1. In this region the derivative is

$$\frac{d}{dz}\arccos z = -i\frac{1}{z+i\sqrt{1-z^2}}\left(1 + \frac{i}{2}\frac{-2z}{\sqrt{1-z^2}}\right)$$

$$= \frac{-1}{\sqrt{z^2-1}}.$$

6.30

(a) The principal branch of $\sqrt{z^2-1} = \exp[\frac{1}{2}\log(z^2 - 1)]$ is a function analytic everywhere in $\mathbb{C}$ except at the points z such that $z^2 - 1 = -u$ with $u \in [0, \infty)$. Depending on the parameter u such points are given by

$$z(u) = \pm\sqrt{1-u}, \qquad u \in [0, 1),$$

$$z(u) = \pm i\sqrt{u-1}, \qquad u \in [1, \infty),$$

and represent the real segment $[-1, 1]$ and the entire imaginary axis, see Fig. 12.2.

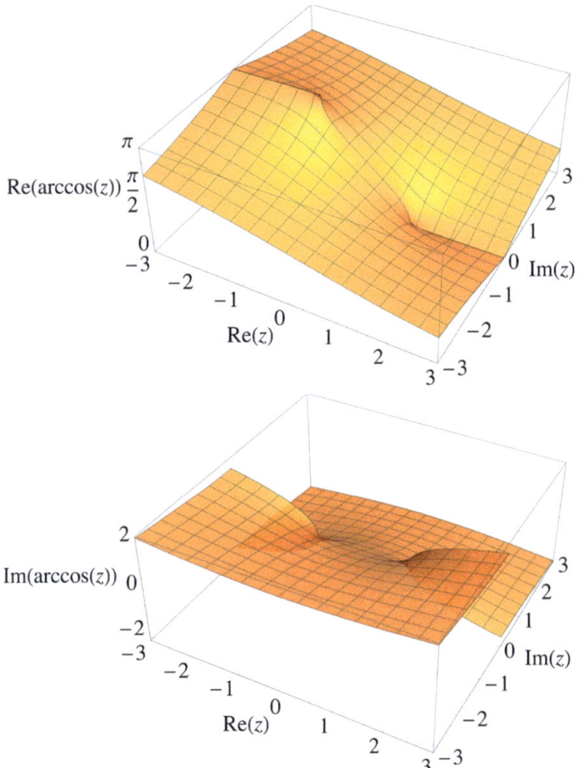

Fig. 12.1 Real and imaginary parts of the principal branch of $\arccos(z)$

(b) The principal branch of $i\sqrt{1-z^2} = i\exp[\frac{1}{2}\log(1-z^2)]$ is a function analytic
everywhere in $\mathbb{C}$ except at the points z such that $1-z^2 = -u$ with $u \in [0,\infty)$.
Depending on the parameter u these points are given by

$$z(u) = \pm\sqrt{1+u}, \qquad u \in [0,\infty),$$

and represent the real semi-axes $(-\infty,-1]$ and $[1,+\infty)$, see Fig. 12.3.

6.31

(a) The function z^n is entire since it is the product of n entire functions. For $n = 17$,
at $z = i$ it has value

$$i^{17} = i^{16}i = (i^2)^8 i = (-1)^8 i = i.$$

(b) If n is a multiple of m we fall into the case (a). Otherwise the function
$z^{n/m} = \exp((n/m)\log z)$ is multi-valued with m distinct branches. The

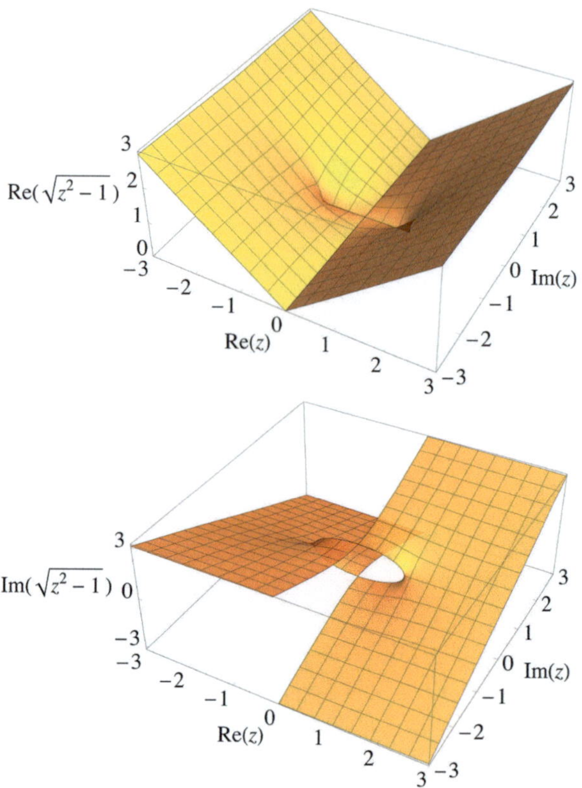

Fig. 12.2 Real and imaginary parts of the principal branch of $\sqrt{z^2-1}$

principal branch, corresponding to the principal branch of $\log z$, is a function analytic everywhere in the complex plane except on the negative real semi-axis, including the origin. For $n = 17$ and $m = 3$, at $z = i$ it has value

$$i^{17/3} = e^{(17/3)\log i} = e^{(17/3)(\ln 1 + i\pi/2)} = e^{17\pi i/6} = e^{12\pi i/6}e^{5\pi i/6}$$

$$= \cos(5\pi/6) + i\sin(5\pi/6) = -\frac{\sqrt{3}}{2} + \frac{i}{2}.$$

(c) The function $z^c = \exp(c \log z)$ is multi-valued with infinitely distinct branches. The principal branch, corresponding to the principal branch of $\log z$, is a function analytic everywhere in the complex plane except on the negative real semi-axis, including the origin. For $c = 1/\pi$, at $z = i$ it has value

$$i^{1/\pi} = e^{(1/\pi)\log i} = e^{(1/\pi)i\pi/2} = e^{i/2} = \cos(1/2) + i\sin(1/2).$$

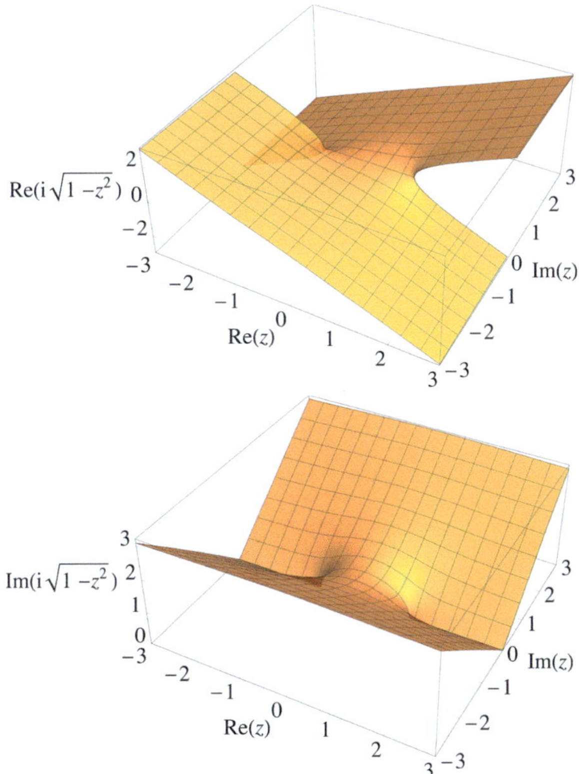

Fig. 12.3 Real and imaginary parts of the principal branch of $i\sqrt{1-z^2}$

6.32

(a) The principal branch of the logarithm is a function analytic everywhere in $\mathbb{C}$ except on the negative real semi-axis, including the origin. Therefore, the function considered is analytic in $\mathbb{C}$ except at the points $z(t) = -t$ with $t \in [0, \infty)$ and those of the type

$$(\log z(u))^2 = -u, \qquad u \in [0, \infty),$$

that is to say

$$z(u) = e^{\log z(u)} = e^{\pm i\sqrt{u}}, \qquad u \in [0, \infty).$$

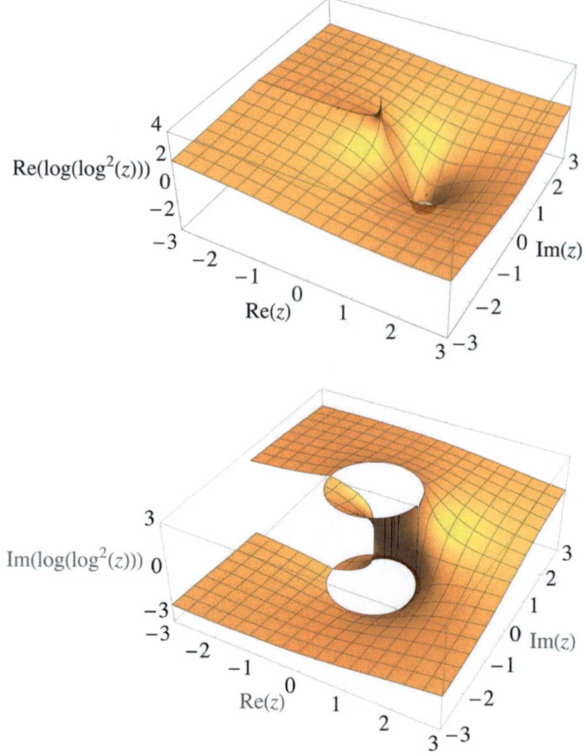

Fig. 12.4 Real and imaginary parts of the principal branch of $\log(\log^2(z))$

Therefore, the domain of analyticity is $D = \{z \in \mathbb{C} : |z| \neq 1, \ z \neq -t, \ t \in [0, \infty)\}$, see Fig. 12.4. For $z \in D$ the derivative is

$$\frac{\mathrm{d}}{\mathrm{d}z} \log((\log z)^2) = \frac{1}{(\log z)^2} \frac{2}{z} \log z = \frac{2}{z \log z}.$$

(b) The function is entire because it is a composition of entire functions. The derivative is

$$\frac{\mathrm{d}}{\mathrm{d}z} e^{z^2+\mathrm{i}} = 2z e^{z^2+\mathrm{i}}.$$

(c) The principal branch of the function considered is analytic in $D = \{z \in \mathbb{C} : z \neq e^{-\mathrm{i}\pi/4}, \ z \neq e^{\mathrm{i}3\pi/4}, \ z \neq 1+t, \ t \in [0, \infty)\}$ since

$$\sqrt{1-z} = e^{\frac{1}{2}\log(1-z)}$$

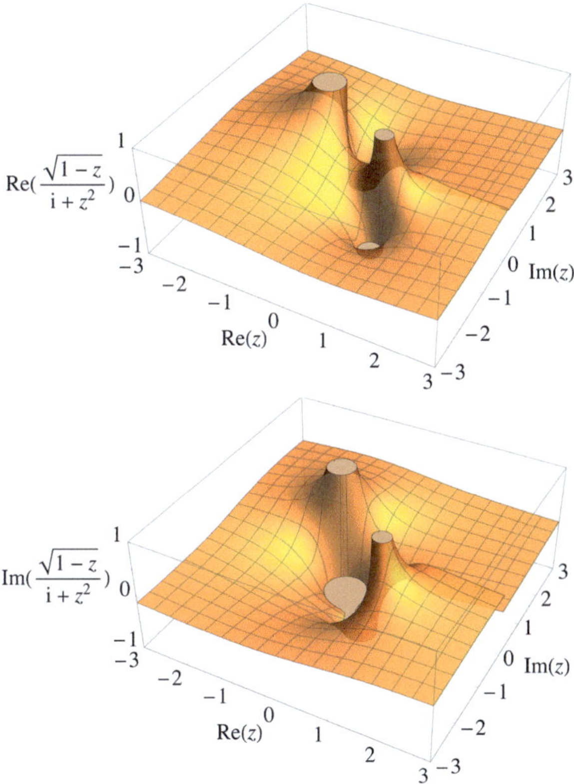

Fig. 12.5 Real and imaginary parts of the principal branch of $\sqrt{1-z}/(z^2+i)$

and $z^2 + i = 0$ has solutions $z = e^{-i\pi/4}$ and $z = e^{i3\pi/4}$, see Fig. 12.5. In D the derivative is

$$\frac{d}{dz}\frac{\sqrt{1-z}}{z^2+i} = \frac{-\frac{1}{2}(1-z)^{-1/2}(z^2+i) - (1-z)^{1/2}2z}{(z^2+i)^2}$$

$$= \frac{-i - 4z + 3z^2}{2\sqrt{1-z}(z^2+i)^2}.$$

6.33 We have

$$f(z) = \sqrt{4 + z^4} = \exp\left(\frac{1}{2}\log(4+z^4)\right),$$

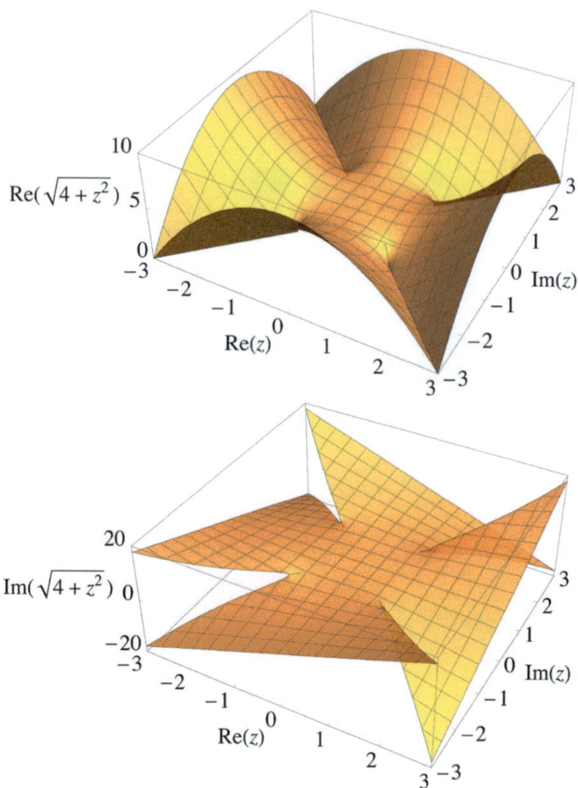

Fig. 12.6 Real and imaginary parts of the principal branch of $\sqrt{4+z^4}$

where the logarithm is intended as the principal branch. This function is analytic everywhere in $\mathbb{C}$ except on the points of the branch line of $\log(4+z^4)$ determined by the equation

$$4 + z^4 = -t, \qquad t \in [0, \infty).$$

These points are

$$z(t) = (-4 - t)^{1/4}$$

$$= ((4+t)e^{i\pi})^{1/4}$$

$$= (4+t)^{1/4}e^{i(\pi+2k\pi)/4}, \qquad t \in [0, \infty), \quad k = 0, 1, 2, 3.$$

These are the semi-axes that bisect the four quadrants starting from the points at distance $\sqrt{2}$ from the origin, see Fig. 12.6.

6.34 The functions $\sin z$ and $\cos z$ are entire, therefore the points of non-analyticity of f are all and only those determined by the principal branch of the square root

$$\sqrt{\cos z} = \exp\left(\frac{1}{2}\log(\cos z)\right).$$

The branch line of the principal branch of $\log(\cos z)$ is determined by the equation

$$\cos z = -t, \qquad 0 \le t < \infty.$$

For $0 \le t \le 1$ this equation admits, as in the real case, the solutions

$$z(t) = \arccos(-t) + 2\pi k, \qquad t \in [0, 1], \qquad k \in \mathbb{Z},$$

which represent the segments of the real axis $[\pi/2 + 2\pi k, 3\pi/2 + 2\pi k]$, $k \in \mathbb{Z}$. For $t > 1$ the equation is equivalent to

$$\left(e^{iz}\right)^2 + 2te^{iz} + 1 = 0,$$

whose solution is

$$e^{iz} = -t \pm \sqrt{t^2 - 1}.$$

Observing that the right-hand side of this last expression is real negative for every $t > 1$, we have

$$z(t) = -i\log\left[\left(t \mp \sqrt{t^2 - 1}\right)e^{i\pi}\right]$$
$$= (2k + 1)\pi - i\ln\left(t \mp \sqrt{t^2 - 1}\right), \qquad t \in (1, \infty), \qquad k \in \mathbb{Z}.$$

These points represent the imaginary axes passing through the points $z = (2k+1)\pi$, see Fig. 12.7.

6.35 The functions $\sin z$ and $\cos z$ are entire, therefore the points of non-analyticity of f are all and only those determined by the square root

$$\sqrt{\cos z} = \exp\left(\frac{1}{2}\log(\cos z)\right).$$

The branch line of the chosen branch of $\log(\cos z)$ is determined by the equation

$$\cos z = t, \qquad 0 \le t < \infty.$$

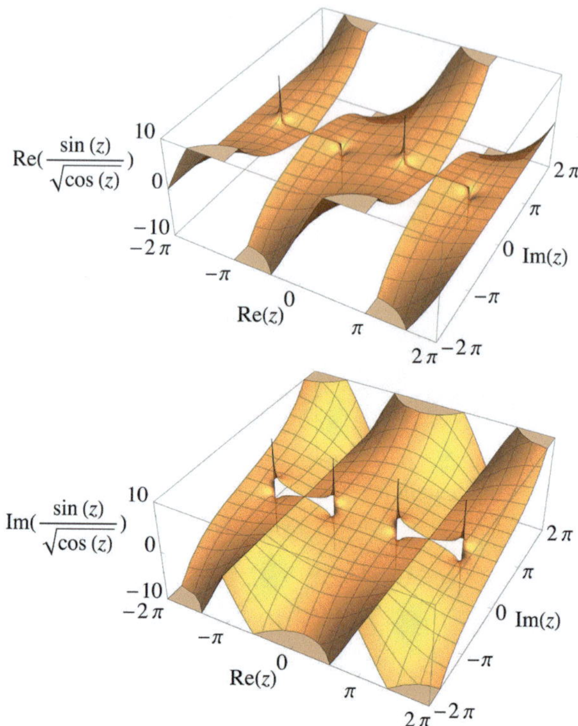

Fig. 12.7 Real and imaginary parts of the principal branch of $\sin(z)/\sqrt{\cos(z)}$

For $0 \le t \le 1$ this equation admits, as in the real case, the solutions

$$z(t) = \arccos(t) + 2\pi k, \qquad t \in [0,1], \qquad k \in \mathbb{Z},$$

which represent the segments of the real axis $[-\pi/2 + 2\pi k, \pi/2 + 2\pi k]$, $k \in \mathbb{Z}$.
For $t > 1$ the equation is equivalent to

$$\left(e^{iz}\right)^2 - 2t e^{iz} + 1 = 0,$$

whose solution is

$$e^{iz} = t \pm \sqrt{t^2 - 1}.$$

Observing that the right-hand side of this last expression is real positive for every
$t > 1$, we have

$$z(t) = -i \log\left[\left(t \pm \sqrt{t^2 - 1}\right) e^{i0}\right]$$

$$= 2k\pi - i \ln\left(t \pm \sqrt{t^2 - 1}\right), \qquad t \in (1, \infty), \qquad k \in \mathbb{Z}.$$

These points represent the imaginary axes passing through the points $z = 2k\pi$.

6.36 The function $g(z) = \log(f(z))$ is a composition of the principal branch of $\log(z)$ and of the function $f(z) = (3z^2 - \bar{z}^2)\bar{z}/2$. Let us study separately the properties of continuity, differentiability and analyticity of these two component functions.

The function $\log$ is continuous, differentiable and analytic everywhere in the complex plane except on the negative real semi-axis including the origin.

The function f has been already studied in the Exercise 5.10, we repeat the analysis here for convenience. As a composition of continuous functions in $\mathbb{C}$, f is continuous in $\mathbb{C}$. To study its differentiability, observe that, setting $z = x + iy$ with $x, y \in \mathbb{R}$, we have

$$f(x + iy) = x^3 + 3xy^2 + i(y^3 + 3x^2y),$$

therefore

$$u(x, y) = \operatorname{Re} f(x + iy) = x^3 + 3xy^2,$$

$$v(x, y) = \operatorname{Im} f(x + iy) = y^3 + 3x^2y.$$

The functions u and v are differentiable throughout $\mathbb{R}^2$ with derivatives

$$u_x(x, y) = 3x^2 + 3y^2, \qquad u_y(x, y) = 6xy,$$

$$v_x(x, y) = 6xy, \qquad v_y(x, y) = 3y^2 + 3x^2,$$

continuous throughout $\mathbb{R}^2$. The Cauchy-Riemann equations, $u_x = v_y$ and $u_y = -v_x$, therefore give

$$3x^2 + 3y^2 = 3x^2 + 3y^2,$$

$$6xy = -6xy.$$

The first equation is always satisfied, the second only if $x = 0$ or $y = 0$. It follows that f is differentiable only at the points of the coordinate axes. At none of these points, however, f is analytic. In fact $\forall \varepsilon > 0$ the ball $B(z, \varepsilon)$, with z real or pure imaginary, contains points w such that $\operatorname{Re} w \neq 0$ and $\operatorname{Im} w \neq 0$ where f is not differentiable.

When composing the continuity domains of $\log$ and f, note that $f(z) = -t$, $t \in [0, \infty)$, is satisfied for the points $z = x + iy$ such that

$$\begin{cases} x^3 + 3xy = -t, \\ y^3 + 3x^2y = 0. \end{cases}$$

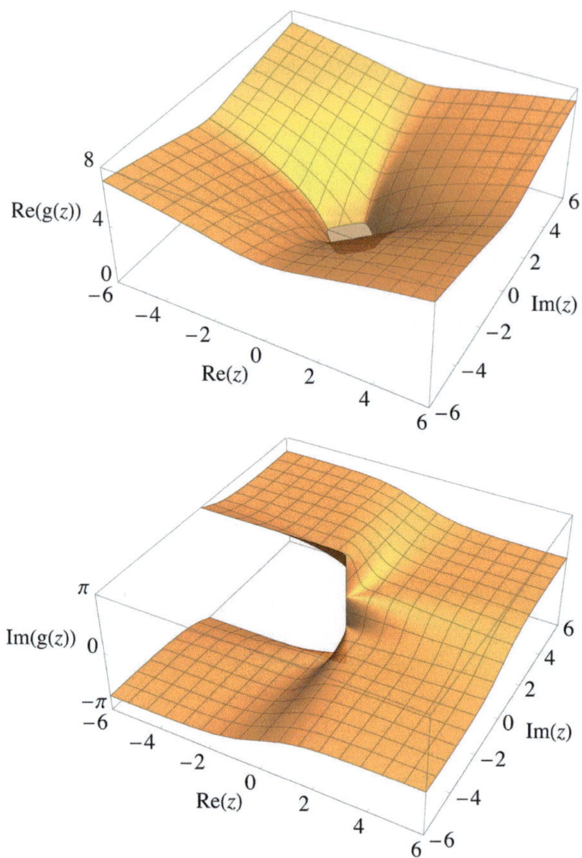

Fig. 12.8 Real and imaginary parts of $g(z) = \log((3z^2 - \bar{z}^2)\bar{z}/2)$. Note the discontinuity along the negative real semi-axis, including the origin. The function is differentiable only along the imaginary axis and the positive real semi-axis, excluding the origin, and nowhere analytic

The second equation has solution $y = 0$. For $y = 0$, the first equation yields $x = -\sqrt[3]{t}$. The continuity domain of $g(z) = (\log \circ f)(z)$ is therefore

$$D_c = \mathbb{C} \setminus \{z(t) = -\sqrt[3]{t},\ t \in [0, \infty)\} = \mathbb{C} \setminus \{z(u) = -u,\ u \in [0, \infty)\},$$

see Fig. 12.8. Since $\log$ is also analytic in the set D_c, the composition $\log \circ f$ is differentiable in the set D_d intersection of D_c and the set of points where f is differentiable, namely,

$$D_d = \{z(u) = u \text{ or } z(u) = \pm iu,\ u \in (0, \infty)\}.$$

In a similar way we conclude that the domain of analyticity of g is empty

$$D_a = \emptyset.$$

6.37 The function $\cos z$ is entire, therefore the points of non-analyticity of f are all and only those determined by the square root defined in terms of the logarithm by the relation

$$\sqrt{\cos z} = \exp\left(\frac{1}{2}\log(\cos z)\right).$$

The principal branch of the logarithm having as branch line $z = it$, $t \geq 0$, is the analytic function in $D = \{z \in \mathbb{C} : z \neq it, \ t \in [0, \infty)\}$ defined by

$$\log z = \ln|z| + i\,\mathrm{Arg}_{(\frac{\pi}{2}, \frac{5\pi}{2}]}(z),$$

where $\pi/2 < \mathrm{Arg}_{(\frac{\pi}{2}, \frac{5\pi}{2}]}(z) \leq 5\pi/2$ can be expressed in terms of the standard principal argument $-\pi < \mathrm{Arg}(z) \leq \pi$ as

$$\mathrm{Arg}_{(\frac{\pi}{2}, \frac{5\pi}{2}]}(z) = \begin{cases} \mathrm{Arg}(z), & \pi/2 < \mathrm{Arg}(z) \leq \pi, \\ \mathrm{Arg}(z) + 2\pi, & -\pi < \mathrm{Arg}(z) \leq \pi/2. \end{cases}$$

The branch line of $\log(\cos z)$ is therefore determined by the equation

$$\cos z = it, \qquad 0 \leq t < \infty.$$

This equation is equivalent to the second degree equation in e^{iz}

$$\left(e^{iz}\right)^2 - 2it\,e^{iz} + 1 = 0,$$

whose solution is

$$e^{iz} = it + \sqrt{(it)^2 - 1} = i\left(t \pm \sqrt{t^2 + 1}\right).$$

Observing that $\mathrm{Arg}_{(\frac{\pi}{2}, \frac{5\pi}{2}]}(i) = 5\pi/2$ while $\mathrm{Arg}_{(\frac{\pi}{2}, \frac{5\pi}{2}]}(-i) = 3\pi/2$ and that e^{iz} is periodic with period 2π, the above solution provides branch lines defined by

$$z(t) = -i\log\left[i\left(t + \sqrt{t^2 + 1}\right)\right]$$

$$= -i\log\left[\left(t + \sqrt{t^2 + 1}\right)e^{i5\pi/2}\right]$$

$$= (5/2 + 2k)\pi - i\ln\left(t + \sqrt{t^2 + 1}\right), \qquad k \in \mathbb{Z}, \quad t \in [0, \infty),$$

$$z(t) = -i\log\left[i\left(t - \sqrt{t^2 + 1}\right)\right]$$

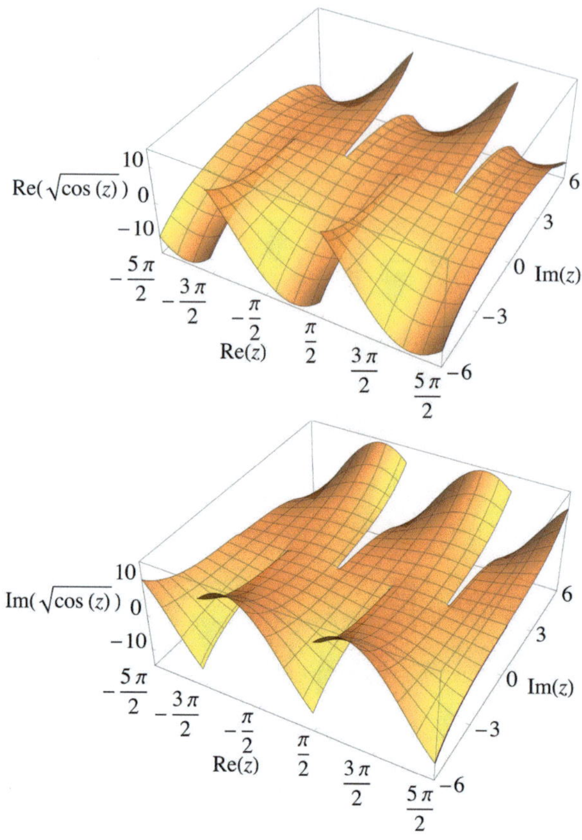

Fig. 12.9 Real and imaginary parts of the branch of $f(z) = \sqrt{\cos z}$ which corresponds to the principal branch of $\log z$ having as branch line the positive imaginary semi-axis

$$= -\mathrm{i} \log\left[\left(\sqrt{t^2 + 1} - t\right) \mathrm{e}^{\mathrm{i}3\pi/2}\right]$$

$$= (3/2 + 2k)\pi - \mathrm{i} \ln\left(\sqrt{t^2 + 1} - t\right), \qquad k \in \mathbb{Z}, \quad t \in [0, \infty).$$

These points represent the negative imaginary semi-axes having the origin at the points $z = (5/2 + 2k)\pi$ and the positive imaginary semi-axes with the origin at $z = (3/2 + 2k)\pi$. The function $f(z)$ is analytic everywhere in the complex plane except on these semi-axes, see Fig. 12.9.

Alternatively, we can solve the equation $\cos z = \mathrm{i}t$ in Cartesian coordinates. Setting $z = x + \mathrm{i}y$, with $x, y \in \mathbb{R}$, and using $\cos z = \cos x \cosh y + \mathrm{i} \sin x \sinh y$, we obtain the system

$$\begin{cases} \cos x \cosh y = 0, \\ \sin x \sinh y = -t. \end{cases}$$

Observing that $\cosh y > 0 \ \forall y \in \mathbb{R}$, the first equation has solution $x = \pi/2 + k\pi$, $k \in \mathbb{Z}$. Since $\sin(\pi/2 + k\pi) = (-1)^k$, the second equation is equivalent to $\sinh y = -t$ for $x = \pi/2 + 2k\pi$ and $\sinh y = t$ for $x = 3\pi/2 + 2k\pi$, with $k \in \mathbb{Z}$. For $t \geq 0$, we therefore obtain that $f(z)$ is non-analytic in the semi-axes $x = \pi/2 + 2k\pi$, $y \leq 0$ and $x = 3\pi/2 + 2k\pi$, $y \geq 0$, with $k \in \mathbb{Z}$.

6.38 The function $f(z)$ is analytic in $\mathbb{C}$ except on the branch line $w(t) = -i + (\pi - t)$, with $t \in [0, \infty)$, and the isolated singularities at the points

$$w_k = (2k+1)\pi/2, \qquad k = 0, \pm 1, \pm 2, \ldots.$$

The radius of convergence R is the radius of the greatest circle centered at $z_0 = 2 + i$ within which $f(z)$ is analytic. Since the minimal distances of z_0 from the singularities of f are

$$\min_{k=0,\pm 1,\pm 2,\ldots} |z_0 - w_k| = |z_0 - w_1| = \sqrt{(2 - \pi/2)^2 + 1},$$

$$\min_{t \in [0,\infty)} |z_0 - w(t)| = |z_0 - w(\pi - 2)| = 2,$$

we obtain that $R = \sqrt{(2 - \pi/2)^2 + 1}$.

6.39 The principal branch of

$$f(z) = 1 + \frac{\log z}{\ln 2} = 1 + \frac{\ln|z| + i \operatorname{Arg} z}{\ln 2}$$

biunivocally transforms A into B. In fact f is analytic in A and, setting $z = re^{i\theta}$, for $1 < r < 2$ and $0 < \theta \leq \pi/2$ we have $1 \leq \operatorname{Re} f \leq 2$ and $0 \leq \operatorname{Im} f \leq \pi/(2 \ln 2)$.

6.40 Since

$$i^{nz^2} = e^{nz^2 \log i} = \left(e^{z^2 \log i}\right)^n,$$

we recognize that the series considered is a geometric series of ratio $e^{z^2 \log i}$. It therefore converges when $|e^{z^2 \log i}| < 1$. Setting $w = z^2$ and using the principal branch of the logarithm, the previous inequality yields

$$\left| e^{w(\ln|i| + i \operatorname{Arg} i)} \right| = e^{-(\pi/2)\operatorname{Im} w} < 1.$$

Therefore, we must have $\operatorname{Im} w > 0$, that is to say $0 < \operatorname{Arg} w < \pi$. Since $w = z^2$ transforms angular sectors into angular sectors of double extension, the given series is convergent when $0 < \operatorname{Arg} z < \pi/2$ or $-\pi < \operatorname{Arg} z < -\pi/2$.

12.7   Exercises of Chap. 7

7.1 We have

$$\int_{\gamma} \frac{z}{\bar{z}}\mathrm{d}z = \int_{\gamma_1} \frac{z}{\bar{z}}\mathrm{d}z + \int_{\gamma_2} \frac{z}{\bar{z}}\mathrm{d}z,$$

where $\gamma_1(t) = 2e^{it}, 0 \le t \le \pi$ and $\gamma_2(t) = t, \ -2 \le t \le 2$. Using the definition of complex integral

$$\int_{\gamma_1} \frac{z}{\bar{z}}\mathrm{d}z = \int_0^{\pi} \frac{2e^{it}}{2e^{-it}} i2e^{it}\,\mathrm{d}t = \int_0^{\pi} 2ie^{i3t}\,\mathrm{d}t = -\frac{4}{3},$$

$$\int_{\gamma_2} \frac{z}{\bar{z}}\mathrm{d}z = \int_{-2}^{2} \frac{t}{t} 1\mathrm{d}t = \int_{-2}^{2} \mathrm{d}t = 4.$$

In conclusion

$$\int_{\gamma} \frac{z}{\bar{z}}\mathrm{d}z = -\frac{4}{3} + 4 = \frac{8}{3}.$$

7.2 Let $\gamma = \gamma_1 + \gamma_2 + \gamma_3$ where $\gamma_1(t) = t$, $\gamma_2(t) = 1 - t + it$ and $\gamma_3(t) = i(1 - t)$, with $0 \le t \le 1$. Observing that $f(z) = -4(\mathrm{Im}\, z)^2$ and using the definition of complex integral, we have

$$\int_{\gamma_1} f(z)\mathrm{d}z = \int_0^1 0 \, 1\mathrm{d}t = 0,$$

$$\int_{\gamma_2} f(z)\mathrm{d}z = \int_0^1 -4t^2(-1 + i)\mathrm{d}t = 4(1 - i)\frac{t^3}{3}\Big|_0^1 = \frac{4}{3} - i\frac{4}{3},$$

$$\int_{\gamma_3} f(z)\mathrm{d}z = \int_0^1 -4(1 - t)^2(-i)\mathrm{d}t = -i\frac{4}{3}(1 - t)^3\Big|_0^1 = i\frac{4}{3}.$$

In conclusion

$$\int_{\gamma} f(z)\mathrm{d}z = \int_{\gamma_1+\gamma_2+\gamma_3} f(z)\mathrm{d}z$$

$$= \int_{\gamma_1} f(z)\mathrm{d}z + \int_{\gamma_2} f(z)\mathrm{d}z + \int_{\gamma_3} f(z)\mathrm{d}z$$

$$= \frac{4}{3}.$$

7.3 The integrand function is

$$f(z) = z^2 - \bar{z}^2 = 2i\,\mathrm{Im}\,z^2$$

and the integration path can be parametrized as $\gamma(t) = 1 - t + it$, $0 \le t \le 1$. Therefore

$$\int_\gamma f(z)\mathrm{d}z = \int_0^1 2i\,\mathrm{Im}\left((1 - t + it)^2\right)(-1+i)\mathrm{d}t$$

$$= \int_0^1 2i2(1-t)t(-1+i)\mathrm{d}t$$

$$= 4i(-1+i)\int_0^1 (t - t^2)\mathrm{d}t$$

$$= (-4+4i)\left(\frac{t^2}{2} - \frac{t^3}{3}\right)\Big|_0^1$$

$$= -\frac{2}{3} - \frac{2}{3}i.$$

7.4 Let $\gamma = \lambda + \gamma_R$ where $\lambda(t) = t - i$, $-1 \le t \le 1$ and $\gamma_R(t) = \sqrt{2}e^{it}$, $-\pi/4 \le t \le 5\pi/4$, we have

$$\int_\gamma \frac{1}{\bar{z}}\mathrm{d}z = \int_\lambda \frac{1}{\bar{z}}\mathrm{d}z + \int_{\gamma_R} \frac{1}{\bar{z}}\mathrm{d}z.$$

Using the definition of complex integral we have

$$\int_\lambda \frac{1}{\bar{z}}\mathrm{d}z = \int_{-1}^1 \frac{1}{t+i}\mathrm{d}t = \log(1+i) - \log(-1+i) = -i\frac{\pi}{2},$$

$$\int_{\gamma_R} \frac{1}{\bar{z}}\mathrm{d}z = \int_{-\pi/4}^{5\pi/4} \frac{1}{\sqrt{2}e^{-it}} i\sqrt{2}e^{it}\,\mathrm{d}t = \frac{1}{2}e^{i5\pi/2} - \frac{1}{2}e^{-i\pi/2} = i.$$

In conclusion

$$\int_\gamma \frac{1}{\bar{z}}\mathrm{d}z = i(1 - \frac{\pi}{2}).$$

7.5 The curves in question can be parametrized as $\gamma_1(y) = iy$, $0 \le y \le 1$, $\gamma_2(x) = x + i$, $0 \le x \le 1$ and $\gamma_3(t) = t + it$, $0 \le t \le 1$. Therefore

$$\int_{\gamma_1} f(z)\mathrm{d}z = \int_0^1 f(\gamma_1(y))\gamma_1'(y)\mathrm{d}y = \int_0^1 y\mathrm{i}\,\mathrm{d}y = \mathrm{i}/2,$$

$$\int_{\gamma_2} f(z)\mathrm{d}z = \int_0^1 f(\gamma_2(x))\gamma_2'(x)\mathrm{d}x = \int_0^1 (1 - x - \mathrm{i}3x^2)\mathrm{i}\,\mathrm{d}x = 1/2 - \mathrm{i},$$

and so

$$\int_{\gamma_1+\gamma_2} f(z)\mathrm{d}z = \int_{\gamma_1} f(z)\mathrm{d}z + \int_{\gamma_2} f(z)\mathrm{d}z = 1/2 - \mathrm{i}/2,$$

while

$$\int_{\gamma_3} f(z)\mathrm{d}z = \int_0^1 f(\gamma_3(t))\gamma_3'(t)\mathrm{d}t = \int_0^1 (t - t - \mathrm{i}3t^2)(1 + \mathrm{i})\mathrm{d}t = 1 - \mathrm{i}.$$

7.6 The square path γ has sides of length $d = \sqrt{2}$. Each point of γ has a distance from the origin which is not smaller than the radius r of the circle inscribed in γ and not larger than the radius R of the circle in which γ is inscribed. Such radii are $r = d/2 = 1/\sqrt{2}$ and $R = 1$. Therefore

$$\left|\frac{\sqrt{z}}{z^2}\right| = \frac{|z|^{1/2}}{|z|^2} \le \frac{R^{1/2}}{r^2} = 2, \qquad \forall z \in \{\gamma\}.$$

Using Darboux's inequality we have

$$\left|\int_\gamma f(z)\mathrm{d}z\right| \le M\,L,$$

where M is a constant such that $|f(z)| \le M \ \forall z \in \{\gamma\}$ and L is the length of the path γ. We can then set $M = 2$ and $L = 4\sqrt{2}$ and get

$$\left|\int_\gamma \frac{\sqrt{z}}{z^2}\mathrm{d}z\right| \le 8\sqrt{2}.$$

7.7 Let $z = r\mathrm{e}^{\mathrm{i}\theta}$, the integrand function is

$$z^{-3/2} \log z = r^{-\frac{3}{2}}\mathrm{e}^{-\mathrm{i}\frac{3}{2}\theta} (\ln r + \mathrm{i}\theta), \qquad r > 0, \qquad -\pi < \theta < \pi.$$

Since on the circle $|z| = R$ we have $r = R$ and $-\pi \le \theta \le \pi$, the modulus of the integrand function can be upper bounded as follows

$$\left|R^{-\frac{3}{2}}\mathrm{e}^{-\mathrm{i}\frac{3}{2}\theta} (\ln R + \mathrm{i}\theta)\right| = R^{-\frac{3}{2}} |\ln R + \mathrm{i}\theta| \le R^{-\frac{3}{2}} (\ln R + \pi).$$

Therefore

$$\left| \int_{|z|=R} z^{-3/2} \log z\, dz \right| \le 2\pi\, R\, R^{-\frac{3}{2}} (\ln R + \pi)$$

$$= 2\pi \frac{\ln R + \pi}{\sqrt{R}} \xrightarrow{R \to \infty} 0.$$

7.8 The continuity of $e^{az} f(z)$ for $\operatorname{Re} z \ge 0$ ensures the existence of the integral of this function on γ_R. By hypothesis, $\forall \varepsilon > 0$ there exists $\delta(\varepsilon) > 0$ such that $|f(z)| < \varepsilon$ whenever $|1/z| < \delta$. Then for $R > \delta^{-1}$ we find

$$\left| \int_{\gamma_R} e^{az} f(z)\,dz \right| \le \int_{-\frac{\pi}{2}}^{\frac{\pi}{2}} \left| e^{a\gamma_R(\theta)} f(\gamma_R(\theta)) \gamma_R'(\theta) \right| d\theta$$

$$= \int_{-\frac{\pi}{2}}^{\frac{\pi}{2}} e^{aR\cos\theta}\, |f(\gamma_R(\theta))|\, R\,d\theta$$

$$< 2\varepsilon R \int_0^{\frac{\pi}{2}} e^{aR\cos\theta}\,d\theta$$

$$= 2\varepsilon R \int_0^{\frac{\pi}{2}} e^{aR\sin\varphi}\,d\varphi \qquad \left(\theta = \frac{\pi}{2} - \varphi\right)$$

$$\le 2\varepsilon R \int_0^{\frac{\pi}{2}} e^{-|a|R2\varphi/\pi}\,d\varphi$$

$$= 2\varepsilon R \frac{\pi}{2\,|a|\,R}\left(1 - e^{-|a|R}\right)$$

$$< \frac{\pi\varepsilon}{|a|},$$

having used the fact that for $0 \le \varphi \le \pi/2$ we have $\sin\varphi \ge 2\varphi/\pi$ and then $\exp(aR\sin\varphi) = \exp(-|a|\,R\sin\varphi) \le \exp(-|a|\,R2\varphi/\pi)$. From the arbitrariness of ε it follows

$$\lim_{R \to \infty} \int_{\gamma_R} e^{az} f(z)\,dz = 0.$$

7.9 The principal branch of $f(z) = e^{(a-1)\log z}$, defined by $\log z = \ln r + i\theta$, where $z = re^{i\theta}$ with $r > 0$ and $-\pi < \theta < \pi$, is an analytic function in $\{z \in \mathbb{C},\ z \ne -t,\ t \in [0,\infty)\}$. Chosen for γ the parametric representation $\gamma(\theta) = Re^{i\theta}$ with $-\pi \le \theta \le \pi$, we have

$$\int_\gamma z^{a-1}\,dz = \int_{-\pi}^{\pi} e^{(a-1)\log\gamma(\theta)} \gamma'(\theta)\,d\theta$$

$$= \int_{-\pi}^{\pi} e^{(a-1)(\ln R + i\theta)} i R e^{i\theta} d\theta$$

$$= i R e^{(a-1)\ln R} \int_{-\pi}^{\pi} e^{ia\theta} d\theta$$

$$= i R^a \frac{e^{ia\pi} - e^{-ia\pi}}{ia}$$

$$= 2\pi i R^a \frac{\sin(\pi a)}{\pi a}.$$

Note that the integral vanishes when a takes on positive or negative integer values excluding zero. As expected, for $a \to 0$ we have

$$\lim_{a \to 0} \int_{\gamma} z^{a-1} dz = 2\pi i = \int_{\gamma} z^{-1} dz.$$

Alternatively, the integral can be calculated by observing that for $a \neq 0$ the function z^{a-1} has primitive $a^{-1} z^a$, for the procedure see other exercises.

7.10 In the open and connected domain D, the function $f(z) = \tan(z)/z^2$, analytic in D, has primitive if and only if $\int_{\gamma} f(z) dz = 0$ for every piecewise regular closed curve γ contained in D. Consider the circle γ centered at the origin, of radius 1, positively oriented. By the Cauchy integral formula applied to the function $\tan z$, analytic on and within γ, we get

$$\int_{\gamma} f(z) dz = \int_{\gamma} \frac{\tan z}{(z-0)^2} dz = 2\pi i \frac{d}{dz} \tan z \Big|_{z=0} = 2\pi i \frac{1}{\cos^2 0} \neq 0.$$

Therefore, f has no primitive in D.

7.11 The function $z^{-1} \log z$ is continuous on $\{\gamma\}$ with the exception of the point $z = 1$ where $\{\gamma\}$ intersects the branch line chosen for the logarithm. The integral in question therefore exists and its value is given by

$$\int_{\gamma} z^{-1} \log z \, dz = \int_{0}^{2\pi} e^{-i\theta} (\ln 1 + i\theta) e^{i\theta} i d\theta$$

$$= \int_{0}^{2\pi} i^2 \theta d\theta$$

$$= -2\pi^2.$$

Alternatively, note that except at the points on the branch line

$$z^{-1} \log z = \frac{d}{dz} \left(\frac{1}{2} \log^2 z \right).$$

Then let $\gamma_\varepsilon(\theta) = \exp(i\theta)$, $\varepsilon \le \theta \le 2\pi - \varepsilon$, we have

$$\int_\gamma z^{-1} \log z\, dz = \lim_{\varepsilon \to 0} \int_{\gamma_\varepsilon} z^{-1} \log z\, dz$$

$$= \lim_{\varepsilon \to 0} \frac{1}{2} \log^2 z \Big|_{z=e^{i\varepsilon}}^{z=e^{i(2\pi-\varepsilon)}}$$

$$= \lim_{\varepsilon \to 0} \frac{1}{2} \left(i^2 (2\pi - \varepsilon)^2 - i^2 \varepsilon^2 \right)$$

$$= -2\pi^2.$$

7.12 Let $z = re^{i\theta}$, the integrand function is defined by

$$z^{3/2} = r^{\frac{3}{2}} e^{i\frac{3}{2}\theta}, \qquad r > 0, \qquad -\pi < \theta < \pi.$$

For every $\varepsilon > 0$ this function is analytic on the path $\gamma_\varepsilon(\theta) = Re^{i\theta}$, $-\pi + \varepsilon \le \theta \le \pi - \varepsilon$, and there it admits as primitive the principal branch of $\frac{2}{5} z^{5/2}$. We thus have

$$\int_\gamma z^{3/2} dz = \lim_{\varepsilon \to 0} \int_{\gamma_\varepsilon} z^{3/2} dz$$

$$= \lim_{\varepsilon \to 0} \frac{2}{5} z^{5/2} \Big|_{z=Re^{i(-\pi+\varepsilon)}}^{z=Re^{i(\pi-\varepsilon)}}$$

$$= \lim_{\varepsilon \to 0} \frac{2}{5} R^{5/2} \left(e^{i\frac{5}{2}(\pi-\varepsilon)} - e^{i\frac{5}{2}(-\pi+\varepsilon)} \right)$$

$$= \lim_{\varepsilon \to 0} i \frac{4}{5} R^{5/2} \cos\left(\frac{5}{2}\varepsilon\right)$$

$$= i \frac{4}{5} R^{5/2}.$$

7.13 Note that the principal branch of $z^{4/3}$ is an analytic function on the path $\gamma_{R,\varepsilon}(\theta) = Re^{i\theta}$, $-\pi + \varepsilon \le \theta \le \pi - \varepsilon$, with arbitrary $\varepsilon > 0$, and there it admits as primitive the principal branch of $\frac{3}{7} z^{7/3}$. We thus have

$$\int_{\gamma_R} z^{4/3} dz = \lim_{\varepsilon \to 0} \int_{\gamma_{R,\varepsilon}} z^{4/3} dz$$

$$= \lim_{\varepsilon \to 0} \frac{3}{7} z^{7/3} \Big|_{z=Re^{i(-\pi+\varepsilon)}}^{z=Re^{i(\pi-\varepsilon)}}$$

$$- \lim_{\varepsilon \to 0} \frac{3}{7} R^{7/3} \left(e^{i\frac{7}{3}(\pi-\varepsilon)} - e^{i\frac{7}{3}(-\pi+\varepsilon)} \right)$$

$$= \lim_{\varepsilon \to 0} \frac{3}{7} R^{7/3} \, 2\mathrm{i} \sin\left[\frac{7}{3}(\pi - \varepsilon)\right]$$

$$= \frac{3}{7} R^{7/3} \, 2\mathrm{i} \sin\left(\frac{\pi}{3}\right)$$

$$= \mathrm{i}\frac{3\sqrt{3}}{7} R^{7/3}.$$

7.14 Let $z = r\mathrm{e}^{\mathrm{i}\theta}$, the principal branch of $F(z) = z^z$ is an analytic function for $r > 0$ and $-\pi < \theta < \pi$, and in this region its derivative is

$$\frac{\mathrm{d}}{\mathrm{d}z} z^z = \frac{\mathrm{d}}{\mathrm{d}z} \mathrm{e}^{z \log z} = \mathrm{e}^{z \log z}(\log z + 1).$$

Let $\gamma_{R,\varepsilon}(\varphi) = R\mathrm{e}^{\mathrm{i}\varphi}$, $-(\pi - \varepsilon) \le \varphi \le \pi + \varepsilon$, with arbitrary $\varepsilon > 0$, for $z \in \{\gamma_{R,\varepsilon}\}$ the function $F(z)$ is a primitive of the integrand function $f(z) = z^z(1 + \log z)$. Therefore

$$\int_{\gamma_R} z^z(1 + \log z)\mathrm{d}z = \lim_{\varepsilon \to 0} \int_{\gamma_{R,\varepsilon}} z^z(1 + \log z)\mathrm{d}z$$

$$= \lim_{\varepsilon \to 0} z^z \Big|_{z=R\mathrm{e}^{-\mathrm{i}(\pi-\varepsilon)}}^{z=R\mathrm{e}^{\mathrm{i}(\pi-\varepsilon)}}$$

$$= \lim_{\varepsilon \to 0} \left(\mathrm{e}^{R\mathrm{e}^{\mathrm{i}(\pi-\varepsilon)}[\ln R + \mathrm{i}(\pi-\varepsilon)]} - \mathrm{e}^{R\mathrm{e}^{-\mathrm{i}(\pi-\varepsilon)}[\ln R - \mathrm{i}(\pi-\varepsilon)]} \right)$$

$$= \mathrm{e}^{-R(\ln R + \mathrm{i}\pi)} - \mathrm{e}^{-R(\ln R - \mathrm{i}\pi)}$$

$$= -2\mathrm{i}\mathrm{e}^{-R \ln R} \sin(\pi R)$$

$$= -\frac{2\mathrm{i} \sin(\pi R)}{R^R}.$$

7.15 Let $z = r\mathrm{e}^{\mathrm{i}\theta}$, the principal branch of $F(z) = 2\sin(z^{1/2}) = 2\sin(r^{1/2}\mathrm{e}^{\mathrm{i}\theta/2})$ is a function analytic for $r > 0$ and $-\pi < \theta < \pi$, and in this region its derivative is

$$\frac{\mathrm{d}}{\mathrm{d}z} 2\sin(z^{1/2}) = \cos(z^{1/2})z^{-1/2}.$$

Let $\gamma_\varepsilon(\varphi) = R\mathrm{e}^{\mathrm{i}\varphi}$, $-(\pi - \varepsilon) \le \varphi \le \pi - \varepsilon$, with arbitrary $\varepsilon > 0$, for $z \in \{\gamma_\varepsilon\}$ the function $F(z)$ is a primitive of the integrand function $\cos\sqrt{z}/\sqrt{z}$. Therefore

$$\int_\gamma \frac{\cos\sqrt{z}}{\sqrt{z}}\mathrm{d}z = \lim_{\varepsilon \to 0} \int_{\gamma_\varepsilon} \frac{\cos\sqrt{z}}{\sqrt{z}}\mathrm{d}z$$

$$= \lim_{\varepsilon \to 0} 2\sin(z^{1/2}) \Big|_{z=R\mathrm{e}^{-\mathrm{i}(\pi-\varepsilon)}}^{z=R\mathrm{e}^{\mathrm{i}(\pi-\varepsilon)}}$$

$$= \lim_{\varepsilon \to 0} \left(2 \sin(R^{1/2} e^{i(\pi-\varepsilon)/2}) - 2 \sin(R^{1/2} e^{-i(\pi-\varepsilon)/2}) \right)$$

$$= 2 \sin(i\sqrt{R}) - 2 \sin(-i\sqrt{R})$$

$$= 4 \sin(i\sqrt{R})$$

$$= 4i \sinh \sqrt{R}.$$

7.16 The integrand function

$$\frac{\log z}{z^2} = \frac{\ln r + i\theta}{r^2 e^{i2\theta}}, \qquad z = re^{i\theta}, \quad r > 0, \quad -\pi < \theta < \pi,$$

is continuous on the whole $\{\gamma_R\}$ except at the point $z = -R$. It is convenient to parametrize the integration path as $\gamma_R(\theta) = Re^{i\theta}, \; -\pi \le \theta \le \pi$. Therefore

$$\int_{\gamma_R} \frac{\log z}{z^2} dz = \int_{-\pi}^{+\pi} \frac{\ln R + i\theta}{R^2 e^{2i\theta}} \, Re^{i\theta} i\, d\theta$$

$$= i\frac{\ln R}{R} \int_{-\pi}^{+\pi} e^{-i\theta} d\theta - \frac{1}{R} \int_{-\pi}^{+\pi} \theta e^{-i\theta} d\theta$$

$$= i\frac{\ln R}{R} \frac{e^{-i\theta}}{-i} \Big|_{-\pi}^{+\pi} - \frac{1}{R} e^{-i\theta} (1 + i\theta) \Big|_{-\pi}^{+\pi}$$

$$= -\frac{\ln R}{R} \left(e^{-i\pi} - e^{i\pi} \right) - \frac{1}{R} \left(e^{-i\pi}(1 + i\pi) - e^{i\pi}(1 - i\pi) \right)$$

$$= 0 - \frac{1}{R}(-1 - i\pi + 1 - i\pi)$$

$$= \frac{2\pi i}{R}.$$

Alternatively, note that for $z \in \{\gamma_{R,\varepsilon}\}$ where $\gamma_{R,\varepsilon}(\theta) = Re^{i\theta}$, with $-(\pi - \varepsilon) \le \theta \le (\pi - \varepsilon)$ and $\varepsilon > 0$ arbitrary, the function $z^{-2} \log z$ admits the primitive $z^{-1}(-1 - \log z)$. Therefore

$$\int_{\gamma_R} \frac{\log z}{z^2} dz = \lim_{\varepsilon \to 0} \int_{\gamma_{R,\varepsilon}} \frac{\log z}{z^2} dz$$

$$= \lim_{\varepsilon \to 0} \frac{-1 - \log z}{z} \Big|_{Re^{-i(\pi-\varepsilon)}}^{Re^{i(\pi-\varepsilon)}}$$

$$= \frac{-1 - (\ln R + i\pi)}{-R} - \frac{-1 - (\ln R - i\pi)}{-R}$$

$$= \frac{2\pi i}{R}.$$

7.17 The integrand function

$$z \log z = r e^{i\theta} \left(\ln r + i\theta \right), \qquad z = r e^{i\theta}, \quad r > 0, \quad -\pi < \theta < \pi,$$

It is continuous on all γ_R except at the point $z = -R$. It is convenient to parametrize the integration path as $\gamma_R(\theta) = R e^{i\theta}$, $-\pi \leq \theta \leq \pi$. From the definition of integral it follows

$$\int_{\gamma_R} z \log z \, dz = \int_{-\pi}^{+\pi} R e^{i\theta} \left(\ln R + i\theta \right) R e^{i\theta} i \, d\theta$$

$$= iR^2 \ln R \int_{-\pi}^{+\pi} e^{2i\theta} \, d\theta - R^2 \int_{-\pi}^{+\pi} \theta e^{2i\theta} \, d\theta$$

$$= iR^2 \ln R \left. \frac{e^{2i\theta}}{2i} \right|_{-\pi}^{+\pi} - R^2 \left. e^{2i\theta} \left(\frac{1}{4} - \frac{i\theta}{2} \right) \right|_{-\pi}^{+\pi}$$

$$= 0 - R^2 \left(\frac{1}{4} - \frac{i\pi}{2} - \frac{1}{4} - \frac{i\pi}{2} \right)$$

$$= i\pi R^2.$$

Alternatively, note that for $z \in \{\gamma_{R,\varepsilon}\}$ where $\gamma_{R,\varepsilon}(\theta) = R e^{i\theta}$, with $-(\pi - \varepsilon) \leq \theta \leq (\pi - \varepsilon)$ and $\varepsilon > 0$ arbitrary, the function $z \log z$ admits the primitive $z^2(-1 + 2 \log z)/4$. Therefore

$$\int_{\gamma_R} z \log z \, dz = \lim_{\varepsilon \to 0} \int_{\gamma_{R,\varepsilon}} z \log z \, dz$$

$$= \lim_{\varepsilon \to 0} \left. \frac{z^2}{4} (-1 + 2 \log z) \right|_{R e^{-i(\pi - \varepsilon)}}^{R e^{i(\pi - \varepsilon)}}$$

$$= \frac{R^2}{4} (-1 + 2(\ln R + i\pi)) - \frac{R^2}{4} (-1 + 2(\ln R - i\pi))$$

$$= i\pi R^2.$$

7.18 Let $z = r e^{i\theta}$, the principal branch of $F(z) = \tan(z^{1/2})$ is an analytic function for $r > 0$ and $-\pi < \theta < \pi$ except at the isolated singularities at points $z = ((2n+1)\pi/2)^2$. In the analyticity domain, its derivative is

$$\frac{d}{dz} \tan(z^{1/2}) = \frac{1}{2z^{1/2} \cos^2(z^{1/2})}.$$

Let $\gamma_\varepsilon(\varphi) = R e^{i\varphi}$, with $-(\pi - \varepsilon) \leq \varphi \leq \pi - \varepsilon$ and $\varepsilon > 0$ arbitrary, for $z \in \{\gamma_\varepsilon\}$ the function $F(z)$ is a primitive of the integrand function $1/(2\sqrt{z} \cos^2(\sqrt{z}))$. Therefore

$$\int_{\gamma} \frac{1}{2\sqrt{z}\cos^2(\sqrt{z})}dz = \lim_{\varepsilon\to 0}\int_{\gamma_\varepsilon} \frac{1}{2\sqrt{z}\cos^2(\sqrt{z})}dz$$

$$= \lim_{\varepsilon\to 0}\tan(z^{1/2})\Big|_{z=Re^{-i(\pi-\varepsilon)}}^{z=Re^{i(\pi-\varepsilon)}}$$

$$= \lim_{\varepsilon\to 0}\left(\tan(R^{1/2}e^{i(\pi-\varepsilon)/2}) - \tan(R^{1/2}e^{-i(\pi-\varepsilon)/2})\right)$$

$$= \tan(i\sqrt{R}) - \tan(-i\sqrt{R})$$

$$= 2\tan(i\sqrt{R})$$

$$= 2i\tanh(\sqrt{R}).$$

7.19 Note that the principal branch of $z^{1/3} = \exp[(1/3)\log z]$ is an analytic function on the path $\gamma_{R,\varepsilon}(\theta) = Re^{i\theta}$, with $-\pi + \varepsilon \le \theta \le \pi - \varepsilon$ and $\varepsilon > 0$ arbitrary, and here admits as primitive the principal branch of $\frac{3}{4}z^{4/3}$. We then have

$$\int_{\gamma_R} z^{1/3}dz = \lim_{\varepsilon\to 0}\int_{\gamma_{R,\varepsilon}} z^{1/3}dz$$

$$= \lim_{\varepsilon\to 0}\frac{3}{4}z^{4/3}\Big|_{z=Re^{i(-\pi+\varepsilon)}}^{z=Re^{i(\pi-\varepsilon)}}$$

$$= \lim_{\varepsilon\to 0}\frac{3}{4}R^{4/3}\left(e^{i\frac{4}{3}(\pi-\varepsilon)} - e^{i\frac{4}{3}(-\pi+\varepsilon)}\right)$$

$$= \lim_{\varepsilon\to 0}\frac{3}{4}R^{4/3}\, 2i\sin\left(\frac{4}{3}(\pi-\varepsilon)\right)$$

$$= \frac{3}{4}R^{4/3}\, 2i\sin\left(\frac{4\pi}{3}\right)$$

$$= -i\frac{3\sqrt{3}}{4}R^{4/3}.$$

7.20 Let $z = re^{i\theta}$, the integrand function is defined by

$$z^{3/2} = r^{\frac{3}{2}}e^{i\frac{3}{2}\theta}, \qquad r > 0, \qquad -\pi < \theta < \pi.$$

This function is analytic on any arbitrary curve $\gamma_\varepsilon = \gamma - [-L + i\varepsilon, -L - i\varepsilon]$ with $\varepsilon > 0$, and there it admits as primitive the principal branch of $\frac{2}{5}z^{5/2}$. Let $\varphi(\varepsilon) = \arctan(\varepsilon/L)$, we have then

$$\int_{\gamma} z^{3/2}dz = \lim_{\varepsilon\to 0}\int_{\gamma_\varepsilon} z^{3/2}dz$$

$$= \lim_{\varepsilon \to 0} \frac{2}{5} z^{5/2} \Big|_{z=Le^{i(-\pi+\varphi(\varepsilon))}}^{z=Le^{i(\pi-\varphi(\varepsilon))}}$$

$$= \lim_{\varepsilon \to 0} \frac{2}{5} L^{5/2} \left(e^{i\frac{5}{2}(\pi-\varphi(\varepsilon))} - e^{i\frac{5}{2}(-\pi+\varphi(\varepsilon))} \right)$$

$$= \lim_{\varepsilon \to 0} i\frac{4}{5} L^{5/2} \cos\left(\frac{5}{2}\varphi(\varepsilon) \right)$$

$$= i\frac{4}{5} L^{5/2}.$$

7.21 The principal branch of the integrand function

$$(z - z_0)^{q-1} = e^{(q-1)\log(z-z_0)}$$

is an analytic function in $D = \mathbb{C} \setminus \sigma$ where $\sigma = \{z(u) = z_0 - u, \ u \in [0, \infty)\}$ is the branch line of the principal branch of $\log(z - z_0)$ that intersects the curve γ at the point $z_0 - R$. Let us suppose, for simplicity, that this is the only point of intersection between γ and σ. In the case of more points of intersection the arguments are similar. Having parametrized the path $\gamma : [a, b] \mapsto \mathbb{C}$ such that $\gamma(a) = \gamma(b) = z_0 - R$, we define a new path $\gamma_\varepsilon : [a + \varepsilon, b - \varepsilon] \mapsto \mathbb{C}$, with arbitrary $\varepsilon > 0$, such that $\gamma_\varepsilon(t) = \gamma(t) \ \forall t \in [a + \varepsilon, b - \varepsilon]$. Evidently we have

$$\int_\gamma (z - z_0)^{q-1} dz = \lim_{\varepsilon \to 0} \int_{\gamma_\varepsilon} (z - z_0)^{q-1} dz.$$

The trace of γ_ε is contained in D and in D the principal branch of the integrand function admits as primitive the principal branch of $q^{-1}(z - z_0)^q$

$$\frac{d}{dz} \frac{e^{q\log(z-z_0)}}{q} = e^{(q-1)\log(z-z_0)}.$$

Therefore, we have

$$\int_{\gamma_\varepsilon} (z - z_0)^{q-1} dz = \frac{e^{q\log(\gamma_\varepsilon(b-\varepsilon)-z_0)}}{q} - \frac{e^{q\log(\gamma_\varepsilon(a+\varepsilon)-z_0)}}{q}$$

$$= \frac{e^{q\log(\gamma(b-\varepsilon)-z_0)}}{q} - \frac{e^{q\log(\gamma(a+\varepsilon)-z_0)}}{q}.$$

Taking the limit $\varepsilon \to 0$ we conclude

$$\int_\gamma (z - z_0)^{q-1} dz = \frac{e^{q\log(Re^{i\pi})}}{q} - \frac{e^{q\log(Re^{-i\pi})}}{q}$$

$$= \frac{e^{q \ln R}}{q} \left(e^{i\pi q} - e^{-i\pi q} \right)$$

$$= 2i \frac{\sin(\pi q)}{q} R^q .$$

7.22 Let $z = re^{i\theta}$, the principal branch of $\sin(\log z) = \sin(\ln r + i\theta)$ is an analytic function for $r > 0$ and $-\pi < \theta < \pi$, and in this region its derivative is

$$\frac{d}{dz} \sin(\log z) = \frac{\cos(\log z)}{z}.$$

Therefore

$$\int_\gamma \frac{\cos(\log z)}{z} dz = \lim_{\varepsilon \to 0} \sin(\log z)|_{z=-a-i\varepsilon}^{z=-a+i\varepsilon}$$

$$= \sin(\ln a + i\pi) + \sin(\ln a - i\pi)$$

$$= \sin(\ln a) \cos(i\pi) - \cos(\ln a) \sin(i\pi)$$

$$- (\sin(\ln a) \cos(i\pi) - \cos(\ln a) \sin(i\pi))$$

$$= 2i \cos(\ln a) \sinh(\pi).$$

7.23 Let $z = re^{i\theta}$, the principal branch of $-\cos(\log z) = -\cos(\ln r + i\theta)$ is an analytic function for $r > 0$ and $-\pi < \theta < \pi$, and in this region its derivative is

$$\frac{d}{dz}(-\cos(\log z)) = \frac{\sin(\log z)}{z}.$$

Therefore

$$\int_\gamma \frac{\sin(\log z)}{z} dz = \lim_{\varepsilon \to 0} (-\cos(\log z))|_{z=-a-i\varepsilon}^{z=-a+i\varepsilon}$$

$$= -\cos(\ln a + i\pi) + \cos(\ln a - i\pi)$$

$$= -(\cos(\ln a) \cos(i\pi) - \sin(\ln a) \sin(i\pi))$$

$$+ (\cos(\ln a) \cos(i\pi) + \sin(\ln a) \sin(i\pi))$$

$$= 2i \sin(\ln a) \sinh(\pi).$$

7.24 The function

$$f(z) = \frac{1}{z-1} + \frac{1}{z-2}$$

is analytic in A but not in $\overline{A}$. It admits as primitive any of the branches of $F(z) = \log(z-1) + \log(z-2)$ in the domain of analyticity of the branch considered. In no case such analyticity domain does cover the entire annulus A.

7.25

(a) The function $e^z/(1-z)^3$ is analytic everywhere within and on the curve γ. From the Cauchy integral formula we have

$$\int_\gamma \frac{e^z}{z(1-z)^3}\,dz = \int_\gamma \frac{e^z/(1-z)^3}{z-0}\,dz$$

$$= 2\pi i \left.\frac{e^z}{(1-z)^3}\right|_{z=0}$$

$$= 2\pi i.$$

(b) The function e^z/z is analytic everywhere within and on the curve γ. From the Cauchy integral formula for the second derivative we have

$$\int_\gamma \frac{e^z}{z(1-z)^3}\,dz = -\int_\gamma \frac{e^z/z}{(z-1)^3}\,dz$$

$$= -\frac{2\pi i}{2!}\left.\frac{d^2}{dz^2}\left(\frac{e^z}{z}\right)\right|_{z=1}$$

$$= -\pi i \left.\left(\frac{e^z}{z} - 2\frac{e^z}{z^2} + 2\frac{e^z}{z^3}\right)\right|_{z=1}$$

$$= -\pi i e.$$

(c) Consider a line that, by joining two points of the curve γ, identifies two closed curves γ_0 and γ_1, containing in their interior respectively the points $z = 0$ and $z = 1$. By traveling the path γ_0 and γ_1 in the positive direction, as stated in points a) and b), we have

$$\int_\gamma \frac{e^z}{z(1-z)^3}\,dz = \int_{\gamma_0} \frac{e^z}{z(1-z)^3}\,dz + \int_{\gamma_1} \frac{e^z}{z(1-z)^3}\,dz$$

$$= 2\pi i \left(1 - \frac{e}{2}\right).$$

7.26 The polynomial in the denominator of the integrand function can be factorized as

$$z^4 - 1 = (z^2 - 1)(z^2 + 1) = (z-1)(z+1)(z-i)(z+i).$$

Using the Cauchy integral formula we have

$$\int_\gamma \frac{z}{z^4 - 1}\, dz = \int_\gamma \frac{z/((z+1)(z^2+1))}{z-1}\, dz$$

$$= 2\pi i \left.\frac{z}{(z+1)(z^2+1)}\right|_{z=1}$$

$$= \frac{\pi i}{2}.$$

7.27 If $z_0 \in \mathrm{Int}(\gamma)$, then the following Cauchy integral formulas hold

$$\int_\gamma \frac{f(z)}{(z-z_0)^2}\, dz = 2\pi i f'(z_0), \qquad \int_\gamma \frac{f'(z)}{z-z_0}\, dz = 2\pi i f'(z_0),$$

hence the statement. If $z_0 \in \mathrm{Ext}(\gamma)$, then the functions

$$\frac{f(z)}{(z-z_0)^2}, \qquad \frac{f'(z)}{z-z_0}$$

are analytic on and within γ, and by the Cauchy-Goursat theorem

$$\int_\gamma \frac{f(z)}{(z-z_0)^2}\, dz = 0, \qquad \int_\gamma \frac{f'(z)}{z-z_0}\, dz = 0,$$

therefore the equality between the two integrals still holds.

7.28 The principal branch of $\log(z + 2\pi)$ is an analytic function everywhere in the complex plane except at the points

$$z(t) = -2\pi - t, \qquad t \in [0, \infty).$$

Therefore, $\log(z + 2\pi)$ is analytic on the circle γ and within it. Since $i\pi$ is a point within γ, for the Cauchy integral representation we have

$$\int_\gamma \frac{\log(z+2\pi)}{(z-i\pi)^3}\, dz = \frac{2\pi i}{2!}\, \frac{d^2}{dz^2}\, \log(z+2\pi)\bigg|_{z=i\pi}$$

$$= \pi i \left.\frac{-1}{(z+2\pi)^2}\right|_{z=i\pi}$$

$$= -\frac{4+3i}{25\pi}.$$

7.29 The principal branch of

$$f(z) = e^z (3 + z)^{-1/2} = \exp\left(z - \frac{1}{2}\log(3 + z)\right)$$

is an analytic function in the whole complex plane except at the points

$$z(t) = -3 - t, \qquad t \in [0, \infty).$$

Therefore, $f(z)$ is analytic on and within the curve γ. Using the Cauchy integral formula we get

$$\begin{aligned}
\int_\gamma \frac{e^z}{z^2\sqrt{3+z}}\,dz &= \int_\gamma \frac{f(z)}{z^2}\,dz \\
&= \frac{2\pi i}{1!} f'(0) \\
&= 2\pi i \left(1 - \frac{1}{2(3+z)}\right)\exp\left(z - \frac{1}{2}\log(3+z)\right)\bigg|_{z=0} \\
&= \frac{5\pi i}{3\sqrt{3}}.
\end{aligned}$$

7.30 As usual, let us assume for $\sqrt{z}$ the principal branch which has a branch line coinciding with the negative real semi-axis and observe that $|2 + 2i - 3| = \sqrt{5} > 2$, that is, the point $z = 2 + 2i$ is outside γ.

(a) Since the integrand function is analytic on and within γ, it turns out

$$\int_\gamma \frac{f(z)\sqrt{z}}{(z+i)^2}\,dz = 0.$$

(b) Since the integrand function has a singularity on γ, the integral

$$\int_\gamma \frac{f(z)\sqrt{z}}{(z-1)^2}\,dz$$

cannot be computed.

(c) The function $g(z) = f(z)\sqrt{z}$ is analytic on and within γ and $z = 4$ is within γ. Then using the Cauchy integral formula we get

$$\int_\gamma \frac{f(z)\sqrt{z}}{(z-4)^2}\,dz = 2\pi i g'(4) = 2\pi i(f(4)/4 + 2f'(4)).$$

7.31 In the open and connected domain $D = \mathbb{C} \setminus \{0\}$ (not simply connected), the function f has primitive if and only if the integral of f along any piecewise regular closed curve contained in D is zero. Since f is analytic in D, this condition is equivalent to the vanishing of the integral of f along a simple closed curve that contains within it the point $z = 0$. Called γ one of these curves, the condition is therefore

$$\int_\gamma a \frac{\sin z}{z^{p+1}} \, \mathrm{d}z = \int_\gamma b \frac{\cos z}{z^{q+1}} \, \mathrm{d}z.$$

Using the Cauchy integral formula, we obtain

$$ap! \left. \frac{\mathrm{d}^p \sin z}{\mathrm{d}z^p} \right|_{z=0} = bq! \left. \frac{\mathrm{d}^q \cos z}{\mathrm{d}z^q} \right|_{z=0}.$$

For $p = q = 0$ the condition gives $b = 0$, while for $p = q = 1$ we have $a = 0$.

7.32 Let $g(z) = \exp(-\mathrm{i} f'(z))$. The function g is analytic in $\mathbb{C}$ and its module

$$|g(z)| = e^{\mathrm{Re}(-\mathrm{i} f'(z))} = e^{\mathrm{Im}(f'(z))}$$

is a bounded function in $\mathbb{C}$. By Liouville's theorem we can then state that $g(z)$ is constant in $\mathbb{C}$. It follows that $f'(z)$ is also constant in $\mathbb{C}$, say $f'(z) = b$ with $b \in \mathbb{C}$. From this we can deduce that $f(z) = bz + h(z)$ with $h'(z) = 0 \; \forall z \in \mathbb{C}$. On the other hand $h(z) = f(z) - bz$ is also analytic in $\mathbb{C}$ and therefore constant in $\mathbb{C}$, say $h(z) = a$ with $a \in \mathbb{C}$.

7.33 It turns out that f is constant. In fact, from the Cauchy integral formula and from the hypothesis on $|f|$ it follows that for $\forall z \in \mathbb{C}$

$$|f'(z)| = \left| \frac{1}{2\pi \mathrm{i}} \int_{|w-z|=R} \frac{f(w)}{(w-z)^2} \mathrm{d}w \right|$$
$$\leq \frac{1}{2\pi} \int_{|w-z|=R} \frac{|f(w)|}{|w-z|^2} \mathrm{d}w$$
$$\leq \frac{1}{2\pi} \, 2\pi R \, \frac{M + \sqrt{R + |z|}}{R^2},$$

where we used $|w| = |w - z + z| \leq |w - z| + |z|$ and therefore $|f(w)| \leq M + \sqrt{R + |z|}$ for $w \in \{\gamma\}$. Since R can be taken arbitrarily large, f being entire, we conclude that $f'(z) = 0 \; \forall z \in \mathbb{C}$. Since f is analytic with zero first derivative in $\mathbb{C}$ it follows that it is constant in $\mathbb{C}$.

7.34 If $\mathrm{Re}(f) \leq a$, $a \in \mathbb{R}$, let $g(z) = \exp(f(z))$. The function g is entire. It is also bounded in $\mathbb{C}$ since $\forall z \in \mathbb{C}$ we have

$$|g(z)| = e^{\operatorname{Re} f(z)} \le e^a.$$

By Liouville's theorem, $g(z)$ is therefore constant in $\mathbb{C}$. On the other hand, if $\operatorname{Im}(f) \le b$, $b \in \mathbb{R}$, let $g(z) = \exp(-if(z))$. The function g is entire and bounded in $\mathbb{C}$ since $\forall z \in \mathbb{C}$ we have

$$|g(z)| = e^{\operatorname{Im} f(z)} \le e^b.$$

By Liouville's theorem, $g(z)$ is therefore constant in $\mathbb{C}$. In both cases if g is constant then f is also constant.

7.35 Since f is analytic in $B(0, R)$, from the Cauchy inequalities we can write

$$\left| f^{(4)}(0) \right| \le \frac{4!M}{R^4}, \qquad M = \sup_{|z| \le R} |f(z)|.$$

On the other hand, by the properties of f we have $|f^{(4)}(0)| = \sqrt{2}$ and

$$M = \sup_{|z| \le R} |f(z)| \le \sup_{|z| \le R} 5|z|^3 = 5R^3.$$

Combining these results we conclude

$$\sqrt{2} \le \frac{24 \, 5R^3}{R^4},$$

that is, $R \le 60\sqrt{2}$.

12.8   Exercises of Chap. 8

8.1 Note that $f(z)$ is analytic everywhere except the point $z = -b/a$. Therefore, $f(z)$ can be expanded in the Taylor series around $z_0 = 0$ for $|z| < |b/a|$. Since

$$f^{(1)}(z) = -a(az + b)^{-2},$$

$$f^{(2)}(z) = +2a^2(az + b)^{-3},$$

$$f^{(3)}(z) = -6a^3(az + b)^{-4},$$

$$f^{(4)}(z) = +24a^4(az + b)^{-5},$$

$$\vdots$$

$$f^{(n)}(z) = (-1)^n n! a^n (az + b)^{-(n+1)},$$

we have

$$\frac{1}{az+b} = \sum_{n=0}^{\infty} \frac{f^{(n)}(0)}{n!} z^n$$

$$= \sum_{n=0}^{\infty} (-1)^n \frac{a^n n!}{n! b^{n+1}} z^n$$

$$= \sum_{n=0}^{\infty} (-1)^n \frac{a^n}{b^{n+1}} z^n.$$

Alternatively, using the notable result valid for the geometric series, we have

$$\frac{1}{az+b} = \frac{1}{b} \frac{1}{1+az/b} = \frac{1}{b} \sum_{n=0}^{\infty} (-1)^n (az/b)^n = \sum_{n=0}^{\infty} (-1)^n \frac{a^n}{b^{n+1}} z^n,$$

which converges for $|az/b| < 1$, that is for $|z| < |b|/|a|$. In both cases, the radius of convergence of the found power series can also be calculated directly as

$$R = \lim_{n\to\infty} \frac{\left|a^n/b^{n+1}\right|}{\left|a^{n+1}/b^{n+2}\right|} = \frac{|b|}{|a|}.$$

8.2 Setting $g(z) = \exp(z^2)$ and changing $z \to z^2$ in the Taylor series expansion of $\exp(z)$, for $|z| < \infty$ we have

$$g(z) = \sum_{n=0}^{\infty} \frac{z^{2n}}{n!}.$$

From the theorem on the integration of power series we obtain

$$f(z) = \int_0^z g(w)\mathrm{d}w$$

$$= \sum_{n=0}^{\infty} \frac{1}{n!} \int_0^z w^{2n}\mathrm{d}w$$

$$= \sum_{n=0}^{\infty} \frac{1}{n!} \frac{w^{2n+1}}{2n+1} \Big|_0^z$$

$$= \sum_{n=0}^{\infty} \frac{z^{2n+1}}{(2n+1)n!}.$$

The radius of convergence of the obtained series is infinite as that of the series representing g.

8.3   The principal branch of $\log z$ is a function analytic everywhere except on the negative real semi-axis including the origin. Therefore, the function under consideration is analytic everywhere except at the point $z = 1$ and the points z that satisfy

$$\frac{1+z}{1-z} = -t, \qquad t \in [0, \infty),$$

that is

$$z(t) = \frac{1+t}{t-1}, \qquad t \in [0, \infty).$$

The set of these points represents the real semi-axes $(-\infty, -1]$ and $[1, +\infty)$. To determine the Taylor expansion around $z = 0$, such expansion exists within the circle centered at the origin and of radius 1, observe that

$$\frac{d}{dz} \log \frac{1+z}{1-z} = \frac{2}{1-z^2} = 2 \sum_{n=0}^{\infty} z^{2n}.$$

Therefore

$$\log \frac{1+z}{1-z} = \log \frac{1+z}{1-z} - \log \frac{1+0}{1-0}$$

$$= \int_0^z \frac{d}{dw} \left(\log \frac{1+w}{1-w} \right) dw$$

$$= \int_0^z 2 \sum_{n=0}^{\infty} w^{2n} \, dw$$

$$= 2 \sum_{n=0}^{\infty} \frac{z^{2n+1}}{2n+1}.$$

8.4   The principal branch of $\log z$

$$\log z = \ln r + i\theta, \qquad z = re^{i\theta}, \qquad r > 0, \qquad -\pi < \theta < \pi,$$

is a function analytic everywhere in $\mathbb{C}$ except on the negative real semi-axis including the origin. It is evident that the maximum circle of center $z_0 = 3$ in which $\log z$ is analytic has radius 3. Let z be any point such that $|z - 3| < 3$, we have

$$\log z - \log 3 = \int_3^z \frac{ds}{s}$$

$$= \int_3^z \frac{ds}{3\left(1 + \frac{s-3}{3}\right)}$$

$$= \int_3^z \frac{ds}{3} \sum_{k=0}^{\infty} (-1)^k \left(\frac{s-3}{3}\right)^k$$

$$= \sum_{k=0}^{\infty} \frac{(-1)^k}{3^{k+1}} \int_3^z ds\,(s-3)^k$$

$$= \sum_{k=0}^{\infty} \frac{(-1)^k}{(k+1)3^{k+1}} (z-3)^{k+1}$$

$$= \sum_{n=1}^{\infty} \frac{(-1)^{n-1}}{n3^n} (z-3)^n.$$

In conclusion

$$\log z = \ln 3 + \sum_{n=1}^{\infty} \frac{(-1)^{n-1}}{n3^n} (z-3)^n$$

$$= \ln 3 + \frac{z-3}{3} - \frac{(z-3)^2}{18} + \frac{(z-3)^3}{81} + \ldots.$$

A direct check confirms that the radius of convergence of the series is $R = 3$. In fact, if a_n is the n-th coefficient of the series, we have

$$\frac{|a_n|}{|a_{n+1}|} = \frac{(n+1)3^{n+1}}{n3^n} = 3\left(1 + \frac{1}{n}\right) \xrightarrow{n \to \infty} 3.$$

8.5 The principal branch of $\log z$ is a function analytic everywhere in $\mathbb{C}$ except on the negative real semi-axis including the origin. Therefore, the function under consideration is analytic everywhere except at the points $z \in \mathbb{C}$ that satisfy

$$1 + z^2 = -t, \qquad t \geq 0,$$

that is

$$z(t) = \pm i\sqrt{1+t}, \qquad t \geq 0.$$

The Taylor series expansion around $z = 0$ exists within the maximum circle of analyticity centered at the origin, which has radius 1. Using the notable expansion,

$$\log(1+z) = \sum_{n=1}^{\infty} \frac{(-1)^{n+1}}{n} z^n,$$

valid for $|z| < 1$, we have

$$(z+1)\log\left(1+z^2\right) = (z+1)\sum_{n=1}^{\infty}\frac{(-1)^{n+1}}{n}z^{2n}$$

$$= \sum_{n=1}^{\infty}\frac{(-1)^{n+1}}{n}z^{2n+1} + \sum_{n=1}^{\infty}\frac{(-1)^{n+1}}{n}z^{2n}$$

$$= z^2 + z^3 - \frac{z^4}{2} - \frac{z^5}{2} + \frac{z^6}{3} + \frac{z^7}{3} - \frac{z^8}{4} - \frac{z^9}{4} + \dots$$

$$= \sum_{k=2}^{\infty} a_k z^k,$$

where

$$a_k = \begin{cases} (2(-1)^{k/2+1})/k, & k \text{ even} \\ (2(-1)^{(k-1)/2+1})/(k-1), & k \text{ odd} \end{cases}.$$

As already observed, the radius of convergence of this series is

$$R = \lim_{k\to\infty}\left|\frac{a_k}{a_{k+1}}\right| = 1.$$

8.6 The principal branch of $\log z$

$$\log z = \ln r + i\theta, \qquad z = re^{i\theta}, \quad r > 0, \quad -\pi < \theta < \pi,$$

is analytic everywhere in $\mathbb{C}$ except on the negative real semi-axis including the origin. Therefore, the function $\log(z^2)$ is analytic everywhere in $\mathbb{C}$ except on the entire imaginary axis. The maximum circle of center $z_0 = -1$ within which $\log(z^2)$ is analytic has radius 1. Let z be any point such that $|z + 1| < 1$, observing that $d\log(z^2)/dz = 2z^{-1}$ we have

$$\log(z^2) - \log((-1)^2) = \int_{-1}^{z}\frac{2}{w}\,dw$$

$$= -2\int_{-1}^{z}\frac{1}{1-(w+1)}\,dw$$

$$= -2\int_{-1}^{z}\sum_{k=0}^{\infty}(w+1)^k\,dw$$

$$= -2\sum_{k=0}^{\infty}\int_{-1}^{z}(w+1)^k\,dw$$

$$= -2 \sum_{k=0}^{\infty} \frac{(z+1)^{k+1}}{k+1}$$

$$= - \sum_{n=1}^{\infty} \frac{2}{n}(z+1)^n.$$

Since $\log((-1)^2) = 0$, we conclude

$$\log(z^2) = - \sum_{n=1}^{\infty} \frac{2}{n}(z+1)^n$$

$$= -2(z+1) - (z+1)^2 - \frac{2}{3}(z+1)^3 - \frac{2}{4}(z+1)^4 + \dots.$$

A direct check confirms that the radius of convergence of this power series is $R = 1$. In fact, calling a_n the n-th coefficient of the series, we have

$$\left| \frac{a_n}{a_{n+1}} \right| = \left| \frac{2}{n} \frac{n+1}{2} \right| = \frac{n+1}{n} \xrightarrow{n \to \infty} 1.$$

8.7 Using the notable expansion

$$\mathrm{e}^z = \sum_{n=0}^{\infty} \frac{z^n}{n!},$$

we have

$$\mathrm{e}^z \sin z = \mathrm{e}^z \frac{\mathrm{e}^{\mathrm{i}z} - \mathrm{e}^{-\mathrm{i}z}}{2\mathrm{i}}$$

$$= \frac{1}{2\mathrm{i}} \left(\mathrm{e}^{(1+\mathrm{i})z} - \mathrm{e}^{(1-\mathrm{i})z} \right)$$

$$= \frac{1}{2\mathrm{i}} \left(\sum_{n=0}^{\infty} \frac{(1+\mathrm{i})^n}{n!} z^n - \sum_{n=0}^{\infty} \frac{(1-\mathrm{i})^n}{n!} z^n \right)$$

$$= \sum_{n=0}^{\infty} \frac{\left(\sqrt{2}\mathrm{e}^{\mathrm{i}\pi/4} \right)^n - \left(\sqrt{2}\mathrm{e}^{-\mathrm{i}\pi/4} \right)^n}{2\mathrm{i}n!} z^n$$

$$= \sum_{n=0}^{\infty} \frac{2^{n/2} \sin(n\pi/4)}{n!} z^n$$

$$= z + z^2 + \frac{z^3}{3} - \frac{z^5}{30} - \frac{z^6}{90} - \frac{z^7}{630} + \dots.$$

8.8 The principal branch of $\log(z^2)$ is an analytic function at all points of the complex plane except those for which $z^2 = -t$ with $t \in [0, +\infty)$, namely, the the points $z(t) = \pm i\sqrt{t}$ with $t \in [0, +\infty)$. Observing that

$$\frac{\mathrm{d}}{\mathrm{d}z}\log(z^2) = \frac{2}{z}$$

$$= 2\,\frac{1}{1+(z-1)}$$

$$= 2\sum_{k=0}^{\infty}(-1)^k(z-1)^k,$$

which holds for $|z-1| < 1$, we have

$$\log(z^2) - \log(1^2) = \int_1^z \frac{\mathrm{d}\log(w^2)}{\mathrm{d}w}\,\mathrm{d}w$$

$$= 2\sum_{k=0}^{\infty}(-1)^k \int_1^z (w-1)^k\,\mathrm{d}w$$

$$= 2\sum_{k=0}^{\infty}(-1)^k\,\frac{(z-1)^{k+1}}{k+1}$$

$$= 2\sum_{n=1}^{\infty}\frac{(-1)^{n-1}}{n}(z-1)^n.$$

Therefore, we conclude that $\forall z \in B(1,1)$

$$\log(z^2) = 2\sum_{n=1}^{\infty}\frac{(-1)^{n-1}}{n}(z-1)^n.$$

Alternatively, note that in the open and connected region $\operatorname{Re} z > 0$ the principal branches of the functions $\log z$ and $\log(z^2)$ are both analytic functions. Furthermore, we have $\log(z^2) = 2\log(z)$ when, for example, $z \in [1, 2]$. Therefore, by the Identity Theorem 9.16 in the whole region $\operatorname{Re} z > 0$ it must be

$$\log(z^2) = 2\log(z).$$

Using the notable expansion

$$\log z = \sum_{n=1}^{\infty}\frac{(-1)^{n-1}}{n}(z-1)^n, \qquad |z-1| < 1,$$

we obtain again

$$\log(z^2) = 2 \sum_{n=1}^{\infty} \frac{(-1)^{n-1}}{n} (z-1)^n, \qquad |z-1| < 1.$$

8.9 Using the notable expansion

$$e^z = \sum_{n=0}^{\infty} \frac{z^n}{n!},$$

we have

$$e^{iz} \sinh z = e^{iz} \frac{e^z - e^{-z}}{2}$$

$$= \frac{1}{2} \left(e^{(1+i)z} - e^{(-1+i)z} \right)$$

$$= \frac{1}{2} \left(\sum_{n=0}^{\infty} \frac{(1+i)^n}{n!} z^n - \sum_{n=0}^{\infty} \frac{(-1+i)^n}{n!} z^n \right)$$

$$= \sum_{n=0}^{\infty} \frac{\left(\sqrt{2} e^{i\pi/4} \right)^n - \left(\sqrt{2} e^{i3\pi/4} \right)^n}{2n!} z^n$$

$$= \sum_{n=0}^{\infty} \frac{2^{n/2-1} \left[\cos(n\pi/4) + i \sin(n\pi/4) \right] (1 - i^n)}{n!} z^n$$

$$= z + iz^2 - \frac{z^3}{3} - \frac{z^5}{30} - \frac{iz^6}{90} + \frac{z^7}{630} + \dots.$$

8.10 Observing that

$$\frac{d}{dz} \arctan z = \frac{1}{1+z^2} = \sum_{n=0}^{\infty} \left(-z^2 \right)^n, \qquad |z| < 1,$$

integrating term by term the above expression we obtain

$$\arctan z = \arctan z - \arctan 0 = \int_0^z \sum_{n=0}^{\infty} (-1)^n w^{2n} \, dw = \sum_{n=0}^{\infty} \frac{(-1)^n}{2n+1} z^{2n+1}, \qquad |z| < 1.$$

The radius of convergence of this series is $R = 1$. The same result is obtained by the Hadamard formula

$$\frac{1}{R} = \limsup_{n \to \infty} |a_n|^{1/n} \, ,$$

where

$$a_n = \begin{cases} (-1)^{(n-1)/2}/n & n \text{ odd} \\ 0 & n \text{ even} \end{cases}$$

is the coefficient of the series rewritten in the form $\sum_{n=0}^{\infty} a_n z^n$.

8.11 The function $f(z)$ is analytic in $\mathbb{C}$ except at the zeros of $z^3 + 1 = 0$, namely, the points

$$w_k = \sqrt[3]{-1} = \left(1 e^{i\pi} \right)^{1/3} = e^{i(\pi + 2k\pi)/3}, \qquad k = 0, 1, 2.$$

Within the greatest circle centered at z_0 where $f(z)$ is analytic, this function has a Taylor series expansion around the point $z_0 = 2$ which coincides with the power series $\sum_{n=0}^{\infty} a_n (z - z_0)^n$. Therefore, the radius of convergence R of the given series is

$$R = \min_{k=0,1,3} |z_0 - w_k|$$

$$= \left| 2 - e^{i\pi/3} \right|$$

$$= \sqrt{5 - 4 \cos \frac{\pi}{3}}$$

$$= \sqrt{3}.$$

8.12 We start writing

$$f(z) = \frac{\cos z - 1}{z^2 \sin z}.$$

Using the Taylor series expansions of $\cos z$ and $\sin z$ around $z = 0$ and the behavior of the geometric series, we have

$$f(z) = \frac{-\frac{z^2}{2} + \frac{z^4}{24} + \cdots}{z^2 \left(z - \frac{z^3}{6} + \cdots \right)}$$

$$= \frac{1}{z} \left(-\frac{1}{2} + \frac{z^2}{24} + \cdots \right) \left(1 - \frac{z^2}{6} + \cdots \right)^{-1}$$

$$= \frac{1}{z}\left(-\frac{1}{2} + \frac{z^2}{24} + \cdots\right)\left[1 + \left(\frac{z^2}{6} - \cdots\right) + \left(\frac{z^2}{6} - \cdots\right)^2 + \cdots\right]$$

$$= \frac{1}{z}\left(-\frac{1}{2} - \frac{z^2}{24} + O(z^4)\right)$$

$$= -\frac{1}{2z} - \frac{z}{24} + O(z^3).$$

Therefore, f has a simple pole at $z = 0$ with

$$\operatorname*{Res}_{z=0} f(z) = -\frac{1}{2}.$$

8.13 The function $f(z)$ is analytic for $z \neq 0$ and $z \notin [1/3, 1/2]$. Therefore, in the annulus $A(0, 0, 1/3)$ admits the Laurent series expansion

$$f(z) = \int_2^3 \frac{t^2 \sin t}{t - 1/z}\, dt$$

$$= \int_2^3 \frac{t^2 \sin t}{-1/z(1 - (tz))}\, dt$$

$$= \int_2^3 \frac{t^2 \sin t}{-1/z}\left(\sum_{k=0}^{\infty}(tz)^k\right) dt$$

$$= \sum_{k=0}^{\infty} z^{k+1}\left(-\int_2^3 t^{k+2} \sin t\, dt\right)$$

$$= \sum_{n=0}^{\infty} z^n\left(-\int_2^3 t^{n+1} \sin t\, dt\right).$$

We thus have

$$\operatorname*{Res}_{z=0} f(z) = 0.$$

Observe that f has a removable singularity at $z = 0$.

8.14 Using the notable Taylor expansions of $\cos z$, $\exp z$ and $\sin z$ around $z = 0$, for $|z| < \infty$ we can write

$$\exp(\cos z) = \exp\left(1 - \frac{z^2}{2} + \frac{z^4}{24} + \cdots\right)$$

$$= \exp(1)\exp\left(-\frac{z^2}{2} + \frac{z^4}{24} + \cdots\right)$$

$$= e\left[1 + \left(-\frac{z^2}{2} + \frac{z^4}{24} + \dots\right) + \frac{1}{2}\left(-\frac{z^2}{2} + \frac{z^4}{24} + \dots\right)^2 + \dots\right]$$

$$= e\left(1 - \frac{z^2}{2} + \frac{z^4}{6} + \dots\right),$$

and for $0 < |z| < \infty$

$$\frac{1}{\sin z} = \frac{1}{z\left(1 - \frac{z^2}{6} + \dots\right)} = \frac{1}{z} + \frac{z}{6} + \dots.$$

In conclusion, in the annulus $0 < |z| < \infty$ for $f(z)$ the following Laurent series expansion holds

$$f(z) = \left[e\left(1 - \frac{z^2}{2} + \frac{z^4}{6} + \dots\right)\left(\frac{1}{z} + \frac{z}{6} + \dots\right)\right]^3$$

$$= e^3\left(\frac{1}{z} - \frac{z}{3} + \dots\right)^3$$

$$= e^3\left(\frac{1}{z^3} - \frac{1}{z} + \dots\right).$$

The function $f(z)$ has at $z = 0$ a pole of order 3 with

$$\operatorname*{Res}_{z=0} f(z) = -e^3.$$

From the same expansion of f it immediately follows

$$\operatorname*{Res}_{z=0} zf(z) = 0, \qquad \operatorname*{Res}_{z=0} z^2 f(z) = e^3.$$

8.15 Since $(z^2 + 1)^2 = (z - i)^2(z + i)^2$, the function $f(z)$ is analytic everywhere except at the points $z = \pm i$. In the region $0 < |z - i| < 2$ for $f(z)$ we have the Laurent series expansion

$$\frac{1}{(z^2 + 1)^2} = \sum_{n=-\infty}^{\infty} c_n(z - i)^n,$$

with

$$c_n = \frac{1}{2\pi i}\int_\gamma \frac{(z^2 + 1)^{-2}}{(z - i)^{n+1}}dz,$$

where γ is an arbitrary piecewise regular simple closed, positively oriented, integration path that contains $z_0 = i$ within it and is contained in $A(i, 0, 2)$. Let $g(z) = (z + i)^{-2}$, analytic for $z \neq -i$, by the Cauchy-Goursat theorem and the Cauchy integral formula, we have

$$c_n = \frac{1}{2\pi i} \int_\gamma \frac{g(z)}{(z - i)^{n+3}} \, dz = \begin{cases} 0 & n \leq -3 \\ g^{(n+2)}(i)/(n + 2)! & n \geq -2 \end{cases}.$$

Observing that

$$g^{(1)}(z) = -2(z + i)^{-3},$$

$$g^{(2)}(z) = +6(z + i)^{-4},$$

$$g^{(3)}(z) = -24(z + i)^{-5},$$

$$\vdots$$

$$g^{(n)}(z) = (-1)^n (n + 1)!(z + i)^{-(n+2)},$$

we find

$$\frac{1}{(z^2 + 1)^2} = \sum_{n=-2}^{\infty} \frac{(-1)^{n+2}(n + 3)!(2i)^{-(n+4)}}{(n + 2)!} (z - i)^n$$

$$= \sum_{n=-2}^{\infty} \frac{i^n (n + 3)}{2^{(n+4)}} (z - i)^n$$

$$= -\frac{1}{4}(z - i)^{-2} - \frac{i}{4}(z - i)^{-1} + \sum_{n=0}^{\infty} \frac{i^n (n + 3)}{2^{(n+4)}} (z - i)^n.$$

It is possible to expand $f(z)$ in the Laurent series also in region $2 < |z - i| < \infty$. Instead of following a procedure similar to the previous one, note that

$$\frac{1}{(z^2 + 1)^2} = \frac{1}{(z - i)^2 (z - i + 2i)^2}$$

$$= \frac{1}{(z - i)^4} \frac{1}{\left[1 + \left(\frac{2i}{z-i} \right) \right]^2}$$

$$= \frac{-1}{(z - i)^4} \frac{d}{dw} \frac{1}{1 + w},$$

where $w = 2i/(z - i)$. For $|w| < 1$, that is, $|z - i| > 2$, we have

$$\frac{1}{(z^2 + 1)^2} = \frac{-1}{(z - i)^4} \frac{d}{dw} \sum_{n=0}^{\infty} (-1)^n w^n$$

$$= \frac{-1}{(z - i)^4} \sum_{n=1}^{\infty} (-1)^n n \left(\frac{2i}{z - i} \right)^{n-1}$$

$$= \sum_{n=1}^{\infty} (-1)^{n+1} n \frac{(2i)^{n-1}}{(z - i)^{n+3}}$$

$$= \sum_{k=0}^{\infty} (-1)^k (k + 1)(2i)^k (z - i)^{-(k+4)}.$$

8.16 For $0 < |z| < 1$ we can write

$$f(z) = \frac{1}{z^2} \frac{d}{dz} \frac{1}{1 - z}$$

$$= \frac{1}{z^2} \frac{d}{dz} \sum_{k=0}^{\infty} z^k$$

$$= \sum_{k=1}^{\infty} k z^{k-3}$$

$$= \sum_{n=-2}^{\infty} (n + 3) z^n$$

$$= \frac{1}{z^2} + \frac{2}{z} + 3 + 4z + 5z^2 + \dots.$$

For $1 < |z| < \infty$, setting $w = z^{-1}$ and noting that $|w| < 1$, we have

$$f(z) = \frac{w^4}{(1 - w)^2}$$

$$= w^4 \frac{d}{dw} \frac{1}{1 - w}$$

$$= w^4 \frac{d}{dw} \sum_{k=0}^{\infty} w^k$$

$$= \sum_{k=1}^{\infty} k w^{k+3}$$

$$= \sum_{n=4}^{\infty}(n-3)z^{-n}$$

$$= \frac{1}{z^4} + \frac{2}{z^5} + \frac{3}{z^6} + \frac{4}{z^7} + \frac{5}{z^8} + \dots$$

8.17 For $0 < |z| < 2$ we can write the function $f(z)$ in the form

$$f(z) = \frac{1}{2z^2}\,\frac{1}{1+z/2}.$$

Setting $w = z/2$ and using the notable expansion

$$\frac{1}{1+w} = \sum_{n=0}^{\infty}(-1)^n w^n, \qquad |w| < 1,$$

we have

$$\frac{1}{z^3 + 2z^2} = \frac{1}{2z^2}\sum_{n=0}^{\infty}(-1)^n\left(\frac{z}{2}\right)^n$$

$$= \sum_{n=0}^{\infty}\frac{(-1)^n}{2^{n+1}}z^{n-2}$$

$$= \frac{1}{2}\frac{1}{z^2} - \frac{1}{2^2}\frac{1}{z} + \frac{1}{2^3} - \frac{1}{2^4}z + \frac{1}{2^5}z^2 + \dots.$$

For $2 < |z| < \infty$ we can instead write

$$f(z) = \frac{1}{z^3}\,\frac{1}{1+2/z}.$$

Setting $w = 2/z$, for $z \in A(0,2,\infty)$ we have $|w| < 1$ and

$$\frac{1}{z^3 + 2z^2} = \frac{1}{z^3}\sum_{n=0}^{\infty}(-1)^n\left(\frac{2}{z}\right)^n$$

$$= \sum_{n=0}^{\infty}(-1)^n 2^n z^{-n-3}$$

$$= \frac{1}{z^3} - \frac{2}{z^4} + \frac{2^2}{z^5} - \frac{2^3}{z^6} + \frac{2^4}{z^7} - \frac{2^5}{z^8} + \dots.$$

8.18 For $0 < |z| < \sqrt{2}$ we can write the function $f(z)$ in the form

$$f(z) = \frac{1}{2z}\frac{1}{1 + z^2/2}.$$

Setting $w = z^2/2$ and using the notable expansion

$$\frac{1}{1 + w} = \sum_{n=0}^{\infty}(-1)^n w^n, \qquad |w| < 1,$$

we have

$$\frac{1}{z^3 + 2z} = \frac{1}{2z}\sum_{n=0}^{\infty}(-1)^n \left(\frac{z^2}{2}\right)^n$$

$$= \sum_{n=0}^{\infty}\frac{(-1)^n}{(2)^{n+1}}z^{2n-1}$$

$$= \frac{1}{2z} - \frac{z}{4} + \frac{z^3}{8} - \frac{z^5}{16} + \frac{z^7}{32} + \dots.$$

For $\sqrt{2} < |z| < \infty$ we can instead write

$$f(z) = \frac{1}{z^3}\frac{1}{1 + 2/z^2}.$$

Setting $w = 2/z^2$, for $z \in A(0, \sqrt{2}, \infty)$ we have $|w| < 1$ and

$$\frac{1}{z^3 + 2z} = \frac{1}{z^3}\sum_{n=0}^{\infty}(-1)^n \left(\frac{2}{z^2}\right)^n$$

$$= \sum_{n=0}^{\infty}(-1)^n 2^n z^{-2n-3}$$

$$= \frac{1}{z^3} - \frac{2}{z^5} + \frac{4}{z^7} - \frac{8}{z^9} + \frac{16}{z^{11}} - \frac{32}{z^{13}} + \dots.$$

8.19 For $0 < |z - i| < 2$ we can write the function $f(z)$ in the form

$$f(z) = \frac{1}{(z - i)(z + i)} = \frac{1}{(z - i)(z - i + 2i)} = \frac{1}{2i(z - i)}\frac{1}{1 + (z - i)/(2i)}$$

Setting $w = (z - \mathrm{i})/(2\mathrm{i})$ and using the notable expansion

$$\frac{1}{1+w} = \sum_{n=0}^{\infty}(-1)^n w^n, \qquad |w| < 1,$$

we have

$$\frac{1}{z^2+1} = \frac{1}{2\mathrm{i}(z-\mathrm{i})} \sum_{n=0}^{\infty}(-1)^n \left(\frac{z-\mathrm{i}}{2\mathrm{i}}\right)^n$$

$$= \sum_{n=0}^{\infty} \frac{(-1)^n}{(2\mathrm{i})^{n+1}} (z-\mathrm{i})^{n-1}$$

$$= -\frac{\mathrm{i}}{2}\frac{1}{z-\mathrm{i}} + \frac{1}{2^2} + \frac{\mathrm{i}}{2^3}(z-\mathrm{i}) - \frac{1}{2^4}(z-\mathrm{i})^2 - \frac{\mathrm{i}}{2^5}(z-\mathrm{i})^3 + \ldots.$$

For $2 < |z - \mathrm{i}| < \infty$ we can instead write

$$f(z) = \frac{1}{(z-\mathrm{i})(z+\mathrm{i})} = \frac{1}{(z-\mathrm{i})(z-\mathrm{i}+2\mathrm{i})} = \frac{1}{(z-\mathrm{i})^2}\frac{1}{1+2\mathrm{i}/(z-\mathrm{i})}.$$

Setting $w = 2\mathrm{i}/(z-\mathrm{i})$, for $z \in A(\mathrm{i}, 2, \infty)$ we have $|w| < 1$ and

$$\frac{1}{z^2+1} = \frac{1}{(z-\mathrm{i})^2} \sum_{n=0}^{\infty}(-1)^n \left(\frac{2\mathrm{i}}{z-\mathrm{i}}\right)^n$$

$$= \sum_{n=0}^{\infty}(-1)^n (2\mathrm{i})^n (z-\mathrm{i})^{-n-2}$$

$$= \frac{1}{(z-\mathrm{i})^2} - \frac{2\mathrm{i}}{(z-\mathrm{i})^3} - \frac{2^2}{(z-\mathrm{i})^4} + \frac{2^3\mathrm{i}}{(z-\mathrm{i})^5} + \frac{2^4}{(z-\mathrm{i})^6} + \ldots.$$

8.20 For $0 < |z| < 1$ we can write

$$f(z) = \frac{1}{z}\frac{1}{1+z^2},$$

and using notable expansion

$$\frac{1}{1+w} = \sum_{n=0}^{\infty}(-1)^n w^n, \qquad |w| < 1,$$

we have

$$\frac{1}{z+z^3} = \frac{1}{z}\sum_{n=0}^{\infty}(-1)^n\left(z^2\right)^n$$

$$= \sum_{n=0}^{\infty}(-1)^n z^{2n-1}$$

$$= \frac{1}{z} - z + z^3 - z^5 + \dots.$$

8.21 For $0 < |z| < 2$ we can write the function $f(z)$ in the form

$$f(z) = \frac{3+z}{2z^2}\,\frac{1}{1+z/2}.$$

Setting $w = z/2$ and using the notable expansion

$$\frac{1}{1+w} = \sum_{n=0}^{\infty}(-1)^n w^n, \qquad |w| < 1,$$

we have

$$\frac{3+z}{z^3+2z^2} = \left(\frac{3}{2z^2}+\frac{1}{2z}\right)\sum_{n=0}^{\infty}(-1)^n\left(\frac{z}{2}\right)^n$$

$$= \sum_{n=0}^{\infty}\frac{(-1)^n 3}{2^{n+1}}z^{n-2} + \sum_{n=0}^{\infty}\frac{(-1)^n}{2^{n+1}}z^{n-1}$$

$$= \sum_{n=0}^{\infty}\frac{(-1)^n 3}{2^{n+1}}z^{n-2} + \sum_{n=1}^{\infty}\frac{(-1)^{n-1}}{2^n}z^{n-2}$$

$$= \frac{3}{2z^2} + \sum_{n=1}^{\infty}\frac{(-1)^n}{2^{n+1}}z^{n-2}$$

$$= \frac{3}{2z^2} - \frac{1}{4z} + \frac{1}{8} - \frac{z}{16} + \frac{z^2}{32} + \dots.$$

For $2 < |z| < \infty$ we can instead write

$$f(z) = \frac{3+z}{z^3}\,\frac{1}{1+2/z}.$$

Setting $w = 2/z$, for $z \in A(0, 2, \infty)$ we have $|w| < 1$ and

$$\frac{3+z}{z^3 + 2z^2} = \left(\frac{3}{z^3} + \frac{1}{z^2}\right) \sum_{n=0}^{\infty} (-1)^n \left(\frac{2}{z}\right)^n$$

$$= \sum_{n=0}^{\infty} 3(-1)^n 2^n z^{-n-3} + \sum_{n=0}^{\infty} (-1)^n 2^n z^{-n-2}$$

$$= \sum_{n=0}^{\infty} 3(-1)^n 2^n z^{-n-3} + \sum_{n=-1}^{\infty} (-1)^{n+1} 2^{n+1} z^{-n-3}$$

$$= \frac{1}{z^2} + \sum_{n=0}^{\infty} (-1)^n 2^n z^{-n-3}$$

$$= \frac{1}{z^2} + \frac{1}{z^3} - \frac{2}{z^4} + \frac{4}{z^5} - \frac{8}{z^6} + \frac{16}{z^7} - \frac{32}{z^8} + \dots.$$

8.22 Using the notable series expansion

$$\frac{1}{1-w} = \sum_{k=0}^{\infty} w^k, \qquad |w| < 1,$$

for $0 < |z - 1| < 1$ we have

$$\frac{1}{(z-1)^2(z-2)^2} = \frac{1}{(z-1)^2} \frac{d}{dz} \frac{-1}{z-2}$$

$$= \frac{1}{(z-1)^2} \frac{d}{dz} \frac{1}{1-(z-1)}$$

$$= \frac{1}{(z-1)^2} \frac{d}{dz} \sum_{k=0}^{\infty} (z-1)^k$$

$$= \frac{1}{(z-1)^2} \sum_{k=1}^{\infty} k(z-1)^{k-1}$$

$$= \sum_{k=1}^{\infty} k(z-1)^{k-3}$$

$$= \sum_{n=-2}^{\infty} (n+3)(z-1)^n$$

$$= \frac{1}{(z-1)^2} + \frac{2}{z-1} + 3 + 4(z-1) + 5(z-1)^2 + \dots.$$

8.23 In the region $0 < |z| < 1$, since $\left|z^2\right| < 1$, we have

$$
\begin{aligned}
\frac{2z}{z^2(1+z^2)^2} &= -\frac{1}{z^2}\frac{d}{dz}\frac{1}{1+z^2} \\
&= -\frac{1}{z^2}\frac{d}{dz}\sum_{k=0}^{\infty}(-1)^k z^{2k} \\
&= \sum_{k=1}^{\infty}(-1)^{k-1}2k z^{2k-3} \\
&= \sum_{n=0}^{\infty}(-1)^n 2(n+1)z^{2n-1} \\
&= \frac{2}{z} - 4z + 6z^3 - 8z^5 + O(z^7).
\end{aligned}
$$

In the region $1 < |z| < \infty$, since $\left|1/z^2\right| < 1$, we have

$$
\begin{aligned}
\frac{2z}{z^2(1+z^2)^2} &= -\frac{1}{z^2}\frac{d}{dz}\frac{1}{z^2}\frac{1}{1+\frac{1}{z^2}} \\
&= -\frac{1}{z^2}\frac{d}{dz}\frac{1}{z^2}\sum_{k=0}^{\infty}(-1)^k z^{-2k} \\
&= \sum_{k=0}^{\infty}(-1)^k(2k+2)z^{-2k-5} \\
&= \frac{2}{z^5} - \frac{4}{z^7} + \frac{6}{z^9} - \frac{8}{z^{11}} + O(z^{-13}).
\end{aligned}
$$

8.24 For $0 < |z-i| < 1$ we can write

$$
\begin{aligned}
f(z) &= \frac{1}{z(z-i)(z+i)} \\
&= \frac{1}{z-i}\frac{1}{i}\frac{1}{1+\frac{z-i}{i}}\frac{1}{2i}\frac{1}{1+\frac{z-i}{2i}} \\
&= \frac{1}{z-i}\frac{1}{2}\left(\frac{1}{1+\frac{z-i}{2i}} - \frac{2}{1+\frac{z-i}{i}}\right)
\end{aligned}
$$

and then

$$\frac{1}{z+z^3} = \frac{1}{2}\frac{1}{z-\mathrm{i}}\left(\sum_{n=0}^{\infty}(-1)^n\frac{(z-\mathrm{i})^n}{2^n\mathrm{i}^n} - 2\sum_{n=0}^{\infty}(-1)^n\frac{(z-\mathrm{i})^n}{\mathrm{i}^n}\right)$$

$$= \sum_{n=0}^{\infty}\mathrm{i}^n\left(2^{-(n+1)}-1\right)(z-\mathrm{i})^{n-1}$$

$$= -\frac{1}{2(z-\mathrm{i})} - \frac{3}{4}\mathrm{i} + \frac{7}{8}(z-\mathrm{i}) + \frac{15}{16}\mathrm{i}(z-\mathrm{i})^2 + \dots.$$

For $2 < |z-\mathrm{i}| < \infty$ we can instead write

$$f(z) = \frac{1}{z(z-\mathrm{i})(z+\mathrm{i})}$$

$$= \frac{1}{(z-\mathrm{i})^3}\,\frac{1}{1+\frac{\mathrm{i}}{z-\mathrm{i}}}\,\frac{1}{1+\frac{2\mathrm{i}}{z-\mathrm{i}}}$$

$$= \frac{1}{(z-\mathrm{i})^3}\left(\frac{2}{1+\frac{2\mathrm{i}}{z-\mathrm{i}}} - \frac{1}{1+\frac{\mathrm{i}}{z-\mathrm{i}}}\right)$$

and then

$$\frac{1}{z+z^3} = \frac{1}{(z-\mathrm{i})^3}\left(2\sum_{n=0}^{\infty}(-1)^n\frac{2^n\mathrm{i}^n}{(z-\mathrm{i})^n} - \sum_{n=0}^{\infty}(-1)^n\frac{\mathrm{i}^n}{(z-\mathrm{i})^n}\right)$$

$$- \sum_{n=0}^{\infty}(-1)^n\mathrm{i}^n\left(2^{(n+1)}-1\right)(z-\mathrm{i})^{-n-3}$$

$$= \frac{1}{(z-\mathrm{i})^3} - \frac{3\mathrm{i}}{(z-\mathrm{i})^4} - \frac{7}{(z-\mathrm{i})^5} + \frac{15\mathrm{i}}{(z-\mathrm{i})^6} + \dots.$$

8.25 The function $f(z)$ as a composition of entire functions is analytic everywhere in $\mathbb{C}$ except at the zeros of the functions in the denominator. It therefore has isolated singularities at the points $z_\pm = \pm\mathrm{i}$ and $z_0 = 0$. Since in a neighborhood of $z_\pm$ we have $f(z) = g_\pm(z)/(z-z_\pm)$ with $g_\pm(z) = (z^5+1)/((z-z_\mp)z)$ analytic and non-zero at $z_\pm$, the singularities at $z_\pm$ are simple poles. Since in a neighborhood of z_0 we have $f(z) = g_0(z)/(z-0)$ with $g_0(z) = (z^5+1)/(z^2+1)$ analytic and non-zero at z_0, the singularity at z_0 is a simple pole. The residue of f at z_0 is

$$\operatorname*{Res}_{z=0} f(z) = g_0(0) = 1.$$

In the annulus $A(0, 1, \infty)$ the function f is analytic and can be there expanded in the Laurent series. Observing that in this annulus we have $|1/z| < 1$ and using the notable geometric series expansion, we have

$$f(z) = \frac{z^5 + 1}{z^3(1 + z^{-2})}$$

$$= \left(z^2 + \frac{1}{z^3}\right) \sum_{n=0}^{\infty} (-1)^n z^{-2n}$$

$$= \sum_{n=0}^{\infty} (-1)^n z^{-2n+2} + \sum_{n=0}^{\infty} (-1)^n z^{-2n-3}$$

$$= z^2 - 1 + \frac{1}{z^2} + \frac{1}{z^3} - \frac{1}{z^4} - \frac{1}{z^5} + \ldots.$$

Note that the coefficient of the term z^{-1} of this series, namely, 0, is not the residue of f at $z = 0$. This residue is found expanding $f(z)$ in the Laurent series in the annulus $A(0, 0, 1)$

$$f(z) = \frac{z^5 + 1}{z(1 + z^2)}$$

$$= \left(z^4 + \frac{1}{z}\right) \sum_{n=0}^{\infty} (-1)^n z^{2n}$$

$$= \sum_{n=0}^{\infty} (-1)^n z^{-2n+4} + \sum_{n=0}^{\infty} (-1)^n z^{-2n-1}$$

$$= z^4 - z^2 + 1 + \frac{1}{z} - \frac{1}{z^2} - \frac{1}{z^3} + \frac{1}{z^4} + \frac{1}{z^5} + \ldots.$$

In agreement with the result found above we have

$$\operatorname*{Res}_{z=0} f(z) = 1.$$

8.26 Using the notable Taylor expansions

$$\cos z = \sum_{n=0}^{\infty} \frac{(-1)^n}{(2n)!} z^{2n}, \qquad |z| < \infty,$$

$$\log(1 + z) = \sum_{n=1}^{\infty} \frac{(-1)^{n-1}}{n} z^n, \qquad |z| < 1,$$

we can write

$$(\cos z)^{1/z^2} = \exp\left(\frac{1}{z^2}\log(\cos z)\right)$$

$$= \exp\left(\frac{1}{z^2}\log\left(1+\left(-\frac{z^2}{2!}+\frac{z^4}{4!}-\frac{z^6}{6!}+\cdots\right)\right)\right)$$

$$= \exp\left(\frac{1}{z^2}\left(\left(-\frac{z^2}{2!}+\frac{z^4}{4!}-\frac{z^6}{6!}+\cdots\right)\right.\right.$$

$$\left.\left.-\frac{1}{2}\left(-\frac{z^2}{2!}+\frac{z^4}{4!}-\frac{z^6}{6!}+\cdots\right)^2+\cdots\right)\right)$$

$$= \exp\left(-\frac{1}{2}-\frac{1}{12}z^2-\frac{1}{45}z^4+\cdots\right).$$

Therefore

$$\lim_{z\to 0}(\cos z)^{1/z^2} = \lim_{z\to 0}\exp\left(-\frac{1}{2}-\frac{1}{12}z^2-\frac{1}{45}z^4+\cdots\right)$$

$$= e^{-1/2}.$$

8.27

(a) Using the Maclaurin series expansions

$$\sin z = \sum_{k=0}^{\infty}\frac{(-1)^k}{(2k+1)!}z^{2k+1}, \qquad \cos z = \sum_{k=0}^{\infty}\frac{(-1)^k}{(2k)!}z^{2k},$$

valid $\forall z \in \mathbb{C}$, for $|z| < \pi/2$ we have $\cos z \neq 0$ and therefore in this region $\tan z$ is analytic and can be expanded in a Taylor series with center at $z_0 = 0$

$$\tan z = \frac{z-\frac{z^3}{3!}+\frac{z^5}{5!}-\frac{z^7}{7!}+\cdots}{1-\frac{z^2}{2!}+\frac{z^4}{4!}-\frac{z^6}{6!}+\cdots}$$

$$= \left(z-\frac{z^3}{3!}+\frac{z^5}{5!}+\cdots\right)\frac{1}{1+\left(-\frac{z^2}{2!}+\frac{z^4}{4!}+\cdots\right)}$$

$$= \left(z-\frac{z^3}{3!}+\frac{z^5}{5!}+\cdots\right)\left[1-\left(-\frac{z^2}{2!}+\frac{z^4}{4!}+\cdots\right)\right.$$

$$\left.+\left(-\frac{z^2}{2!}+\frac{z^4}{4!}+\cdots\right)^2\cdots\right]$$

$$= z + \left(-\frac{1}{3!} + \frac{1}{2!}\right) z^3 + \left(\frac{1}{5!} - \frac{1}{4!} - \frac{1}{3!2!} + \left(\frac{1}{2!}\right)^2\right) z^5 + \dots$$

$$= z + \frac{1}{3} z^3 + \frac{2}{15} z^5 + \dots$$

(b) The function $\tan z$ is analytic in the annular region $0 < |z - \pi/2| < \pi$ and in this region it can be expanded in the Laurent series. Setting $z = w + \pi/2$ we have

$$\sin z = \sin(w + \pi/2) = \cos w = \sum_{k=0}^{\infty} \frac{(-1)^k}{(2k)!} w^{2k},$$

$$\cos z = \cos(w + \pi/2) = -\sin w = \sum_{k=0}^{\infty} \frac{(-1)^{k+1}}{(2k+1)!} w^{2k+1},$$

and then

$$\tan z = \frac{1 - \frac{w^2}{2!} + \frac{w^4}{4!} - \frac{w^6}{6!} + \dots}{-w + \frac{w^3}{3!} - \frac{w^5}{5!} + \frac{w^7}{7!} + \dots}$$

$$= -\frac{1}{w} \left(1 - \frac{w^2}{2!} + \frac{w^4}{4!} + \dots\right) \frac{1}{1 + \left(-\frac{w^2}{3!} + \frac{w^4}{5!} + \dots\right)}$$

$$= \left(-\frac{1}{w} + \frac{w}{2!} - \frac{w^3}{4!} + \dots\right) \left[1 - \left(-\frac{w^2}{3!} + \frac{w^4}{5!} + \dots\right) + \left(-\frac{w^2}{3!} + \frac{w^4}{5!} + \dots\right)^2 + \dots\right]$$

$$= -\frac{1}{w} + \left(\frac{1}{2!} - \frac{1}{3!}\right) w + \left(-\frac{1}{4!} + \frac{1}{2!3!} + \frac{1}{5!} - \left(\frac{1}{3!}\right)^2\right) w^3 + \dots$$

$$= -\frac{1}{z - \frac{\pi}{2}} + \frac{z - \frac{\pi}{2}}{3} + \frac{(z - \frac{\pi}{2})^3}{45} + \dots.$$

8.28 Using the notable expansions

$$\sinh z = \sum_{n=0}^{\infty} \frac{z^{2n+1}}{(2n+1)!}, \qquad \frac{1}{1+z} = \sum_{n=0}^{\infty} (-1)^n z^n,$$

we have

$$
\frac{1}{z^2 \sinh z} = \frac{1}{z^2 \left(z + \frac{z^3}{3!} + \frac{z^5}{5!} + \cdots \right)}
$$

$$
= \frac{1}{z^3} \frac{1}{1 + \left(\frac{z^2}{3!} + \frac{z^4}{5!} + \cdots \right)}
$$

$$
= \frac{1}{z^3} \left[1 - \left(\frac{z^2}{3!} + \frac{z^4}{5!} + \cdots \right) + \left(\frac{z^2}{3!} + \frac{z^4}{5!} + \cdots \right)^2 + \cdots \right]
$$

$$
= \frac{1}{z^3} \left[1 - \frac{z^2}{3!} - \frac{z^4}{5!} + \left(\frac{z^2}{3!} \right)^2 + \cdots \right]
$$

$$
= \frac{1}{z^3} - \frac{1}{6z} + \frac{7z}{360} + \cdots.
$$

8.29 The function $f(z)$ is analytic for $|z| < \pi/2$ and in this region it can be expanded in the Maclaurin series

$$
f(z) = \sum_{k=0}^{\infty} \frac{f^{(k)}(0)}{k!} z^k.
$$

On the other hand, using the notable expansion of $\cos z$ around $z = 0$, we have

$$
f(z) = \frac{1}{1 - \frac{z^2}{2} + \frac{z^4}{24} - \frac{z^6}{720} + \cdots}
$$

$$
= 1 - \left(-\frac{z^2}{2} + \frac{z^4}{24} + \cdots \right) + \left(-\frac{z^2}{2} + \frac{z^4}{24} + \cdots \right)^2 + \cdots
$$

$$
= 1 + \frac{1}{2} z^2 + \left(-\frac{1}{24} + \frac{1}{4} \right) z^4 + \cdots.
$$

For the uniqueness of the Taylor series expansion, it must be

$$
f^{(4)}(0) = \left(-\frac{1}{24} + \frac{1}{4} \right) 4! = 5.
$$

8.30 In the common region of convergence and assuming that the sum of the series in the denominator does not vanish, the ratio of two power series is still a power series given by

$$\frac{\sum_{n=0}^{\infty} a_n z^n}{\sum_{n=0}^{\infty} b_n z^n} = \sum_{n=0}^{\infty} d_n z^n, \qquad d_n = \left(a_n - \sum_{k=0}^{n-1} d_k b_{n-k}\right)\frac{1}{b_0}.$$

In the case considered where $a_n = 1$, $b_n = 2^n$, $n = 0, 1, 2, \ldots$, the iterative equation for the coefficients d_n gives

$$d_0 = 1,$$

$$d_1 = 1 - d_0 2^1 = 1 - 2 = -1,$$

$$d_2 = 1 - d_0 2^2 - d_1 2^1 = 1 - 4 + 2 = -1,$$

$$d_3 = 1 - d_0 2^3 - d_1 2^2 - d_2 2^1 = 1 - 8 + 4 + 2 = -1,$$

$$\vdots$$

$$d_n = 1 - \sum_{k=0}^{n-1} d_k 2^{n-k} = -2^n + \sum_{j=0}^{n-1} 2^j = -2^n + \frac{1 - 2^n}{1 - 2} = -1, \qquad n > 1.$$

The radius of convergence R is given by the Hadamard formula

$$\frac{1}{R} = \limsup_{n \to \infty} |d_n|^{1/n} = 1.$$

Alternatively, observe that for $|z| < 1$ we have $\sum_{n=0}^{\infty} z^n = 1/(1 - z)$, while for $|z| < 1/2$ it holds $\sum_{n=0}^{\infty} 2^n z^n = 1/(1 - 2z)$. Therefore

$$\frac{\sum_{n=0}^{\infty} z^n}{\sum_{n=0}^{\infty} 2^n z^n} = 1 - z\frac{1}{1 - z} = 1 - z\sum_{n=0}^{\infty} z^n = 1 - \sum_{n=1}^{\infty} z^n.$$

This expansion is valid for $|z| < 1$ and coincides with the result found previously.

Note that the theorem on the ratio of two power series ensures the convergence of the resulting series within the circle of radius $R_{\text{th}} = \min\{R_1, R_2\}$, where R_1 and R_2 are the radii of convergence of the series in the numerator and denominator (assumed to have the same center). In general, we will have $R_{\text{th}} \leq R$ where R is the effective radius of convergence. In the present example $R_1 = 1$, $R_2 = 1/2$ and $R_{\text{th}} = 1/2 < R = 1$.

8.31 Observing that

$$\frac{d^n}{dw^n}(1 - w)^{-1/2} = \frac{1}{2}\left(\frac{1}{2} + 1\right)\cdots\left(\frac{1}{2} + n - 1\right)(1 - w)^{-1/2-n},$$

for $|w| < 1$ the principal branch of $(1 - w)^{-1/2}$ admits the Taylor series expansion

$$(1 - w)^{-1/2} = \sum_{n=0}^{\infty} \binom{\frac{1}{2} + n - 1}{n} w^n,$$

where the binomial coefficient is defined by

$$\binom{a}{n} = \frac{a(a-1)(a-2)\ldots(a-n+1)}{n!}.$$

Let $w = 2xz - z^2$, for $|z|$ sufficiently small we have $|w| < 1$ and therefore it is possible to use the previous expansion to obtain

$$\frac{1}{\sqrt{1 - 2xz + z^2}} = 1 + \frac{1}{2}(2xz - z^2) + \frac{3}{8}(2xz - z^2)^2 + \frac{5}{16}(2xz - z^2)^3 + \ldots$$

$$= 1 + xz + \left(-\frac{1}{2} + \frac{3}{2}x^2\right) z^2 + \left(-\frac{3}{2}x + \frac{5}{2}x^3\right) z^3 + \ldots.$$

From this expression we derive

$$P_1(x) = x,$$

$$P_2(x) = -\frac{1}{2} + \frac{3}{2}x^2,$$

$$P_3(x) = -\frac{3}{2}x + \frac{5}{2}x^3.$$

8.32 Using the notable expansions of $\cos z$, $\sin z$, e^z and $1/(1 \pm z)$, we have

$$f(z) = \frac{1 - \frac{z^2}{2} + \frac{z^4}{24} + \ldots}{\left(z^2 - \frac{z^6}{6} + \frac{z^{10}}{120} + \ldots\right)\left(1 + z + \frac{z^2}{2} + \frac{z^3}{6} + \ldots - 1\right)}$$

$$= \frac{1 - \frac{z^2}{2} + \frac{z^4}{24} + \ldots}{z^3 \left(1 + \frac{z}{2} + \frac{z^2}{6} + \ldots\right)}$$

$$= \frac{1}{z^3} \left(1 - \frac{z^2}{2} + \frac{z^4}{24} + \ldots\right) \left[1 - \left(\frac{z}{2} + \frac{z^2}{6} + \ldots\right)\right.$$

$$\left. + \left(\frac{z}{2} + \frac{z^2}{6} + \ldots\right)^2 + \ldots\right]$$

$$= \frac{1}{z^3}\left(1 - \frac{z}{2} - \frac{5z^2}{12} + \dots\right)$$

$$= \frac{1}{z^3} - \frac{1}{2z^2} - \frac{5}{12z} + O(z^0).$$

8.33 Set $g(w) = \sin(\pi w^2/2)$, from the Taylor series expansion of $\sin(z)$, valid for $|z| < \infty$, changing $z \to \pi w^2/2$ we have

$$g(w) = \sum_{n=0}^{\infty} \frac{(-1)^n}{(2n+1)!} (\pi w^2/2)^{2n+1} = \sum_{n=0}^{\infty} \frac{(-1)^n \pi^{2n+1}}{2^{2n+1}(2n+1)!} w^{4n+2}.$$

From the theorem on the integration of power series we obtain

$$S(z) = \int_0^z \sum_{n=0}^{\infty} \frac{(-1)^n \pi^{2n+1}}{2^{2n+1}(2n+1)!} w^{4n+2} dw$$

$$= \sum_{n=0}^{\infty} \frac{(-1)^n \pi^{2n+1}}{2^{2n+1}(2n+1)!} \int_0^z w^{4n+2} dw$$

$$= \sum_{n=0}^{\infty} \frac{(-1)^n \pi^{2n+1}}{2^{2n+1}(2n+1)!} \frac{w^{4n+3}}{4n+3}\Bigg|_0^z$$

$$= \sum_{n=0}^{\infty} \frac{(-1)^n \pi^{2n+1}}{(4n+3)2^{2n+1}(2n+1)!} z^{4n+3}$$

$$= \frac{\pi}{6} z^3 - \frac{\pi^3}{336} z^7 + O(z^{11}).$$

The radius of convergence of the obtained series is infinite as that of the series representing g. The function $S(z)$ is entire and has a zero at $z = 0$. Therefore, $1/S(z)$ can be expanded in the Laurent series in $A(0, 0, \infty)$

$$\frac{1}{S(z)} = \frac{1}{\frac{\pi}{6} z^3 + O(z^7)} = \frac{1}{\frac{\pi}{6} z^3} \frac{1}{1 + O(z^4)} = \frac{6}{\pi z^3} + O(z).$$

8.34 First of all, note that f is analytic in a neighborhood of $z_0 = 0$, precisely in the ball $B(0, \pi/2)$ as seen in the previous exercise, and in such a neighborhood it is

$$f(z) = -2\frac{d}{dz}\sqrt{\cos z}.$$

Therefore, it is sufficient to determine the Taylor series expansion with center at $z_0 = 0$ of the principal branch of $\sqrt{\cos z}$. Using notable expansions

$$\cos z = \sum_{k=0}^{\infty} \frac{(-1)^k}{(2k)!} z^{2k}, \qquad |z| < \infty,$$

$$\sqrt{1+z} = \sum_{k=0}^{\infty} \binom{\frac{1}{2}}{k} z^k, \qquad |z| < 1,$$

we have

$$\sqrt{\cos z} = \sqrt{1 + (\cos z - 1)}$$

$$= 1 + \frac{1}{2}\left(-\frac{z^2}{2!} + \frac{z^4}{4!} - \frac{z^6}{6!} + \dots\right) - \frac{1}{8}\left(-\frac{z^2}{2!} + \frac{z^4}{4!} - \frac{z^6}{6!} + \dots\right)^2$$

$$+ \frac{1}{16}\left(-\frac{z^2}{2!} + \frac{z^4}{4!} - \frac{z^6}{6!} + \dots\right)^3 + \dots$$

$$= 1 - \frac{1}{2}\frac{1}{2!}z^2 + \left(\frac{1}{2}\frac{1}{4!} - \frac{1}{8}\frac{1}{(2!)^2}\right)z^4 + \left(-\frac{1}{2}\frac{1}{6!} + \frac{1}{8}\frac{2}{2!4!} - \frac{1}{16}\frac{1}{(2!)^3}\right)z^6 + O(z^8)$$

$$= 1 - \frac{1}{4}z^2 - \frac{1}{96}z^4 - \frac{19}{5760}z^6 + O(z^8).$$

This expansion is valid within the largest circle centered at $z_0 = 0$ and such that for every z within it we get $\mathrm{Re}(\cos z - 1) > -1$. This condition is equivalent to $|z| < \pi/2$ and we can conclude

$$\frac{\sin z}{\sqrt{\cos z}} = z + \frac{1}{12}z^3 + \frac{19}{480}z^5 + O(z^7), \qquad |z| < \frac{\pi}{2}.$$

12.9 Exercises of Chap. 9

9.1 The function $f(z) = 1/\sin(z^{-1})$ is analytic everywhere except for the isolated singularities at the points $\pm z_n = (n\pi)^{-1}$, $n = 1, 2, \dots$, and the non-isolated singularity at $z - 0$. Therefore, the function $1/\sin(z^{-1})$ is analytic on the closed path γ and on its outside except at the points $\pm z_1$. By the Theorem 9.5 of the residue at infinity the integral of f along a simple closed piecewise regular positively oriented curve η that contains within it all the singularities of f is

$$\int_{\eta} f(z)\mathrm{d}z = 2\pi\,\mathrm{i}\,\underset{z=0}{\mathrm{Res}}\,\frac{1}{z^2}\frac{1}{\sin(z)}.$$

This residue can be calculated considering that for $0 < |z| < \infty$ the following Laurent series expansion holds

$$\frac{1}{z^2}\frac{1}{\sin(z)} = \frac{1}{z^2}\frac{1}{z - \frac{z^3}{3!} + \frac{z^5}{5!} - \frac{z^7}{7!} + \ldots}$$

$$= \frac{1}{z^3}\frac{1}{1 - \left(\frac{z^2}{3!} - \frac{z^4}{5!} + \ldots\right)}$$

$$= \frac{1}{z^3}\left[1 + \left(\frac{z^2}{3!} - \frac{z^4}{5!} + \ldots\right) + \left(\frac{z^2}{3!} - \frac{z^4}{5!} + \ldots\right)^2 + \ldots\right]$$

$$= z^{-3} + \frac{1}{3!}z^{-1} + \left(-\frac{1}{5!} + \frac{1}{(3!)^2}\right)z + \ldots,$$

therefore

$$\int_\eta f(z)\mathrm{d}z = 2\pi\mathrm{i}\frac{1}{3!} = \frac{\pi}{3}\mathrm{i}.$$

By the path deformation principle and the residue theorem, we have

$$\int_\gamma f(z)\mathrm{d}z = \int_\eta f(z)\mathrm{d}z - 2\pi\mathrm{i}\left(\operatorname*{Res}_{z=z_1} f(z) + \operatorname*{Res}_{z=-z_1} f(z)\right).$$

Since $\sin(z^{-1})$ has simple zeros at the points $\pm z_1$, we have

$$\operatorname*{Res}_{z=\pm z_1} f(z) = -\frac{z_1^2}{\cos(z_1^{-1})} = \frac{1}{\pi^2}.$$

In conclusion

$$\int_\gamma \frac{1}{\sin(z^{-1})}\mathrm{d}z = \left(\frac{\pi}{3} - \frac{4}{\pi}\right)\mathrm{i}.$$

9.2 To calculate the integral on γ_1 consider a path γ_4 homotopic to γ_1 that contains within it both z_1 and z_3. The integral of f along γ_4 is calculated by the residue at infinity since f is analytic on γ_4 and on its outside. By the path deformation principle it follows immediately

$$\int_{\gamma_1} f(z)\mathrm{d}z = 2\pi\mathrm{i}\left(\operatorname*{Res}_{z=0}\frac{1}{z^2}f\left(\frac{1}{z}\right) - \operatorname*{Res}_{z=z_3} f(z)\right).$$

The function f is not analytic outside γ_2. Within this curve we do not know if f has only isolated singularities. Therefore

$$\int_{\gamma_2} f(z)\,dz \ \text{is not computable.}$$

The curve γ_3 is not simple. Let us break it down into the sum of two simple closed curves on and within which f is analytic except for the isolated singularities z_2 and z_3, respectively. Paying attention to how the two simple closed curves are oriented, we have

$$\int_{\gamma_3} f(z)\,dz = 2\pi i \left(\operatorname*{Res}_{z=z_3} f(z) - \operatorname*{Res}_{z=z_2} f(z) \right).$$

9.3   The function $1/\sin(z^{-2})$ is analytic everywhere except for the isolated singularities at the points

$$z_{n,k} = \frac{e^{i\pi k}}{\sqrt{n\pi}}, \qquad n = 1, 2, 3, \ldots, \qquad k = 0, 1,$$

and the non-isolated singularity at $z = 0$. Therefore, the function $1/\sin(z^{-2})$ is analytic on the closed path γ and on its outside. By the theorem of the residue at infinity we have

$$\int_\gamma \frac{1}{\sin(z^{-2})}\,dz = 2\pi i \operatorname*{Res}_{z=0} \frac{1}{z^2} \frac{1}{\sin(z^2)}.$$

This residue can be calculated considering that for $0 < |z| < \sqrt{\pi}$ the following Laurent series expansion holds

$$
\begin{aligned}
\frac{1}{z^2}\frac{1}{\sin(z^2)} &= \frac{1}{z^2}\frac{1}{z^2 - \frac{z^6}{3!} + \frac{z^{10}}{5!} - \frac{z^{14}}{7!} + \cdots} \\[2mm]
&= \frac{1}{z^4}\frac{1}{1 - \left(\frac{z^4}{3!} - \frac{z^8}{5!} + \cdots\right)} \\[2mm]
&= \frac{1}{z^4}\left[1 + \left(\frac{z^4}{3!} - \frac{z^8}{5!} + \cdots\right) + \left(\frac{z^4}{3!} - \frac{z^8}{5!} + \cdots\right)^2 + \cdots \right] \\[2mm]
&= z^{-4} + \frac{1}{3!} + \left(-\frac{1}{5!} + \frac{1}{(3!)^2}\right) z^4 + O(z^8).
\end{aligned}
$$

So we have

$$\operatorname*{Res}_{z=0} \frac{1}{z^2} \frac{1}{\sin(z^2)} = 0$$

and, in conclusion,

$$\int_\gamma \frac{1}{\sin(z^{-2})}\, dz = 0.$$

9.4 The function $1/\sin(z^{-1})$ is analytic everywhere except for the simple poles at the points $z_n = (n\pi)^{-1}$, $n = \pm 1, \pm 2, \ldots$, and the non-isolated singularity at $z = 0$. Called γ_1 the circle centered at the origin, of radius 1, traveled anti-clockwise, the function $1/\sin(z^{-1})$ is analytic on γ_1 and on its outside, and by the Theorem 9.5 of the residue at infinity we have

$$\int_{\gamma_1} \frac{1}{\sin(z^{-1})}\, dz = 2\pi i \operatorname*{Res}_{z=0} \frac{1}{z^2} \frac{1}{\sin(z)}.$$

This residue can be calculated considering that for $0 < |z| < \infty$ the following Laurent series expansion holds

$$
\begin{aligned}
\frac{1}{z^2} \frac{1}{\sin(z)} &= \frac{1}{z^2} \frac{1}{z - \frac{z^3}{3!} + \frac{z^5}{5!} - \frac{z^7}{7!} + \cdots} \\
&= \frac{1}{z^3} \frac{1}{1 - \left(\frac{z^2}{3!} - \frac{z^4}{5!} + \cdots\right)} \\
&= \frac{1}{z^3}\left[1 + \left(\frac{z^2}{3!} - \frac{z^4}{5!} + \cdots\right) + \left(\frac{z^2}{3!} - \frac{z^4}{5!} + \cdots\right)^2 + \cdots\right] \\
&= z^{-3} + \frac{1}{3!}z^{-1} + \left(-\frac{1}{5!} + \frac{1}{(3!)^2}\right)z + \cdots.
\end{aligned}
$$

Therefore

$$\int_{\gamma_1} \frac{1}{\sin(z^{-1})}\, dz = 2\pi i \frac{1}{3!} = \frac{\pi}{3}i.$$

By the path deformation principle, it follows

$$\int_{\gamma_1} \frac{1}{\sin(z^{-1})}\, dz = \int_\gamma \frac{1}{\sin(z^{-1})}\, dz + \int_{\gamma_+} \frac{1}{\sin(z^{-1})}\, dz + \int_{\gamma_-} \frac{1}{\sin(z^{-1})}\, dz,$$

where $\gamma_\pm$ are circles, centered at the points $\pm\,\pi^{-1}$, positively oriented and with radius smaller than $\min(1/\pi - 1/5, 1 - 1/\pi)$. Since

$$\operatorname*{Res}_{z=\pm\pi^{-1}} \frac{1}{\sin(z^{-1})} = \frac{1}{\sin'(z^{-1})}\Bigg|_{z=\pm\pi^{-1}} = \frac{1}{\pi^2},$$

we conclude

$$\int_\gamma \frac{1}{\sin(z^{-1})}\,dz = \frac{\pi}{3}i - 2\pi i\frac{1}{\pi^2} - 2\pi i\frac{1}{\pi^2} = i\left(\frac{\pi}{3} - \frac{4}{\pi}\right).$$

9.5 The function $f(z) = z/((z-2)(z^2+1))$ has 3 simple poles at $z = 2$ and $z = \pm i$. The poles at $z = 2$ and $z = i$ are within the circle γ while the one at $z = -i$ is on its outside, therefore

$$\int_\gamma \frac{z\,dz}{(z-2)(z^2+1)} = 2\pi i\left(\operatorname*{Res}_{z=2} f(z) + \operatorname*{Res}_{z=i} f(z)\right)$$

$$= 2\pi i\left(\frac{2}{5} + \frac{i}{2i(i-2)}\right)$$

$$= \frac{\pi}{5}(1+2i).$$

9.6 Note that the integration path is the circle of radius 2, centered at the point $1+2i$ and positively oriented. The integrand function has simple poles corresponding to the simple zeros of $e^{\pi z} + 1$. These zeros are $z_k = i(1 + 2k)$, $k \in \mathbb{Z}$. Observing that z_0 and z_1 are within the integration path while all the other z_k are external, by the residue theorem we conclude that

$$\int_\gamma \frac{z^2}{e^{\pi z} + 1}\,dz = 2\pi i\left(\operatorname*{Res}_{z=z_0} \frac{z^2}{e^{\pi z}+1} + \operatorname*{Res}_{z=z_1} \frac{z^2}{e^{\pi z}+1}\right)$$

$$= 2\pi i\left(\frac{z_0^2}{\pi e^{\pi z_0}} + \frac{z_1^2}{\pi e^{\pi z_1}}\right)$$

$$= 20i.$$

9.7 It is sufficient to show that the modulus of the integral tends to 0 for $n \to \infty$. By the Darboux inequality

$$\left|\int_{\gamma_n} \frac{1}{z^2 \sin z}\,dz\right| \le L_{\gamma_n} \sup_{z\in\{\gamma_n\}} \left|\frac{1}{z^2 \sin z}\right|,$$

where $L_{\gamma_n} = 8a_n$ is the length of γ_n. For $z \in \{\gamma_n\}$ it is evident that $|z|^2 \geq a_n^2$. Furthermore, $\forall x, y \in \mathbb{R}$ we have

$$|\sin(x + iy)|^2 = |\sin x \cos(iy) + \cos x \sin(iy)|^2$$
$$= |\sin x \cosh y - i \cos x \sinh y|^2$$
$$= \sin^2 x \cosh^2 y + \cos^2 x \sinh^2 y$$
$$= \sin^2 x + \sinh^2 y.$$

Therefore, along the two sides of the square γ_n where $x = \pm a_n$ and $-a_n \leq y \leq a_n$ it turns out $|\sin z|^2 \geq \sin^2 a_n = 1$ while along the other two sides where $-a_n \leq x \leq a_n$ and $y = \pm a_n$ we have $|\sin z|^2 \geq \sinh^2 a_n$. It follows that $|\sin z| \geq \min(1, \sinh a_n) = 1$ for $z \in \{\gamma_n\}$. We can then conclude that

$$\sup_{z \in \{\gamma_n\}} \left| \frac{1}{z^2 \sin z} \right| \leq \frac{1}{a_n^2}$$

which entails

$$\left| \int_{\gamma_n} \frac{1}{z^2 \sin z} \, dz \right| \leq 8a_n \frac{1}{a_n^2} = \frac{8}{\pi(n + 1/2)} \xrightarrow{n \to \infty} 0.$$

For a generic n the integral can be evaluated with the residue theorem. The integrand function is analytic on and within γ_n except for the simple poles at points $z_k = k\pi$ with $k = \pm 1, \pm 2, \ldots, \pm n$ and the pole of order 3 at point $z_0 = 0$. Observing that each z_k is a simple zero of $\sin z$, it follows

$$\operatorname*{Res}_{z = z_k} \frac{1}{z^2 \sin z} = \frac{z_k^{-2}}{\cos z_k} = \frac{(-1)^k}{\pi^2 k^2}, \qquad k = \pm 1, \pm 2, \ldots, \pm n.$$

To calculate the residue at $z_0 = 0$, note that for $z \in A(0, 0, \pi)$ the following Laurent series expansion holds

$$\frac{1}{z^2 \sin z} = \frac{1}{z^3} \frac{1}{1 - \left(\frac{z^2}{3!} - \frac{z^4}{5!} + \ldots \right)}$$
$$= \frac{1}{z^3} \left[1 + \left(\frac{z^2}{3!} - \frac{z^4}{5!} + \ldots \right) + \left(\frac{z^2}{3!} - \frac{z^4}{5!} + \ldots \right)^2 + \ldots \right]$$
$$= \frac{1}{z^3} + \frac{1}{6} \frac{1}{z} + \frac{7}{360} z + O(z^3),$$

from which it follows

$$\operatorname*{Res}_{z=z_0} \frac{1}{z^2 \sin z} = \frac{1}{6}.$$

By the Theorem (9.3) we conclude

$$\int_{\gamma_n} \frac{1}{z^2 \sin z}\, dz = 2\pi i \sum_{k=-n}^{n} \operatorname*{Res}_{z=z_k} \frac{1}{z^2 \sin z} = 2\pi i \left(\frac{1}{6} + 2 \sum_{k=1}^{n} \frac{(-1)^k}{\pi^2 k^2} \right).$$

Taking the limit $n \to \infty$, from the previous two results it follows

$$\sum_{k=1}^{\infty} \frac{(-1)^k}{\pi^2 k^2} = -\frac{1}{12}.$$

9.8

(a) The function is analytic everywhere except for the isolated singularity at $z = 2$.
Recalling that

$$\cos w = \sum_{n=0}^{\infty} \frac{(-1)^n w^{2n}}{(2n)!}, \qquad |w| < \infty,$$

and letting $w = 1/(z-2)$, for $0 < |z-2| < \infty$ we have

$$\cos\left(\frac{1}{z-2} \right) = \sum_{n=0}^{\infty} \frac{(-1)^n}{(2n)!(z-2)^{2n}}.$$

From this expression it follows that the singularity is essential and

$$\operatorname*{Res}_{z=2} \cos\left(\frac{1}{z-2} \right) = 0.$$

(b) Also in this case the function is analytic everywhere except for the isolated
singularity at $z = 2$. For $0 < |z-2| < \infty$ we have

$$z^3 \cos\left(\frac{1}{z-2} \right) = (z-2+2)^3 \sum_{n=0}^{\infty} \frac{(-1)^n}{(2n)!(z-2)^{2n}}$$

$$= \left((z-2)^3 + 6(z-2)^2 + 12(z-2) + 8 \right)$$

$$\times \left(1 - \frac{1}{2(z-2)^2} + \frac{1}{24(z-2)^4} \right.$$

$$\left. - \frac{1}{120(z-2)^6} + \cdots \right).$$

From this expression it follows that the singularity is essential and

$$\operatorname*{Res}_{z=2} z^3 \cos\left(\frac{1}{z-2}\right) = \frac{1}{24} - \frac{12}{2} = -\frac{143}{24}.$$

9.9 Expanding the exponential and sine functions in the Taylor series, we write

$$f(z) = \frac{1}{\left(z + \frac{z^2}{2} + \frac{z^3}{6} + \dots\right)\left(z - \frac{z^3}{6} + \frac{z^5}{120} + \dots\right)} = \frac{g(z)}{z^2},$$

with

$$g(z) = \frac{1}{\left(1 + \frac{z}{2} + \frac{z^2}{6} + \dots\right)\left(1 - \frac{z^2}{6} + \frac{z^4}{120} + \dots\right)}$$

analytic and non-zero at $z = 0$. The function $f(z)$ therefore has a double pole at $z = 0$ with

$$\operatorname*{Res}_{z=0} f(z) = g'(0)$$

$$= -\frac{\frac{1}{2} + \frac{z}{3} + \dots}{\left(1 + \frac{z}{2} + \frac{z^2}{6} + \dots\right)^2 \left(1 - \frac{z^2}{6} + \frac{z^4}{120} + \dots\right)}$$

$$- \frac{-\frac{z}{3} + \frac{z^3}{30} + \dots}{\left(1 + \frac{z}{2} + \frac{z^2}{6} + \dots\right)\left(1 - \frac{z^2}{6} + \frac{z^4}{120} + \dots\right)^2}\Bigg|_{z=0}$$

$$= -\frac{1}{2}.$$

9.10 Note that $f(z)$ is analytic everywhere except for the isolated singularity at $z = 0$ and the branch line $\cos(z(t)) = -t$ with $t \in [0, \infty)$, see Exercise 9.19 (b). It follows that the function $f(z)$ is analytic in the annulus $A(0, 0, \pi/2)$ and here can be expanded in the Laurent series. Using the well known Taylor series expansions

$$\cos z = \sum_{n=0}^{\infty} \frac{(-1)^n}{(2n)!} z^{2n}, \qquad |z| < \infty,$$

$$\log(1 + z) = \sum_{n=1}^{\infty} \frac{(-1)^{n-1}}{n} z^n, \qquad |z| < 1,$$

we have

$$\frac{1}{z^2}\log(\cos z) = \frac{1}{z^2}\log\left[1 + \left(-\frac{z^2}{2!} + \frac{z^4}{4!} - \frac{z^6}{6!} + \cdots\right)\right]$$

$$= \frac{1}{z^2}\left[\left(-\frac{z^2}{2!} + \frac{z^4}{4!} - \frac{z^6}{6!} + \cdots\right)\right.$$

$$\left. - \frac{1}{2}\left(-\frac{z^2}{2!} + \frac{z^4}{4!} - \frac{z^6}{6!} + \cdots\right)^2 + \cdots\right]$$

$$= -\frac{1}{2} - \frac{1}{12}z^2 - \frac{1}{45}z^4 + \cdots.$$

The expansion thus found is the Laurent series expansion of $f(z)$ valid for $|z| > 0$ and $|\cos z - 1| < 1$, that is, for $0 < |z| < \pi/2$, which coincides with the region of analyticity of $f(z)$ around $z = 0$. The function $f(z)$ has a singularity that can be removed at $z = 0$. The function

$$g(z) = \begin{cases} f(z) & z \neq 0 \\ -1/2 & z = 0 \end{cases}$$

is analytic for $|z| < \pi/2$. The Laurent series expansion found above for $f(z)$ coincides with the Taylor series expansion for $g(z)$.

9.11

(a) Regardless of the branch chosen, the multi-valued function

$$\frac{\sqrt{z}}{z} = z^{-1/2} = e^{-\frac{1}{2}\log z}$$

has a non-isolated singularity (branch point) at $z = 0$. Therefore, there is no annular region $0 < |z| < \varepsilon$ where it is possible to expand it in the Laurent series.

(b) The function $1 - \cos z$ is entire and at $z = 0$ has a double zero

$$1 - \cos z = 1 - \sum_{k=0}^{\infty} \frac{(-1)^k}{(2k)!} z^{2k}$$

$$= \frac{z^2}{2!}\left(1 - \frac{2!}{4!}z^2 + \frac{2!}{6!}z^4 + \cdots\right).$$

It follows that the function $z/(1-\cos z)$ has a simple pole at $z = 0$. In a suitable annular region $0 < |z| < \varepsilon$, we have

$$\frac{z}{1-\cos z} = \frac{2!}{z}\frac{1}{1 + \left(-\frac{2!}{4!}z^2 + \frac{2!}{6!}z^4 + \cdots\right)}$$

$$
= \frac{2!}{z}\left[1 - \left(-\frac{2!}{4!}z^2 + \frac{2!}{6!}z^4 + \dots\right)\right.
$$

$$
\left. + \left(-\frac{2!}{4!}z^2 + \frac{2!}{6!}z^4 + \dots\right)^2 + \dots\right]
$$

$$
= \frac{2!}{z} + \frac{(2!)^2}{4!}z + O\left(z^3\right).
$$

Therefore

$$
\operatorname*{Res}_{z=0} \frac{z}{1 - \cos z} = 2.
$$

(c) In this case for $0 < |z| < \infty$ we have

$$
\left(z^3 + 3\right)e^{1/z} = \left(z^3 + 3\right)\sum_{k=0}^{\infty}\frac{1}{k!}z^{-k}
$$

$$
= \sum_{k=0}^{\infty}\frac{1}{k!}z^{-k+3} + \sum_{k=0}^{\infty}\frac{3}{k!}z^{-k}
$$

$$
= z^3 + z^2 + \frac{1}{2}z + \sum_{n=0}^{\infty}\left(\frac{1}{(n+3)!} + \frac{3}{n!}\right)z^{-n}.
$$

So $z = 0$ is an essential singularity of $\left(z^3 + 3\right)e^{1/z}$ and we have

$$
\operatorname*{Res}_{z=0}\left(z^3 + 3\right)e^{1/z} = \frac{1}{(1+3)!} + \frac{3}{1!} = \frac{1}{24} + 3 = \frac{73}{24}.
$$

9.12 Expanding the exponential, sine and cosine functions in the Maclaurin series we have

$$
f(z) = \frac{\left(1 + z + \frac{z^2}{2!} + \frac{z^3}{3!} + \dots\right)\left(z - \frac{z^3}{3!} + \frac{z^5}{5!} + \dots\right)}{z\left(\frac{z^2}{2!} - \frac{z^4}{4!} + \frac{z^6}{6!} + \dots\right)}
$$

$$
= \frac{2}{z^3}\,\frac{z + z^2 + \frac{1}{3}z^3 - \frac{1}{30}z^5 + \dots}{1 - \left(\frac{1}{12}z^2 - \frac{1}{360}z^4 + \dots\right)}
$$

$$
= \frac{2}{z^3}\left(z + z^2 + \frac{1}{3}z^3 - \frac{1}{30}z^5 + \dots\right)\left[1 + \left(\frac{1}{12}z^2 - \frac{1}{360}z^4 + \dots\right)\right.
$$

$$+ \left(\frac{1}{12}z^2 - \frac{1}{360}z^4 + \dots \right)^2 + \dots \Bigg]$$

$$= \frac{2}{z^3}\left(z + z^2 + \frac{1}{3}z^3 - \frac{1}{30}z^5 + \dots \right)\left(1 + \frac{1}{12}z^2 + \frac{7}{720}z^4 + \dots \right)$$

$$= \frac{2}{z^3}\left(z + z^2 + \frac{5}{12}z^3 + \frac{1}{12}z^4 - \frac{1}{720}z^5 + \dots \right)$$

$$= \frac{2}{z^2} + \frac{2}{z} + \frac{5}{6} + \frac{z}{6} - \frac{z^2}{360} + \dots .$$

Therefore, $f(z)$ has at $z = 0$ a pole of order 2 with

$$\operatorname*{Res}_{z=0} f(z) = 2.$$

9.13

(a) Using the well known expansion $\sin z = \sum_{n=0}^{\infty} \frac{(-1)^n}{(2n+1)!} z^{2n+1}$ we have

$$z \sin\left(\frac{1}{z-3} \right) = ((z-3)+3) \sum_{n=0}^{\infty} \frac{(-1)^n}{(2n+1)!} \frac{1}{(z-3)^{2n+1}}$$

$$= \sum_{n=0}^{\infty} \frac{(-1)^n}{(2n+1)!} \frac{1}{(z-3)^{2n}}$$

$$+ 3 \sum_{n=0}^{\infty} \frac{(-1)^n}{(2n+1)!} \frac{1}{(z-3)^{2n+1}}$$

$$= 1 + \frac{3}{z-3} - \frac{1}{3!(z-3)^2} - \frac{3}{3!(z-3)^3} + \dots .$$

Therefore, $z = 3$ is an essential isolated singularity with

$$\operatorname*{Res}_{z=3} z \sin\left(\frac{1}{z-3} \right) = 3.$$

(b) We now write

$$\frac{z}{(z-3)\sin(z-3)} = \frac{z}{(z-3)\left((z-3) - \frac{(z-3)^3}{6} + \frac{(z-3)^5}{120} + \dots \right)}$$

$$= \frac{(z-3)+3}{(z-3)^2\left[1 - \left(\frac{(z-3)^2}{6} - \frac{(z-3)^4}{120} + \dots \right) \right]}$$

$$= \frac{(z-3)+3}{(z-3)^2}\left[1 + \left(\frac{(z-3)^2}{6} + \cdots\right)\right.$$

$$\left. + \left(\frac{(z-3)^2}{6} + \cdots\right)^2 + \cdots\right]$$

$$= \frac{3}{(z-3)^2} + \frac{1}{z-3} + \frac{1}{2} + \frac{1}{6}(z-3) + \cdots .$$

The isolated singularity at $z = 3$ is in this case a double pole with

$$\operatorname*{Res}_{z=3} \frac{z}{(z-3)\sin(z-3)} = 1.$$

9.14

(a) Chosen a generic branch of the logarithm (the principal branch corresponds to have $\varphi = -\pi$)

$$\log z = \ln r + i\theta, \qquad z = re^{i\theta}, \qquad r > 0, \qquad \varphi < \theta < \varphi + 2\pi,$$

the function $\log(z^2)$ is analytic everywhere except for the branch points $z = \pm\sqrt{t}e^{i\varphi/2}$ with $t \geq 0$. Ti follows that this function has a non-isolated singularity at $z = 0$ (branch point) and there is no annular region $0 < |z| < \varepsilon$ where we can expand it in the Laurent series.

(b) Remembering that

$$\cos w = \sum_{n=0}^{\infty} \frac{(-1)^n}{(2n)!} w^{2n}, \qquad |w| < \infty,$$

and setting $w = z^{-2}$, for $0 < |z| < \infty$ we have

$$z^3 \cos\left(z^{-2}\right) = z^3 \sum_{n=0}^{\infty} \frac{(-1)^n}{(2n)!} z^{-4n}$$

$$= z^3 \left(1 - \frac{z^{-4}}{2!} + \frac{z^{-8}}{4!} - \frac{z^{-12}}{6!} + \cdots\right)$$

$$= z^3 - \frac{1}{2!}\frac{1}{z} + \frac{1}{4!}\frac{1}{z^5} - \frac{1}{6!}\frac{1}{z^9} + \cdots .$$

Therefore, the function $z^3 \cos(z^{-2})$ has at $z = 0$ an essential singularity with

$$\operatorname*{Res}_{z=0} z^3 \cos\left(z^{-2}\right) = -\frac{1}{2}.$$

(c) In this case for $0 < |z| < \infty$ we have

$$\frac{1}{z^4 \sinh z} = \frac{1}{z^4} \frac{1}{z + \frac{z^3}{3!} + \frac{z^5}{5!} + \frac{z^7}{7!} + \cdots}$$

$$= \frac{1}{z^5} \frac{1}{1 + \left(\frac{z^2}{3!} + \frac{z^4}{5!} + \frac{z^6}{7!} + \cdots \right)}$$

$$= \frac{1}{z^5} \left[1 - \left(\frac{z^2}{3!} + \frac{z^4}{5!} + \frac{z^6}{7!} + \cdots \right) \right.$$

$$+ \left(\frac{z^2}{3!} + \frac{z^4}{5!} + \frac{z^6}{7!} + \cdots \right)^2$$

$$\left. - \left(\frac{z^2}{3!} + \frac{z^4}{5!} + \frac{z^6}{7!} + \cdots \right)^3 + \cdots \right]$$

$$= \frac{1}{z^5} \left[1 - \frac{1}{3!} z^2 + \left(-\frac{1}{5!} + \frac{1}{(3!)^2} \right) z^4 \right.$$

$$\left. + \left(-\frac{1}{7!} + \frac{2}{3!5!} - \frac{1}{(3!)^3} \right) z^6 + \cdots \right]$$

$$= \frac{1}{z^5} - \frac{1}{6} \frac{1}{z^3} + \frac{7}{360} \frac{1}{z} - \frac{31}{15120} z + \cdots$$

Therefore, $z = 0$ is a pole of order 5 of $1/(z^4 \sinh z)$ and we have

$$\operatorname*{Res}_{z=0} \frac{1}{z^4 \sinh z} = \frac{7}{360}.$$

9.15

(a) The function $1 - \cos z$ is entire and at $z = 0$ has an isolated double zero

$$1 - \cos z = 1 - \sum_{k=0}^{\infty} \frac{(-1)^k}{(2k)!} z^{2k}$$

$$= \frac{z^2}{2!} \left(1 - \frac{2!}{4!} z^2 + \frac{2!}{6!} z^4 + \cdots \right).$$

It follows that the function $1/(1 - \cos z)$ has at $z = 0$ a double pole. In the annular region $0 < |z| < \infty$ we have

$$\frac{1}{1 - \cos z} = \frac{2!}{z^2} \frac{1}{1 + \left(-\frac{2!}{4!}z^2 + \frac{2!}{6!}z^4 + \dots \right)}$$

$$= \frac{2!}{z^2}\left[1 - \left(-\frac{2!}{4!}z^2 + \frac{2!}{6!}z^4 + \dots \right) \right.$$

$$\left. + \left(-\frac{2!}{4!}z^2 + \frac{2!}{6!}z^4 + \dots \right)^2 + \dots \right]$$

$$= \frac{2!}{z^2} + \frac{(2!)^2}{4!} + O\left(z^2 \right)$$

and therefore

$$\operatorname*{Res}_{z=0} \frac{1}{1 - \cos z} = 0.$$

(b) Regardless of the branch chosen, the multi-valued function

$$\frac{\sqrt{z}}{z^2} = z^{-3/2} = \mathrm{e}^{-\frac{3}{2}\log z}$$

has a non-isolated singularity (branch point) at $z = 0$. Therefore, there is no annular region $0 < |z| < \varepsilon$ where it is possible to expand it in the Laurent series.

(c) In this case for $0 < |z| < \infty$ we have

$$z\mathrm{e}^{1/z^2} = z \sum_{k=0}^{\infty} \frac{1}{k!} \left(\frac{1}{z^2} \right)^k$$

$$= \sum_{k=0}^{\infty} \frac{1}{k!} z^{-2k+1}$$

$$= z + \frac{1}{z} + \frac{1}{2z^3} + \frac{1}{6z^5} + \dots$$

Therefore, $z = 0$ is an essential singularity of $z\mathrm{e}^{1/z^2}$ with

$$\operatorname*{Res}_{z=0} z\mathrm{e}^{1/z^2} = 1.$$

9.16

(a) The function can be rewritten as

$$\frac{3+z}{z^3+2z^2} = \frac{g(z)}{z^2},$$

with $g(z) = (3+z)/(2+z)$ analytic and non-zero at $z = 0$. Therefore the considered function has at $z = 0$ a double pole and

$$\operatorname*{Res}_{z=0} \frac{3+z}{z^3+2z^2} = \frac{g'(0)}{1!} = -\frac{1}{4}.$$

(b) Regardless of the branch chosen, the multi-valued function

$$z \log z$$

has a non-isolated singularity (branch point) at $z = 0$. Therefore, there is no annular region $0 < |z| < \varepsilon$ where it is possible to expand it in the Laurent series.

(c) Using the well known expansion

$$\log(1+z) = \sum_{n=1}^{\infty} \frac{(-1)^{n-1}}{n} z^n, \qquad |z| < 1,$$

we have

$$\begin{aligned}
\frac{1}{\log(1+z)} &= \frac{1}{z - \frac{z^2}{2} + \frac{z^3}{3} - \frac{z^4}{4} + \cdots} \\
&= \frac{1}{z} \frac{1}{1 + \left(-\frac{z}{2} + \frac{z^2}{3} - \frac{z^3}{4} + \cdots\right)} \\
&= \frac{1}{z} \left[1 - \left(-\frac{z}{2} + \frac{z^2}{3} - \frac{z^3}{4} + \cdots\right) \right. \\
&\qquad \left. + \left(-\frac{z}{2} + \frac{z^2}{3} - \frac{z^3}{4} + \cdots\right)^2 + \cdots \right] \\
&= \frac{1}{z} + \frac{1}{2} - \frac{z}{12} + \frac{z^2}{24} + \cdots.
\end{aligned}$$

Therefore, the function considered has a simple pole at $z = 0$ with

$$\operatorname*{Res}_{z=0} \frac{1}{\log(1+z)} = 1.$$

9.17

(a) The function is analytic everywhere except for the isolated singularity at $z = 0$.
Using the notable expansion

$$\cos w = \sum_{n=0}^{\infty} \frac{(-1)^n w^{2n}}{(2n)!}, \qquad |w| < \infty,$$

and letting $w = 1/z$, for $0 < |z| < \infty$ we have

$$z \cos \left(z^{-1} \right) = z \sum_{n=0}^{\infty} \frac{(-1)^n}{(2n)! z^{2n}}$$

$$= z - \frac{1}{2! z} + \frac{1}{4! z^3} - \frac{1}{6! z^5} + \ldots .$$

From this expression it follows that the singularity is essential and

$$\operatorname*{Res}_{z=0} z \cos \left(z^{-1} \right) = -\frac{1}{2}.$$

(b) Also in this case the function is analytic everywhere except for the isolated
singularity at $z = 0$. For $0 < |z| < \infty$ we obtain

$$z^{-1} \cos (z) = z^{-1} \sum_{n=0}^{\infty} \frac{(-1)^n z^{2n}}{(2n)!}$$

$$= \frac{1}{z} - \frac{z}{2!} + \frac{z^3}{4!} - \frac{z^5}{6!} + \ldots .$$

From this expression it follows that the singularity is a simple pole with

$$\operatorname*{Res}_{z=0} z^{-1} \cos (z) = 1.$$

9.18

(a) The function $\tan(1/z)$ has singular points at $z = 0$ and at

$$z_n = \pm \frac{2}{\pi (2n + 1)}, \qquad n \in \mathbb{N},$$

solutions of $\cos(1/z) = 0$. It follows that $z = 0$ is a non-isolated singular point.
Note that the function can be expanded in the Laurent series in the annulus
$2/\pi < |z| < \infty$, but the coefficient of $1/z$ of this expansion is not the residue
of the function at $z = 0$.

(b) Using the notable expansion

$$\sin w = \sum_{k=0}^{\infty} \frac{(-1)^k}{(2k+1)!} w^{2k+1}, \qquad |w| < \infty,$$

and setting $w = 1/z$, for $0 < |z| < \infty$ we have

$$z^n \sin \frac{1}{z} = \sum_{k=0}^{\infty} \frac{(-1)^k}{(2k+1)!} \frac{1}{z^{2k-n+1}}.$$

The singularity at $z = 0$ is therefore essential and we have

$$\operatorname*{Res}_{z=0} z^n \sin \frac{1}{z} = \begin{cases} (-1)^{n/2}/(n+1)! & n \text{ even} \\ 0 & n \text{ odd} \end{cases}.$$

(c) The function

$$\sqrt{1 + \frac{1}{z}} = \exp\left(\frac{1}{2}\log(1 + 1/z)\right)$$

has a singularity at $z = 0$ and a branch line corresponding to the points $1+1/z = -t$, that is $z(t) = -1/(1+t)$, with $t \in [0, \infty)$. Therefore, this function is analytic everywhere in the complex plane except on the real segment $[-1, 0]$. The point $z = 0$ is a non-isolated singularity. Note that the function can be expanded in the Laurent series in the annulus $1 < |z| < \infty$

$$\sqrt{1 + \frac{1}{z}} = \sum_{k=0}^{\infty} \binom{\frac{1}{2}}{k} \frac{1}{z^k} = 1 + \frac{1}{2z} - \frac{1}{8z^2} + \frac{1}{16z^3} + \dots.$$

but the coefficient of $1/z$ of this expansion is not the residue of the function at $z = 0$.

9.19

(a) Recalling the notable Taylor series expansions of $\exp z$ and $\sin z$, for $0 < |z| < \infty$ we have

$$\frac{\exp(\sin z)}{z^2} = \frac{1}{z^2} \sum_{k=0}^{\infty} \frac{1}{k!} \left(\sum_{n=0}^{\infty} \frac{(-1)^n}{(2n+1)!} z^{2n+1}\right)^k$$

$$= \frac{1}{z^2}\left[1 + \left(z - \frac{z^3}{6} + \frac{z^5}{120} + \dots\right)\right.$$

$$+ \frac{1}{2}\left(z - \frac{z^3}{6} + \frac{z^5}{120} + \dots\right)^2 + \dots\Big]$$

$$= \frac{1}{z^2} + \frac{1}{z} + \frac{1}{2} + \left(-\frac{1}{6} + \frac{1}{6}\right)z + \left(-\frac{1}{6} + \frac{1}{24}\right)z^2 + \dots.$$

Therefore, the singularity at $z = 0$ is a double pole with

$$\operatorname*{Res}_{z=0} \frac{\exp(\sin z)}{z^2} = 1.$$

(b) Recalling the notable expansions of $\log(1 + z)$, $\cos z$ and $\sin z$, and observing that $\log(\cos z)$ is analytic in a neighborhood of $z = 0$, we have

$$\log(\cos z) = \log(1 + (\cos z - 1))$$

$$= \left(-\frac{z^2}{2} + \frac{z^4}{24} + \dots\right) - \frac{1}{2}\left(-\frac{z^2}{2} + \frac{z^4}{24} + \dots\right)^2 + \dots$$

$$= -\frac{z^2}{2} - \frac{z^4}{12} - \frac{z^6}{45} + \dots$$

and then

$$\frac{\sin z}{\log(\cos z)} = \frac{z - \frac{z^3}{6} + \dots}{-\frac{z^2}{2}\left(1 + \frac{z^2}{6} + \frac{2z^4}{45} \dots\right)}$$

$$= -\frac{2}{z^2}\left(z - \frac{z^3}{6} + \dots\right)\left[1 - \left(\frac{z^2}{6} + \frac{2z^4}{45} + \dots\right) + \dots\right]$$

$$= -\frac{2}{z} + \frac{2z}{3} + \dots.$$

It follows that the singularity at $z = 0$ is a simple pole and

$$\operatorname*{Res}_{z=0} \frac{\sin z}{\log(\cos z)} = -2.$$

Note that the function $\log(\cos z)$ is analytic in the whole complex plane except at the points $z(t)$ such that $\cos(z(t)) = -t$ with $t \in [0, \infty)$. Setting $w(t) = \exp(iz(t))$, we have

$$w(t) = \begin{cases} -t \pm i\sqrt{1 - t^2} & t \in [0, 1) \\ -t \pm \sqrt{t^2 - 1} & t \in [1, \infty) \end{cases}.$$

For $t \in [0, 1)$, we have

$$z(t) = -i \log \left(-t \pm i\sqrt{1 - t^2} \right)$$
$$= -i \left(\ln \sqrt{t^2 + (1 - t^2)} + i(\varphi(t) + 2k\pi) \right)$$
$$= \varphi(t) + 2k\pi, \qquad k \in \mathbb{Z},$$

where $\pi/2 \le \varphi(t) \le 3\pi/2$ ($\cos \varphi$ negative and $\sin \varphi$ positive or negative), that is, the segments of the real axis $[(2k + 1/2)\pi, (2k + 3/2)\pi]$ with $k \in \mathbb{Z}$. For $t \in [1, \infty)$, we have

$$z(t) = -i \log \left(-t \pm \sqrt{t^2 - 1} \right)$$
$$= -i \left(\ln \left(t \mp \sqrt{t^2 - 1} \right) + i(\pi + 2k\pi) \right)$$
$$= (2k + 1)\pi - i \ln \left(t \mp \sqrt{t^2 - 1} \right), \qquad k \in \mathbb{Z},$$

which represent the imaginary axes passing through $z = (2k + 1)\pi$ with $k \in \mathbb{Z}$, see Fig. 12.10.

9.20 First note that we can write $f(z) = h(z)/(z - z_0)$ with h analytic and non-zero at z_0. Therefore

$$\operatorname*{Res}_{z=z_0} f(z) = h(z_0).$$

Furthermore, note that fg is analytic in an annulus $0 < |z - z_0| < r$ and can be there expanded in the Laurent series. Let γ be a simple closed piecewise regular positively oriented curve, contained in $A(z_0, 0, r)$ and containing z_0 within it, the residue of fg at z_0 is by definition

$$\operatorname*{Res}_{z=z_0} f(z)g(z) = \frac{1}{2\pi i} \int_\gamma f(z)g(z)\,dz$$
$$= \frac{1}{2\pi i} \int_\gamma \frac{h(z)g(z)}{z - z_0}\,dz$$
$$= h(z_0)g(z_0),$$

having calculated the integral using the Cauchy formula since hg is an analytic function on and within the integration path γ. Combining the two results we get the statement.

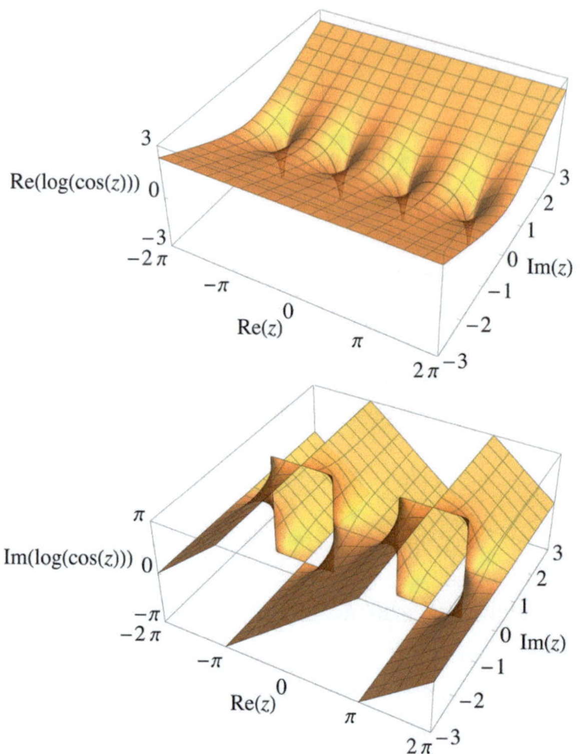

Fig. 12.10 Real and imaginary parts of the principal branch of $\log(\cos(z))$

9.21 Note that $\sin z = 0$ at points $z_n = \pi n$, $n \in \mathbb{Z}$. For $n \neq 0$, $z^2 \sin z$ has a simple zero at z_n, moreover e^{z_n} is different from zero. So f has at z_n a simple pole with

$$\operatorname*{Res}_{z=z_n} f(z) = \left.\frac{e^z}{2z \sin z + z^2 \cos z}\right|_{z=z_n} = (-1)^n \frac{e^{n\pi}}{n^2 \pi^2}.$$

At $z_0 = 0$, $z^2 \sin z$ has a triple zero, moreover $e^{z_0} \neq 0$. Therefore, f has at z_0 a pole of order 3. The corresponding residue is found by expanding f in the Laurent series in the region $0 \leq |z| \leq \pi$

$$f(z) = \frac{1 + z + \frac{z^2}{2} + \dots}{z^2(z - \frac{z^3}{6} + \dots)}$$

$$= \frac{1}{z^3}\left(1 + z + \frac{z^2}{2} + \dots\right)\left(1 + \frac{z^2}{6} + \dots\right)$$

$$= \frac{1}{z^3} + \frac{1}{z^2} + \frac{2}{3z} + \dots,$$

from which we conclude

$$\operatorname*{Res}_{z=z_0} f(z) = \frac{2}{3}.$$

9.22 The function has only one isolated singular point at $z = 2$. Using the notable Taylor series expansion

$$\cos w = \sum_{k=0}^{\infty} \frac{(-1)^k}{(2k)!} w^{2k}, \qquad |w| < \infty,$$

and setting $w = 1/(z-2)$, in the annular region $0 < |z-2| < \infty$ we have

$$f(z) = ((z-2)+2)^3 \sum_{k=0}^{\infty} \frac{(-1)^k}{(2k)!} \frac{1}{(z-2)^{2k}}$$

$$= \left((z-2)^3 + 6(z-2)^2 + 12(z-2) + 8 \right) \sum_{k=0}^{\infty} \frac{(-1)^k}{(2k)!} \frac{1}{(z-2)^{2k}}.$$

From this expression it follows that the singularity is essential and

$$\operatorname*{Res}_{z=2} f(z) = \frac{(-1)^2}{4!} + 12\frac{(-1)^1}{2!} = \frac{1}{24} - 6 = -\frac{143}{24}.$$

9.23 The principal branch of $\log(1+z)$ is a function analytic everywhere in $\mathbb{C}$ except on the real semi-axis $\{z = -1 - t, \ t \in [0, \infty)\}$. The denominator of $f(z)$ vanishes at the points $z = k\pi$, $k \in \mathbb{Z}$. Therefore, f has isolated singularities only at the non-negative points $z = k\pi$ with $k = 0, 1, 2 \ldots$. Let us first consider the singular point $z = 0$. In the annulus $0 < |z| < 1$, f can be expanded in the Laurent series and the first terms of this expansion are

$$f(z) = \frac{z - \frac{z^2}{2} + \frac{z^3}{3} + \ldots}{z\left(z - \frac{z^3}{3!} + \ldots\right)}$$

$$= \frac{1}{z} \frac{1 - \frac{z}{2} + \frac{z^2}{3} + \ldots}{1 - \frac{z^2}{3!} + \ldots}$$

$$= \frac{1}{z}\left(1 - \frac{z}{2} + \frac{z^2}{3} + \ldots\right)\left(1 + \frac{z^2}{3!} + \ldots\right)$$

$$= \frac{1}{z} - \frac{1}{2} + \frac{z}{2} + \ldots.$$

From this expression we deduce that $z = 0$ is a simple pole and

$$\operatorname*{Res}_{z=0} f(z) = 1.$$

As for the singular points $z = k\pi$, $k = 1, 2, \ldots$, observe that in a neighborhood of these points we can set $f(z) = p(z)/q(z)$ with $p(z) = z^{-1}\log(1+z)$ and $q(z) = \sin z$. The functions p and q are analytic at $z = k\pi$, moreover $p(k\pi) = (k\pi)^{-1}\ln(1+k\pi) \neq 0$, $q(k\pi) = 0$ and $q'(k\pi) = (-1)^k$. We conclude that $z = k\pi$ is a simple pole and

$$\operatorname*{Res}_{z=k\pi} f(z) = \frac{p(k\pi)}{q'(k\pi)} = (-1)^k \frac{\ln(1+k\pi)}{k\pi}, \qquad k = 1, 2, \ldots.$$

9.24 For $n \geq 1$, the function $f_n(z)$ has a simple pole at $z = 2$, a double pole at $z = 1$, a pole of order n at $z = 0$ and is elsewhere analytic. In the region $0 < |z| < 1$ it can be expanded in the Laurent series and, recalling the notable geometric series, we have

$$
\begin{aligned}
f_n(z) &= \frac{1}{z^n}\left(\frac{1}{1-z}\right)^2\left(-\frac{1}{2}\frac{1}{1-z/2}\right) \\
&= -\frac{1}{2z^n}\left(1 + z + z^2 + z^3 + \ldots\right)^2\left(1 + \frac{z}{2} + \frac{z^2}{4} + \frac{z^3}{8} + \ldots\right) \\
&= -\frac{1}{2z^n}\left(1 + 2z + 3z^2 + 4z^3 + \ldots\right)\left(1 + \frac{z}{2} + \frac{z^2}{4} + \frac{z^3}{8} + \ldots\right) \\
&= -\frac{1}{2z^n}\left(1 + \frac{5}{2}z + \frac{17}{4}z^2 + \frac{49}{8}z^3 + \ldots\right) \\
&= -\frac{1}{2}z^{-n} - \frac{5}{4}z^{1-n} - \frac{17}{8}z^{2-n} - \frac{49}{16}z^{3-n} + O(z^{4-n}).
\end{aligned}
$$

From this expression it follows

$$\operatorname*{Res}_{z=0} f_1(z) = -\frac{1}{2}, \qquad \operatorname*{Res}_{z=0} f_2(z) = -\frac{5}{4},$$

$$\operatorname*{Res}_{z=0} f_3(z) = -\frac{17}{8}, \qquad \operatorname*{Res}_{z=0} f_4(z) = -\frac{49}{16}.$$

If $g(z)$ is analytic at $z = 0$, it can be expanded in a series of powers around $z = 0$, that is, $\exists r > 0$ such that $\forall z \in B(0, r)$

$$g(z) = \sum_{k=0}^{\infty} a_k z^k.$$

Then, for $0 < |z| < \min(1, r)$ we have

$$f_2(z)g(z) = \left(-\frac{1}{2z^2} - \frac{5}{4z} - \frac{17}{8} - \frac{49}{16}z + \ldots\right)\left(a_0 + a_1 z + a_2 z^2 + \ldots\right)$$

$$= -\frac{a_0}{2}\frac{1}{z^2} + \left(-\frac{5}{4}a_0 - \frac{1}{2}a_1\right)\frac{1}{z} + O(z^0).$$

A double pole is obtained if $a_0 \neq 0$ and a zero residue if $a_1/a_0 = -5/2$. The multi-valued function $g(z) = (1+z)^{-5/2}$ satisfies the required conditions. Note that the principal branch of g, whose branch line is $z(t) = -1 - t$, $t \geq 0$, is analytic at $z = 0$.

9.25 The function $f(z)$ has an essential singularity at $z = -2$ and a simple pole at $z = 3$, elsewhere being analytic. In the annulus $A(-2, 0, 5)$ it can be expanded in the Laurent series. Remembering the notable expansions

$$\sin z = \sum_{k=0}^{\infty} \frac{(-1)^k}{(2k+1)!} z^{2k+1}, \qquad |z| < \infty,$$

$$\frac{1}{1-z} = \sum_{k=0}^{\infty} z^k, \qquad |z| < 1,$$

for $0 < |(z+2)/3| < 1$ we can write

$$f(z) = -\frac{1}{5}\frac{1}{1 - \frac{z+2}{5}} (z+2)^4 \sin\left(\frac{1}{z+2}\right)$$

$$= -\frac{1}{5}\left(\sum_{k=0}^{\infty} \frac{1}{5^k}(z+2)^{k+4}\right)\left(\sum_{n=0}^{\infty} \frac{(-1)^n}{(2n+1)!}\frac{1}{(z+2)^{2n+1}}\right).$$

The residue of $f(z)$ at $z = -2$ is the sum of the coefficients of the terms obtained from the previous series product whose exponents satisfy $k + 4 - (2n+1) = -1$. This condition is equivalent to $n = k/2 + 2$ and can be satisfied only for even k. Setting $k = 2j$, we have

$$\operatorname*{Res}_{z=-2} f(z) = -\frac{1}{5}\sum_{j=0}^{\infty} \frac{1}{5^{2j}}\frac{(-1)^{j+2}}{(2j+5)!}$$

$$= \sum_{j=0}^{\infty} \frac{(-1)^{j+1}}{5^{2j+1}(2j+5)!}$$

$$= -\frac{1}{600} + \frac{1}{630000} + \ldots.$$

9.26 By hypothesis, for $0 < |z - z_0| < \varepsilon$ we can write

$$f(z) = (z - z_0)^m g(z),$$

with $g(z)$ analytic and non-zero at z_0. Differentiating this relationship we find

$$f'(z) = m(z - z_0)^{m-1} g(z) + (z - z_0)^m g'(z)$$

and

$$\frac{f'(z)}{f(z)} p(z) = \left(\frac{m}{z - z_0} + \frac{g'(z)}{g(z)} \right) p(z)$$

$$= \left(\frac{m}{z - z_0} + \frac{g'(z)}{g(z)} \right) \sum_{k=0}^{\infty} \frac{p^{(k)}(z_0)}{k!} (z - z_0)^k.$$

Since $g'(z)/g(z)$ is analytic at z_0, it immediately follows

$$\operatorname*{Res}_{z=z_0} \left[\frac{f'(z)}{f(z)} p(z) \right] = mp(z_0).$$

9.27 By hypothesis $\exists \varepsilon > 0$ such that for $|z - z_0| < \varepsilon$ the following Taylor series expansion holds

$$q(z) = \sum_{k=1}^{\infty} \frac{q^{(k)}(z_0)}{k!} (z - z_0)^k$$

$$= (z - z_0) \sum_{n=0}^{\infty} \frac{q^{(n+1)}(z_0)}{(n + 1)!} (z - z_0)^n$$

$$= (z - z_0)h(z).$$

The function $h(z)$ being the sum of a convergent power series is analytic in z_0. Moreover $h(z_0) \neq 0$. In fact, differentiating $q(z) = (z - z_0)h(z)$ with respect to z we have

$$q'(z) = h(z) + (z - z_0)h'(z) \qquad \Rightarrow \qquad h(z_0) = q'(z_0).$$

Differentiating again we obtain

$$q''(z) = 2h'(z) + (z - z_0)h''(z) \qquad \Rightarrow \qquad 2h'(z_0) = q''(z_0).$$

Set

$$f(z) = \frac{1}{q(z)^2} = \frac{g(z)}{(z-z_0)^2}, \qquad g(z) = h(z)^{-2}.$$

Using the properties of h, we can say that the function g is analytic and non-zero at z_0. So $f(z)$ has a pole of order 2 in z_0 with

$$\operatorname*{Res}_{z=z_0} f(z) = g'(z_0) = -\frac{2h'(z_0)}{h(z_0)^3} = -\frac{q''(z_0)}{q'(z_0)^3}.$$

9.28   Since $f(z)$ has a pole of order m at z_0, there exists $\varepsilon > 0$ such that for $0 < |z - z_0| < \varepsilon$ the following Laurent series expansion holds

$$f(z) = \sum_{k=0}^{\infty} a_k(z-z_0)^k + \frac{b_1}{z-z_0} + \frac{b_2}{(z-z_0)^2} + \cdots + \frac{b_m}{(z-z_0)^m},$$

with $b_m \neq 0$. Therefore

$$(z-z_0)^m f(z) = \sum_{k=0}^{\infty} a_k(z-z_0)^{k+m}$$
$$+ b_1(z-z_0)^{m-1} + b_2(z-z_0)^{m-2} + \cdots + b_m.$$

Differentiating this expression member by member $m-1$ times with respect to z, we obtain

$$\frac{d^{m-1}}{dz^{m-1}}\left((z-z_0)^m f(z)\right)$$
$$= \sum_{k=0}^{\infty} a_k(k+m)(k+m-1)\ldots(k+2)(z-z_0)^{k+1}$$
$$+ b_1(m-1)(m-2)\ldots 1.$$

Dividing by $(m-1)!$ and taking the limit $z \to z_0$, we conclude

$$\lim_{z \to z_0} \frac{1}{(m-1)!} \frac{d^{m-1}}{dz^{m-1}}\left((z-z_0)^m f(z)\right) = b_1.$$

9.29   Set $f(z) = \sin^2 z + \cos^2 z - 1$ analytic throughout $\mathbb{C}$. Since $f(z) = 0 \; \forall z \in \mathbb{R}$ and $\mathbb{C}$ is open and connected, by Theorem 9.15 f is identically null in $\mathbb{C}$.

9.30   By hypothesis p and q are analytic at z_0 and we have $p(z_0) \neq 0$, $q(z_0) = q'(z_0) = 0$ and $q''(z_0) \neq 0$. This implies that in a neighbourhood of z_0 we have

$q(z) = (z - z_0)^2 g(z)$, with g analytic and non-zero at z_0. In the same neighborhood we have therefore

$$f(z) = \frac{h(z)}{(z - z_0)^2},$$

with $h(z) = p(z)/g(z)$ analytic and non-zero at z_0. This is enough to conclude that f has a double pole at z_0 and

$$\operatorname*{Res}_{z=z_0} f(z) = h'(z_0) = \frac{p'(z_0)g(z_0) - p(z_0)g'(z_0)}{g(z_0)^2}.$$

From the relationships

$$q(z) = (z - z_0)^2 g(z),$$
$$q'(z) = 2(z - z_0)g(z) + (z - z_0)^2 g'(z),$$
$$q''(z) = 2g(z) + 4(z - z_0)g'(z) + (z - z_0)^2 g''(z),$$
$$q'''(z) = 6g'(z) + 6(z - z_0)g''(z) + (z - z_0)g'''(z),$$

for $z = z_0$ we obtain

$$q''(z_0) = 2g(z_0),$$
$$q'''(z_0) = 6g'(z_0),$$

and so

$$\operatorname*{Res}_{z=z_0} f(z) = \frac{p'(z_0)q''(z_0)/2 - p(z_0)q'''(z_0)/6}{(q''(z_0)/2)^2}.$$

9.31 By the Cauchy theorem we have

$$\int_{|z|=1} \frac{f(z)}{nz - 1}\,\mathrm{d}z = \frac{2\pi\mathrm{i}}{n} f(1/n)$$

and therefore

$$f(1/n) = 0, \qquad \forall n \in \mathbb{N}.$$

In the open and connected domain $B(0, 2)$ where f is analytic, the zeros of f admit the limit point $z = 0$. By the identity theorem this implies that f is identically zero at $B(0, 2)$.

9.32 Consider the two functions

$$f(z) = \log \frac{1+z}{1-z}, \qquad g(z) = \log(1+z) - \log(1-z),$$

where log denotes the principal branch of the logarithm. These functions are analytic everywhere in $\mathbb{C}$ except on the real semi-axes $(-\infty, -1]$ and $[1, +\infty)$, see Exercise 8.3. Therefore, in the open ball $B(0, 1)$, where it is possible to expand $f(z)$ in the Taylor series, also $g(z)$ is analytic and $f(x) = g(x)$ holds, for example, when $x \in [0, 1/2]$. By the identity theorem this is sufficient to have $f(z) = g(z)$ $\forall z \in B(0, 1)$. Using the notable expansions

$$\log(1+z) = \sum_{k=1}^{\infty} \frac{(-1)^{k-1}}{k} z^k,$$

$$\log(1-z) = \sum_{k=1}^{\infty} \frac{-1}{k} z^k,$$

valid for $|z| < 1$, in the same region we obtain

$$\log \frac{1+z}{1-z} = \sum_{k=1}^{\infty} \frac{(-1)^{k-1}+1}{k} z^k$$

$$= \sum_{n=0}^{\infty} \frac{2}{2n+1} z^{2n+1}.$$

12.10 Exercises of Chap. 10

10.1 Setting $e^{i\theta} = z$ we have $d\theta = dz/iz$, $\cos\theta = (z + z^{-1})/2$ and

$$\int_0^{2\pi} \frac{d\theta}{(a + b\cos\theta)^2} = \int_{|z|=1} f(z)dz,$$

where

$$f(z) = \frac{-4iz}{\left(bz^2 + 2az + b\right)^2}.$$

and the circle $|z| = 1$ is positively oriented. Since the roots of $bz^2 + 2az + b = 0$ are the real numbers

$$z_{\pm} = -\frac{a}{b} \pm \sqrt{\left(\frac{a}{b}\right)^2 - 1},$$

with $z_- < -1$ and $-1 < z_+ < 0$ (note that $z_+ z_- = 1$), the function $f(z)$ has within the circle $|z| = 1$ an isolated singularity at $z = z_+$. This singularity is a double pole since

$$f(z) = \frac{-4iz}{b^2(z - z_-)^2(z - z_+)^2} = \frac{g(z)}{(z - z_+)^2},$$

with

$$g(z) = \frac{-4iz}{b^2(z - z_-)^2}$$

analytic and non-zero at z_+. The corresponding residue is

$$\operatorname*{Res}_{z=z_+} f(z) = g'(z_+)$$

$$= \frac{-4i}{b^2} \frac{(z_+ - z_-) - 2z_+}{(z_+ - z_-)^3}$$

$$= \frac{-ia}{(a^2 - b^2)^{3/2}}.$$

In conclusion

$$\int_0^{2\pi} \frac{d\theta}{(a + b\cos\theta)^2} = 2\pi i \operatorname*{Res}_{z=z_+} f(z)$$

$$= \frac{2\pi a}{(a^2 - b^2)^{3/2}}.$$

10.2 Setting $e^{i\theta} = z$ we have $d\theta = dz/iz$, $\cos\theta = (z + z^{-1})/2$ and

$$\int_0^{2\pi} \cos^2\theta \, d\theta = \int_\gamma \left(\frac{z + z^{-1}}{2}\right)^2 \frac{dz}{iz} = \int_\gamma \frac{(z^2 + 1)^2}{4iz^3} \, dz,$$

where $\gamma(\theta) = e^{i\theta}$, $0 \le \theta \le 2\pi$. The function to be integrated

$$f(z) = \frac{(z^2 + 1)^2}{4iz^3}$$

is analytic on and within γ except for the pole of order 3 at $z = 0$. By the residue theorem

$$\int_\gamma f(z)\, dz = 2\pi i \operatorname*{Res}_{z=0} f(z) = 2\pi i \, \frac{g''(0)}{2!},$$

where $g(z) = \left(z^2 + 1\right)^2 / 4i$. Since $g''(z) = \left(3z^2 + 1\right)/i$ we conclude

$$\int_0^{2\pi} \cos^2\theta \, d\theta = 2\pi i \, \frac{g''(0)}{2!} = 2\pi i \, \frac{1}{2i} = \pi.$$

10.3 Setting $e^{i\theta} = z$ we have $d\theta = dz/iz$, $\cos\theta = (z + z^{-1})/2$ and

$$\int_0^\pi \frac{d\theta}{a + \cos\theta} = \frac{1}{2} \int_0^{2\pi} \frac{d\theta}{a + \cos\theta} = \int_\gamma f(z)dz,$$

where $\gamma(\theta) = e^{i\theta}$, $0 \le \theta \le 2\pi$, and $f(z) = -i/(z^2 + 2az + 1)$. The function f has two simple poles at $z_\pm = -a \pm \sqrt{a^2 - 1}$. Note that $|z_-| > 1$ and, since $z_+ z_- = 1$, $|z_+| < 1$. Therefore, the pole at z_+ is internal to the path γ while the pole at z_- is external, and by the residue theorem we conclude

$$\int_0^\pi \frac{d\theta}{a + \cos\theta} = 2\pi i \operatorname*{Res}_{z=z_+} f(z)$$

$$= 2\pi i \, \left. \frac{-i}{z - z_-} \right|_{z=z_+}$$

$$= \frac{\pi}{\sqrt{a^2 - 1}}.$$

10.4 Setting $z = e^{i\theta}$ and calling γ the closed path $|z| = 1$ traveled counterclockwise, we have

$$\int_0^{2\pi} (\sin\theta)^{2n} d\theta = \int_\gamma \left(\frac{z - z^{-1}}{2i} \right)^{2n} \frac{dz}{iz} = \int_\gamma \frac{(z^2 - 1)^{2n}}{i(-1)^n 2^{2n} z^{2n+1}} dz.$$

The integrand function is analytic on and within γ except for the pole of order $2n+1$ at $z = 0$. Using the expansion of Newton's binomial

$$(z^2 - 1)^{2n} = \sum_{k=0}^{2n} \frac{(2n)!}{k!(2n - k)!} (z^2)^k (-1)^{2n-k},$$

we obtain the Laurent series expansion of $(z^2 - 1)^{2n}/z^{2n+1}$ in the annulus $0 < |z| < \infty$

$$\frac{(z^2 - 1)^{2n}}{z^{2n+1}} = \sum_{k=0}^{2n} \frac{(-1)^{2n-k}(2n)!}{k!(2n - k)!} z^{2k-1-2n}.$$

The term $k = n$ of this expansion, proportional to z^{-1}, gives

$$\operatorname*{Res}_{z=0} \frac{(z^2 - 1)^{2n}}{z^{2n+1}} = \frac{(-1)^n (2n)!}{n!n!}.$$

By the residue theorem we conclude

$$\int_0^{2\pi} (\sin\theta)^{2n} d\theta = \frac{1}{i(-1)^n 2^{2n}} 2\pi i \operatorname*{Res}_{z=0} \frac{(z^2 - 1)^{2n}}{z^{2n+1}}$$

$$= \frac{\pi(2n)!}{2^{2n-1}(n!)^2}.$$

10.5 Setting $e^{i\theta} = z$ we have $d\theta = dz/iz$, $\cos\theta = (z + z^{-1})/2$ and

$$\int_0^\pi \frac{d\theta}{(a + \cos\theta)^2} = \frac{1}{2} \int_0^{2\pi} \frac{d\theta}{(a + \cos\theta)^2} = \int_\gamma f(z)dz,$$

where $\gamma(\theta) = e^{i\theta}$, $0 \le \theta \le 2\pi$, and

$$f(z) = \frac{-2iz}{(z^2 + 2az + 1)^2}.$$

The function f has two double poles at $z_\pm = -a \pm \sqrt{a^2 - 1}$. Note that $|z_-| > 1$ and, since $z_+ z_- = 1$, $|z_+| < 1$. Therefore, the pole at z_+ is internal to the path γ while the one at z_- is external. The residue theorem provides

$$\int_0^\pi \frac{d\theta}{(a + \cos\theta)^2} = 2\pi i \operatorname*{Res}_{z=z_+} f(z)$$

$$= 2\pi i \frac{d}{dz} \frac{-2iz}{(z - z_-)^2}\bigg|_{z=z_+}$$

$$= 2\pi i \frac{2i(z_+ + z_-)}{(z_+ - z_-)^3}$$

$$= \frac{\pi a}{(a^2 - 1)^{3/2}}.$$

10.6 Setting $e^{i\theta} = z$ we have $d\theta = dz/iz$, $\cos\theta = (z + z^{-1})/2$ and

$$\int_0^{2\pi} \frac{d\theta}{2 + \cos\theta} = \int_\gamma f(z)\,dz,$$

where $\gamma(\theta) = e^{i\theta}$, $0 \le \theta \le 2\pi$, and

$$f(z) = \frac{-2i}{z^2 + 4z + 1}.$$

The function f has two simple poles at $z_\pm = -2 \pm \sqrt{3}$. Note that $|z_-| > 1$ and, since $z_+ z_- = 1$, $|z_+| < 1$. Therefore, the pole at z_+ is internal to the path γ while the one at z_- is external. From the residue theorem we then obtain

$$\int_0^{2\pi} \frac{d\theta}{2 + \cos\theta} = 2\pi i \operatorname*{Res}_{z=z_+} f(z)$$

$$= 2\pi i \left. \frac{-2i}{z - z_-} \right|_{z=z_+}$$

$$= \frac{2\pi}{\sqrt{3}}.$$

10.7 Setting $z = e^{i\theta} = \cos\theta + i\sin\theta$ we have $dz = i e^{i\theta}\, d\theta$ and

$$\int_0^{2\pi} \sin(\exp(\cos\theta + i\sin\theta))\,d\theta = \int_\gamma \sin(\exp(z))\frac{dz}{iz},$$

where $\gamma(\theta) = e^{i\theta}$, $0 \le \theta \le 2\pi$. The integrand function $\sin(\exp(z))/(iz)$ is analytic on and within γ except for the simple pole at $z = 0$. By the residue theorem we therefore have

$$\int_\gamma \sin(\exp(z))\frac{dz}{iz} = 2\pi i \operatorname*{Res}_{z=0} \frac{\sin(\exp(z))}{iz}$$

$$= 2\pi i \frac{\sin(\exp(0))}{i}$$

$$= 2\pi \sin(1).$$

10.8 The proposed integral is equivalent to the integral of the complex function $e^z/(iz)$ along the path $\gamma(\theta) = e^{i\theta}$, $0 \le \theta \le 2\pi$, in fact,

$$\int_0^{2\pi} e^{e^{i\theta}}\,d\theta = \int_0^{2\pi} e^{\gamma(\theta)}\frac{\gamma'(\theta)}{i\gamma(\theta)}\,d\theta = \int_\gamma \frac{e^z}{iz}\,dz.$$

The last integral is immediately evaluated by the residue theorem

$$\int_\gamma \frac{e^z}{iz}\,dz = 2\pi i \operatorname*{Res}_{z=0} \frac{e^z}{iz} = 2\pi i \frac{e^0}{i} = 2\pi.$$

10.9 First note that the improper integral exists and we have

$$\int_0^\infty \frac{1}{(x^2+1)^2}\,dx = \frac{1}{2}\int_{-\infty}^{+\infty} \frac{1}{(x^2+1)^2}\,dx = \frac{1}{2}\lim_{R\to\infty}\int_{-R}^{+R} \frac{1}{(x^2+1)^2}\,dx.$$

The function

$$f(z) = \frac{1}{(z^2+1)^2} = \frac{1}{(z-i)^2(z+i)^2}$$

is analytic everywhere except for the two double poles at $z = \pm i$. The integral of f along the positively oriented closed path $\gamma = \lambda_R + \gamma_R$, where $\lambda_R(x) = x$, $-R \le x \le R$ and $\gamma_R(\theta) = Re^{i\theta}, 0 \le \theta \le \pi$, gives, for $R > 1$,

$$\begin{aligned}
\int_\gamma f(z)dz &= \int_{-R}^{R} \frac{1}{(x^2+1)^2}\,dx + \int_{\gamma_R} f(z)dz \\
&= 2\pi i \operatorname*{Res}_{z=i} f(z) \\
&= 2\pi i g'(i) \\
&= \frac{\pi}{2},
\end{aligned}$$

having used for the calculation of the residue the decomposition $f(z) = g(z)/(z-i)^2$ with $g(z) = 1/(z+i)^2$ analytic on and within γ and non-zero at $z = i$. Since

$$\lim_{R\to\infty}\int_{\gamma_R} f(z)dz = 0,$$

taking the limit $R \to \infty$ of the integral over γ we obtain

$$\int_{-\infty}^{+\infty} \frac{1}{(x^2+1)^2}\,dx = \frac{\pi}{2},$$

that is,

$$\int_0^\infty \frac{1}{(x^2+1)^2}\,dx = \frac{\pi}{4}.$$

10.10 The function

$$f(z) = \frac{z}{(z^2 + 4z + 13)^2} = \frac{z}{(z - z_+)^2(z - z_-)^2}$$

is analytic everywhere except for the two double poles at $z_\pm = -2\pm 3\mathrm{i}$. The integral of f along the closed path $\gamma = \lambda_R + \gamma_R$, where $\lambda_R(x) = x$, $-R \le x \le R$, and $\gamma_R(\theta) = R\mathrm{e}^{\mathrm{i}\theta}$, $0 \le \theta \le \pi$, by the residue theorem gives

$$\int_\gamma f(z)\mathrm{d}z = \int_{-R}^{R} \frac{x}{(x^2 + 4x + 13)^2}\mathrm{d}x + \int_{\gamma_R} f(z)\mathrm{d}z$$

$$= 2\pi\mathrm{i}\,\operatorname*{Res}_{z=z_+} f(z)$$

$$= 2\pi\mathrm{i}\,\frac{\mathrm{d}}{\mathrm{d}z}\frac{z}{(z - z_-)^2}\bigg|_{z=z_+}$$

$$= 2\pi\mathrm{i}\,\frac{-z_+ - z_-}{(z_+ - z_-)^3}$$

$$= -\frac{\pi}{27}.$$

Since for $z \in \{\gamma_R\}$ with R large we have $|f(z)| \le R/(R^2 - 4R - 13)^2$, it follows that

$$\left|\int_{\gamma_R} f(z)\mathrm{d}z\right| \le \frac{\pi R^2}{(R^2 - 4R - 13)^2} \xrightarrow{R\to\infty} 0.$$

Taking the limit $R \to \infty$ of the integral over γ we thus conclude

$$\int_{-\infty}^{+\infty} \frac{x}{(x^2 + 4x + 13)^2}\mathrm{d}x = \mathrm{PV}\int_{-\infty}^{+\infty} \frac{x}{(x^2 + 4x + 13)^2}\mathrm{d}x = -\frac{\pi}{27}.$$

10.11 The function

$$f(z) = \frac{z^2}{(z^2 + 1)(z^2 + 9)} = \frac{z^2}{(z - \mathrm{i})(z + \mathrm{i})(z - 3\mathrm{i})(z + 3\mathrm{i})}$$

is analytic everywhere except for the four simple poles at $\pm\mathrm{i}$ and $\pm 3\mathrm{i}$. The integral of f on the closed path $\gamma = \lambda_R + \gamma_R$, where $\lambda_R(x) = x$, $-R \le x \le R$, and $\gamma_R(\theta) = R\mathrm{e}^{\mathrm{i}\theta}$, $0 \le \theta \le \pi$, gives, for $R > 3$,

$$\int_\gamma f(z)\mathrm{d}z = \int_{-R}^{R} \frac{x^2}{(x^2 + 1)(x^2 + 9)}\mathrm{d}x + \int_{\gamma_R} f(z)\mathrm{d}z$$

$$= 2\pi i \left(\operatorname*{Res}_{z=i} f(z) + \operatorname*{Res}_{z=3i} f(z) \right)$$

$$= 2\pi i \left(-\frac{1}{16i} + \frac{3}{16i} \right)$$

$$= \frac{\pi}{4}.$$

Taking the limit $R \to \infty$ of the integral over γ and observing that

$$\lim_{R \to \infty} \int_{\gamma_R} f(z)\mathrm{d}z = 0,$$

we find

$$\int_{-\infty}^{+\infty} \frac{x^2}{(x^2+1)(x^2+9)}\,\mathrm{d}x = \mathrm{PV} \int_{-\infty}^{+\infty} \frac{x^2}{(x^2+1)(x^2+9)}\,\mathrm{d}x = \frac{\pi}{4}.$$

10.12 The function

$$f(z) = \frac{z^2}{(z^2+16)^2}$$

has double poles at $z = \pm 4i$. Put $\gamma = \lambda_R + \gamma_R$ the closed integration path defined by $\lambda_R(x) = x, \ -R \le x \le R$, and $\gamma_R(\theta) = Re^{i\theta}, 0 \le \theta \le \pi$, since f is, for $R > 4$, analytic on and within γ except for the double pole at $z = 4i$, the residue theorem gives

$$\int_{\gamma} f(z)\mathrm{d}z = \int_{\lambda_R} f(z)\mathrm{d}z + \int_{\gamma_R} f(z)\mathrm{d}z$$

$$= 2\pi i \operatorname*{Res}_{z=4i} f(z)$$

$$= 2\pi i \left. \frac{g'(z)}{1!} \right|_{z=4i}$$

$$= \frac{\pi}{8},$$

where $g(z) = z^2/(z+4i)^2$. For the integrals over the paths that compose γ we have

$$\int_{\lambda_R} f(z)\mathrm{d}z = \int_{-R}^{R} \frac{x^2}{(x^2+16)^2}\,\mathrm{d}x,$$

$$\left| \int_{\gamma_R} f(z)\mathrm{d}z \right| \le \pi R \, \frac{R^2}{(R^2-16)^2} \xrightarrow{R \to \infty} 0.$$

Therefore, in the limit $R \to \infty$ we obtain

$$\int_{-\infty}^{+\infty} \frac{x^2}{(x^2 + 16)^2} \, dx = \frac{\pi}{8}.$$

By the parity of the integrand function we conclude

$$\int_0^\infty \frac{x^2}{(x^2 + 16)^2} \, dx = PV \int_0^\infty \frac{x^2}{(x^2 + 16)^2} \, dx = \frac{\pi}{16}.$$

10.13 Setting $f(z) = z/(z^2 + b^2)$ and calling $\gamma = \lambda + \gamma_R$ the closed path of integration with $\lambda(x) = x,\ -R \le x \le R$, and $\gamma_R(\theta) = R \exp(i\theta),\ 0 \le \theta \le \pi$, for $R > b$ we have

$$\int_\gamma f(z) e^{iaz} dz = \int_\lambda f(z) e^{iaz} dz + \int_{\gamma_R} f(z) e^{iaz} dz$$

$$= \int_{-R}^{R} \frac{x e^{iax}}{x^2 + b^2} \, dx + \int_{\gamma_R} f(z) e^{iaz} dz$$

$$= 2\pi i \operatorname*{Res}_{z=ib} \left[f(z) e^{iaz} \right]$$

$$= 2\pi i \left. \frac{z e^{iaz}}{z + ib} \right|_{z=ib}$$

$$= \pi i e^{-ab}.$$

For $z \in \{\gamma_R\}$, we have that $|f(z)| \le R/(R^2 - b^2)$ is bounded by a quantity that vanishes for $R \to \infty$. Hence by Jordan's lemma

$$\lim_{R \to \infty} \int_{\gamma_R} f(z) e^{iaz} dz = 0.$$

In conclusion

$$PV \int_{-\infty}^{+\infty} \frac{x e^{iax}}{x^2 + b^2} \, dx = \pi i e^{-ab},$$

from which, taking the imaginary part,

$$PV \int_{-\infty}^{+\infty} \frac{x \sin(ax)}{x^2 + b^2} \, dx = \pi e^{-ab}.$$

10.14 Let $f(z) = z/((z-b)^2 + c^2)$ and $\gamma = \lambda_R + \gamma_R$ the closed integration path with $\lambda_R(x) = x,\ -R \le x \le R$, and $\gamma_R(\theta) = R e^{i\theta},\ 0 \le \theta \le \pi$. For $R > \sqrt{b^2 + c^2}$ the function $f(z)e^{iaz}$ is analytic on and within γ except for the simple pole at $z = b+ic$. By the residue theorem we can write

$$\int_\gamma f(z)e^{iaz}dz = \int_{\lambda_R} f(z)e^{iaz}dz + \int_{\gamma_R} f(z)e^{iaz}dz$$

$$= \int_{-R}^{R} \frac{xe^{iax}}{(x-b)^2+c^2}dx + \int_{\gamma_R} f(z)e^{iaz}dz$$

$$= 2\pi i \operatorname*{Res}_{z=b+ic} f(z)e^{iaz}$$

$$= \frac{\pi}{c}e^{-ac}(b+ic)e^{iab}.$$

For $z \in \{\gamma_R\}$, we have that $|f(z)| \leq R/((R-b)^2 - c^2)$ is bounded by a quantity which vanishes for $R \to \infty$. Hence by Jordan's lemma

$$\lim_{R\to\infty} \int_{\gamma_R} f(z)e^{iaz}dz = 0.$$

The limit $R \to \infty$ of the integral over γ in conclusion gives

$$\lim_{R\to\infty} \int_{-R}^{+R} \frac{xe^{iax}}{(x-b)^2+c^2}dx = \frac{\pi}{c}e^{-ac}(b+ic)e^{iab},$$

from which, taking the imaginary part,

$$\mathrm{PV} \int_{-\infty}^{+\infty} \frac{x\sin(ax)}{(x-b)^2+c^2}dx = \frac{\pi}{c}e^{-ac}\left(b\sin(ab)+c\cos(ab)\right).$$

10.15 Consider the complex function $f(z) = z^2 e^{iz}/((z-1)(z^2+1))$ which has simple poles at $z = 1$ and $z = \pm i$ and integrate it along the positively oriented closed path $\gamma = \lambda_1^l + \gamma_r^- + \lambda_1^r + \lambda_2 + \lambda_3 + \lambda_4$, where $\lambda_1^l(x) = x, -R_1 \leq x \leq 1-r$, $\gamma_r^-(\theta) = 1 + re^{-i\theta}, -\pi \leq \theta \leq 0$, $\lambda_1^r(x) = x, 1+r \leq x \leq R_2$, $\lambda_2(y) = R_2 + iy$, $0 \leq y \leq R_1 + R_2$, $\lambda_3(x) = x + i(R_1 + R_2)$, $R_2 \geq x \geq -R_1$, and $\lambda_4(y) = -R_1 + iy$, $R_1 + R_2 \geq y \geq 0$. By the residue theorem we have

$$\int_\gamma f(z)dz = 2\pi i \operatorname*{Res}_{z=i} f(z).$$

Observing that the integrals along the components of γ give

$$\int_{\lambda_1^l} f(z)dz = \int_{-R_1}^{1-r} \frac{x^2\sin x}{(x-1)(x^2+1)}dx,$$

$$\int_{\gamma_r^-} f(z)\mathrm{d}z \xrightarrow{r\to 0} -\pi i \operatorname*{Res}_{z=1} f(z),$$

$$\int_{\lambda_1^r} f(z)\mathrm{d}z = \int_{1+r}^{R_2} \frac{x^2 \sin x}{(x-1)(x^2+1)}\mathrm{d}x,$$

$$\int_{\lambda_2} f(z)\mathrm{d}z = \int_0^{R_1+R_2} \frac{(R_2+iy)e^{iR_2-y}}{(R_2+iy-1)((R_2+iy)^2+1)}\mathrm{d}y \xrightarrow{R_1,R_2\to\infty} 0,$$

$$\int_{\lambda_3} f(z)\mathrm{d}z = \int_{R_2}^{-R_1} \frac{(x+i(R_1+R_2))e^{ix-(R_1+R_2)}}{(x+i(R_1+R_2)-1)((x+i(R_1+R_2))^2+1)}\mathrm{d}x \xrightarrow{R_1,R_2\to\infty} 0,$$

$$\int_{\lambda_4} f(z)\mathrm{d}z = \int_{R_1+R_2}^0 \frac{(-R_1+iy)e^{-iR_1-y}}{(-R_1+iy-1)((-R_1+iy)^2+1)}\mathrm{d}y \xrightarrow{R_1,R_2\to\infty} 0,$$

we conclude

$$\mathrm{PV}\int_{-\infty}^{+\infty} \frac{x^2 \sin x}{(x-1)(x^2+1)}\mathrm{d}x = \operatorname{Im}\left(2\pi i \operatorname*{Res}_{z=i} f(z) + \pi i \operatorname*{Res}_{z=1} f(z)\right)$$

$$= \operatorname{Im}\left(2\pi i \frac{i^2 e^{-1}}{(i-1)2i} + \pi i \frac{1 e^i}{2}\right)$$

$$= \operatorname{Im}\left(\frac{\pi}{2e}(1+i) + \frac{\pi}{2}i(\cos 1 + i \sin 1)\right)$$

$$= \frac{\pi}{2e} + \frac{\pi}{2}\cos 1.$$

10.16 Consider the function $f(z) = e^{iz}/z$ which has a simple pole at $z = 0$ and integrate it along the positively oriented closed path $\gamma = \lambda_1^l + \gamma_r^- + \lambda_1^r + \lambda_2 + \lambda_3 + \lambda_4$, where $\lambda_1^l(x) = x$, $-R_1 \le x \le -r$, $\gamma_r^-(\theta) = re^{-i\theta}$, $-\pi \le \theta \le 0$, $\lambda_1^r(x) = x$, $r \le x \le R_2$, $\lambda_2(y) = R_2 + iy$, $0 \le y \le R_1 + R_2$, $\lambda_3(x) = x + i(R_1 + R_2)$, $R_2 \ge x \ge -R_1$, and $\lambda_4(y) = -R_1 + iy$, $R_1 + R_2 \ge y \ge 0$. Since the pole is external to the path γ, by the residue theorem we have

$$\int_\gamma f(z)\mathrm{d}z = 0.$$

Observing that the integrals along the components of γ are

$$\int_{\lambda_1^l} f(z)\mathrm{d}z = \int_{-R_1}^{-r} \frac{e^{ix}}{x}\mathrm{d}x,$$

$$\int_{\gamma_r^-} f(z)\mathrm{d}z \xrightarrow{r\to 0} -\pi\mathrm{i}\operatorname*{Res}_{z=0} f(z) = -\pi\mathrm{i},$$

$$\int_{\lambda_1^r} f(z)\mathrm{d}z = \int_r^{R_2} \frac{\mathrm{e}^{\mathrm{i}x}}{x}\mathrm{d}x,$$

$$\int_{\lambda_2} f(z)\mathrm{d}z = \int_0^{R_1+R_2} \frac{\mathrm{e}^{\mathrm{i}R_2-y}}{R_2+\mathrm{i}y}\mathrm{d}y \xrightarrow{R_1,R_2\to\infty} 0,$$

$$\int_{\lambda_3} f(z)\mathrm{d}z = \int_{R_2}^{-R_1} \frac{\mathrm{e}^{\mathrm{i}x-(R_1+R_2)}}{x+\mathrm{i}(R_1+R_2)}\mathrm{d}x \xrightarrow{R_1,R_2\to\infty} 0,$$

$$\int_{\lambda_4} f(z)\mathrm{d}z = \int_{R_1+R_2}^0 \frac{\mathrm{e}^{-\mathrm{i}R_1-y}}{-R_1+\mathrm{i}y}\mathrm{d}y \xrightarrow{R_1,R_2\to\infty} 0,$$

we conclude

$$\mathrm{PV}\int_{-\infty}^{+\infty} \frac{\mathrm{e}^{\mathrm{i}x}}{x}\mathrm{d}x = \pi\mathrm{i}.$$

10.17 Let $f(z) = z\mathrm{e}^{\mathrm{i}z}/((z-b)^2+c^2)$ and call γ the boundary, positively oriented, of the square with vertices $-R_1, R_2, R_2+\mathrm{i}(R_1+R_2), -R_1+\mathrm{i}(R_1+R_2)$. For $R_1 > 0$ and $R_2 > 0$ sufficiently large the function $f(z)$ is analytic on and within γ except for the simple pole at $z = b+\mathrm{i}c$. By the residue theorem we have

$$\int_\gamma f(z)\mathrm{d}z = 2\pi\mathrm{i}\operatorname*{Res}_{z=b+\mathrm{i}c} \frac{z\mathrm{e}^{\mathrm{i}az}/(z-(b-\mathrm{i}c))}{z-(b+\mathrm{i}c)}$$

$$= 2\pi\mathrm{i}\left.\frac{z\mathrm{e}^{\mathrm{i}az}}{z-(b-\mathrm{i}c)}\right|_{z=b+\mathrm{i}c}$$

$$= \frac{\pi}{c}\mathrm{e}^{-ac}(b+\mathrm{i}c)\mathrm{e}^{\mathrm{i}ab}.$$

On the other hand

$$\int_\gamma f(z)\mathrm{d}z = \sum_{k=1}^4 \int_{\lambda_k} f(z)\mathrm{d}z,$$

where $\lambda_1(x) = x, -R_1 \le x \le R_2$, $\lambda_2(y) = R_2+\mathrm{i}y, 0 \le y \le R_1+R_2$, $\lambda_3(x) = x+\mathrm{i}(R_1+R_2), R_2 \ge x \ge -R_1$, and $\lambda_4(y) = -R_1+\mathrm{i}y, R_1+R_2 \ge y \ge 0$.

The integrals along the paths that compose γ give

$$\int_{\lambda_1} f(z)\mathrm{d}z = \int_{-R_1}^{R_2} \frac{x\mathrm{e}^{iax}}{(x-b)^2+c^2}\mathrm{d}x,$$

$$\int_{\lambda_2} f(z)\mathrm{d}z = \int_0^{R_1+R_2} \frac{(R_2+iy)\mathrm{e}^{iaR_2-ay}}{(R_2+iy-b)^2+c^2}i\,\mathrm{d}y \xrightarrow{R_1,R_2\to\infty} 0,$$

$$\int_{\lambda_3} f(z)\mathrm{d}z = \int_{R_2}^{-R_1} \frac{x\mathrm{e}^{iax-a(R_1+R_2)}}{(x+i(R_1+R_2)-b)^2+c^2}\mathrm{d}x \xrightarrow{R_1,R_2\to\infty} 0,$$

$$\int_{\lambda_4} f(z)\mathrm{d}z = \int_{R_1+R_2}^0 \frac{(-R_1+iy)\mathrm{e}^{-iaR_1-ay}}{(-R_1+iy-b)^2+c^2}i\,\mathrm{d}y \xrightarrow{R_1,R_2\to\infty} 0.$$

Therefore

$$\int_{-\infty}^{+\infty} \frac{x\mathrm{e}^{iax}}{(x-b)^2+c^2}\mathrm{d}x = \frac{\pi}{c}\mathrm{e}^{-ac}(b+ic)\mathrm{e}^{iab}.$$

Taking the real and imaginary parts of this expression we have

$$\int_{-\infty}^{+\infty} \frac{x\cos(ax)}{(x-b)^2+c^2}\mathrm{d}x = \frac{\pi}{c}\mathrm{e}^{-ac}(b\cos(ab)-c\sin(ab)),$$

$$\int_{-\infty}^{+\infty} \frac{x\sin(ax)}{(x-b)^2+c^2}\mathrm{d}x = \frac{\pi}{c}\mathrm{e}^{-ac}(b\sin(ab)+c\cos(ab)).$$

10.18 Let $f(z) = \mathrm{e}^{iz}/(z(z^2+1))$ and call $\gamma = \lambda_1+\gamma_R+\lambda_2+\gamma_\rho$ the closed path of integration, where $\lambda_1(x) = x$, $\rho \le x \le R$, $\gamma_R(\theta) = R\mathrm{e}^{i\theta}$, $0 \le \theta \le \pi$, $\lambda_2(x) = -x$, $R \ge x \ge \rho$, and $\gamma_\rho(\theta) = \rho\mathrm{e}^{i\theta}$, $\pi \ge \theta \ge 0$. Since f is, for $R > 1$ and $\rho < 1$, analytic on and within γ except for the simple pole at $z = i$, by the residue theorem we find

$$\int_\gamma f(z)\mathrm{d}z = \int_{\lambda_1} f(z)\mathrm{d}z + \int_{\gamma_R} f(z)\mathrm{d}z + \int_{\lambda_2} f(z)\mathrm{d}z + \int_{\gamma_\rho} f(z)\mathrm{d}z$$

$$= 2\pi i \operatorname*{Res}_{z=i} f(z)$$

$$= 2\pi i \left.\frac{\mathrm{e}^{iz}}{z(z+i)}\right|_{z=i}$$

$$= -i\pi\mathrm{e}^{-1}.$$

The integrals on the paths that compose γ give

$$\int_{\lambda_1} f(z)\mathrm{d}z = \int_{\rho}^{R} \frac{\mathrm{e}^{\mathrm{i}x}}{x(x^2+1)}\mathrm{d}x,$$

$$\int_{\lambda_2} f(z)\mathrm{d}z = \int_{R}^{\rho} \frac{\mathrm{e}^{-\mathrm{i}x}}{x(x^2+1)}\mathrm{d}x = -\int_{\rho}^{R} \frac{\mathrm{e}^{-\mathrm{i}x}}{x(x^2+1)}\mathrm{d}x,$$

$$\left|\int_{\gamma_R} f(z)\mathrm{d}z\right| \leq \frac{\pi R}{R(R^2-1)} \xrightarrow{R\to\infty} 0.$$

To evaluate the integral on γ_ρ, note that, having f a simple pole at $z = 0$, for $0 < |z| < 1$ we can set $f(z) = g(z) + z^{-1}\operatorname{Res}_{z=0} f(z)$ with $g(z)$ analytic in $|z| < 1$. Furthermore

$$\left|\int_{\gamma_\rho} g(z)\mathrm{d}z\right| \leq \pi\rho \sup_{z\in\{\gamma_\rho\}} |g(z)| \xrightarrow{\rho\to 0} 0,$$

$$\int_{\gamma_\rho} \frac{1}{z}\mathrm{d}z = \int_{\pi}^{0} \frac{1}{\mathrm{e}^{\mathrm{i}\theta}}\mathrm{i}\mathrm{e}^{\mathrm{i}\theta}\,\mathrm{d}\theta = -\mathrm{i}\pi,$$

and so

$$\int_{\gamma_\rho} f(z)\mathrm{d}z \xrightarrow{\rho\to 0} -\mathrm{i}\pi \operatorname*{Res}_{z=0} f(z) = -\mathrm{i}\pi \left.\frac{\mathrm{e}^{\mathrm{i}z}}{z^2+1}\right|_{z=0} = -\mathrm{i}\pi.$$

In conclusion, in the limit $\rho \to 0$ and $R \to \infty$ we have

$$\int_{0}^{\infty} \frac{\mathrm{e}^{\mathrm{i}x} - \mathrm{e}^{-\mathrm{i}x}}{x(x^2+1)}\mathrm{d}x - \mathrm{i}\pi = -\mathrm{i}\pi\mathrm{e}^{-1},$$

that is

$$\int_{0}^{\infty} \frac{\sin x}{x(x^2+1)}\mathrm{d}x = \frac{\pi}{2}(1 - \mathrm{e}^{-1}).$$

10.19 First of all note that since the improper integral exists, we can more simply calculate its Cauchy principal value. Let $f(z) = \mathrm{e}^{-\mathrm{i}\omega z}/(z^2+a^2)$ and call $\gamma_\pm = \lambda_R + \gamma_{R_\pm}$ the closed paths of integration with $\lambda_R(x) = x, -R \leq x \leq R$, $\gamma_{R_\pm}(\theta) = R\mathrm{e}^{\pm\mathrm{i}\theta}, 0 \leq \theta \leq \pi$. Observe that for $R \to \infty$ the function $f(z)$ vanishes on $\{\gamma_{R_+}\}$ if $\omega < 0$ and on $\{\gamma_{R_-}\}$ if $\omega > 0$. Furthermore f is analytic on and within $\gamma_\pm$ except for the simple pole at $z = \pm\mathrm{i}a$. By the residue theorem, for $\omega < 0$ we have

$$\int_{\gamma_+} f(z)\mathrm{d}z = \int_{\lambda_R} f(z)\mathrm{d}z + \int_{\gamma_{R_+}} f(z)\mathrm{d}z$$

$$= 2\pi i \operatorname*{Res}_{z=ia} f(z)$$

$$= 2\pi i \left. \frac{e^{-i\omega z}}{z + ia} \right|_{z=ia}$$

$$= \pi \frac{e^{\omega a}}{a},$$

while for $\omega > 0$ we get (note the negative sign due to the negative orientation of γ_-)

$$\int_{\gamma_-} f(z)dz = \int_{\lambda_R} f(z)dz + \int_{\gamma_{R_-}} f(z)dz$$

$$= -2\pi i \operatorname*{Res}_{z=-ia} f(z)$$

$$= -2\pi i \left. \frac{e^{-i\omega z}}{z - ia} \right|_{z=-ia}$$

$$= \pi \frac{e^{-\omega a}}{a}.$$

The integrals over the components of $\gamma_\pm$ give

$$\int_{\lambda_R} f(z)dz = \int_{-R}^{R} \frac{e^{-i\omega x}}{x^2 + a^2} dx,$$

$$\int_{\gamma_{R_\pm}} f(z)dz = \int_0^\pi \frac{e^{-i\omega R(\cos\theta \pm i\sin\theta)}}{R^2 e^{\pm i2\theta} + a^2} \left(\pm i R e^{\pm i\theta} \right) d\theta,$$

and therefore

$$\left| \int_{\gamma_{R_\pm}} f(z)dz \right| \leq \pi \frac{R}{R^2 - a^2} \xrightarrow{R \to \infty} 0.$$

In conclusion, in the limit $R \to \infty$ for every value of ω, including the trivial case $\omega = 0$, we have

$$\int_{-\infty}^{+\infty} \frac{e^{-i\omega x}}{x^2 + a^2} dx = \frac{\pi}{a} e^{-|\omega|a}.$$

10.20 Assuming that the improper integral exists, we have

$$\int_{-\infty}^{+\infty} \frac{x \sin(\pi x)}{x^2 + 2x + 5} dx = \lim_{R \to \infty} \int_{-R}^{+R} \frac{x \sin(\pi x)}{x^2 + 2x + 5} dx.$$

Let $f(z) = z/(z^2 + 2z + 5)$ and $\gamma = \lambda_R + \gamma_R$ the closed integration path with $\lambda_R(x) = x, -R \leq x \leq R$, and $\gamma_R(\theta) = Re^{i\theta}, 0 \leq \theta \leq \pi$. For $R > \sqrt{5}$ the function $f(z)e^{i\pi z}$ is analytic on and within γ except for the simple pole at $z = -1 + 2i$. By the residue theorem we have

$$\int_\gamma f(z)e^{i\pi z}dz = \int_{\lambda_R} f(z)e^{i\pi z}dz + \int_{\gamma_R} f(z)e^{i\pi z}dz$$

$$= 2\pi i \operatorname*{Res}_{z=-1+2i} f(z)e^{i\pi z}$$

$$= \frac{\pi}{2}(1 - 2i)e^{-2\pi}.$$

For $z \in \{\gamma_R\}$, we have that $|f(z)| \leq R/(R^2 - 2R - 5)$ is bounded by a quantity which vanishes for $R \to \infty$ and therefore by Jordan's lemma

$$\lim_{R \to \infty} \int_{\gamma_R} f(z)e^{i\pi z}dz = 0.$$

Furthermore

$$\int_{\lambda_R} f(z)e^{i\pi z}dz = \int_{-R}^{+R} \frac{xe^{i\pi x}}{x^2 + 2x + 5}dx.$$

In the limit $R \to \infty$ we thus have

$$\int_{-\infty}^{+\infty} \frac{x(\cos(\pi x) + i\sin(\pi x))}{x^2 + 2x + 5}dx = \frac{\pi}{2}(1 - 2i)e^{-2\pi},$$

from which, taking the imaginary part,

$$\int_{-\infty}^{+\infty} \frac{x\sin(\pi x)}{x^2 + 2x + 5}dx = -\pi e^{-2\pi}.$$

The same result is reached, without having to assume the existence of the improper integral, by integrating the function $f(z)e^{i\pi z}$ along a square path as in the Exercise 10.17.

10.21   First note that, since the improper integral exists, we can more simply calculate its Cauchy principal value. Integrate the function $f(z) = e^{iz}/(z^2 - 6z + 10)$ along the path $\gamma = \lambda + \gamma_R$, where $\lambda(x) = x, -R \leq x \leq R$, and $\gamma_R(\theta) = Re^{i\theta}$, $0 \leq \theta \leq \pi$. Since f has two simple poles at $3 \pm i$, for $R > \sqrt{10}$ by the residue theorem we have

$$\int_\gamma f(z)dz = \int_\lambda f(z)dz + \int_{\gamma_R} f(z)dz$$

$$= 2\pi i \operatorname*{Res}_{z=3+i} f(z)$$

$$= 2\pi i \left. \frac{e^{iz}}{z - (3 - i)} \right|_{z=3+i}$$

$$= \frac{\pi e^{3i}}{e}.$$

Taking the imaginary part of this expression and noting that

$$\int_{\lambda} f(z)dz = \int_{-R}^{R} \frac{e^{ix}\,dx}{x^2 - 6x + 10} \xrightarrow{R\to\infty} \int_{-\infty}^{+\infty} \frac{e^{ix}\,dx}{x^2 - 6x + 10},$$

$$\int_{\gamma_R} f(z)dz \xrightarrow{R\to\infty} 0,$$

we conclude

$$\int_{-\infty}^{+\infty} \frac{\sin x\,dx}{x^2 - 6x + 10} = \frac{\pi \sin 3}{e}.$$

10.22 Let I be the value of the integral, for the formula (10.15) we have

$$I = \frac{1}{2} \int_{-\infty}^{+\infty} \frac{x \sin x}{x^4 + 1}\,dx$$

$$= \frac{1}{2} \operatorname{Im} \int_{-\infty}^{+\infty} \frac{x e^{ix}}{x^4 + 1}\,dx$$

$$= \frac{1}{2} \operatorname{Im} \left[2\pi i \left(\operatorname*{Res}_{z=z_0} \frac{z e^{iz}}{z^4 + 1} + \operatorname*{Res}_{z=z_1} \frac{z e^{iz}}{z^4 + 1} \right) \right],$$

where $z_k = e^{i(\pi + 2\pi k)/4}$, $k = 0, 1, 2, 3$, are the four distinct roots of $(-1)^{1/4}$. At the simple poles z_0 and z_1, we have the following residues

$$\operatorname*{Res}_{z=z_0} \frac{z e^{iz}}{z^4 + 1} = \frac{z_0 e^{iz_0}}{4z_0^3} = \frac{e^{-1/\sqrt{2}}}{4i}(\cos(1/\sqrt{2}) + i\sin(1/\sqrt{2})),$$

$$\operatorname*{Res}_{z=z_1} \frac{z e^{iz}}{z^4 + 1} = \frac{z_1 e^{iz_0}}{4z_1^3} = -\frac{e^{-1/\sqrt{2}}}{4i}(\cos(1/\sqrt{2}) - i\sin(1/\sqrt{2})).$$

Therefore

$$I = \frac{1}{2}\,\mathrm{Im}\left[2\pi i\,\frac{e^{-1/\sqrt{2}}}{4i}\left(\cos(1/\sqrt{2}) + i\sin(1/\sqrt{2})\right.\right.$$

$$\left.\left. - \cos(1/\sqrt{2}) + i\sin(1/\sqrt{2})\right)\right]$$

$$= \frac{\pi}{2}\,\sin(1/\sqrt{2})e^{-1/\sqrt{2}}.$$

10.23 Consider the complex function $f(z) = (ze^{iz} + 3i)/(z^2 + 1)$ that has simple poles at $z = \pm i$ and integrate it along the positively oriented boundary γ of the square with vertices $-R_1$, R_2, $R_2 + i(R_1 + R_2)$, $-R_1 + i(R_1 + R_2)$, with $R_1 > 0$, $R_2 > 0$ and $R_1 + R_2 > 1$. By the residue theorem we have

$$\int_\gamma f(z)dz = 2\pi i \operatorname*{Res}_{z=i} f(z)$$

$$= 2\pi i\,\frac{ie^{-1} + 3i}{2i}$$

$$= i\pi(3 + 1/e).$$

On the other hand

$$\int_\gamma f(z)dz = \sum_{k=1}^{4} \int_{\lambda_k} f(z)dz,$$

where $\lambda_1(x) = x$, $-R_1 \le x \le R_2$, $\lambda_2(y) = R_2 + iy$, $0 \le y \le R_1 + R_2$, $\lambda_3(x) = x + i(R_1 + R_2)$, $R_2 \ge x \ge -R_1$, and $\lambda_4(y) = -R_1 + iy$, $R_1 + R_2 \ge y \ge 0$. The integrals along the paths that compose γ are given by

$$\int_{\lambda_1} f(z)dz = \int_{-R_1}^{R_2} \frac{xe^{ix} + 3i}{x^2 + 1}dx,$$

$$\int_{\lambda_2} f(z)dz = \int_0^{R_1+R_2} \frac{(R_2 + iy)e^{iR_2 - y} + 3i}{(R_2 + iy)^2 + 1}idy \xrightarrow{R_1,R_2 \to \infty} 0,$$

$$\int_{\lambda_3} f(z)dz = \int_{R_2}^{-R_1} \frac{xe^{ix-(R_1+R_2)} + 3i}{(x + i(R_1 + R_2))^2 + 1}dx \xrightarrow{R_1,R_2 \to \infty} 0,$$

$$\int_{\lambda_4} f(z)dz = \int_{R_1+R_2}^{0} \frac{(-R_1 + iy)e^{-iR_1 - y} + 3i}{(-R_1 + iy)^2 + 1}idy \xrightarrow{R_1,R_2 \to \infty} 0.$$

Therefore

$$\int_{-\infty}^{+\infty} \frac{xe^{ix} + 3i}{x^2 + 1}\,dx = i\pi(3 + 1/e).$$

Taking the imaginary part of this expression, we conclude

$$\int_{-\infty}^{+\infty} \frac{x\sin(x) + 3}{x^2 + 1}\,dx = \pi(3 + 1/e).$$

10.24 Consider the complex function $f(z) = (ze^{iz} + 7)/(z^2 + 1)^2$ that has two double poles at $z = \pm i$ and integrate it along the positively oriented boundary γ of the square with vertices $-R_1,\ R_2,\ R_2 + i(R_1 + R_2),\ -R_1 + i(R_1 + R_2)$, with $R_1 > 0$, $R_2 > 0$ and $R_1 + R_2 > 1$. By the residue theorem we have

$$\int_\gamma f(z)dz = 2\pi i \operatorname*{Res}_{z=i} f(z)$$

$$= 2\pi i\, \frac{d}{dz} \frac{ze^{iz} + 7}{(z + i)^2}\bigg|_{z=i}$$

$$= 2\pi i\, \frac{(e^{iz} + ize^{iz})(z + i)^2 - (ze^{iz} + 7)2(z + i)}{(z + i)^4}\bigg|_{z=i}$$

$$= 2\pi i\, \frac{-4i(7 + ie^{-1})}{16}$$

$$= \frac{\pi}{2}(7 + ie^{-1}).$$

On the other hand

$$\int_\gamma f(z)dz = \sum_{k=1}^{4} \int_{\lambda_k} f(z)dz,$$

where $\lambda_1(x) = x,\ -R_1 \le x \le R_2$, $\lambda_2(y) = R_2 + iy,\ 0 \le y \le R_1 + R_2$, $\lambda_3(x) = x + i(R_1 + R_2),\ R_2 \ge x \ge -R_1$, and $\lambda_4(y) = -R_1 + iy,\ R_1 + R_2 \ge y \ge 0$. The integrals along the paths that compose γ are given by

$$\int_{\lambda_1} f(z)dz = \int_{-R_1}^{R_2} \frac{xe^{ix} + 7}{(x^2 + 1)^2}\,dx,$$

$$\int_{\lambda_2} f(z)dz = \int_0^{R_1+R_2} \frac{(R_2 + iy)e^{iR_2-y} + 7}{((R_2 + iy)^2 + 1)^2}\,idy \xrightarrow{R_1,R_2 \to \infty} 0,$$

$$\int_{\lambda_3} f(z)dz = \int_{R_2}^{-R_1} \frac{xe^{ix-(R_1+R_2)}+7}{((x+i(R_1+R_2))^2+1)^2}dx \xrightarrow{R_1,R_2\to\infty} 0,$$

$$\int_{\lambda_4} f(z)dz = \int_{R_1+R_2}^{0} \frac{(-R_1+iy)e^{-iR_1-y}+7}{((-R_1+iy)^2+1)^2}idy \xrightarrow{R_1,R_2\to\infty} 0.$$

Therefore

$$\int_{-\infty}^{+\infty} \frac{xe^{ix}+7}{(x^2+1)^2}dx = \frac{\pi}{2}(7+ie^{-1}).$$

Taking the real and imaginary parts of this expression we have

$$\int_{-\infty}^{+\infty} \frac{x\cos(x)+7}{(x^2+1)^2}dx = \int_{-\infty}^{+\infty} \frac{7}{(x^2+1)^2}dx = \frac{\pi}{2}7,$$

$$\int_{-\infty}^{+\infty} \frac{x\sin(x)}{(x^2+1)^2}dx = \frac{\pi}{2}e^{-1},$$

from which we conclude

$$\int_{-\infty}^{+\infty} \frac{x\sin(x)+7}{(x^2+1)^2}dx = \frac{\pi}{2}(7+e^{-1}).$$

10.25 The integral is convergent and can be evaluated by looking at its Cauchy principal value. Due to the parity of the cosine function we can assume p,q non negative. Consider the complex function $f(z) = \left(e^{ipz}-e^{iqz}\right)/z^2$ which has a simple pole at $z=0$ with

$$\operatorname*{Res}_{z=0} f(z) = i(p-q).$$

Integrate $f(z)$ along the positively oriented closed path $\gamma = \lambda_1^l + \gamma_r^- + \lambda_1^r + \gamma_R$, where $\lambda_1^l(x) = x,\ -R \le x \le -r,\ \gamma_r^-(\theta) = re^{-i\theta},\ -\pi \le \theta \le 0,\ \lambda_1^r(x) = x,$ $r \le x \le R,\ \gamma_R(\theta)) = Re^{i\theta},\ 0 \le \theta \le \pi$. Since the pole of f is external to the path γ, by the residue theorem we have

$$\int_{\gamma} f(z)dz = 0.$$

The integrals along the components of γ are

$$\int_{\lambda_1^l} f(z)dz = \int_{-R}^{-r} \frac{e^{ipx}-e^{iqx}}{x^2}dx,$$

$$\int_{\gamma_r^-} f(z)\mathrm{d}z \xrightarrow{r\to 0} -\pi\mathrm{i}\operatorname*{Res}_{z=0} f(z) = \pi(p-q),$$

$$\int_{\lambda_1^r} f(z)\mathrm{d}z = \int_r^R \frac{\mathrm{e}^{\mathrm{i}px} - \mathrm{e}^{\mathrm{i}qx}}{x^2}\,\mathrm{d}x,$$

$$\int_{\gamma_R} f(z)\mathrm{d}z = \int_0^\pi \frac{\mathrm{e}^{\mathrm{i}pR(\cos\theta+\mathrm{i}\sin\theta)} - \mathrm{e}^{\mathrm{i}qR(\cos\theta+\mathrm{i}\sin\theta)}}{R^2\mathrm{e}^{2\mathrm{i}\theta}} R\mathrm{i}\mathrm{e}^{\mathrm{i}\theta}\,\mathrm{d}\theta \xrightarrow{R\to\infty} 0.$$

Note that to determine the last limit we used the properties $\mathrm{e}^{-pR\sin\theta} \leq 1$ and $\mathrm{e}^{-qR\sin\theta} \leq 1$ valid for $0 \leq \theta \leq \pi$ and p, q non-negative. We conclude

$$\int_{-\infty}^{+\infty} \frac{\cos(px) - \cos(qx)}{x^2}\,\mathrm{d}x = \pi(q-p).$$

10.26 Let $f(z) = (\log z)/(z^2 + 1) = (\log z)/((z+\mathrm{i})(z-\mathrm{i}))$, where

$$\log z = \ln r + \mathrm{i}\theta, \qquad z = r\mathrm{e}^{\mathrm{i}\theta}, \qquad r > 0, \qquad -\frac{\pi}{2} < \theta < \frac{3\pi}{2},$$

and call $\gamma = \lambda_1 + \gamma_R + \lambda_2 + \gamma_\rho$ the positively oriented closed path of integration with $\lambda_1(x) = x$, $\rho \leq x \leq R$, $\gamma_R(\theta) = R\mathrm{e}^{\mathrm{i}\theta}$, $0 \leq \theta \leq \pi$, $\lambda_2(x) = x\mathrm{e}^{\mathrm{i}\pi}$, $R \geq x \geq \rho$, and $\gamma_\rho(\theta) = \rho\mathrm{e}^{\mathrm{i}\theta}$, $\pi \geq \theta \geq 0$. Since f is analytic on and within γ except for the simple pole at $z = \mathrm{i}$, for $R > 1$ and $\rho < 1$ the residue theorem gives

$$\int_\gamma f(z)\mathrm{d}z = \int_{\lambda_1} f(z)\mathrm{d}z + \int_{\gamma_R} f(z)\mathrm{d}z + \int_{\lambda_2} f(z)\mathrm{d}z + \int_{\gamma_\rho} f(z)\mathrm{d}z$$

$$= 2\pi\mathrm{i}\operatorname*{Res}_{z=\mathrm{i}} f(z)$$

$$= 2\pi\mathrm{i}\frac{\log\left(1\mathrm{e}^{\mathrm{i}\pi/2}\right)}{\mathrm{i}+\mathrm{i}}$$

$$= \mathrm{i}\frac{\pi^2}{2}.$$

For the integrals along the paths composing γ we have

$$\int_{\lambda_1} f(z)\mathrm{d}z = \int_\rho^R \frac{\ln x}{x^2 + 1}\,\mathrm{d}x,$$

$$\int_{\lambda_2} f(z)\mathrm{d}z = \int_R^\rho \frac{\ln x + \mathrm{i}\pi}{x^2 + 1}\mathrm{e}^{\mathrm{i}\pi}\,\mathrm{d}x = \int_\rho^R \frac{\ln x}{x^2 + 1}\,\mathrm{d}x + \mathrm{i}\pi\int_\rho^R \frac{1}{x^2 + 1}\,\mathrm{d}x,$$

$$\int_{\gamma_R} f(z)\mathrm{d}z = \int_0^\pi \frac{\ln R + \mathrm{i}\theta}{R^2 e^{2\mathrm{i}\theta} + 1}\, \mathrm{i}R e^{\mathrm{i}\theta}\mathrm{d}\theta \xrightarrow{R\to\infty} 0,$$

$$\int_{\gamma_\rho} f(z)\mathrm{d}z = \int_\pi^0 \frac{\ln \rho + \mathrm{i}\theta}{\rho^2 e^{2\mathrm{i}\theta} + 1}\, \mathrm{i}\rho e^{\mathrm{i}\theta}\mathrm{d}\theta \xrightarrow{\rho\to 0} 0.$$

So in the limit $\rho \to 0$ and $R \to \infty$ we get

$$2\int_0^\infty \frac{\ln x}{x^2 + 1}\mathrm{d}x + \mathrm{i}\pi \int_0^\infty \frac{1}{x^2 + 1}\mathrm{d}x = \mathrm{i}\frac{\pi^2}{2}.$$

Taking the real and imaginary parts of the above expression we conclude

$$\int_0^\infty \frac{\ln x}{x^2 + 1}\mathrm{d}x = 0, \qquad \int_0^\infty \frac{1}{x^2 + 1}\mathrm{d}x = \frac{\pi}{2}.$$

10.27 Consider the function

$$f(z) = \frac{e^{\frac{1}{2}\log z}}{z^2 + 1},$$

$$\log z = \ln r + \mathrm{i}\theta, \qquad z = r e^{\mathrm{i}\theta}, \qquad r > 0, \qquad -\frac{\pi}{2} < \theta < \frac{3\pi}{2},$$

and call $\gamma = \lambda_1 + \gamma_R + \lambda_2 + \gamma_\rho$ the positively oriented closed path of integration, where $\lambda_1(x) = x$, $\rho \le x \le R$, $\gamma_R(\theta) = R e^{\mathrm{i}\theta}$, $0 \le \theta \le \pi$, $\lambda_2(x) = x e^{\mathrm{i}\pi}$, $R \ge x \ge \rho$, and $\gamma_\rho(\theta) = \rho e^{\mathrm{i}\theta}$, $\pi \ge \theta \ge 0$. Since f is, for $R > 1$ and $\rho < 1$, analytic on and within γ except for the simple pole at $z = \mathrm{i}$, the residue theorem gives

$$\int_\gamma f(z)\mathrm{d}z = \int_{\lambda_1} f(z)\mathrm{d}z + \int_{\gamma_R} f(z)\mathrm{d}z + \int_{\lambda_2} f(z)\mathrm{d}z + \int_{\gamma_\rho} f(z)\mathrm{d}z$$

$$= 2\pi\mathrm{i}\operatorname*{Res}_{z=\mathrm{i}} f(z)$$

$$= 2\pi\mathrm{i}\, \frac{e^{\frac{1}{2}(\ln 1 + \mathrm{i}\frac{\pi}{2})}}{\mathrm{i} + \mathrm{i}}$$

$$= \pi e^{\mathrm{i}\frac{\pi}{4}}.$$

For the integrals along the paths composing γ we have

$$\int_{\lambda_1} f(z)\mathrm{d}z = \int_\rho^R \frac{\sqrt{x}}{x^2 + 1}\mathrm{d}x,$$

$$\int_{\lambda_2} f(z)dz = \int_R^\rho \frac{e^{\frac{1}{2}(\ln x + i\pi)}}{x^2 + 1}\, e^{i\pi}\, dx = i \int_\rho^R \frac{\sqrt{x}}{x^2 + 1}\, dx,$$

$$\int_{\gamma_R} f(z)dz = \int_0^\pi \frac{e^{\frac{1}{2}(\ln R + i\theta)}}{R^2 e^{2i\theta} + 1}\, iRe^{i\theta}\, d\theta \xrightarrow{R \to \infty} 0,$$

$$\int_{\gamma_\rho} f(z)dz = \int_\pi^0 \frac{e^{\frac{1}{2}(\ln \rho + i\theta)}}{\rho^2 e^{2i\theta} + 1}\, i\rho e^{i\theta}\, d\theta \xrightarrow{\rho \to 0} 0.$$

Therefore, in the limit $\rho \to 0$ and $R \to \infty$ we obtain

$$(1 + i) \int_0^\infty \frac{\sqrt{x}}{x^2 + 1}\, dx = \pi e^{i\frac{\pi}{4}} = \pi \left(\cos \frac{\pi}{4} + i \sin \frac{\pi}{4} \right) = \frac{\pi}{\sqrt{2}}(1 + i)$$

and thus

$$\int_0^\infty \frac{\sqrt{x}}{x^2 + 1}\, dx = \frac{\pi}{\sqrt{2}}.$$

10.28 Define

$$f(z) = \frac{\log z}{z^2 + 4},$$

$$\log z = \ln r + i\theta, \qquad z = re^{i\theta}, \quad r > 0, \quad -\frac{\pi}{2} < \theta < \frac{3\pi}{2},$$

and call $\gamma = \lambda_1 + \gamma_R + \lambda_2 + \gamma_\rho$ the positively oriented closed path of integration, where $\lambda_1(x) = x$, $\rho \le x \le R$, $\gamma_R(\theta) = Re^{i\theta}$, $0 \le \theta \le \pi$, $\lambda_2(x) = xe^{i\pi}$, $R \ge x \ge \rho$, and $\gamma_\rho(\theta) = \rho e^{i\theta}$, $\pi \ge \theta \ge 0$. Since f is, for $R > 2$ and $\rho < 2$, analytic on and within γ except for the simple pole at $z = 2i$, the residue theorem gives

$$\int_\gamma f(z)dz = \int_{\lambda_1} f(z)dz + \int_{\gamma_R} f(z)dz + \int_{\lambda_2} f(z)dz + \int_{\gamma_\rho} f(z)dz$$

$$- 2\pi i \operatorname*{Res}_{z=2i} f(z)$$

$$= 2\pi i \frac{\log \left(2e^{i\pi/2} \right)}{2i + 2i}$$

$$= \frac{\pi}{2} \left(\ln 2 + i\frac{\pi}{2} \right).$$

For the integrals along the paths that compose γ we have

$$\int_{\lambda_1} f(z)\mathrm{d}z = \int_\rho^R \frac{\ln x}{x^2+4}\mathrm{d}x,$$

$$\int_{\lambda_2} f(z)\mathrm{d}z = \int_R^\rho \frac{\ln x + i\pi}{x^2+4}\, e^{i\pi}\,\mathrm{d}x = \int_\rho^R \frac{\ln x}{x^2+4}\mathrm{d}x + i\pi \int_\rho^R \frac{1}{x^2+4}\mathrm{d}x,$$

$$\int_{\gamma_R} f(z)\mathrm{d}z = \int_0^\pi \frac{\ln R + i\theta}{R^2 e^{2i\theta}+4}\, iRe^{i\theta}\,\mathrm{d}\theta \xrightarrow{R\to\infty} 0,$$

$$\int_{\gamma_\rho} f(z)\mathrm{d}z = \int_\pi^0 \frac{\ln \rho + i\theta}{\rho^2 e^{2i\theta}+4}\, i\rho e^{i\theta}\,\mathrm{d}\theta \xrightarrow{\rho\to0} 0.$$

Therefore, in the limit $\rho \to 0$ and $R \to \infty$ we obtain

$$2\int_0^\infty \frac{\ln x}{x^2+4}\mathrm{d}x + i\pi \int_0^\infty \frac{1}{x^2+4}\mathrm{d}x = \frac{\pi}{2}\left(\ln 2 + i\frac{\pi}{2}\right).$$

Taking the real and imaginary parts of the above expression we conclude

$$\int_0^\infty \frac{\ln x}{x^2+4}\mathrm{d}x = \frac{\pi}{4}\ln 2, \qquad \int_0^\infty \frac{1}{x^2+4}\mathrm{d}x = \frac{\pi}{4}.$$

10.29 Let

$$f(z) = \frac{z^a}{z^2+1} = \frac{e^{a\log z}}{z^2+1},$$

$$\log z = \ln r + i\theta, \qquad z = re^{i\theta}, \quad r > 0, \quad 0 < \theta < 2\pi,$$

with $a = -1/3$. The following solution is valid for every $-1 < a < 1$. In the specified domain f is analytic everywhere except for the two simple poles at $z = \pm i$ where it has residues

$$\operatorname*{Res}_{z=i} f(z) = \left.\frac{e^{a\log z}}{z+i}\right|_{z=i} = \frac{e^{a(\ln 1 + i\pi/2)}}{i+i} = \frac{e^{i\pi a/2}}{2i},$$

$$\operatorname*{Res}_{z=-i} f(z) = \left.\frac{e^{a\log z}}{z-i}\right|_{z=-i} = \frac{e^{a(\ln 1 + i3\pi/2)}}{-i-i} = \frac{e^{i3\pi a/2}}{-2i}.$$

The integral of f along the positively oriented closed path $\gamma = \lambda_1 + \gamma_R + \lambda_2 + \gamma_\rho$, where $\lambda_1(x) = x + \mathrm{i}0$, $\rho \le x \le R$, $\gamma_R(\theta) = R\mathrm{e}^{\mathrm{i}\theta}$, $0 \le \theta \le 2\pi$, $\lambda_2(x) = x - \mathrm{i}0$, $R \ge x \ge \rho$, and $\gamma_\rho(\theta) = \rho\mathrm{e}^{\mathrm{i}\theta}$, $2\pi \ge \theta \ge 0$, for $R > 1$ and $\rho < 1$ gives

$$\int_\gamma f(z)\mathrm{d}z = \int_{\lambda_1} f(z)\mathrm{d}z + \int_{\gamma_R} f(z)\mathrm{d}z + \int_{\lambda_2} f(z)\mathrm{d}z + \int_{\gamma_\rho} f(z)\mathrm{d}z$$

$$= 2\pi\mathrm{i}\left(\operatorname*{Res}_{z=\mathrm{i}} f(z) + \operatorname*{Res}_{z=-\mathrm{i}} f(z) \right)$$

$$= \pi\left(\mathrm{e}^{\mathrm{i}\pi a/2} - \mathrm{e}^{\mathrm{i}3\pi a/2} \right).$$

The integrals along the individual paths that compose γ are

$$\int_{\lambda_1} f(z)\mathrm{d}z = \int_\rho^R \frac{\mathrm{e}^{a(\ln x + \mathrm{i}0)}}{(x\mathrm{e}^{\mathrm{i}0})^2 + 1}\mathrm{e}^{\mathrm{i}0}\mathrm{d}x = \int_\rho^R \frac{\mathrm{e}^{a\ln x}}{x^2 + 1}\mathrm{d}x,$$

$$\int_{\lambda_2} f(z)\mathrm{d}z = \int_R^\rho \frac{\mathrm{e}^{a(\ln x + \mathrm{i}2\pi)}}{(x\mathrm{e}^{\mathrm{i}2\pi})^2 + 1}\mathrm{e}^{\mathrm{i}2\pi}\mathrm{d}x = -\mathrm{e}^{\mathrm{i}2\pi a}\int_\rho^R \frac{\mathrm{e}^{a\ln x}}{x^2 + 1}\mathrm{d}x,$$

$$\left| \int_{\gamma_R} f(z)\mathrm{d}z \right| \le \frac{\mathrm{e}^{a\ln R}}{R^2 - 1}\, 2\pi R \xrightarrow{R\to\infty} 0, \qquad a < 1,$$

$$\left| \int_{\gamma_\rho} f(z)\mathrm{d}z \right| \le \frac{\mathrm{e}^{a\ln \rho}}{1 - \rho^2}\, 2\pi\rho \xrightarrow{\rho\to 0} 0, \qquad a > -1.$$

In conclusion, taking the limits $R \to \infty$ and $\rho \to 0$, for $-1 < a < 1$ we get

$$\int_0^\infty \frac{x^a}{x^2 + 1}\mathrm{d}x = \frac{\pi\left(\mathrm{e}^{\mathrm{i}\pi a/2} - \mathrm{e}^{\mathrm{i}3\pi a/2} \right)}{1 - \mathrm{e}^{\mathrm{i}2\pi a}}$$

$$= \frac{\pi\mathrm{e}^{\mathrm{i}\pi a}\left(\mathrm{e}^{-\mathrm{i}\pi a/2} - \mathrm{e}^{\mathrm{i}\pi a/2} \right)}{\mathrm{e}^{\mathrm{i}\pi a}\left(\mathrm{e}^{-\mathrm{i}\pi a} - \mathrm{e}^{\mathrm{i}\pi a} \right)}$$

$$= \frac{\pi \sin(\pi a/2)}{\sin(\pi a)}$$

$$= \frac{\pi/2}{\cos(\pi a/2)}.$$

10.30 To evaluate the improper integral

$$I = \int_0^\infty \frac{\ln x}{x^2 - 2x + 2}\mathrm{d}x$$

let's try to integrate the function

$$f(z) = \frac{\log z}{z^2 - 2z + 2} = \frac{\log z}{(z - (1 + \mathrm{i}))(z - (1 - \mathrm{i}))},$$

$$\log z = \ln r + \mathrm{i}\theta, \qquad z = r\mathrm{e}^{\mathrm{i}\theta}, \quad r > 0, \quad -\frac{\pi}{2} < \theta < \frac{3\pi}{2},$$

as in exercise 10.26. The integral of f along the positively oriented closed path $\gamma = \lambda_1 + \gamma_R + \lambda_2 + \gamma_\rho$, where $\lambda_1(x) = x\mathrm{e}^{\mathrm{i}0}$, $\rho \leq x \leq R$, $\gamma_R(\theta) = R\mathrm{e}^{\mathrm{i}\theta}$, $0 \leq \theta \leq \pi$, $\lambda_2(x) = x\mathrm{e}^{\mathrm{i}\pi}$, $R \geq x \geq \rho$, and $\gamma_\rho(\theta) = \rho\mathrm{e}^{\mathrm{i}\theta}$, $\pi \geq \theta \geq 0$, by the residue theorem gives

$$\int_\gamma f(z)\mathrm{d}z = \int_{\lambda_1} f(z)\mathrm{d}z + \int_{\gamma_R} f(z)\mathrm{d}z + \int_{\lambda_2} f(z)\mathrm{d}z + \int_{\gamma_\rho} f(z)\mathrm{d}z$$

$$= 2\pi\mathrm{i} \operatorname*{Res}_{z=1+\mathrm{i}} f(z)$$

$$= \pi \ln\sqrt{2} + \mathrm{i}\frac{\pi^2}{4}.$$

We assumed that ρ and R are arbitrary with $\rho < \sqrt{2} < R$. The integrals along the individual paths that compose γ are

$$\int_{\lambda_1} f(z)\mathrm{d}z = \int_\rho^R \frac{\ln x + \mathrm{i}0}{(x\mathrm{e}^{\mathrm{i}0})^2 - 2(x\mathrm{e}^{\mathrm{i}0}) + 2}\mathrm{e}^{\mathrm{i}0}\mathrm{d}x = \int_\rho^R \frac{\ln x}{x^2 - 2x + 2}\mathrm{d}x,$$

$$\int_{\lambda_2} f(z)\mathrm{d}z = \int_R^\rho \frac{\ln x + \mathrm{i}\pi}{(x\mathrm{e}^{\mathrm{i}\pi})^2 - 2(x\mathrm{e}^{\mathrm{i}\pi}) + 2}\mathrm{e}^{\mathrm{i}\pi}\mathrm{d}x$$

$$= \int_\rho^R \frac{\ln x}{x^2 + 2x + 2}\mathrm{d}x + \mathrm{i}\pi \int_\rho^R \frac{1}{x^2 + 2x + 2}\mathrm{d}x,$$

$$\left|\int_{\gamma_R} f(z)\mathrm{d}z\right| \leq \frac{\ln R + \pi}{R^2 - 2R - 2}\,\pi R \xrightarrow{R\to\infty} 0,$$

$$\left|\int_{\gamma_\rho} f(z)\mathrm{d}z\right| \leq \frac{\pi - \ln\rho}{2 - \rho^2 - 2\rho}\,\pi\rho \xrightarrow{\rho\to 0} 0.$$

Taking the limits $R \to \infty$ and $\rho \to 0$, we obtain

$$I + \int_0^\infty \frac{\ln x}{x^2 + 2x + 2}\mathrm{d}x + \mathrm{i}\pi \int_0^\infty \frac{1}{x^2 + 2x + 2}\mathrm{d}x = \pi \ln\sqrt{2} + \mathrm{i}\frac{\pi^2}{4},$$

whose real part supplies

$$I + \int_0^\infty \frac{\ln x}{x^2 + 2x + 2}\,dx = \pi \ln \sqrt{2}.$$

Using identity

$$\frac{1}{x^2 + 2x + 2} = \frac{1}{x^2 - 2x + 2} + \left(\frac{1}{x^2 + 2x + 2} - \frac{1}{x^2 - 2x + 2} \right)$$

$$= \frac{1}{x^2 - 2x + 2} - \frac{4x}{x^4 + 4},$$

the previous expression allows us to derive

$$I = \frac{\pi}{2} \ln \sqrt{2} + 2 \int_0^\infty \frac{x \ln x}{x^4 + 4}\,dx.$$

The new integral can be calculated by integrating the function $z \log(z)/(z^4 + 4)$ along the closed path $\gamma = \lambda_1 + \gamma_R + \lambda_2 + \gamma_\rho$, where $\lambda_1(x) = x e^{i0}$, $\rho \le x \le R$, $\gamma_R(\theta) = R e^{i\theta}$, $0 \le \theta \le \pi/4$, $\lambda_2(x) = x e^{i\pi/4}$, $R \ge x \ge \rho$, and $\gamma_\rho(\theta) = \rho e^{i\theta}$, $\pi/4 \ge \theta \ge 0$. Using the residue theorem and sending $\rho \to 0$ and $R \to \infty$, we have

$$\int_0^\infty \frac{x \ln x}{x^4 + 4}\,dx + \int_0^\infty \frac{x(\ln x + i\pi/4)}{x^4 + 4}\,dx = 2\pi i \operatorname*{Res}_{z=1+i} \frac{z \log z}{z^4 + 4}$$

$$= \frac{\pi}{8} \ln 2 + i\frac{\pi^2}{16}.$$

We conclude that

$$I = \frac{\pi}{2} \ln \sqrt{2} + \frac{\pi}{8} \ln 2 = \frac{3\pi}{8} \ln 2.$$

10.31 To evaluate the improper integral

$$I = \int_0^\infty \frac{\ln x}{x^2 - 2x + 2}\,dx$$

let's try to integrate the function

$$f(z) = \frac{\log z}{z^2 - 2z + 2} = \frac{\log z}{(z - (1 + i))(z - (1 - i))},$$

$$\log z = \ln r + i\theta, \qquad z = r e^{i\theta}, \qquad r > 0, \quad 0 < \theta < 2\pi,$$

as in the Exercise 10.29. In the specified domain f is analytic everywhere except for the two simple poles at $z = 1 \pm \mathrm{i}$ where it has residues

$$\operatorname*{Res}_{z=1+\mathrm{i}} f(z) = \left.\frac{\log z}{z - (1 - \mathrm{i})}\right|_{z=1+\mathrm{i}} = \frac{\ln\sqrt{2} + \mathrm{i}\pi/4}{2\mathrm{i}},$$

$$\operatorname*{Res}_{z=1-\mathrm{i}} f(z) = \left.\frac{\log z}{z - (1 + \mathrm{i})}\right|_{z=1-\mathrm{i}} = -\frac{\ln\sqrt{2} + \mathrm{i}7\pi/4}{2\mathrm{i}}.$$

The integral of f along the positively oriented closed path $\gamma = \lambda_1 + \gamma_R + \lambda_2 + \gamma_\rho$, where $\lambda_1(x) = x\mathrm{e}^{\mathrm{i}0}$, $\rho \leq x \leq R$, $\gamma_R(\theta) = R\mathrm{e}^{\mathrm{i}\theta}$, $0 \leq \theta \leq 2\pi$, $\lambda_2(x) = x\mathrm{e}^{\mathrm{i}2\pi}$, $R \geq x \geq \rho$, and $\gamma_\rho(\theta) = \rho\mathrm{e}^{\mathrm{i}\theta}$, $2\pi \geq \theta \geq 0$, gives, for arbitrary ρ and R with $\rho < \sqrt{2} < R$,

$$\int_\gamma f(z)\mathrm{d}z = \int_{\lambda_1} f(z)\mathrm{d}z + \int_{\gamma_R} f(z)\mathrm{d}z + \int_{\lambda_2} f(z)\mathrm{d}z + \int_{\gamma_\rho} f(z)\mathrm{d}z$$

$$= 2\pi\mathrm{i}\left(\operatorname*{Res}_{z=1+\mathrm{i}} f(z) + \operatorname*{Res}_{z=1-\mathrm{i}} f(z)\right)$$

$$= -\mathrm{i}\frac{3\pi^2}{2}.$$

The integrals along the individual paths that compose γ are

$$\int_{\lambda_1} f(z)\mathrm{d}z = \int_\rho^R \frac{\ln x + \mathrm{i}0}{(x\mathrm{e}^{\mathrm{i}0})^2 - 2(x\mathrm{e}^{\mathrm{i}0}) + 2}\mathrm{e}^{\mathrm{i}0}\mathrm{d}x = \int_\rho^R \frac{\ln x}{x^2 - 2x + 2}\mathrm{d}x,$$

$$\int_{\lambda_2} f(z)\mathrm{d}z = \int_R^\rho \frac{\ln x + \mathrm{i}2\pi}{(x\mathrm{e}^{\mathrm{i}2\pi})^2 - 2(x\mathrm{e}^{\mathrm{i}2\pi}) + 2}\mathrm{e}^{\mathrm{i}2\pi}\mathrm{d}x$$

$$= -\int_\rho^R \frac{\ln x}{x^2 - 2x + 2}\mathrm{d}x - \mathrm{i}2\pi \int_\rho^R \frac{1}{x^2 - 2x + 2}\mathrm{d}x,$$

$$\left|\int_{\gamma_R} f(z)\mathrm{d}z\right| \leq \frac{\ln R + 2\pi}{R^2 - 2R - 2}\, 2\pi R \xrightarrow{R\to\infty} 0,$$

$$\left|\int_{\gamma_\rho} f(z)\mathrm{d}z\right| \leq \frac{2\pi - \ln\rho}{2 - \rho^2 - 2\rho}\, 2\pi\rho \xrightarrow{\rho\to 0} 0.$$

Taking the limits $R \to \infty$ and $\rho \to 0$ we obtain

$$I - I - \mathrm{i}2\pi \int_0^\infty \frac{1}{x^2 - 2x + 2}\mathrm{d}x = -\mathrm{i}\frac{3\pi^2}{2}.$$

Therefore, a cancellation of I occurs and the proposed method only allows the evaluation of the integral

$$\int_0^\infty \frac{1}{x^2 - 2x + 2}\,dx = \frac{3\pi}{4}.$$

Let's then try to obtain I as $I = \lim_{a \to 0} I_a$, where

$$I_a = \int_0^\infty \frac{x^a \ln x}{x^2 - 2x + 2}\,dx, \qquad 0 < a < 1,$$

This integral can be calculated as above by changing $f(z) \to f_a(z)$ defined by

$$f_a(z) = \frac{z^a \log z}{z^2 - 2z + 2} = \frac{e^{a \log z} \log z}{(z - (1 + i))(z - (1 - i))},$$

$$\log z = \ln r + i\theta, \qquad z = re^{i\theta}, \qquad r > 0, \qquad 0 < \theta < 2\pi.$$

In the specified connected open domain the function $f_a(z)$ is analytic everywhere except for the two simple poles at $z = 1 \pm i$ where it has residues

$$\operatorname*{Res}_{z=1+i} f_a(z) = \left. \frac{e^{a \log z} \log z}{z - (1 - i)} \right|_{z=1+i} = e^{a(\ln \sqrt{2} + i\pi/4)} \frac{\ln \sqrt{2} + i\pi/4}{2i},$$

$$\operatorname*{Res}_{z=1-i} f_a(z) = \left. \frac{e^{a \log z} \log z}{z - (1 + i)} \right|_{z=1-i} = -e^{a(\ln \sqrt{2} + i7\pi/4)} \frac{\ln \sqrt{2} + i7\pi/4}{2i}.$$

We thus find

$$\int_\gamma f_a(z)dz = \pi e^{a \ln \sqrt{2}} \left(e^{ia\pi/4}(\ln \sqrt{2} + i\pi/4) - e^{ia7\pi/4}(\ln \sqrt{2} + i7\pi/4) \right),$$

$$\int_{\lambda_1} f_a(z)dz = \int_\rho^R \frac{e^{a(\ln x + i0)}(\ln x + i0)}{(xe^{i0})^2 - 2(xe^{i0}) + 2} e^{i0}dx = \int_\rho^R \frac{x^a \ln x}{x^2 - 2x + 2}\,dx,$$

$$\int_{\lambda_2} f_a(z)dz = \int_R^\rho \frac{e^{a(\ln x + i2\pi)}(\ln x + i2\pi)}{(xe^{i2\pi})^2 - 2(xe^{i2\pi}) + 2} e^{i2\pi}dx$$

$$= -e^{i2\pi a} \int_\rho^R \frac{x^a \ln x}{x^2 - 2x + 2}\,dx - i2\pi e^{i2\pi a} \int_\rho^R \frac{x^a}{x^2 - 2x + 2}\,dx,$$

$$\int_{\gamma_R} f_a(z)dz \xrightarrow{R \to \infty} 0, \qquad \int_{\gamma_\rho} f_a(z)dz \xrightarrow{\rho \to 0} 0.$$

Taking the limits $R \to \infty$ and $\rho \to 0$, we obtain

$$\left(1 - e^{i2\pi a}\right) I_a = Q_a,$$

where

$$Q_a = \pi e^{a \ln \sqrt{2}} \left(e^{ia\pi/4}(\ln \sqrt{2} + i\pi/4) - e^{ia7\pi/4}(\ln \sqrt{2} + i7\pi/4)\right)$$

$$+ i2\pi e^{i2\pi a} \int_0^\infty \frac{x^a}{x^2 - 2x + 2} dx.$$

To evaluate $\lim_{a \to 0} Q_a/(1 - e^{i2\pi a})$, note that $Q_a \xrightarrow{a \to 0} 0$ and $(1 - e^{i2\pi a}) \xrightarrow{a \to 0} 0$. It is therefore necessary to calculate the first non-zero order of the Taylor series expansion around $a = 0$ of these two quantities

$$Q_a = \pi a \ln \sqrt{2} \left((\ln \sqrt{2} + i\pi/4) - (\ln \sqrt{2} + i7\pi/4)\right)$$

$$+ \pi \left(ia\pi/4(\ln \sqrt{2} + i\pi/4) - ia7\pi/4(\ln \sqrt{2} + i7\pi/4)\right)$$

$$+ (i2\pi)^2 a \int_0^\infty \frac{1}{x^2 - 2x + 2} dx$$

$$+ i2\pi \int_0^\infty \frac{a \ln x}{x^2 - 2x + 2} dx + O(a^2)$$

$$= -i3\pi^2 a \ln \sqrt{2} + 3\pi^3 a - 4\pi^2 a \int_0^\infty \frac{1}{x^2 - 2x + 2} dx + i2\pi a I + O(a^2)$$

$$= -i2\pi a \frac{3\pi}{2} \ln \sqrt{2} + i2\pi a I + O(a^2),$$

$$1 - e^{i2\pi a} = -i2\pi a + O(a^2).$$

From the ratio of the two terms $O(a)$ obtained above, we find

$$I = \lim_{a \to 0} Q_a/(1 - e^{i2\pi a}) = \frac{3\pi}{2} \ln \sqrt{2} - I$$

and so

$$I = \frac{3\pi}{4} \ln \sqrt{2} = \frac{3\pi}{8} \ln 2.$$

10.32 Integrating by parts we have

$$\int_0^\infty \frac{\log(1+x^2)}{x^{a+1}}\,dx = \frac{2}{a}\int_0^\infty \frac{x^{1-a}}{1+x^2}\,dx.$$

Let $f(z) = e^{(1-a)\log z}/(1+z^2)$, assuming for the logarithm the branch

$$\log z = \ln r + i\theta, \qquad z = re^{i\theta}, \qquad r > 0, \quad 0 < \theta < 2\pi.$$

Observe that f has simple poles at $z = \pm i$ and a branch line coinciding with the positive real semi-axis. Integrating f along the positively oriented closed path $\gamma = \lambda_1 + \gamma_R + \lambda_2 + \gamma_r$, where $\lambda_1(x) = x + i0$, $r \leq x \leq R$, $\gamma_R(\theta) = Re^{i\theta}$, $0 \leq \theta \leq \pi$, $\lambda_2(x) = xe^{i\pi}$, $R \geq x \geq r$, and $\gamma_r(\theta) = re^{i\theta}$, $\pi \geq \theta \geq 0$, for $R > 1$ and $r < 1$ we obtain

$$\int_\gamma f(z)dz = \int_{\lambda_1} f(z)dz + \int_{\gamma_R} f(z)dz + \int_{\lambda_2} f(z)dz + \int_{\gamma_r} f(z)dz$$

$$= 2\pi i \operatorname*{Res}_{z=i} f(z)$$

$$= \pi e^{i(1-a)\pi/2}.$$

The integrals over the paths that compose γ are

$$\int_{\lambda_1} f(z)dz = \int_r^R \frac{e^{(1-a)(\ln x + i0)}}{(xe^{i0})^2 + 1}e^{i0}dx = \int_r^R \frac{e^{(1-a)\ln x}}{x^2 + 1}\,dx,$$

$$\int_{\lambda_2} f(z)dz = \int_R^r \frac{e^{(1-a)(\ln x + i\pi)}}{(xe^{i\pi})^2 + 1}e^{i\pi}dx = e^{i(1-a)\pi}\int_r^R \frac{e^{(1-a)\ln x}}{x^2 + 1}\,dx,$$

$$\left|\int_{\gamma_R} f(z)dz\right| \leq \frac{e^{(1-a)\ln R}}{R^2 - 1}\,\pi R = \frac{\pi R^{2-a}}{R^2 - 1} \xrightarrow{R\to\infty} 0, \qquad a > 0,$$

$$\left|\int_{\gamma_r} f(z)dz\right| \leq \frac{e^{(1-a)\ln r}}{1 - r^2}\,\pi r = \frac{\pi r^{2-a}}{1 - r^2} \xrightarrow{r\to 0} 0, \qquad a < 2.$$

Taking the limits $R \to \infty$ and $r \to 0$, for $0 < a < 2$ we conclude

$$\int_0^\infty \frac{x^{1-a}}{x^2 + 1}\,dx = \frac{\pi e^{i(1-a)\pi/2}}{1 + e^{i(1-a)\pi}}$$

$$= \frac{\pi i}{e^{ia\pi/2} - e^{-ia\pi/2}}$$

$$= \frac{\pi/2}{\sin(\pi a/2)}.$$

and thus

$$\int_0^\infty \frac{\log(1+x^2)}{x^{a+1}}\,dx = \frac{\pi/a}{\sin(\pi a/2)}, \qquad 0 < a < 2.$$

10.33 Let $f(z) = z^{\frac{2}{3}}/(z^2+2\sqrt{3}z+4) = e^{\frac{2}{3}\log z}/(z^2+2\sqrt{3}z+4)$, assuming for the logarithm that branch whose branch line coincides with the positive real semi-axis

$$\log z = \ln r + i\theta, \qquad z = re^{i\theta}, \qquad r > 0, \quad 0 < \theta < 2\pi.$$

The function f is analytic everywhere except on the positive real semi-axis and at the two simple poles at $z = -\sqrt{3} \pm i$ where has residues

$$\operatorname*{Res}_{z=-\sqrt{3}+i} f(z) = \left.\frac{e^{\frac{2}{3}\log z}}{z + \sqrt{3} + i}\right|_{z=-\sqrt{3}+i} = \frac{e^{\frac{2}{3}\ln 2 + i\frac{5}{9}\pi}}{2i},$$

$$\operatorname*{Res}_{z=-\sqrt{3}-i} f(z) = \left.\frac{e^{\frac{2}{3}\log z}}{z + \sqrt{3} - i}\right|_{z=-\sqrt{3}-i} = \frac{e^{\frac{2}{3}\ln 2 + i\frac{7}{9}\pi}}{-2i}.$$

Integrate f along the positively oriented closed path $\gamma = \lambda_1 + \gamma_R + \lambda_2 + \gamma_r$, where $\lambda_1(x) = x+i0, r \le x \le R, \gamma_R(\theta) = Re^{i\theta}, 0 \le \theta \le 2\pi, \lambda_2(x) = x-i0, R \ge x \ge r$, and $\gamma_r(\theta) = re^{i\theta}, 2\pi \ge \theta \ge 0$. As usual, $x \pm i0$ stands for $x \pm i\varepsilon$ with ε positive infinitesimal so that in polar representation $x + i0 = xe^{i\varepsilon}$ while $x - i0 = xe^{i(2\pi-\varepsilon)}$. For $r < 2 < R$, we have

$$\int_\gamma f(z)dz = \int_{\lambda_1} f(z)dz + \int_{\gamma_R} f(z)dz + \int_{\lambda_2} f(z)dz + \int_{\gamma_r} f(z)dz$$

$$= 2\pi i \left(\operatorname*{Res}_{z=-\sqrt{3}+i} f(z) + \operatorname*{Res}_{z=-\sqrt{3}-i} f(z) \right)$$

$$= \pi e^{\frac{2}{3}\ln 2} \left(e^{i\frac{5}{9}\pi} - e^{i\frac{7}{9}\pi} \right).$$

The integrals along the paths that compose γ give

$$\int_{\lambda_1} f(z)dz = \int_r^R \frac{e^{\frac{2}{3}(\ln x + i0)}}{(xe^{i0})^2 + 2\sqrt{3}(xe^{i0}) + 4} e^{i0}dx$$

$$= \int_r^R \frac{e^{\frac{2}{3}\ln x}}{x^2 + 2\sqrt{3}x + 4}\,dx,$$

$$\int_{\lambda_2} f(z)\mathrm{d}z = \int_R^r \frac{e^{\frac{2}{3}(\ln x + i2\pi)}}{(xe^{i2\pi})^2 + 2\sqrt{3}(xe^{i2\pi}) + 4} e^{i2\pi}\,\mathrm{d}x$$

$$= -e^{i\frac{4}{3}\pi}\int_r^R \frac{e^{\frac{2}{3}\ln x}}{x^2 + 2\sqrt{3}x + 4}\,\mathrm{d}x,$$

$$\left|\int_{\gamma_R} f(z)\mathrm{d}z\right| \lesssim \frac{e^{\frac{2}{3}\ln R}}{R^2}\, R = \frac{R^{\frac{4}{3}}}{R^2} \xrightarrow{R\to\infty} 0,$$

$$\left|\int_{\gamma_r} f(z)\mathrm{d}z\right| \lesssim \frac{e^{\frac{2}{3}\ln r}}{1}\, r = r^{\frac{4}{3}} \xrightarrow{r\to 0} 0.$$

Therefore, in the limit $r \to 0$ and $R \to \infty$ we find

$$\int_0^\infty \frac{x^{\frac{2}{3}}}{x^2 + 2\sqrt{3}x + 4}\,\mathrm{d}x = \pi e^{\frac{2}{3}\ln 2}\frac{e^{i\frac{5}{9}\pi} - e^{i\frac{7}{9}\pi}}{1 - e^{i\frac{4}{3}\pi}}$$

$$= \pi 2^{\frac{2}{3}}\frac{e^{i\frac{6}{9}\pi}}{e^{i\frac{2}{3}\pi}}\frac{e^{-i\frac{1}{9}\pi} - e^{i\frac{1}{9}\pi}}{e^{-i\frac{2}{3}\pi} - e^{i\frac{2}{3}\pi}}$$

$$= \pi 2^{\frac{2}{3}}\frac{\sin(\pi/9)}{\sin(2\pi/3)}$$

$$= \frac{\pi 2^{\frac{5}{3}}}{3^{\frac{1}{2}}}\sin(\pi/9).$$

10.34 Let

$$f(z) = \frac{(z+1)e^{-\frac{1}{2}\log z}}{z^2 - 2z + 2},$$

where for the logarithm we assume the branch

$$\log z = \ln r + i\theta, \qquad z = re^{i\theta}, \quad r > 0, \quad 0 < \theta < 2\pi.$$

The function f is analytic everywhere except on the positive real semi-axis and at the two simple poles at $z = 1 \pm i$ where it has residues

$$\operatorname*{Res}_{z=1+i} f(z) = \left.\frac{(z+1)e^{-\frac{1}{2}\log z}}{z - 1 + i}\right|_{z=1+i} = \frac{2+i}{2i}e^{-\frac{1}{2}(\ln\sqrt{2}+i\frac{\pi}{4})},$$

$$\operatorname*{Res}_{z=1-i} f(z) = \left.\frac{(z+1)e^{-\frac{1}{2}\log z}}{z - 1 - i}\right|_{z=1-i} = \frac{2-i}{-2i}e^{-\frac{1}{2}(\ln\sqrt{2}+i\frac{7\pi}{4})}.$$

Integrate f along the positively oriented closed path $\gamma = \lambda_1 + \gamma_R + \lambda_2 + \gamma_r$, where $\lambda_1(x) = x + \mathrm{i}0$, $r \leq x \leq R$, $\gamma_R(\theta) = R\mathrm{e}^{\mathrm{i}\theta}$, $0 \leq \theta \leq 2\pi$, $\lambda_2(x) = x - \mathrm{i}0$, $R \geq x \geq r$, and $\gamma_r(\theta) = r\mathrm{e}^{\mathrm{i}\theta}$, $2\pi \geq \theta \geq 0$. For $r < \sqrt{2} < R$, by the residue theorem we have

$$\int_\gamma f(z)\mathrm{d}z = \int_{\lambda_1} f(z)\mathrm{d}z + \int_{\gamma_R} f(z)\mathrm{d}z + \int_{\lambda_2} f(z)\mathrm{d}z + \int_{\gamma_r} f(z)\mathrm{d}z$$

$$= 2\pi\mathrm{i}\left(\operatorname*{Res}_{z=1+\mathrm{i}} f(z) + \operatorname*{Res}_{z=1-\mathrm{i}} f(z) \right)$$

$$= 2\pi\mathrm{i}\left(\frac{2+\mathrm{i}}{2\mathrm{i}} 2^{-1/4} \mathrm{e}^{-\mathrm{i}\frac{\pi}{8}} - \frac{2-\mathrm{i}}{2\mathrm{i}} 2^{-1/4} \mathrm{e}^{-\mathrm{i}\frac{7\pi}{8}} \right)$$

$$= \pi 2^{-1/4}\left((2+\mathrm{i})\mathrm{e}^{-\mathrm{i}\frac{\pi}{8}} + (2-\mathrm{i})\mathrm{e}^{\mathrm{i}\frac{\pi}{8}} \right)$$

$$= \frac{2\pi}{\sqrt[4]{2}}\left(2\cos\frac{\pi}{8} + \sin\frac{\pi}{8} \right).$$

In the limit $r \to 0$ and $R \to \infty$ the integrals over γ_r and γ_R vanish,

$$\left| \int_{\gamma_R} f(z)\mathrm{d}z \right| \leq \pi R \frac{R+1}{\sqrt{R}(R^2 + 2R + 2)} \xrightarrow{R\to\infty} 0,$$

$$\left| \int_{\gamma_r} f(z)\mathrm{d}z \right| \leq \pi r \frac{1-r}{\sqrt{r}(2 - 2r - r^2)} \xrightarrow{r\to 0} 0,$$

while those on λ_1 and λ_2 tend both to the real integral to be calculated

$$\int_{\lambda_1} f(z)\mathrm{d}z = \int_r^R \frac{(x\mathrm{e}^{\mathrm{i}0} + 1)\mathrm{e}^{-\frac{1}{2}(\ln x + \mathrm{i}0)}}{(x\mathrm{e}^{\mathrm{i}0})^2 - 2x\mathrm{e}^{\mathrm{i}0}) + 2}\mathrm{e}^{\mathrm{i}0}\mathrm{d}x = \int_r^R \frac{(x+1)}{\sqrt{x}(x^2 - 2x + 2)}\mathrm{d}x,$$

$$\int_{\lambda_2} f(z)\mathrm{d}z = \int_R^r \frac{(x\mathrm{e}^{\mathrm{i}2\pi} + 1)\mathrm{e}^{-\frac{1}{2}(\ln x + \mathrm{i}2\pi)}}{(x\mathrm{e}^{\mathrm{i}2\pi})^2 - 2x\mathrm{e}^{\mathrm{i}2\pi} + 2}\mathrm{e}^{\mathrm{i}2\pi}\mathrm{d}x = \int_r^R \frac{(x+1)}{\sqrt{x}(x^2 - 2x + 2)}\mathrm{d}x.$$

In conclusion

$$\int_0^\infty \frac{x+1}{\sqrt{x}(x^2 - 2x + 2)}\mathrm{d}x = \frac{\pi}{\sqrt[4]{2}}\left(2\cos\frac{\pi}{8} + \sin\frac{\pi}{8} \right).$$

10.35 Let $f(z) = \mathrm{e}^{(1-a)\log z}/(1+z^2)^2$, having assumed for the logarithm the branch

$$\log z = \ln r + \mathrm{i}\theta, \qquad z = r\mathrm{e}^{\mathrm{i}\theta}, \quad r > 0, \quad 0 < \theta < 2\pi.$$

Observe that f has double poles at $z = \pm i$ and a branch line coinciding with the positive real semi-axis. Integrate f along the positively oriented closed path $\gamma = \lambda_1 + \gamma_R + \lambda_2 + \gamma_r$, where $\lambda_1(x) = x + i0$, $r \leq x \leq R$, $\gamma_R(\theta) = Re^{i\theta}$, $0 \leq \theta \leq \pi$, $\lambda_2(x) = xe^{i\pi}$, $R \geq x \geq r$, and $\gamma_r(\theta) = re^{i\theta}$, $\pi \geq \theta \geq 0$. For $R > 1$ and $r < 1$ we obtain

$$\int_\gamma f(z)dz = \int_{\lambda_1} f(z)dz + \int_{\gamma_R} f(z)dz + \int_{\lambda_2} f(z)dz + \int_{\gamma_r} f(z)dz$$

$$= 2\pi i \operatorname*{Res}_{z=i} \frac{e^{(1-a)\log z}/(z+i)^2}{(z-i)^2}$$

$$= 2\pi i \left. \frac{d}{dz} \frac{e^{(1-a)\log z}}{(z+i)^2} \right|_{z=i}$$

$$= \frac{ia\pi}{2} e^{-ia\pi/2}.$$

The integrals over the paths that compose γ are

$$\int_{\lambda_1} f(z)dz = \int_r^R \frac{e^{(1-a)(\ln x + i0)}}{((xe^{i0})^2 + 1)^2} e^{i0}dx = \int_r^R \frac{e^{(1-a)\ln x}}{(x^2 + 1)^2}dx,$$

$$\int_{\lambda_2} f(z)dz = \int_R^r \frac{e^{(1-a)(\ln x + i\pi)}}{((xe^{i\pi})^2 + 1)^2} e^{i\pi}dx = -e^{-ia\pi} \int_r^R \frac{e^{(1-a)\ln x}}{(x^2 + 1)^2}dx,$$

$$\left| \int_{\gamma_R} f(z)dz \right| \leq \frac{e^{(1-a)\ln R}}{(R^2 - 1)^2} \pi R = \frac{\pi R^{2-a}}{(R^2 - 1)^2} \xrightarrow{R \to \infty} 0, \qquad a > -2,$$

$$\left| \int_{\gamma_r} f(z)dz \right| \leq \frac{e^{(1-a)\ln r}}{(1 - r^2)^2} \pi r = \frac{\pi r^{2-a}}{(1 - r^2)^2} \xrightarrow{r \to 0} 0, \qquad a < 2.$$

Taking the limits $R \to \infty$ and $r \to 0$, for $-2 < a < 2$ we conclude

$$\int_0^\infty \frac{x^{1-a}}{(x^2 + 1)^2}dx = \frac{ia\pi}{2} \frac{e^{-ia\pi/2}}{1 - e^{-ia\pi}}$$

$$= \frac{a\pi}{4} \frac{2i}{e^{ia\pi/2} - e^{-ia\pi/2}}$$

$$= \frac{a\pi/4}{\sin(a\pi/2)}.$$

10.36 Let $f(z) = e^{(1-a)\log z}/(1+z^3)$, having assumed for the logarithm the branch

$$\log z = \ln r + i\theta, \qquad z = re^{i\theta}, \qquad r > 0, \qquad 0 < \theta < 2\pi.$$

Note that f has simple poles at $z = -1$ and $z = e^{\pm i\pi/3}$ and a branch line coinciding with the positive real semi-axis. Integrate f along the positively oriented closed path $\gamma = \lambda_1 + \gamma_R + \lambda_2 + \gamma_r$, where $\lambda_1(x) = x + i0$, $r \le x \le R$, $\gamma_R(\theta) = Re^{i\theta}$, $0 \le \theta \le 3\pi/2$, $\lambda_2(x) = xe^{i3\pi/2}$, $R \ge x \ge r$, and $\gamma_r(\theta) = re^{i\theta}$, $3\pi/2 \ge \theta \ge 0$. For $R > 1$ and $r < 1$ by the residue theorem we get

$$\int_\gamma f(z)dz = \int_{\lambda_1} f(z)dz + \int_{\gamma_R} f(z)dz + \int_{\lambda_2} f(z)dz + \int_{\gamma_r} f(z)dz$$

$$= 2\pi i \operatorname*{Res}_{z=e^{i\pi/3}} \frac{e^{(1-a)\log z}}{1+z^3}$$

$$= 2\pi i \left. \frac{e^{(1-a)\log z}}{3z^2} \right|_{z=e^{i\pi/3}}$$

$$= 2\pi i \frac{1}{3} e^{-i(1+a)\pi/3}.$$

The integrals over the paths that compose γ give

$$\int_{\lambda_1} f(z)dz = \int_r^R \frac{e^{(1-a)(\ln x + i0)}}{(xe^{i0})^3 + 1} e^{i0} dx = \int_r^R \frac{e^{(1-a)\ln x}}{x^3 + 1} dx,$$

$$\int_{\lambda_2} f(z)dz = \int_R^r \frac{e^{(1-a)(\ln x + i2\pi/3)}}{(xe^{i2\pi/3})^3 + 1} e^{i2\pi/3} dx = -e^{i(2-a)2\pi/3} \int_r^R \frac{e^{(1-a)\ln x}}{x^3 + 1} dx,$$

$$\left| \int_{\gamma_R} f(z)dz \right| \le \frac{e^{(1-a)\ln R}}{R^3 - 1} \frac{3}{2}\pi R = \frac{3\pi R^{2-a}}{2(R^3 - 1)} \xrightarrow{R\to\infty} 0, \qquad \text{if } a > -1,$$

$$\left| \int_{\gamma_r} f(z)dz \right| \le \frac{e^{(1-a)\ln r}}{1 - r^3} \frac{3}{2}\pi r = \frac{3\pi r^{2-a}}{2(1 - r^3)} \xrightarrow{r\to 0} 0, \qquad \text{if } a < 2.$$

Taking the limits $R \to \infty$ and $r \to 0$, for $-1 < a < 2$ we conclude

$$\int_0^\infty \frac{x^{1-a}}{x^3 + 1} dx = 2\pi i \frac{1}{3} \frac{e^{-i(1+a)\pi/3}}{1 - e^{i(2-a)2\pi/3}}$$

$$= \frac{\pi}{3} \frac{2i}{e^{i(1+a)\pi/3} - e^{-i(1+a)\pi/3}}$$

$$= \frac{\pi}{3} \frac{1}{\sin((1 + a)\pi/3)}.$$

10.37 Let $f(z) = e^{(1-a)\log z}/(4 + z^2)^2$, having assumed for the logarithm the branch

$$\log z = \ln r + i\theta, \qquad z = re^{i\theta}, \quad r > 0, \quad 0 < \theta < 2\pi.$$

Observe that f has double poles at $z = \pm 2i$ and a branch line coinciding with the positive real semi-axis. Integrate f along the positively oriented closed path $\gamma = \lambda_1 + \gamma_R + \lambda_2 + \gamma_r$, where $\lambda_1(x) = x + i0$, $r \leq x \leq R$, $\gamma_R(\theta) = Re^{i\theta}$, $0 \leq \theta \leq \pi$, $\lambda_2(x) = xe^{i\pi}$, $R \geq x \geq r$, and $\gamma_r(\theta) = re^{i\theta}$, $\pi \geq \theta \geq 0$. For $R > 2$ and $r < 2$ we obtain

$$\int_\gamma f(z)dz = \int_{\lambda_1} f(z)dz + \int_{\gamma_R} f(z)dz + \int_{\lambda_2} f(z)dz + \int_{\gamma_r} f(z)dz$$

$$= 2\pi i \operatorname*{Res}_{z=2i} \frac{e^{(1-a)\log z}/(z + 2i)^2}{(z - 2i)^2}$$

$$= 2\pi i \left. \frac{d}{dz} \frac{e^{(1-a)\log z}}{(z + 2i)^2} \right|_{z=2i}$$

$$= 2\pi i\, a 2^{-4-a} e^{-ia\pi/2}.$$

The integrals over the paths that compose γ give

$$\int_{\lambda_1} f(z)dz = \int_r^R \frac{e^{(1-a)(\ln x + i0)}}{((xe^{i0})^2 + 4)^2} e^{i0} dx = \int_r^R \frac{e^{(1-a)\ln x}}{(x^2 + 4)^2} dx,$$

$$\int_{\lambda_2} f(z)dz = \int_R^r \frac{e^{(1-a)(\ln x + i\pi)}}{((xe^{i\pi})^2 + 4)^2} e^{i\pi} dx = -e^{-ia\pi} \int_r^R \frac{e^{(1-a)\ln x}}{(x^2 + 4)^2} dx,$$

$$\left| \int_{\gamma_R} f(z)dz \right| \leq \frac{e^{(1-a)\ln R}}{(R^2 - 4)^2} \pi R = \frac{\pi R^{2-a}}{(R^2 - 4)^2} \xrightarrow{R\to\infty} 0, \qquad \text{if } a > -2,$$

$$\left| \int_{\gamma_r} f(z)dz \right| \leq \frac{e^{(1-a)\ln r}}{(4 - r^2)^2} \pi r = \frac{\pi r^{2-a}}{(4 - r^2)^2} \xrightarrow{r\to 0} 0, \qquad \text{if } a < 2.$$

Taking the limits $R \to \infty$ and $r \to 0$, for $-2 < a < 2$ we conclude

$$\int_0^\infty \frac{x^{1-a}}{(x^2 + 4)^2} dx = 2\pi i\, a 2^{-4-a} \frac{e^{-ia\pi/2}}{1 - e^{-ia\pi}}$$

$$= \pi a 2^{-4-a} \frac{2i}{e^{ia\pi/2} - e^{-ia\pi/2}}$$

$$= \frac{a\pi 2^{-4-a}}{\sin(a\pi/2)}.$$

10.38 Let $f(z) = z^a/(z^2 + 2z + 1) = e^{a \log z}/(z + 1)^2$ and assume the logarithm with branch line coinciding with the positive real semi-axis

$$\log z = \ln r + i\theta, \qquad z = re^{i\theta}, \quad r > 0, \quad 0 < \theta < 2\pi.$$

The function f is analytic everywhere except on the positive real semi-axis and at the double pole at $z = -1$ where it has residue

$$\operatorname*{Res}_{z=-1} f(z) = \frac{d}{dz} e^{a \log z}\Big|_{z=-1} = e^{a \log z} \frac{a}{z}\Big|_{z=-1} = -ae^{a(\ln 1 + i\pi)} = -ae^{i\pi a}.$$

Integrate f along the positively oriented closed path $\gamma = \lambda_1 + \gamma_R + \lambda_2 + \gamma_r$, where

$$\lambda_1(x) = x + i0, \qquad r \le x \le R,$$
$$\gamma_R(\theta) = Re^{i\theta}, \qquad 0 \le \theta \le 2\pi,$$
$$\lambda_2(x) = x - i0, \qquad R \ge x \ge r,$$
$$\gamma_r(\theta) = re^{i\theta}, \qquad 2\pi \ge \theta \ge 0.$$

As usual, $x \pm i0$ stands for $x \pm i\varepsilon$ with ε positive infinitesimal so that in polar representation $x + i0 = xe^{i\varepsilon}$ while $x - i0 = xe^{i(2\pi - \varepsilon)}$. For $r < 1 < R$ we have

$$\int_\gamma f(z)dz = \int_{\lambda_1} f(z)dz + \int_{\gamma_R} f(z)dz + \int_{\lambda_2} f(z)dz + \int_{\gamma_r} f(z)dz$$
$$= 2\pi i \operatorname*{Res}_{z=-1} f(z)$$
$$= -2\pi i a e^{i\pi a}.$$

The integrals along the individual paths that compose γ give

$$\int_{\lambda_1} f(z)dz = \int_r^R \frac{e^{a(\ln x + i0)}}{((xe^{i0}) + 1)^2} e^{i0} dx = \int_r^R \frac{e^{a \ln x}}{(x + 1)^2} dx,$$

$$\int_{\lambda_2} f(z)dz = \int_R^r \frac{e^{a(\ln x + i2\pi)}}{((xe^{i2\pi}) + 1)^2} e^{i2\pi} dx = -e^{i2\pi a} \int_r^R \frac{e^{a \ln x}}{(x + 1)^2} dx,$$

$$\left| \int_{\gamma_R} f(z)dz \right| \le \frac{e^{a \ln R}}{(R - 1)^2} 2\pi R = \frac{2\pi R^{a+1}}{(R - 1)^2} \xrightarrow{R \to \infty} 0, \qquad \text{if } a + 1 < 2,$$

$$\left| \int_{\gamma_r} f(z)dz \right| \le \frac{e^{a \ln r}}{(1 - r)^2} 2\pi r = \frac{2\pi r^{a+1}}{(1 - r)^2} \xrightarrow{r \to 0} 0, \qquad \text{if } a + 1 > 0.$$

Therefore, taking the limits $r \to 0$ and $R \to \infty$ we conclude that for $-1 < a < 1$ the assigned integral is convergent and its value is

$$\int_0^\infty \frac{x^a}{x^2 + 2x + 1}\,dx = \frac{-2\pi i a e^{i\pi a}}{1 - e^{i2\pi a}}$$

$$= \frac{-2\pi i a e^{i\pi a}}{e^{i\pi a}\left(e^{-i\pi a} - e^{i\pi a}\right)}$$

$$= \frac{\pi a}{\sin(\pi a)}.$$

10.39   Choose $f(z) = e^{az}/\cosh z$ and the positively oriented closed path of integration $\gamma = \lambda_1 + \lambda_2 + \lambda_3 + \lambda_4$, where $\lambda_1(x) = x,\ -R \le x \le R,\ \lambda_2(y) = R + iy,$ $0 \le y \le \pi,\ \lambda_3(x) = x + i\pi,\ R \ge x \ge -R,$ and $\lambda_4(y) = -R + iy,\ \pi \ge y \ge 0.$ Since f is analytic on and within γ except for the simple pole at $z = i\pi/2$, the residue theorem gives

$$\int_\gamma f(z)dz = \sum_{k=1}^4 \int_{\lambda_k} f(z)dz$$

$$= 2\pi i \operatorname*{Res}_{z=i\pi/2} f(z)$$

$$= 2\pi i \left.\frac{e^{az}}{\sinh z}\right|_{z=i\pi/2}$$

$$= 2\pi e^{ia\pi/2}.$$

For the integrals along the paths $\lambda_k,\ k = 1, 2, 3, 4,$ we have

$$\int_{\lambda_1} f(z)dz = \int_{-R}^R \frac{e^{ax}}{\cosh x}\,dx,$$

$$\int_{\lambda_3} f(z)dz = \int_R^{-R} \frac{e^{a(x+i\pi)}}{\cosh(x + i\pi)}\,dx = e^{ia\pi}\int_{-R}^R \frac{e^{ax}}{\cosh x}\,dx,$$

having used the anti-periodicity property $\cosh(x + i\pi) = -\cosh x$,

$$\left|\int_{\lambda_2} f(z)dz\right| \le \pi\, \frac{e^{aR}}{\frac{1}{2}\left(e^R - 1\right)} \xrightarrow{R\to\infty} 0, \qquad a < 1,$$

$$\left|\int_{\lambda_4} f(z)dz\right| \le \pi\, \frac{e^{-aR}}{\frac{1}{2}\left(e^R - 1\right)} \xrightarrow{R\to\infty} 0, \qquad a > -1.$$

In conclusion, in the limit $R \to \infty$ for $-1 < a < 1$ we have

$$\int_{-\infty}^{+\infty} \frac{e^{ax}}{\cosh x}\,dx = \frac{2\pi\,e^{ia\pi/2}}{1+e^{ia\pi}} = \frac{\pi}{\cos(a\pi/2)}.$$

10.40 Let $f(z) = 1/(1+z^{2n+1})$ which is analytic everywhere in $\mathbb{C}$ except for the $2n+1$ simple poles at the points

$$z_k = e^{i\pi/(2n+1)+i2\pi k/(2n+1)}, \qquad k = 0, 1, \ldots, 2n.$$

Consider the positively oriented closed path of integration $\gamma = \lambda_1 + \lambda_2 + \lambda_3$, where $\lambda_1(x) = x, 0 \le x \le R$, $\lambda_2(\theta) = Re^{i\theta}, 0 \le \theta \le 2\pi/(2n+1)$, $\lambda_3(x) = xe^{i2\pi/(2n+1)}$, $R \ge x \ge 0$. The function f is, for $R > 1$, analytic on and within γ except for the simple pole at $z = z_0$. Therefore

$$\int_\gamma f(z)dz = \sum_{j=1}^{3} \int_{\lambda_j} f(z)dz$$

$$= 2\pi i \operatorname*{Res}_{z=z_0} f(z)$$

$$= 2\pi i \frac{1}{(2n+1)z_0^{2n}}.$$

For the integrals along the paths that compose γ we have

$$\int_{\lambda_1} f(z)dz = \int_0^R \frac{1}{1+x^{2n+1}}\,dx,$$

$$\left| \int_{\lambda_2} f(z)dz \right| \le \frac{2\pi R}{2n+1} \frac{1}{R^{2n+1}-1} \xrightarrow{R\to\infty} 0,$$

$$\int_{\lambda_3} f(z)dz = -e^{i2\pi/(2n+1)} \int_0^R \frac{1}{1+x^{2n+1}}\,dx.$$

In conclusion, in the limit $R \to \infty$ we obtain

$$\int_0^\infty \frac{1}{1+x^{2n+1}}\,dx = \frac{2\pi i/(2n+1)}{e^{i2n\pi/(2n+1)}} \frac{1}{1-e^{i2\pi/(2n+1)}}$$

$$= \frac{2\pi i/(2n+1)}{e^{i2n\pi/(2n+1)}} \frac{1}{\left[e^{-i\pi/(2n+1)} - e^{i\pi/(2n+1)}\right]e^{i\pi/(2n+1)}}$$

$$= \frac{\pi/(2n+1)}{\sin[\pi/(2n+1)]}.$$

10.41 Choose $f(z) = e^{iz}/(e^z + e^{-z})$ and consider the positively oriented closed path $\gamma = \lambda_1 + \lambda_2 + \lambda_3 + \lambda_4$, where $\lambda_1(x) = x$, $-R \le x \le R$, $\lambda_2(y) = R + iy$, $0 \le y \le \pi$, $\lambda_3(x) = x + i\pi$, $R \ge x \ge -R$, and $\lambda_4(y) = -R + iy$, $\pi \ge y \ge 0$. Since $f(z) = p(z)/q(z)$ with $p(z) = e^{iz}$ and $q(z) = e^z + e^{-z}$ is a function analytic everywhere in $\mathbb{C}$ except for the simple poles corresponding to the simple zeros of $q(z)$, namely, the points $z = \pm i\pi/2 + i2\pi k$, $k \in \mathbb{Z}$, the residue theorem gives, for $R > 0$,

$$\int_\gamma f(z)dz = 2\pi i \operatorname*{Res}_{z=i\pi/2} f(z)$$

$$= 2\pi i \frac{p(i\frac{\pi}{2})}{q'(i\frac{\pi}{2})}$$

$$= \pi e^{-\pi/2}.$$

On the other hand

$$\int_\gamma f(z)dz = \int_{\lambda_1} f(z)dz + \int_{\lambda_2} f(z)dz + \int_{\lambda_3} f(z)dz + \int_{\lambda_4} f(z)dz,$$

with

$$\int_{\lambda_1} f(z)dz = \int_{-R}^{R} \frac{e^{ix}}{e^x + e^{-x}}dx,$$

$$\int_{\lambda_3} f(z)dz = \int_{R}^{-R} \frac{e^{i(x+i\pi)}}{e^{x+i\pi} + e^{-(x+i\pi)}}dx = e^{-\pi} \int_{-R}^{R} \frac{e^{ix}}{e^x + e^{-x}}dx.$$

By Darboux's inequality

$$\left| \int_{\lambda_2} f(z)dz \right| \le \frac{\pi}{e^R - e^{-R}} \xrightarrow{R \to \infty} 0,$$

$$\left| \int_{\lambda_4} f(z)dz \right| \le \frac{\pi}{e^R - e^{-R}} \xrightarrow{R \to \infty} 0,$$

so in the limit $R \to \infty$ we have

$$\left(1 + e^{-\pi}\right) \int_{-\infty}^{+\infty} \frac{e^{ix}}{e^x + e^{-x}}dx = \pi e^{-\pi/2}.$$

Taking the real part of the above expression we conclude

$$\int_{-\infty}^{+\infty} \frac{\cos x}{e^x + e^{-x}}dx = \frac{\pi/2}{\cosh(\pi/2)}.$$

10.42 Setting $w = |w|e^{i\alpha}$, since $\operatorname{Re} w \geq 0$ it must be $-\pi/2 \leq \alpha \leq \pi/2$. Integrate the function $f(z) = e^{-wz^2}$, analytic throughout $\mathbb{C}$, along the simple closed path $\gamma = \lambda_1 + \gamma_R + \lambda_2$, where

$$\lambda_1(x) = x, \qquad 0 \leq x \leq R,$$

$$\gamma_R(\theta) = Re^{i\theta}, \qquad 0 \leq \theta \leq -\alpha/2 \text{ if } \alpha < 0, \qquad 0 \geq \theta \geq -\alpha/2 \text{ if } \alpha > 0,$$

$$\lambda_2(r) = re^{-i\alpha/2}, \qquad R \geq r \geq 0.$$

Note that the path is positively oriented if $\alpha < 0$ and negatively oriented if $\alpha > 0$ while for $\alpha = 0$, that is when $\operatorname{Im} w = 0$ and therefore $w = a > 0$, the integral in question is known and is equal to $\sqrt{\pi}/(2\sqrt{a})$. Since f is analytic on and within γ, we have

$$\int_\gamma f(z)\mathrm{d}z = \int_{\lambda_1} f(z)\mathrm{d}z + \int_{\gamma_R} f(z)\mathrm{d}z + \int_{\lambda_2} f(z)\mathrm{d}z = 0.$$

The integrals along the paths λ_1 and λ_2 are given by

$$\int_{\lambda_1} f(z)\mathrm{d}z = \int_0^R e^{-wx^2}\mathrm{d}x,$$

$$\int_{\lambda_2} f(z)\mathrm{d}z = \int_R^0 e^{-|w|e^{i\alpha}(re^{-i\alpha/2})^2} e^{-i\alpha/2}\mathrm{d}r$$

$$= -|w|^{-1/2}e^{-i\alpha/2} \int_0^{R|w|^{1/2}} e^{-u^2}\mathrm{d}u,$$

where we put $u = |w|^{1/2} r$. For the integral along γ_R we find

$$\left| \int_{\gamma_R} f(z)\mathrm{d}z \right| \leq -\operatorname{sgn}(\alpha) \int_0^{-\alpha/2} \left| e^{-|w|e^{i\alpha} R^2 e^{i2\theta}} \, iRe^{i\theta} \right| \mathrm{d}\theta$$

$$= -\operatorname{sgn}(\alpha) R \int_0^{-\alpha/2} e^{-|w|R^2 \cos(\alpha+2\theta)} \mathrm{d}\theta$$

$$= \frac{R}{2} \int_{\pi/2-|\alpha|}^{\pi/2} e^{-|w|R^2 \sin\varphi} \mathrm{d}\varphi,$$

where we put $\alpha + 2\theta = (\pi/2 - \varphi)\operatorname{sgn}(\alpha)$ and used the relation

$$\cos((\pi/2 - \varphi)\operatorname{sgn}(\alpha)) = \cos(\pi/2 - \varphi) = \sin\varphi.$$

Using Jordan's inequality, $\sin\varphi \geq 2\varphi/\pi$, valid for $0 \leq \varphi \leq \pi/2$, we get

$$\left| \int_{\gamma_R} f(z)\mathrm{d}z \right| \leq \frac{R}{2} \int_{\pi/2-\alpha}^{\pi/2} e^{-|w|R^2 2\varphi/\pi}\,\mathrm{d}\varphi$$

$$= \frac{\pi}{4|w|R} \left(e^{-|w|R^2(1-2|\alpha|/\pi)} - e^{-|w|R^2} \right) \xrightarrow{R\to\infty} 0.$$

In the limit $R \to \infty$ we conclude that $\forall w \in \mathbb{C}$ with $w \neq 0$ and $\mathrm{Re}\,w \geq 0$

$$\int_0^\infty e^{-wx^2}\mathrm{d}x = |w|^{-1/2}e^{-i\alpha/2} \int_0^\infty e^{-u^2}\mathrm{d}u = \frac{\sqrt{\pi}}{2\sqrt{w}},$$

where $\sqrt{w} = |w|^{1/2}e^{i\alpha/2}$. Note that in the case $\mathrm{Re}\,w = 0$, that is when $w = ib$ with b real non-zero, we have

$$\int_0^\infty e^{-ibx^2}\mathrm{d}x = \frac{\sqrt{\pi}}{2\sqrt{ib}} = \frac{\sqrt{\pi}}{2\sqrt{|b|}}e^{-i\frac{\pi}{4}\,\mathrm{sgn}(b)},$$

from which we obtain the Fresnel integrals

$$\int_0^\infty \cos(bx^2)\mathrm{d}x = \frac{1}{2}\sqrt{\frac{\pi}{2|b|}}, \qquad \int_0^\infty \sin(bx^2)\mathrm{d}x = \frac{1}{2}\sqrt{\frac{\pi}{2|b|}}\,\mathrm{sgn}(b).$$

10.43 Setting $e^{i\theta} = z$, we have $\mathrm{d}\theta = \mathrm{d}z/iz$, $\cos\theta = (z+z^{-1})/2$ and

$$\int_0^{2\pi} \log\left(a + a^{-1} - 2\cos\theta\right)\mathrm{d}\theta = \int_\gamma \frac{1}{iz} \log\left(a + a^{-1} - z - z^{-1}\right)\mathrm{d}z,$$

where γ is the positively oriented circle with center 0 and radius 1. If we choose for the logarithm the principal branch

$$\log z = \ln r + i\theta, \qquad z = re^{i\theta}, \quad r > 0, \quad -\pi < \theta < \pi,$$

the integrand function

$$f(z) = \frac{1}{iz} \log\left(a + a^{-1} - z - z^{-1}\right)$$

is analytic everywhere in $\mathbb{C}$ except at the point $z = 0$ and the points $z(t)$ such that

$$a + a^{-1} - z(t) - z(t)^{-1} = -t, \qquad t \geq 0.$$

These points satisfy the second degree equation

$$z(t)^2 - \left(a + a^{-1} + t\right)z(t) + 1 = 0,$$

Fig. 12.11 Closed path
$\sigma = \gamma + \lambda_1 + \lambda_2 + \gamma_\varepsilon + \lambda_3 + \lambda_4$,
used to integrate the function
$f(z) = (iz)^{-1}\log(a + a^{-1} - z - z^{-1})$,
with $a > 1$, along the
positively oriented circle
$|z| = 1$

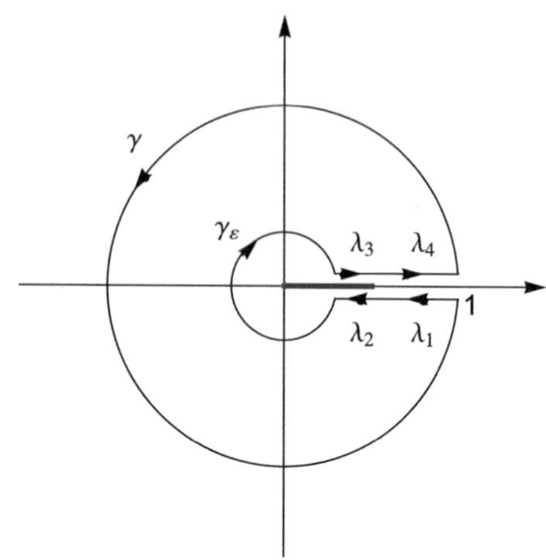

which has solutions

$$z_\pm(t) = \frac{1}{2}\left(\left(a + a^{-1} + t\right) \pm \sqrt{\left(a + a^{-1} + t\right)^2 - 4}\right).$$

For $t \in [0, \infty)$, the solution $z_+(t)$ describes the real semi-axis from a to $+\infty$, while $z_-(t)$ the segment from a^{-1} to 0. Consider the closed path $\sigma = \gamma + \lambda_1 + \lambda_2 + \gamma_\varepsilon + \lambda_3 + \lambda_4$, where (Fig. 12.11)

$$\gamma(\theta) = e^{i\theta}, \qquad 0 \le \theta \le 2\pi,$$
$$\lambda_1(x) = x - i0^+, \qquad 1 \ge x \ge a^{-1},$$
$$\lambda_2(x) = x - i0^+, \qquad a^{-1} \ge x \ge \varepsilon,$$
$$\gamma_\varepsilon(\theta) = \varepsilon e^{i\theta}, \qquad 2\pi \ge \theta \ge 0,$$
$$\lambda_3(x) = x + i0^+, \qquad \varepsilon \le x \le a^{-1},$$
$$\lambda_4(x) = x + i0^+, \qquad a^{-1} \le x \le 1.$$

For every $\varepsilon > 0$ the function f is analytic on σ and within it, therefore

$$\int_\sigma f(z)\mathrm{d}z = \int_\gamma f(z)\mathrm{d}z + \sum_{k=1}^{4} \int_{\lambda_k} f(z)\mathrm{d}z + \int_{\gamma_\varepsilon} f(z)\mathrm{d}z$$

$$= 0.$$

Since f is analytic on the segment of the real axis $(a^{-1}, 1]$ and for $\varepsilon \to 0$ turns out to be $\lambda_1 = -\lambda_4$, we have

$$\lim_{\varepsilon \to 0} \left(\int_{\lambda_1} f(z)\mathrm{d}z + \int_{\lambda_4} f(z)\mathrm{d}z \right) = 0.$$

On the segment $[\varepsilon, a^{-1}]$ at the branch line we have

$$\int_{\lambda_2} f(z)\mathrm{d}z = \int_{a^{-1}}^{\varepsilon} \frac{1}{\mathrm{i}x} \log\left(t(x)\mathrm{e}^{-\mathrm{i}\pi}\right) \mathrm{d}x = \int_{a^{-1}}^{\varepsilon} \frac{1}{\mathrm{i}x} \left(\ln(t(x)) - \mathrm{i}\pi\right) \mathrm{d}x,$$

$$\int_{\lambda_3} f(z)\mathrm{d}z = \int_{\varepsilon}^{a^{-1}} \frac{1}{\mathrm{i}x} \log\left(t(x)\mathrm{e}^{\mathrm{i}\pi}\right) \mathrm{d}x = \int_{\varepsilon}^{a^{-1}} \frac{1}{\mathrm{i}x} \left(\ln(t(x)) + \mathrm{i}\pi\right) \mathrm{d}x,$$

where $t(x) = x + x^{-1} - a - a^{-1} \geq 0$ for $\varepsilon \leq x \leq a^{-1}$. The sum of these two integrals gives

$$\int_{\lambda_2} f(z)\mathrm{d}z + \int_{\lambda_3} f(z)\mathrm{d}z = 2 \int_{\varepsilon}^{a^{-1}} \frac{\pi}{x} \mathrm{d}x = -2\pi \ln a - 2\pi \ln \varepsilon.$$

Finally, on the path γ_ε for ε infinitesimal we can put $a + a^{-1} - z - z^{-1} \simeq -z^{-1}$ and therefore

$$\int_{\gamma_\varepsilon} f(z)\mathrm{d}z = \int_{\gamma_\varepsilon} \frac{1}{\mathrm{i}z} \log\left(-z^{-1}\right) \mathrm{d}z$$

$$= \int_{2\pi}^{0} \frac{1}{\mathrm{i}\varepsilon\mathrm{e}^{\mathrm{i}\theta}} \log\left(\varepsilon^{-1}\mathrm{e}^{-\mathrm{i}(\theta-\pi)}\right) \varepsilon\mathrm{e}^{\mathrm{i}\theta} \mathrm{i}\mathrm{d}\theta$$

$$= \int_{2\pi}^{0} \left(\ln\left(\varepsilon^{-1}\right) - \mathrm{i}(\theta - \pi)\right) \mathrm{d}\theta$$

$$= \theta \ln\left(\varepsilon^{-1}\right) - \mathrm{i}\left(\frac{\theta^2}{2} - \pi\theta\right)\bigg|_{2\pi}^{0}$$

$$= 2\pi \ln \varepsilon.$$

Note that the argument of $-z^{-1}$, namely, $-(\theta - \pi)$, has been chosen to fall within the range $(-\pi, \pi)$ of the principal branch of the logarithm when θ varies in the domain of integration $2\pi \geq \theta \geq 0$. In conclusion

$$\int_{\gamma} f(z)\mathrm{d}z = \lim_{\varepsilon \to 0} \left(-\int_{\lambda_2} f(z)\mathrm{d}z - \int_{\lambda_3} f(z)\mathrm{d}z - \int_{\gamma_\varepsilon} f(z)\mathrm{d}z \right)$$

$$= \lim_{\varepsilon \to 0} (2\pi \ln a + 2\pi \ln \varepsilon - 2\pi \ln \varepsilon)$$

$$= 2\pi \ln a.$$

10.44 Consider the complex function $f(z) = (ze^{i3z} + 3i)/(z^2 - 2z + 2)$ that has simple poles at $z = -1 \pm i$ and integrate it along the positively oriented boundary γ of the square with vertices $-R_1, R_2, R_2 + i(R_1 + R_2), -R_1 + i(R_1 + R_2)$, with $R_1 > 0$, $R_2 > 0$ and $R_1 + R_2 > 1$. By the residue theorem we have

$$\int_\gamma f(z)dz = 2\pi i \operatorname*{Res}_{z=1+i} f(z)$$

$$= 2\pi i \frac{(1+i)e^{-3+3i} + 3i}{2i}$$

$$= \pi e^{-3}(\cos 3 - \sin 3) + i\pi(3 + e^{-3}(\cos 3 + \sin 3)).$$

On the other hand we have

$$\int_\gamma f(z)dz = \sum_{k=1}^{4} \int_{\lambda_k} f(z)dz,$$

where $\lambda_1(x) = x, -R_1 \le x \le R_2$, $\lambda_2(y) = R_2 + iy, 0 \le y \le R_1 + R_2$, $\lambda_3(x) = x + i(R_1 + R_2), R_2 \ge x \ge -R_1$, and $\lambda_4(y) = -R_1 + iy, R_1 + R_2 \ge y \ge 0$. The integrals along the paths that compose γ are given by

$$\int_{\lambda_1} f(z)dz = \int_{-R_1}^{R_2} \frac{xe^{i3x} + 3i}{x^2 - 2x + 2}dx,$$

$$\int_{\lambda_2} f(z)dz = \int_0^{R_1+R_2} \frac{(R_2 + iy)e^{i3R_2 - 3y} + 3i}{(R_2 + iy)^2 - 2(R_2 + iy) + 2}idy \xrightarrow{R_1,R_2 \to \infty} 0,$$

$$\int_{\lambda_3} f(z)dz = \int_{R_2}^{-R_1} \frac{xe^{i3x - 3(R_1+R_2)} + 3i}{(x + i(R_1 + R_2))^2 - 2(x + i(R_1 + R_2)) + 2}dx \xrightarrow{R_1,R_2 \to \infty} 0,$$

$$\int_{\lambda_4} f(z)dz = \int_{R_1+R_2}^0 \frac{(-R_1 + iy)e^{-i3R_1 - 3y} + 3i}{(-R_1 + iy)^2 - 2(-R_1 + iy) + 2}idy \xrightarrow{R_1,R_2 \to \infty} 0.$$

Therefore, the following limit exists and has value

$$\lim_{R_1,R_2 \to \infty} \int_{-R_1}^{R_2} \frac{xe^{i3x} + 3i}{x^2 - 2x + 2}dx = \int_{-\infty}^{+\infty} \frac{xe^{i3x} + 3i}{x^2 - 2x + 2}dx$$

$$= \pi e^{-3}(\cos 3 - \sin 3) + i\pi(3 + e^{-3}(\cos 3 + \sin 3)).$$

Taking the imaginary and real parts of the above expression we respectively have

$$\int_{-\infty}^{+\infty} \frac{x\sin(3x)+3}{x^2-2x+2}\,dx = \pi(3+e^{-3}(\cos 3 + \sin 3)),$$

$$\int_{-\infty}^{+\infty} \frac{x\cos(3x)}{x^2-2x+2}\,dx = \pi e^{-3}(\cos 3 - \sin 3).$$

10.45 Consider the complex function $f(z) = (ze^{i3z}+3i)/(z^2+1)$ that has simple poles at $z = \pm i$ and integrate it along the positively oriented boundary γ of the square with vertices $-R_1$, R_2, $R_2+i(R_1+R_2)$, $-R_1+i(R_1+R_2)$, with $R_1 > 0$, $R_2 > 0$ and $R_1+R_2 > 1$. By the residue theorem we have

$$\int_\gamma f(z)dz = 2\pi i \operatorname*{Res}_{z=i} f(z)$$

$$= 2\pi i \frac{ie^{-3}+3i}{2i}$$

$$= i\pi(3+e^{-3}).$$

On the other hand

$$\int_\gamma f(z)dz = \sum_{k=1}^{4}\int_{\lambda_k} f(z)dz,$$

where $\lambda_1(x) = x$, $-R_1 \le x \le R_2$, $\lambda_2(y) = R_2 + iy$, $0 \le y \le R_1+R_2$, $\lambda_3(x) - x+i(R_1+R_2)$, $R_2 \ge x \ge -R_1$, and $\lambda_4(y) = -R_1+iy$, $R_1+R_2 \ge y \ge 0$. The integrals along the paths that compose γ provide

$$\int_{\lambda_1} f(z)dz = \int_{-R_1}^{R_2} \frac{xe^{i3x}+3i}{x^2+1}\,dx,$$

$$\int_{\lambda_2} f(z)dz = \int_{0}^{R_1+R_2} \frac{(R_2+iy)e^{i3R_2-3y}+3i}{(R_2+iy)^2+1}\,idy \xrightarrow{R_1,R_2\to\infty} 0,$$

$$\int_{\lambda_3} f(z)dz = \int_{R_2}^{-R_1} \frac{xe^{i3x-3(R_1+R_2)}+3i}{(x+i(R_1+R_2))^2+1}\,dx \xrightarrow{R_1,R_2\to\infty} 0,$$

$$\int_{\lambda_4} f(z)dz = \int_{R_1+R_2}^{0} \frac{(-R_1+iy)e^{-i3R_1-3y}+3i}{(-R_1+iy)^2+1}\,idy \xrightarrow{R_1,R_2\to\infty} 0.$$

Therefore, the following limit exists and has value

$$\lim_{R_1,R_2\to\infty}\int_{-R_1}^{R_2}\frac{xe^{i3x}+3i}{x^2+1}\,dx=\int_{-\infty}^{+\infty}\frac{xe^{i3x}+3i}{x^2+1}\,dx=i\pi(3+e^{-3}).$$

Taking the imaginary part of this expression we conclude

$$\int_{-\infty}^{+\infty}\frac{x\sin(3x)+3}{x^2+1}\,dx=\pi(3+e^{-3}).$$

Note that, taking the real part of the same expression we obtain

$$\int_{-\infty}^{+\infty}\frac{x\cos(3x)}{x^2+1}\,dx=0,$$

in accordance with the fact that the integrand function is odd.

10.46 To begin with, complete the square at the exponent of the function to be integrated

$$-\frac{x^2}{\sigma^2}-i\lambda x=-\left(\left(\frac{x}{\sigma}\right)^2+2\frac{x}{\sigma}\frac{i\lambda\sigma}{2}+\left(\frac{i\lambda\sigma}{2}\right)^2-\left(\frac{i\lambda\sigma}{2}\right)^2\right)$$

$$=-\left(\frac{x}{\sigma}+\frac{i\lambda\sigma}{2}\right)^2-\frac{\lambda^2\sigma^2}{4}.$$

With the change of variable $t=x/\sigma$, the integral to be calculated becomes

$$\int_{-\infty}^{+\infty}e^{-x^2/\sigma^2}e^{-i\lambda x}\,dx=\sigma e^{-\frac{\lambda^2\sigma^2}{4}}\int_{-\infty}^{+\infty}e^{-(t+i\lambda\sigma/2)^2}\,dt.$$

The integral in t can be evaluated by integrating the entire function $f(z)=e^{-z^2}$ along the negatively oriented closed path $\gamma=\lambda_1+\lambda_2+\lambda_3+\lambda_4$, where

$$\lambda_1(u)=u,\qquad A\ge u\ge-A,$$

$$\lambda_2(y)=-A+iy,\qquad 0\le y\le\lambda\sigma/2,$$

$$\lambda_3(t)=t+i\lambda\sigma/2,\qquad -A\le t\le A,$$

$$\lambda_4(y)=A+iy,\qquad \lambda\sigma/2\ge y\ge0.$$

In these expressions A is an arbitrary positive real number. By the residue theorem it follows

$$\int_\gamma f(z)\,dz=\sum_{k=1}^{4}\int_{\lambda_k}f(z)\,dz=0.$$

The integrals along the four paths that compose γ provide

$$\int_{\lambda_1} f(z)\mathrm{d}z = -\int_{-A}^{A} \mathrm{e}^{-u^2}\mathrm{d}u,$$

$$\int_{\lambda_2} f(z)\mathrm{d}z = \int_{0}^{\lambda\sigma/2} \mathrm{e}^{-(-A+\mathrm{i}y)^2}\mathrm{i}\mathrm{d}y = \int_{0}^{\lambda\sigma/2} \mathrm{e}^{-A^2+2\mathrm{i}Ay+y^2}\mathrm{i}\mathrm{d}y,$$

$$\int_{\lambda_3} f(z)\mathrm{d}z = \int_{-A}^{A} \mathrm{e}^{-(t+\mathrm{i}\lambda\sigma/2)^2}\mathrm{d}t,$$

$$\int_{\lambda_4} f(z)\mathrm{d}z = \int_{\lambda\sigma/2}^{0} \mathrm{e}^{-(A+\mathrm{i}y)^2}\mathrm{i}\mathrm{d}y = -\int_{0}^{\lambda\sigma/2} \mathrm{e}^{-A^2-2\mathrm{i}Ay+y^2}\mathrm{i}\mathrm{d}y.$$

The modulus of the integrals over λ_2 and λ_4 can be upper bounded as follows

$$\left| \int_{\lambda_2} f(z)\mathrm{d}z \right| \le \mathrm{e}^{-A^2} \int_{0}^{\lambda\sigma/2} \mathrm{e}^{y^2}\mathrm{d}y \xrightarrow{A\to\infty} 0,$$

$$\left| \int_{\lambda_4} f(z)\mathrm{d}z \right| \le \mathrm{e}^{-A^2} \int_{0}^{\lambda\sigma/2} \mathrm{e}^{y^2}\mathrm{d}y \xrightarrow{A\to\infty} 0.$$

Therefore, for $A \to \infty$ we obtain

$$\int_{-\infty}^{+\infty} \mathrm{e}^{-(t+\mathrm{i}\lambda\sigma/2)^2}\mathrm{d}t = \int_{-\infty}^{+\infty} \mathrm{e}^{-u^2}\mathrm{d}u = \sqrt{\pi}.$$

In conclusion,

$$\int_{-\infty}^{+\infty} \mathrm{e}^{-x^2/\sigma^2}\mathrm{e}^{-\mathrm{i}\lambda x}\mathrm{d}x = \sigma\sqrt{\pi}\,\mathrm{e}^{-\frac{\lambda^2\sigma^2}{4}}.$$

10.47 Let $f(z) = (z-z_0)^{-1}(z-z_1)^{-1}$ and choose the positively oriented closed path $\gamma = \lambda_R + \gamma_R$ with λ_R given and $\gamma_R(\theta) = a + R\mathrm{e}^{\mathrm{i}\theta}$, $\pi/2 \le \theta \le 3\pi/2$. For $z_0 \ne z_1$ the function $f(z)\mathrm{e}^{zt}$ is analytic on and within γ except for the simple poles at $z = z_0$ and $z = z_1$. By the residue theorem

$$\int_{\gamma} \mathrm{e}^{zt} f(z)\mathrm{d}z = \int_{\lambda_R} \mathrm{e}^{zt} f(z)\mathrm{d}z + \int_{\gamma_R} \mathrm{e}^{zt} f(z)\mathrm{d}z$$

$$= 2\pi\mathrm{i}\left(\operatorname*{Res}_{z=z_0} \mathrm{e}^{zt} f(z) + \operatorname*{Res}_{z=z_1} \mathrm{e}^{zt} f(z) \right).$$

Note that for $0 < |z - z_0| < |z_0 - z_1|$ we have

$$
e^{zt} f(z) = \frac{1}{z - z_0} \frac{1}{z_0 - z_1} \frac{1}{1 + \frac{z - z_0}{z_0 - z_1}} e^{z_0 t} e^{(z - z_0)t}
$$

$$
= \frac{1}{z - z_0} \frac{1}{z_0 - z_1} \sum_{k=0}^{\infty} (-1)^k \left(\frac{z - z_0}{z_0 - z_1} \right)^k e^{z_0 t} \sum_{n=0}^{\infty} \frac{((z - z_0)t)^n}{n!}
$$

which implies

$$
\operatorname*{Res}_{z=z_0} e^{zt} f(z) = \frac{e^{z_0 t}}{z_0 - z_1}.
$$

Exchanging z_0 with z_1 we get

$$
\operatorname*{Res}_{z=z_1} e^{zt} f(z) = \frac{e^{z_1 t}}{z_1 - z_0}.
$$

Using Darboux's and Jordan's inequalities, the modulus of the integral on γ_R can be upper bounded as follows

$$
\left| \int_{\gamma_R} e^{zt} f(z) \mathrm{d}z \right| \le \int_{\pi/2}^{3\pi/2} \frac{e^{(a + R\cos\theta)t}}{(R + |z_0|)(R + |z_1|)} R\mathrm{d}\theta
$$

$$
= \frac{R e^{at}}{(R + |z_0|)(R + |z_1|)} \int_0^{\pi} e^{-Rt \sin\varphi} \mathrm{d}\varphi
$$

$$
\le \frac{R e^{at}}{(R + |z_0|)(R + |z_1|)} \frac{\pi}{Rt},
$$

which provides

$$
\lim_{R \to \infty} \int_{\gamma_R} e^{zt} f(z) \mathrm{d}z = 0.
$$

In conclusion

$$
\frac{1}{2\pi i} \lim_{R \to \infty} \int_{\lambda_R} e^{zt} \frac{1}{(z - z_0)(z - z_1)} \mathrm{d}z = \frac{e^{z_0 t} - e^{z_1 t}}{z_0 - z_1}.
$$

10.48 Given $g(z) = 1/(z + 5)^3$, we need to determine the function $f(t)$ whose Laplace transform is $g(z)$. This is equivalent to evaluating the Bromwich integral

$$
f(t) = \frac{1}{2\pi i} \lim_{y_0 \to \infty} \int_{\lambda_{y_0}} e^{zt} g(z) \mathrm{d}z, \qquad t > 0,
$$

where $\lambda_{y_0}(y) = x_0 + iy$ with $-y_0 \le y \le y_0$, $y_0 > 0$, and $x_0 > -5$. The integral can be evaluated by the residue theorem. Consider the closed path $\lambda_{y_0} + \gamma_{y_0}$ where $\gamma_{y_0}(\theta) = x_0 + y_0 e^{i\theta}$ with $\pi/2 \le \theta \le 3\pi/2$. For $z \in \{\gamma_{y_0}\}$ we find

$$|g(z)| = \frac{1}{|z + 5|^3} \le \frac{1}{||z| + 5|^3} \le \frac{1}{(y_0 - |x_0| + 5)^3} \xrightarrow{y_0 \to \infty} 0.$$

By Jordan's lemma we then have

$$\left| \int_{\gamma_{y_0}} e^{zt} g(z) dz \right| \xrightarrow{y_0 \to \infty} 0$$

and so

$$f(t) = \operatorname*{Res}_{z=-5} \left(e^{zt} g(z) \right), \qquad t > 0.$$

The function $g(z)$ has a triple pole at $z = -5$ and in the annulus $0 < |z + 5| < \infty$ the following Laurent series expansion hold

$$g(z) = \frac{1}{(z + 5)^3},$$

$$e^{zt} g(z) = \sum_{n=0}^{\infty} \frac{e^{-5t} t^n}{n!} (z + 5)^{(n-3)}.$$

In conclusion

$$f(t) = \frac{e^{-5t} t^2}{2}, \qquad t > 0.$$

It's easy to verify that

$$\int_0^{\infty} e^{-zt} \frac{e^{-5t} t^2}{2} dt = \frac{1}{(z + 5)^3}, \qquad z \in \mathbb{C}, \quad \operatorname{Re} z > -5.$$

10.49 Given $g(z) = z/(z - 3)^2$, we need to determine the function $f(t)$ whose Laplace transform is $g(z)$. This is equivalent to evaluating the Bromwich integral

$$f(t) = \frac{1}{2\pi i} \lim_{y_0 \to \infty} \int_{\lambda_{y_0}} e^{zt} g(z) dz, \qquad t > 0,$$

where $\lambda_{y_0}(y) = x_0 + iy$ with $-y_0 \le y \le y_0$, $y_0 > 0$, and $x_0 > 3$. The integral can be evaluated by the residue theorem. Consider the closed path $\lambda_{y_0} + \gamma_{y_0}$ where

$\gamma_{y_0}(\theta) = x_0 + y_0 e^{i\theta}$ with $\pi/2 \leq \theta \leq 3\pi/2$. For $z \in \{\gamma_{y_0}\}$ we find

$$|g(z)| = \frac{|z|}{|z-3|^2} \leq \frac{|x_0| + y_0}{||z| - 3|^2} \leq \frac{|x_0| + y_0}{(y_0 - |x_0| - 3)^2} \xrightarrow{y_0 \to \infty} 0.$$

By Jordan's lemma we then have

$$\left| \int_{\gamma_{y_0}} e^{zt} g(z) dz \right| \xrightarrow{y_0 \to \infty} 0$$

and so

$$f(t) = \operatorname*{Res}_{z=3} \left(e^{zt} g(z) \right), \qquad t > 0.$$

The function $g(z)$ has a double pole at $z = 3$ and in the annulus $0 < |z - 3| < \infty$ the following Laurent series expansions hold

$$g(z) = \frac{z - 3 + 3}{(z-3)^2} = \frac{1}{z-3} + \frac{3}{(z-3)^2},$$

$$e^{zt} g(z) = \sum_{n=0}^{\infty} \frac{e^{3t} t^n}{n!} (z-3)^n \left(\frac{1}{z-3} + \frac{3}{(z-3)^2} \right).$$

In conclusion

$$f(t) = e^{3t}(1 + 3t), \qquad t > 0.$$

It's easy to verify that

$$\int_0^\infty e^{-zt} e^{3t}(1 + 3t) dt = \frac{z}{(z-3)^2}, \qquad z \in \mathbb{C}, \quad \operatorname{Re} z > 3.$$

12.11 Exercises of Chap. 11

11.1 Note that g is an analytic function in E. As seen in Exercise 10.42, we have $g(z) = f(z) \; \forall z \in D \subset E$. Therefore, g is the analytic continuation of f from D to E.

11.2 Since $f(z) = \sinh z$ is continuous in R, closed and bounded, and is analytic and not constant in its interior, by the maximum modulus theorem $|f(z)|$ has a maximum on the boundary of R. Recalling that

$$|\sinh(z)|^2 = \sinh^2 x + \sin^2 y$$

and noting that $\sinh^2 x$ is monotonically increasing and $\sin^2 y$ has maxima for $y = \pi/2 + n\pi$, $n \in \mathbb{Z}$, the maximum of $|f(z)|$ in R is reached at the point $z = 3 + i\pi/2$ and is

$$|\sinh(3 + i\pi/2)| = \sqrt{\sinh^2 3 + 1} = \cosh 3.$$

11.3 Set $z^4 - 5z + 1 = f(z) + g(z)$, with

$$f(z) = -5z + 1,$$

$$g(z) = z^4.$$

For $|z| = 1$ we have

$$|f(z)| \geq ||-5z| - |1|| = 4,$$

$$|g(z)| = |z^4| = 1,$$

that is $|f(z)| > |g(z)|$. Since $f(z)$ has a simple zero in $z = 1/5$, by Rouché theorem, $f(z) + g(z)$ also has a zero in $|z| < 1$. Always with the same functions f and g, for $|z| = 2$ we have

$$|f(z)| \leq |-5z| + |1| = 11,$$

$$|g(z)| = |z^4| = 16,$$

that is, $|g(z)| > |f(z)|$. Since $g(z)$ has a zero with multiplicity four at $z = 0$, by Rouché theorem, $g(z) + f(z)$ has four zeros in $|z| < 2$. We conclude that in the region $1 \leq |z| < 2$, the equation $z^4 - 5z + 1 = 0$ has 3 solutions.

11.4 Let $f(z) = az^n$ and $g(z) = -\sin z$. Both these functions are analytic in Q. Furthermore $|f(z)| > |g(z)|$ for $z \in \partial Q$. In fact, for z on the boundary of the square Q we have $|z| \geq 1$ and therefore

$$\left|az^n\right|^2 = |a|^2 \left(|z|^2\right)^n \geq |a|^2 > e^2.$$

On the other hand, for $|x| \leq 1$ and $|y| \leq 1$ it turns out

$$|-\sin(x + iy)|^2 = \sin^2 x + \sinh^2 y \leq 1 + \left(\frac{e - e^{-1}}{2}\right)^2 < 1 + \frac{e^2}{4} < e^2.$$

Therefore, by Rouché theorem, the function $f(z) + g(z)$ has in Q° the same number of zeros, counting their multiplicities, as the function $f(z)$, which evidently has n zeros (one zero of multiplicity n at $z = 0$). In conclusion, the equation $az^n = \sin z$ has n solutions for $z \in Q^\circ$.

11.5 Let $f(z) = 4z^5$ and $g(z) = z^8 - z^2 + 1$. On the boundary of the region $|z| < 1$, that is for $|z| = 1$, we have

$$|f(z)| = 4,$$

$$|g(z)| = \left| z^8 - z^2 + 1 \right| \leq \left| z^8 \right| + \left| -z^2 \right| + |1| = 3,$$

therefore $|f(z)| > |g(z)|$ for $|z| = 1$. By Rouché theorem, f and $f + g$ then have the same number of zeros in the region $|z| < 1$. Since $f(z)$ has a zero of order five at $z = 0$, the equation considered has five solutions in $|z| < 1$.

11.6 To answer the question, we separately determine the number of zeros of the given equation in the regions $|z| < 1$ and $|z| < \sqrt{2}$.

In the region $|z| < 1$, set $f(z) = 4z^3$ and $g(z) = z^8 - z^2 + 1$. On the boundary of this region, namely, for $|z| = 1$, we have

$$|f(z)| = \left| 4z^3 \right| = 4,$$

$$|g(z)| = \left| z^8 - z^2 + 1 \right| \leq \left| z^8 \right| + \left| -z^2 \right| + |1| = 3,$$

therefore $|f(z)| > |g(z)|$ for $|z| = 1$. By Rouché theorem, f and $f + g$ then have the same number of zeros in the region $|z| < 1$. Since $f(z)$ has a zero of order three at $z = 0$, the given equation has three zeros in $|z| < 1$.

In the region $|z| < \sqrt{2}$ set $f(z) = z^4 + 4z^3$ and $g(z) = -z^2 + 1$. On the border of this region, that is to say $|z| = \sqrt{2}$, we have

$$|f(z)| = \left| z^8 + 4z^3 \right| \geq \left| \left| z^8 \right| - \left| 4z^3 \right| \right| = \left| 16 - 8\sqrt{2} \right| = 8(2 - \sqrt{2}) > 4,$$

$$|g(z)| = \left| -z^2 + 1 \right| \leq \left| -z^2 \right| + |1| = 2 + 1 = 3,$$

therefore $|f(z)| > |g(z)|$ for $|z| = \sqrt{2}$. By Rouché theorem, f and $f + g$ then have the same number of zeros in the region $|z| < \sqrt{2}$. Since $f(z)$ has a zero of order three at $z = 0$ and five simple zeros corresponding to the five distinct roots of $(-4)^{1/5}$, the given equation has eight zeros in $|z| < \sqrt{2}$.

We conclude that $z^8 + 4z^3 - z^2 + 1 = 0$ has $8 - 3 = 5$ solutions in the region $1 \leq |z| < \sqrt{2}$.

11.7 Let $f(z) = az^n$ and $g(z) = -\cos z$. Both these functions are analytic in Q. Moreover $|f(z)| > |g(z)|$ for $z \in \partial Q$. In fact, for z on the boundary of the square Q we have $|z| \geq 1$ and therefore

$$\left| az^n \right|^2 = |a|^2 \left(|z|^2 \right)^n \geq |a|^2 > e^2.$$

On the other hand, for $|x| \leq 1$ and $|y| \leq 1$ it turns out

$$|-\cos(x+iy)|^2 = \cos^2(x) + \sinh^2(y) \leq 1 + \left(\frac{e - e^{-1}}{2}\right)^2 < 1 + \frac{e^2}{4} < e^2.$$

Therefore, by Rouché theorem, the function $f(z) + g(z)$ has in Q° the same number of zeros, counting their multiplicities, as the function $f(z)$, which has a zero of order n at $z = 0$. In conclusion, the equation $az^n = \cos z$ has n solutions for $z \in Q^\circ$.

11.8 In the region considered, v is harmonic conjugate to u if and only if $f = u + iv$ is analytic. Therefore, the Cauchy-Riemann equations $u_x = v_y$ and $u_y = -v_x$ must be satisfied. From the second one we have

$$v_x(x, y) = -u_y(x, y) = 2y,$$

which, integrated with respect to x, gives

$$v(x, y) = 2xy + \phi(y).$$

From the first one it then follows

$$v_y(x, y) = 2x + \phi'(y) = u_x(x, y) = 2x + 1,$$

that is $\phi'(y) = 1$, which is solved by

$$\phi(y) = y + \text{constant}.$$

In conclusion, up to a constant, we have

$$v(x, y) = 2xy + y.$$

Note that, as it should be,

$$f(z) = u(x, y) + iv(x, y) = z^2 + z,$$

where $z = x + iy$, is entire.

11.9 In the region considered, v is harmonic conjugate to u if and only if $f = u + iv$ is analytic. Therefore, the Cauchy-Riemann equations $u_x = v_y$ and $u_y = -v_x$ must be satisfied. From the second one we have

$$v_x(x, y) = -u_y(x, y)$$
$$= 2y \cos(x^2 - y^2) \cosh(2xy) - 2x \sin(x^2 - y^2) \sinh(2xy)$$
$$= \frac{d}{dx}\left(\cos(x^2 - y^2) \sinh(2xy)\right),$$

which, integrated with respect to x, gives

$$v(x, y) = \cos(x^2 - y^2)\sinh(2xy) + \phi(y).$$

From the first one it then follows

$$v_y(x, y) = 2y\sin(x^2 - y^2)\sinh(2xy) + 2x\cos(x^2 - y^2)\cosh(2xy) + \phi'(y)$$

$$= u_x(x, y)$$

$$= 1 + 2x\cos(x^2 - y^2)\cosh(2xy) + 2y\sin(x^2 - y^2)\sinh(2xy),$$

that is, $\phi'(y) = 1$, which is solved by

$$\phi(y) = y + \text{constant}.$$

In conclusion, up to a constant, we have

$$v(x, y) = y + \cos(x^2 - y^2)\sinh(2xy).$$

Note that, as it should be,

$$f(z) = u(x, y) + iv(x, y) = z + \sin\left(z^2\right),$$

where $z = x + iy$, is entire.

11.10 For every $(x, y) \in \mathbb{R}^2$ we have

$$u_x(x, y) = 2x - 2y - 6xy, \qquad u_{xx}(x, y) = 2 - 6y,$$

$$u_y(x, y) = -2y - 2x - 3x^2 + 3y^2, \qquad u_{yy}(x, y) = -2 + 6y,$$

and therefore $u_{xx} + u_{yy} = 0$, so u is harmonic in $\mathbb{R}^2$. The function v is harmonic conjugate to u in $\mathbb{R}^2$ if and only if the function $f(x + iy) = u(x, y) + iv(x, y)$ is analytic in $\mathbb{C}$. From the first of the Cauchy-Riemann equations, $u_x = v_y$, we have

$$2x - 2y - 6xy = v_y,$$

which integrated with respect to y gives

$$v(x, y) = 2xy - y^2 - 3xy^2 + \phi(x).$$

By imposing the second Cauchy-Riemann equation, $u_y = -v_x$, we have

$$-2y - 2x - 3x^2 + 3y^2 = 2y - 3y^2 + \phi'(x),$$

which integrated with respect to x gives

$$\phi(x) = x^2 + x^3 + \text{constant}.$$

In conclusion, up to a constant,

$$v(x, y) = 2xy - y^2 + x^2 - 3xy^2 + x^3.$$

Setting $z = x + iy$, we have

$$u(x, y) + iv(x, y) = (1 + i)z^2 + iz^3 = f(z),$$

with $f(z)$ analytic in $\mathbb{C}$.

11.11 For every $(x, y) \in \mathbb{R}^2$ we have

$$u_x(x, y) = 2x - e^{-y}\sin x, \qquad u_{xx}(x, y) = 2 - e^{-y}\cos x,$$

$$u_y(x, y) = -2y - e^{-y}\cos x, \qquad u_{yy}(x, y) = -2 + e^{-y}\cos x,$$

and therefore $u_{xx}(x, y) + u_{yy}(x, y) = 0$, so u is harmonic in $\mathbb{R}^2$. The function v is harmonic conjugate to u in $\mathbb{R}^2$ if and only if the function $f(x + iy) = u(x, y) + iv(x, y)$ is analytic in $\mathbb{C}$. From the first of the Cauchy-Riemann equations, $u_x = v_y$, we have

$$2x - e^{-y}\sin x = v_y,$$

which integrated with respect to y yields

$$v(x, y) = 2xy + e^{-y}\sin x + \phi(x).$$

By imposing the second Cauchy-Riemann equation, $u_y = -v_x$, we have

$$2y + e^{-y}\cos x = 2y + e^{-y}\cos x + \phi'(x),$$

that is, $\phi'(x) = 0$, which integrated with respect to x gives

$$\phi(x) - \text{constant}.$$

In conclusion, up to a constant,

$$v(x, y) = 2xy + e^{-y}\sin x.$$

Let $z = x + iy$, we have

$$u(x, y) + iv(x, y) = z^2 + e^{iz} = f(z),$$

with $f(z)$ analytic in $\mathbb{C}$.

11.12 The function v is harmonic conjugate to u in $\mathbb{R}^2$ if and only if the function $f(z) = f(x + iy) = u(x, y) + iv(x, y)$ is analytic in $\mathbb{C}$. Observing that

$$z^2 = (x + iy)^2 = x^2 - y^2 + i2xy$$

and therefore

$$e^{iz^2} = e^{-2xy}\left(\cos(x^2 - y^2) + i\sin(x^2 - y^2)\right),$$

we obtain

$$u(x, y) = \operatorname{Re} f(z), \qquad f(z) = -ie^{iz^2},$$

with $f(z)$ entire. Therefore, up to an arbitrary constant,

$$v(x, y) = \operatorname{Im} f(z) = -e^{-2xy}\cos(x^2 - y^2).$$

11.13 In the region considered, v is harmonic conjugate to u if and only if $f = u + iv$ is analytic. Therefore, the Cauchy-Riemann equations $u_x = v_y$ and $u_y = -v_x$ must be satisfied. From the second one we have

$$\begin{aligned}
v_x(x, y) &= -u_y(x, y)\\[4pt]
&= 2y\cos(x^2 - y^2)\cosh(2xy) - 2x\sin(x^2 - y^2)\sinh(2xy)\\[4pt]
&= \frac{d}{dx}\left(\cos(x^2 - y^2)\sinh(2xy)\right),
\end{aligned}$$

which integrated gives

$$v(x, y) = \cos(x^2 - y^2)\sinh(2xy) + \phi(y).$$

From the first one it then follows

$$\begin{aligned}
v_y(x, y) &= 2y\sin(x^2 - y^2)\sinh(2xy) + 2x\cos(x^2 - y^2)\cosh(2xy) + \phi'(y)\\[4pt]
&= u_x(x, y)\\[4pt]
&= 2x\cos(x^2 - y^2)\cosh(2xy) + 2y\sin(x^2 - y^2)\sinh(2xy),
\end{aligned}$$

or

$$\phi'(y) = 0,$$

which solved gives

$$\phi(y) = \text{constant}.$$

In conclusion, up to a constant,

$$v(x, y) = \cos(x^2 - y^2) \sinh(2xy).$$

Note that, as it should be,

$$f(z) = u(x, y) + iv(x, y) = \sin\left(z^2\right),$$

where $z = x + iy$, is entire.

11.14 Let $v : \mathbb{C} \mapsto \mathbb{R}$ be the harmonic conjugate of u. The function $f = u + iv$ is analytic in $\mathbb{C}$. It follows that the function $g = \exp(-f)$ is entire. Furthermore, since $|\exp(-f)| = \exp(-u)$ and $u \geq 3$, the function g is also bounded in $\mathbb{C}$. By Liouville's theorem it is therefore constant in $\mathbb{C}$. We conclude that $u = -\ln|g|$ is also constant in $\mathbb{C}$.

11.15 The function $u(x, y)$ has second derivatives in $D = \mathbb{R}^2 \setminus \{(1, 0)\}$ and in this domain it follows that

$$u_x(x, y) = \frac{2(x - 1)}{(x - 1)^2 + y^2}, \qquad u_{xx}(x, y) = \frac{-2(x - 1)^2 + 2y^2}{((x - 1)^2 + y^2)^2},$$

$$u_y(x, y) = \frac{2y}{(x - 1)^2 + y^2}, \qquad u_{yy}(x, y) = \frac{2(x - 1)^2 - 2y^2}{((x - 1)^2 + y^2)^2},$$

from which we see that $u_{xx} + u_{yy} = 0$. Let $z = x + iy$, we immediately recognize that

$$\ln((x - 1)^2 + y^2) = \text{Re}(2 \log(z - 1)),$$

where $\log$ is any branch of the logarithm. Note that, in agreement with the fact that D is not a simply connected domain, the function f such that $\text{Re } f = u$ is multi-valued. Moreover, f is analytic only in a subset of D. Choosing for f the principal branch, the domain of analyticity is $\mathbb{C} \setminus \{(-t, 0), \ t \in [0, \infty)\}$. The harmonic conjugate to u is the function

$$v(x, y) = \text{Im}(2 \log(z - 1))$$

$$= 2 \, \text{Arg}(z - 1)$$

$$= \begin{cases} 2(\arctan(y/(x-1)) - \pi) & x < 1, \ y < 0 \\ 2\arctan(y/(x-1)) & x \geq 1, \ (x,y) \neq (1,0) \ . \\ 2(\arctan(y/(x-1)) + \pi) & x < 1, \ y \geq 0 \end{cases}$$

11.16 The integral is of the type

$$I(\lambda) = \int_{\gamma} g(z) e^{\lambda f(z)} \, dz$$

with $g(z) = z^{-1}$ and $f(z) = i(z^3/3 - z)$. It turns out $f'(z) = i(z^2 - 1)$ and $f''(z) = i2z$. The function f has therefore two simple critical points, $z = \pm 1$, in which we have

$$f(\pm 1) = \mp \frac{2}{3} i, \quad f''(\pm 1) = \pm 2i, \quad \mathrm{Arg}(f''(\pm 1)) = \pm \frac{\pi}{2}, \quad g(\pm 1) = \pm 1.$$

Since the function $g(z) e^{\lambda f(z)}$ is analytic everywhere in $\mathbb{C}$ except at the point $z = 0$, it is possible, without changing the value of $I(\lambda)$, to deform the integration path γ so that it passes through the paths of steepest descent at points $z = \pm 1$ (paths orthogonal to the real axis). In conclusion, using the formula (11.21), we obtain

$$I(\lambda) = e^{\lambda f(1)} \left(\sqrt{\frac{2\pi}{\lambda \, |f''(1)|}} g(1) e^{i(\frac{\pi}{2} - \frac{1}{2} \mathrm{Arg}\, f''(1))} + O(\lambda^{-3/2}) \right)$$

$$+ e^{\lambda f(-1)} \left(\sqrt{\frac{2\pi}{\lambda \, |f''(-1)|}} g(-1) e^{i(\frac{\pi}{2} - \frac{1}{2} \mathrm{Arg}\, f''(-1))} + O(\lambda^{-3/2}) \right)$$

$$= \sqrt{\frac{2\pi}{2\lambda}} e^{\lambda 2i/3} e^{i(\pi/2 - \pi/4)} + \sqrt{\frac{2\pi}{2\lambda}} e^{-\lambda 2i/3} e^{i(\pi/2 + \pi/4)} + O(\lambda^{-3/2})$$

$$= i \sqrt{\frac{2\pi}{\lambda}} (\sin(2\lambda/3) + \cos(2\lambda/3)) + O(\lambda^{-3/2}).$$

11.17 The integral is of the type

$$I(\lambda) = \int_{\gamma} g(z) e^{\lambda f(z)} \, dz$$

with $g(z) = z^{-1/2}$ and $f(z) = i(z^2/2 - z)$. It turns out $f'(z) = i(z - 1)$ and $f''(z) = i$. The function f has therefore only one simple critical point, $z = 1$, in which we have

$$f(1) = -\frac{1}{2} i, \quad f''(1) = i, \quad \mathrm{Arg}(f''(1)) = \frac{\pi}{2}, \quad g(1) = 1.$$

Since the function $g(z)e^{\lambda f(z)}$ (assuming the principal branch for g) is analytic everywhere in $\mathbb{C}$ except on the negative real semi-axis including the origin, it is possible, without changing the value of $I(\lambda)$, to deform the integration path γ so that it passes through the path of steepest descent at point $z = 1$ (path orthogonal to the real axis). In conclusion, using the formula (11.21), we obtain

$$I(\lambda) = e^{\lambda f(1)}\left(\sqrt{\frac{2\pi}{\lambda\,|f''(1)|}}\,g(1)e^{i(\frac{\pi}{2}-\frac{1}{2}\,\mathrm{Arg}\,f''(1))} + O(\lambda^{-3/2})\right)$$

$$= \sqrt{\frac{2\pi}{\lambda}}\,e^{-\lambda i/2}e^{i(\pi/2-\pi/4)} + O(\lambda^{-3/2})$$

$$= \sqrt{\frac{2\pi}{\lambda}}\,e^{i\pi/4}\,(\cos(\lambda/2) + i\sin(\lambda/2)) + O(\lambda^{-3/2}).$$

11.18 The integral is of the type

$$I(\lambda) = \int_{\gamma} g(z)e^{\lambda f(z)}dz$$

with $g(z) = z^{3/2}$ and $f(z) = i(z^2/2-z)$. It turns out $f'(z) = i(z-1)$ and $f''(z) = i$. The function f has therefore only one simple critical point, $z = 1$, in which we have

$$f(1) = -\frac{1}{2}i, \quad f''(1) = i, \quad \mathrm{Arg}(f''(1)) = \frac{\pi}{2}, \quad g(1) = 1.$$

Since the function $g(z)e^{\lambda f(z)}$ (assuming for g the principal branch) is analytic everywhere in $\mathbb{C}$ except on the negative real semi-axis including the origin, it is possible, without changing the value of $I(\lambda)$, to deform the integration path γ so that it passes through the path of steepest descent at point $z = 1$ (path orthogonal to the real axis). In conclusion, using the formula (11.21), we obtain

$$I(\lambda) = e^{\lambda f(1)}\left(\sqrt{\frac{2\pi}{\lambda\,|f''(1)|}}\,g(1)e^{i(\frac{\pi}{2}-\frac{1}{2}\,\mathrm{Arg}\,f''(1))} + O(\lambda^{-3/2})\right)$$

$$= \sqrt{\frac{2\pi}{\lambda}}\,e^{-\lambda i/2}e^{i(\pi/2-\pi/4)} + O(\lambda^{-3/2})$$

$$= \sqrt{\frac{2\pi}{\lambda}}\,e^{i\pi/4}\,(\cos(\lambda/2) + i\sin(\lambda/2)) + O(\lambda^{-3/2}).$$

11.19 In the limit $N \to \infty$ we can estimate the dominant term in $\Omega(E)$ by the saddle-point method. We use formula (11.21) changing $\lambda \to N$, $z \to k$ and assuming k_0 solution of $f'(k_0) = 0$. In the present case we have $g(k) = 1$ and

$$f(k) = ik\frac{E}{N} + \log(\cos(kJ)),$$

$$f'(k) = i\frac{E}{N} - J\tan(kJ),$$

$$f''(k) = -\frac{J^2}{\cos^2(kJ)},$$

therefore the equation $f'(k_0) = 0$ gives

$$\tan(k_0 J) = i\frac{E}{NJ}.$$

The critical point is simple and given by, see formula (6.50),

$$k_0 J = \frac{1}{2i}\log\left(\frac{1 + i\frac{iE}{NJ}}{1 - i\frac{iE}{NJ}}\right) = -\frac{i}{2}\ln\left(\frac{1 - \frac{E}{NJ}}{1 + \frac{E}{NJ}}\right).$$

Observe that $-NJ < E < NJ$ (the cases $E = -NJ$ or $E = NJ$ correspond to configurations of zero probability for $N \to \infty$), it then follows that the quantity $(1 - \frac{E}{NJ})/(1 + \frac{E}{NJ})$ is positive smaller than 1 and the corresponding logarithm is real negative. Since k_0 is imaginary, let us put $k_0 = -ix_0$, $x_0 < 0$, given by

$$x_0 = \frac{1}{2J}\ln\left(\frac{1 - \frac{E}{NJ}}{1 + \frac{E}{NJ}}\right).$$

At the critical point we have

$$f(k_0) = i(-ix_0)\frac{E}{N} + \log(\cos(-ix_0 J)) = x_0\frac{E}{N} + \ln(\cosh(x_0 J)),$$

$$f''(k_0) = -\frac{J^2}{\cos^2(-ix_0 J)} = -\frac{J^2}{\cosh^2(x_0 J)}.$$

In terms of the dimensionless quantity $\epsilon = E/(NJ)$, which becomes a constant for $N \to \infty$, we rewrite

$$x_0 = \frac{1}{2J}[\ln(1 - \epsilon) - \ln(1 + \epsilon)],$$

$$\cosh(x_0 J) = \frac{1}{\sqrt{1 + \epsilon}\sqrt{1 - \epsilon}}$$

$$f(k_0) = -\frac{1}{2}(1 + \epsilon)\ln(1 + \epsilon) - \frac{1}{2}(1 - \epsilon)\ln(1 - \epsilon),$$

$$f''(k_0) = -J^2(1 + \epsilon)(1 - \epsilon).$$

Noticing that $\operatorname{Arg} f''(k_0) = \pi$, we conclude

$$\Omega(E) = \frac{2^N}{2\pi} e^{Nf(k_0)}\left(\sqrt{\frac{2\pi}{N\,|f''(k_0)|}}\,e^{i\left(\frac{\pi}{2} - \frac{1}{2}\operatorname{Arg} f''(k_0)\right)} + O\left(N^{-3/2}\right)\right)$$

$$= \exp\left(N\left(\ln 2 - \frac{1}{2}(1 + \epsilon)\ln(1 + \epsilon) - \frac{1}{2}(1 - \epsilon)\ln(1 - \epsilon)\right)\right)$$

$$\times \frac{1}{J}\left(\frac{1}{\sqrt{2\pi N}\sqrt{(1 + \epsilon)(1 - \epsilon)}} + O\left(N^{-3/2}\right)\right).$$

Just for statistical mechanics considerations, observe that

$$\ln\left(\Omega(E)J\right) = N\left(\ln 2 - \frac{1}{2}(1 + \epsilon)\ln(1 + \epsilon) - \frac{1}{2}(1 - \epsilon)\ln(1 - \epsilon)\right)$$

$$+ O(\ln N),$$

which is proportional to the entropy, vanishes at both $\epsilon = \pm 1$, that is, in the configurations of minimum and maximum energy.

11.20   In the limit $N \to \infty$ we can estimate the dominant term in $\Omega(E)$ by the saddle-point method. We use formula (11.21) changing $\lambda \to N$, $z \to k$ and assuming k_0 solution of $f'(k_0) = 0$. In the present case we have $g(k) = 1$ and

$$f(k) = ik\frac{E}{N} - \log(2i\sin(k\hbar\omega/2)),$$

$$f'(k) = i\frac{E}{N} - \frac{\hbar\omega/2}{\tan(k\hbar\omega/2)},$$

$$f''(k) = \frac{(\hbar\omega/2)^2}{\sin^2(k\hbar\omega/2)},$$

therefore the equation $f'(k_0) = 0$ gives

$$\tan(k_0\hbar\omega/2) = \frac{N\hbar\omega}{2iE}.$$

The critical point is simple and given by, see formula (6.50),

$$k_0 \hbar\omega/2 = \frac{1}{2i} \log\left(\frac{1 + i\frac{N\hbar\omega}{2iE}}{1 - i\frac{N\hbar\omega}{2iE}}\right) = -\frac{i}{2} \ln\left(\frac{\frac{2E}{N\hbar\omega} + 1}{\frac{2E}{N\hbar\omega} - 1}\right).$$

Observe that $E > N\hbar\omega/2$ (the case $E = N\hbar\omega/2$ corresponds to a configuration of zero probability for $N \to \infty$), it then follows that the quantity $(\frac{2E}{N\hbar\omega} + 1)/(\frac{2E}{N\hbar\omega} - 1)$ is positive larger than 1 and the corresponding logarithm is real positive. Since k_0 is imaginary, let us put $k_0 = -ix_0$, $x_0 > 0$, given by

$$x_0 = \frac{1}{\hbar\omega} \ln\left(\frac{\frac{2E}{N\hbar\omega} + 1}{\frac{2E}{N\hbar\omega} - 1}\right).$$

At the critical point we have

$$f(k_0) = i(-ix_0)\frac{E}{N} - \log(2i\sin(-ix_0\hbar\omega/2)) = x_0\frac{E}{N} - \ln(2\sinh(x_0\hbar\omega/2)),$$

$$f''(k_0) = \frac{(\hbar\omega/2)^2}{\sin^2(-ix_0\hbar\omega/2)} = -\frac{(\hbar\omega/2)^2}{\sinh^2(x_0\hbar\omega/2)}.$$

In terms of the dimensionless quantity $\epsilon = (2E)/(N\hbar\omega)$, which becomes a constant for $N \to \infty$, we rewrite

$$x_0 = \frac{1}{\hbar\omega}[\ln(\epsilon + 1) - \ln(\epsilon - 1)],$$

$$\sinh(x_0\hbar\omega/2) = \frac{1}{\sqrt{\epsilon + 1}\sqrt{\epsilon - 1}},$$

$$f(k_0) = -\ln 2 + \frac{1}{2}(\epsilon + 1)\ln(\epsilon + 1) - \frac{1}{2}(\epsilon - 1)\ln(\epsilon - 1),$$

$$f''(k_0) = -(\hbar\omega/2)^2(\epsilon + 1)(\epsilon - 1).$$

Noticing that $\mathrm{Arg}\, f''(k_0) = \pi$, we conclude

$$\Omega(E) = \frac{1}{2\pi} e^{Nf(k_0)} \left(\sqrt{\frac{2\pi}{N|f''(k_0)|}}\, e^{i(\frac{\pi}{2} - \frac{1}{2}\mathrm{Arg}\, f''(k_0))} + O\left(N^{-3/2}\right)\right)$$

$$= \exp\left(N\left(-\ln 2 + \frac{1}{2}(\epsilon + 1)\ln(\epsilon + 1) - \frac{1}{2}(\epsilon - 1)\ln(\epsilon - 1)\right)\right)$$

$$\times \frac{2}{\hbar\omega}\left(\frac{1}{\sqrt{2\pi N}\sqrt{(\epsilon + 1)(\epsilon - 1)}} + O\left(N^{-3/2}\right)\right).$$

Just for statistical mechanics considerations, observe that

$$\ln\left(\Omega(E)\frac{\hbar\omega}{2}\right) = N\left(-\ln 2 + \frac{1}{2}(\epsilon + 1)\ln(\epsilon + 1) - \frac{1}{2}(\epsilon - 1)\ln(\epsilon - 1)\right)$$
$$+ O(\ln N),$$

which is proportional to the entropy, vanishes at $\epsilon = 1$, that is, in the configuration of minimum energy.

References

1. Ahlfors, L.V.: Complex Analysis. McGraw-Hill, New York (1979)
2. Bleistein, N., Handelsman, R.A.: Asymptotic Expansions of Integrals. Dover, New York (1986)
3. Brown, J.W., Churchill, R.V.: Complex Variables and Applications. McGraw-Hill, New York (2003)
4. Caratheodory, C.: Theory of Functions of a Complex Variable. Chelsea, New York (19541)
5. Conway, J.B.: Functions of One Complex Variable. Springer, New York (1978)
6. Lang, S.: Complex Analysis. Springer, New York (1999)
7. Markushevich, A.I.: Theory of Functions of a Complex Variable, 2nd edn. (3 vol. set). Chelsea, New York (1977)
8. Narasimhan, R.: Complex Analysis in One Variable. Springer, New York (2001)
9. Needham, T.: Visual Complex Analysis. Oxford University Press, London (1998)
10. Silverman, R.A.: Introductory Complex Analysis. Dover, New York (1984)
11. Spiegel, M.R.: Schaum's Outline Series: Theory and Problems of Complex Variables. McGraw-Hill, New York (1964)
12. Taylor, A.E., Mann W.R.: Advanced Calculus. Wiley, New York (1983)
13. Titchmarsh, E.C.: The Theory of Functions. Oxford University Press, London (1939)
14. Tverberg, H.: A proof of the Jordan curve theorem. Bull. Lond. Math. Soc. **12**(1), 34 (1980)

Index

FSC
www.fsc.org
MIX
Papier aus verantwortungsvollen Quellen
Paper from responsible sources
FSC® C105338